W0231975

Path Integrals

And Their Applications in Quantum, Statistical, and Solid State Physics

NATO ADVANCED STUDY INSTITUTES SERIES

A series of edited volumes comprising multifaceted studies of contemporary scientific issues by some of the best scientific minds in the world, assembled in cooperation with NATO Scientific Affairs Division.

Series B: Physics

RECENT VOLUMES IN THIS SERIES

Volume 27 – Topics in Theoretical and Experimental Gravitation Physics
edited by V. De Sabbata and J. Weber

Volume 28 – Material Characterization Using Ion Beams
edited by J. P. Thomas and A. Cachard

Volume 29 – Electron–Phonon Interactions and Phase Transitions
edited by Tormod Riste

Volume 30 – Many Degrees of Freedom in Field Theory
edited by L. Streit

Volume 31 – Many Degrees of Freedom in Particle Theory
edited by H. Satz

Volume 32 – Electron and Ion Spectroscopy of Solids
edited by L. Fiermans, J. Vennik, and W. Dekeyser

Volume 33 – Microscopic Structure and Dynamics of Liquids
edited by J. Dupuy and A. J. Dianoux

Volume 34 – Path Integrals and Their Applications in Quantum, Statistical, and Solid State Physics
edited by George J. Papadopoulos and J. T. Devreese

Volume 35 – Correlation Functions and Quasiparticle Interactions in Condensed Matter
edited by J. Woods Halley

Volume 36 – Strongly Coupled Plasmas
edited by Gabor Kalman

Volume 37 – Coherence in Spectroscopy and Modern Physics
edited by F. T. Arecchi, R. Bonifacio, and M. O. Scully

The series is published by an international board of publishers in conjunction with NATO Scientific Affairs Division

A	Life Sciences	Plenum Publishing Corporation
B	Physics	New York and London
C	Mathematical and Physical Sciences	D. Reidel Publishing Company Dordrecht and Boston
D	Behavioral and Social Sciences	Sijthoff International Publishing Company Leiden

Path Integrals

And Their Applications in Quantum, Statistical, and Solid State Physics

Edited by
George J. Papadopoulos
University of Athens
Athens, Greece

and

J.T. Devreese
State University of Antwerp
Antwerp, Belgium

Springer Science+Business Media, LLC

Library of Congress Cataloging in Publication Data

NATO Advanced Study Institute on Path Integrals and their Applications in Quantum, Statistical, and Solid State Physics, Rijksuniversitair Centrum Antwerpen, 1977.
Path integrals and their applications in quantum, statistical, and solid state physics.

(NATO advanced study institutes series: Series B, Physics; 34)
Proceedings of a NATO Advanced Study Institute on Path Integrals and their Applications in Quantum, Statistical, and Solid State Physics held at the State University of Antwerp, July 17–30, 1977.
Includes index.
1. Integral, Path–Congresses. 2. Quantum statistics–Congresses. 3. Solid state physics – Congresses. I. Papadopoulos, George J. II. Devreese, Jozef T. III. Title. IV. Series.
QC174.17.P27N18 1977 530.1'2 78-18225

ISBN 978-1-4684-9142-5 ISBN 978-1-4684-9140-1 (eBook)
DOI 10.1007/978-1-4684-9140-1

Proceedings of a NATO Advanced Study Institute on Path Integrals and Their Applications in Quantum, Statistical, and Solid State Physics held at the State University of Antwerp (R.U.C.A.), Antwerp, Belgium, July 17–30, 1977

This volume is dedicated to

Professor R. P. Feynman,

whose path integral has profoundly influenced quantum mechanics,
both in enhancing our understanding
and in extending the scope of application.

Preface

The Advanced Study Institute on "Path Integrals and Their Applications in Quantum, Statistical, and Solid State Physics" was held at the University of Antwerpen (R.U.C.A.), July 17-30, 1977. The Institute was sponsored by NATO. Co-sponsors were: A.C.E.C. (Belgium), Agfa-Gevaert (Belgium), l'Air Liquide Belge (Belgium), Belgonucléaire (Belgium), Bell Telephone Mfg. Co. (Belgium), Boelwerf (Belgium), Generale Bankmaatschappij (Belgium), I.B.M. (Belgium), Kredietbank (Belgium), National Science Foundation (U.S.A.), Siemens (Belgium). A total of 100 lecturers and participants attended the Institute.

The development of path (or functional) integrals in relation to problems of stochastic nature dates back to the early 20's. At that time, Wiener succeeded in obtaining the fundamental solution of the diffusion equation using Einstein's joint probability of finding a Brownian particle in a succession of space intervals during a corresponding succession of time intervals. Dirac in the early 30's sowed the seeds of the path integral formulation of quantum mechanics. However, the major and decisive step in this direction was taken with Feynman's works in quantum and statistical physics, and quantum electrodynamics. The applications now extend to areas such as continuous mechanics, and recently functional integration methods have been employed by Edwards for the study of polymerized matter. In fact the method is of such versatility that one can hardly imagine a branch of theoretical physics which could not proceed via path integration.

Today path integrals play an ever increasing role in providing solutions to problems not so accessible by other methods. Thus, no other approach has been able to match the success of Feynman's treatment of the polaron problem.

From a methodological point of view, the path integral approach serves as a unifying entity among apparently disconnected disciplines. In the first part of the proceedings, a fairly general introduction consisting of concepts and methods of evaluation interwoven with illustrating applications from brownian motion, quantum mechanics, quantum field theory and classical mechanics is given. In the second part, more specialized topics stemming from condensed matter and statistical mechanics, together with recent developments in the field, are considered.

I should like to thank the lecturers for their collaboration in preparing their manuscripts in time which enabled the participants to attend the lectures comfortably and more beneficially. The invited speaker for the opening session of the Institute was Professor Sir S.F. Edwards, Chairman of the Science Research Council in London, whom I should like to take this opportunity to thank once again. My thanks are also due to the members of the International Advisory Committee: Professors W.E. Brittin, J.R. Klauder, J.M. Luttinger and K.K. Thornber. Further I am much obliged to Professor G.J. Papadopoulos, who acted as co-director.

It gives me great pleasure to thank the Rector of the Rijksuniversitair Centrum Antwerpen (R.U.C.A.), Professor Dr. L. Massart, for his warm and continuous interest in the Institute, and for making available so promptly and effectively the university facilities. I should also like to thank Mr. A. Drubbel, Chief Administrator of the University, and through him his staff, for contributing to the success of the Institute.

Finally, I wish to thank Dr. J. De Sitter, Dr. V. Van Doren, Dr. L. Lemmens, Mrs. H. Van Mierlo and Miss H. Evans for their dedication in carrying out the enormous secretarial task which made possible the smooth operation of the Summer School. My thanks are also due to Dr. F. Brosens, Dr. P. Van Camp and Dr. J. Van Royen for reading the proofs, and Mrs. N. Dillen-Verbiest, Mr. M. De Moor, Mr. Z. De Bliek and Mr. R. Verdyck for technical assistance.

Jozef T. Devreese
Professor of Theoretical Physics
University of Antwerp (R.U.C.A. and U.I.A.)
Director of the Advanced Study Institute

April 1978

Contents

Part II

APPLICATIONS OF THE PATH INTEGRAL APPROACH

Part III

SEMINARS

AN INTRODUCTION TO FEYNMAN PATH INTEGRALS

K. K. Thornber

Bell Laboratories

Murray Hill, New Jersey 07974

ABSTRACT

This short introductory article is intended to serve as a guide to readily available, published works which can be read in order to obtain a working knowledge of Feynman path integrals.

TEXT

In order to more fully understand the lectures presented in the following chapters, it will be necessary to have some acquaintance with the path-integral approach to quantum mechanics. The essence of the method is to calculate the quantum mechanical amplitude K(b,a) to go from a point $\underset{\sim}{x}_a$ at time t_a to a point $\underset{\sim}{x}_b$ at time t_b. The $\underset{\sim}{x}_a$, $\underset{\sim}{x}_b$ can correspond to the coordinates of any number of particles as is of interest in the problem. This amplitude is equated to the sum over all paths $\underset{\sim}{x}(t)$ of the amplitude $[\underset{\sim}{x}(t)]$ assigned to each path according to

$$\phi\ [\underset{\sim}{x}(t)] = C \exp\left[(i/\hbar)\ S\ (\underset{\sim}{x}(t))\right] \qquad (1)$$

where $S(\underset{\sim}{x})$ is the classical action of the system, and C is a constant independent of $\underset{\sim}{x}(t)$ chosen so that the total amplitude K(b,a) is properly normalized. In the more familiar Schroedinger or Heisenberg theory K(b,a), a Green function, is often calculated by solving a differential equation, solving an integral equation, or summing eigenstates. The method is most readily applied to systems coupled to harmonic oscillators and arbitrarily time-varying probe fields. In the classical limit of $\hbar \to 0$, Eq.(1) tells us that the only paths we need consider are those for which $\delta S(\underset{\sim}{x}) = O(\hbar)$, that is paths essentially identical with the classical path determined by $\delta S(\underset{\sim}{x}(t)) = 0$. The path-integral approach is equivalent to the Schroedinger or Heisenberg approach to quantum mechanics in that each can be derived from the other. While path integral methods have been extended to problems in relativistic quantum mechanics and quantum field theory, the basic idea of summing over all paths retains the flavor it acquires in the nonrelativistic problems to which it is usually applied.

By far the most comprehensive tutorial on path integrals is the book by Feynman and Hibbs[1] in which path integrals are introduced and applied not only to calculating and analyzing quantum mechanical quantities but also to determining the partition function in statistical mechanics and the characteristic function in probability theory. A shortened treatment in the form of a review devoted solely to quantum aspects and entitled "Space-Time Approach to Non-Relativistic Quantum Mechanics" is also available.[2] A more detailed discussion of applications in statistical mechanics is to be found in a set of lectures[3](SM). The derivation of the path-integral representation from the Schroedinger-Heisenberg representation can be found in a technical article[4] entitled "An Operator Calculus Having Applications in Quantum Electrodynamics". While any of these may serve to adequately acquaint the student with the method, the FM book[1] is certainly the best choice for the uninitiated. The more mathematically inclined will probably prefer the article on operator calculus,[4] especially the appendices, and a book by Norbert Wiener.[5]

There is, of course, little need to excite the prospective reader hereof concerning the power of interpretation given the physicist by the model of the quantum, an idea now in its eighth decade. Nor need I mention any aspect of quantum electrodynamics, for which the path-integral approach was originally intended. Rather what is so surprising is the very limited degree to which both the path integral and the operator calculus have become part of the repertoire of the working theorist. To view of problem from a totally different angle or to express a result in a quite different manner has always been one of the delights of Prof. Feynman. Perhaps this is one source of his admirable creativity, insight, and understanding. Need one doubt that such rarely used methods of his as these can also be put to good use by others.

REFERENCES

1. R. P. Feynman, A. R. Hibbs, Quantum Mechanics and Path Integrals, New York: McGraw Hill (1965).
2. R. P. Feynman, Rev. Mod. Phys. 20, 367 (1948).
3. R. P. Feynman, Statistical Mechanics, Reading, Mass.: W. A. Benjamin (1972).
4. R. P. Feynman, Phys. Rev. 84, 108 (1951).
5. N. Wiener, Nonlinear Problems in Random Theory, Cambridge, Mass.: MIT (1958).

Part I

Methods in Functional Integration

CONTINUOUS REPRESENTATIONS AND PATH INTEGRALS, REVISITED

John R. Klauder

Bell Laboratories

Murray Hill, New Jersey 07974 USA

ABSTRACT

The relation of classical and quantum theories, and the use of semi-classical approximations in quantum problems are topics that pervade all branches of physics. Continuous representations, which are generalizations of coherent states, are ideally suited for the formulation of quantum theory, especially for phase-space, and related, formulations. In addition, it is particularly natural to formulate the path integral in terms of continuous representations. Examples and applications of this approach are presented.

INTRODUCTION

The formulation of quantum theory draws heavily on classical notions, and it is natural that this is so. The various representations of quantum theory--such as the x- or the p-representation of quantum mechanics--are connected with the diagonalization of one or another operator, the physical significance of which is presumed known from a classical context. While classically both x and p are needed to specify a system, the Heisenberg uncertainty relation precludes the simultaneous specification of both variables, and we are familiar with quantum representations that use only half the classical Cauchy data. Nevertheless, various phase-space representations, such as that proposed by Wigner and Weyl, and developed by Moyal,[1] employ functions of phase-space variables x and p in their quantum formulations. The price paid for this particular kind of generalization is the inability to specify states of the system sharp in both x and p that are in violation of the uncertainty relation.

Once this kind of limitation is understood and accepted it is no more difficult to use such quantum descriptions, and indeed sometimes it is of considerable conceptual and practical value. In particular, alternative formulations often lend themselves to new and unconventional approximation schemes, which may have application for a wide variety of problems.

In a general sense these lectures deal with representations of quantum theory different from the conventional ones, with their role in conceptual and formulational aspects of the theory, and especially with their role in path integral analyses. A number of the basic ideas herein contained can be found in publications of the author, notably Ref. 2, and in a series of papers entitled Continuous Representation Theory.[3] Because of this close connection our choice of notation generally follows that of Ref. 3. In addition, unpublished notes based on lectures given in Bern in 1962 contain the essence of some of this material. In view of recent interest in semi-classical methods of analysis in field theory, among other reasons, it is appropriate to re-examine these early ideas for contemporary relevance. In the course of this re-examination several new concepts and approximations will be developed, notably the notion of stationary-phase approximations for path integrals expressed in continuous representations.

CONTINUOUS REPRESENTATIONS--WHAT ARE THEY?

To illustrate the general idea of a continuous representation we appeal to an important special case, which we term the canonical phase-space continuous representation. Let Q and P denote an irreducible pair of self-adjoint operators acting in a (separable) Hilbert space $\mathcal{H}$ that obey the canonical (Heisenberg) commutation relation

$$[Q,P] = i\hbar$$

on a suitable dense domain. As such these operators generate a two-parameter, unitary family of (Weyl) operators

$$U[p,q] \equiv e^{-iqP/\hbar} e^{ipQ/\hbar}$$

that satisfy the composition law

$$U[p,q]U[p',q'] = e^{ipq'/2\hbar} U[p+p',q+q'] \ ,$$

as well as the basic relations

$$U^{-1}[p,q] = U^{\dagger}[p,q] = e^{-ipQ/\hbar} e^{iqP/\hbar}$$

$$= e^{ipq/\hbar} U[-p,-q] \ ,$$

$$U^{\dagger}[p,q]\{\alpha P+\beta Q\}U[p,q] = \alpha(P+p) + \beta(Q+q) \ .$$

Here p and q denote two arbitrary real parameters, and the operators U[p,q] constitute a unitary representation of the two-parameter additive group up to a phase factor.

Let now a fiducial unit vector $\Phi_0 \varepsilon H$ be chosen, and for each p,q define the unit vector

$$\Phi[p,q] \equiv U[p,q]\Phi_0 \ .$$

Such vectors enjoy a number of basic properties, some of which are elementary consequences of the weak continuity property of unitary one-parameter groups of operators.[3] Three basic properties are:

i) The vector-valued function Φ[p,q] is continuous in p and q; namely if p,q → p',q' in the ordinary sense, then the norm

$$||\Phi[p,q] - \Phi[p',q']|| \to 0 \ .$$

ii) The complex-valued function

$$\psi(p,q) \equiv (\Phi[p,q],\Psi) \ ,$$

linearly defined for each $\Psi\varepsilon H$, is bounded and a continuous function of p and q; specifically the Schwarz inequality leads to

$$|\psi(p,q)| \leq ||\Psi|| \ ,$$

$$|\psi(p,q)-\psi(p',q')| \leq ||\Phi[p,q]-\Phi[p',q']||\cdot||\Psi|| \ .$$

iii) $\psi(p,q) \equiv 0$ if and only if $\Psi \equiv 0$. The last property asserts that there is a one-to-one relation between functions ψ(p,q) and vectors Ψ, and as a consequence the continuous function ψ(p,q) can be taken as a representative of the abstract vector Ψ. Therefore we deal with a continuous representation of each element Ψ, and thus of the entire Hilbert space H itself, and we denote the space of continuous functions ψ(p,q) by C.

From a physical point of view a representation by continuous functions is eminently reasonable. In addition the fiducial vector Φ_0 may still be chosen to ensure some desirable features. For example, Φ_0 may be chosen to minimize the uncertainty product which results directly in Φ[p,q] being one of the conventional minimum uncertainty packets, now commonly called coherent states (see below). Alternatively, Φ_0 may be chosen as the ground state of whatever dynamical system is under study, thus providing a direct link between the representation space of functions and the specific dynamics. For field systems, i.e., systems of an infinite number of degrees of freedom, such an approach is essentially mandated.

When Ψ is a normalized state the function $\psi(p,q)$ is the probability amplitude for the state Ψ to be found as the state $\Phi[p,q]$ when a measurement is made. Different values of p and q do not label mutually exclusive events, but that is not essential. The simplest interpretation of p and q arises if we impose on the fiducial vector the conditions

$$(\Phi_o,Q\Phi_o) = 0 \ ,$$

$$(\Phi_o,P\Phi_o) = 0$$

for then

$$(\Phi[p,q],Q\Phi[p,q]) = q \ ,$$

$$(\Phi[p,q],P\Phi[p,q]) = p \ .$$

Hence q and p denote the mean position and momentum of the state $\Phi[p,q]$. Furthermore, whenever the fiducial vector fulfills

$$(\Phi_o,Q^2\Phi_o) + (\Phi_o,P^2\Phi_o) = O(\hbar) \ ,$$

then the state $\Phi[p,q]$ becomes dispersionless about both q and p as $\hbar \to 0$. We shall find it convenient to assume these conditions on the fiducial vector Φ_o.

Clearly the interpretation of q and p, and thus of $\psi(p,q)$, given above critically depends on the special parameterization of the operators U[p,q] in terms of so-called canonical coordinates.[4] If new parameters are introduced that are functions of the canonical coordinates, for example through a classical canonical transformation, then in general the new parameters do not correspond to the mean of either Q or P. The new parameters and the associated continuous representation thus have a different but well-defined interpretation. These matters are fully discussed in Ref. 3.

Resolution of Unity

One is familiar with the resolution of unity provided by the spectral resolution of a self-adjoint operator so commonly used in quantum theory, especially in the elegantly suggestive bra-ket notation of Dirac. In this standard case the decomposition of the unit operator I is onto mutually orthogonal, one-dimensional projection operators (or idealized "projection operators"). In the present context, on the other hand, the decomposition of the unit operator onto nonmutually orthogonal, one-dimensional projection operators given by

$$I = \int \Phi[p,q]\Phi[p,q]^{\dagger}(dpdq/2\pi\hbar)$$

holds true as well.[3] The resolution of unity is here interpreted to mean

$$\begin{aligned}(X,\Psi) &= \int (X,\Phi[p,q])(\Phi[p,q],\Psi)(dpdq/2\pi\hbar) \\ &= \int \chi^*(p,q)\psi(p,q)(dpdq/2\pi\hbar) \\ &\equiv (\chi,\psi)_C ,\end{aligned}$$

which both defines and evaluates the inner product in the space C.[3] Observe if $X \equiv \Phi[p',q']$, then it follows that

$$\psi(p',q') = \int K(p',q';p,q)\psi(p,q)(dpdq/2\pi\hbar) ,$$

where

$$K(p',q';p,q) \equiv (\Phi[p',q'],\Phi[p,q])$$

is a continuous function satisfying

$$K(p,q;p,q) \equiv 1$$

which is called the reproducing kernel. Moreover, it follows that

$$K(p',q';p'',q'') = \int K(p',q';p,q)K(p,q;p'',q'')(dpdq/2\pi\hbar)$$

which shows that K acts as a projection operator in $L^2(R^2)$, projecting out the particular space C of continuous functions that are spanned by finite linear sums of the form

$$\sum_{j=1}^{J} \alpha_j K(p,q;p_j,q_j)$$

for arbitrary real p_j,q_j, complex α_j, and integral J.

Operator Representation: I

The continuous representation of an operator is given as an integral kernel, as with any conventional representation. Specifically, if B denotes an operator, then the function

$$B(p,q;p',q') \equiv (\Phi[p,q],B\Phi[p',q'])$$

represents the operator (at least formally) inasmuch as

$$(\Phi[p,q],B\Psi) = \int B(p,q;p',q')\psi(p',q')(dp'dq'/2\pi\hbar) .$$

Just as the functions $\psi(p,q)$ are not arbitrary but satisfy a reproducing equation, it is clear that

$$\begin{aligned}\mathcal{B}(p',q';p'',q'') &= \int \mathcal{K}(p',q';p,q)\mathcal{B}(p,q;p'',q'')(dpdq/2\pi\hbar) \\ &= \int \mathcal{B}(p',q';p,q)\mathcal{K}(p,q;p'',q'')(dpdq/2\pi\hbar) \ .\end{aligned}$$

Although each function $\psi(p,q)$ is invariably a bounded, continuous function, an operator representative $\mathcal{B}(p,q;p',q')$ is well defined only for an operator $\mathcal{B}$ for which every $\Phi[p,q]$ lies in the operator domain, and generally is an unbounded function of its arguments. Sometimes for an important unbounded operator, or a class of unbounded operators, one can use the flexibility to choose Φ_0 to ensure the domain condition. For example, if $\mathcal{B}$ is one of the operators $\exp(\alpha Q^4)$ for any α, then (in a Schrödinger representation) it would suffice to choose $\Phi_0 = \exp(-x^6)$ up to normalization. If sufficient vectors are not in the domain of $\mathcal{B}$, one may approximate $\mathcal{B}$ by a sequence of operators $\mathcal{B}_m$ (a sequence that approaches $\mathcal{B}$ in some way) and use the approximating sequence of integral kernels $\mathcal{B}_m(p,q;p',q')$ each of which has been arranged to be well defined.

Diagonal Matrix Elements

The functional form of the diagonal matrix elements of a general operator $\mathcal{H}(P,Q)$ are readily determined. In particular

$$\begin{aligned}H(p,q) &\equiv \mathcal{H}(p,q;p,q) \\ &= (\Phi[p,q],\mathcal{H}(P,Q)\Phi[p,q]) \\ &= (\Phi_0,U^\dagger[p,q]\mathcal{H}(P,Q)U[p,q]\Phi_0) \\ &= (\Phi_0,\mathcal{H}(P+p,Q+q)\Phi_o) \ .\end{aligned}$$

For a polynomial $\mathcal{H}(P,Q)$ it follows that

$$H(p,q) = \mathcal{H}(p,q) + \mathcal{O}(p,q;\hbar)$$

where, generally speaking, $\mathcal{O}$ vanishes as $\hbar \to 0$.

Evidently the general matrix elements $\mathcal{B}(p,q;p',q')$ for arbitrary arguments uniquely specify an operator $\mathcal{B}$, but what about just the diagonal elements

$$B(p,q) \equiv \mathcal{B}(p,q;p,q) = (\Phi[p,q],\mathcal{B}\Phi[p,q]) \ .$$

Does B(p,q) = 0 for all p and q imply that $\mathcal{B} = 0$? To answer this question[3] let $\mathcal{B}$ be expressed in a Weyl form, that is

$$\mathcal{B} = \int b(p,q)U[p,q](dpdq/2\pi\hbar) ,$$

in which case B(r,s) = 0 for all r and s implies that

$$\begin{aligned} 0 &= \int b(p,q)(\Phi[r,s],U[p,q]\Phi[r,s])(dpdq/2\pi\hbar) \\ &= \int b(p,q)K(o,o;p,q)e^{i(pr-qs)/\hbar}(dpdq/2\pi\hbar) . \end{aligned}$$

Consequently the distribution b(p,q) has support only on the set where K(o,o;p,q) = 0; and if K(o,o;p,q) never vanishes then b(p,q) = 0. In other words, the diagonal phase-space continuous representation matrix elements uniquely determine the operator whenever the reproducing kernel never vanishes. This condition holds for a harmonic oscillator ground state as the fiducial vector, and for many other choices as well.[3] If $\mathcal{B}$ is restricted to a polynomial in P and Q, then b(p,q) is a distribution with support at the origin, namely at p = q = 0. But every reproducing kernel is unity at the origin and nonzero in a small neighborhood by continuity. Thus it follows that the diagonal phase-space continuous representation matrix elements of an arbitrary polynomial operator uniquely determine the operator for any choice of the fiducial vector.

Operator Representation: II

Besides the integral kernel representation for various operators there is frequently a differential operator representation. To see this first note that

$$\begin{aligned} \psi(p,q) &= (\Phi[p,q],\Psi) \\ &= (\Phi_o,e^{-ipQ/\hbar}e^{iqP/\hbar}\Psi) \\ &= e^{ipq/\hbar}(\Phi_o,e^{iqP/\hbar}e^{-ipQ/\hbar}\Psi) . \end{aligned}$$

Evidently, then

$$\left(-i\hbar\frac{\partial}{\partial q}\right)\psi(p,q) = (\Phi[p,q],P\Psi) ,$$

$$\left(q+i\hbar\frac{\partial}{\partial p}\right)\psi(p,q) = (\Phi[p,q],Q\Psi) ,$$

which states that the action of P on $\psi(p,q)$ is represented by $-i\hbar\partial/\partial q$ and that of P is represented by $q+i\hbar\partial/\partial p$. For a general operator function $H(P,Q)$ it then follows that

$$H(-i\hbar\partial/\partial q, q+i\hbar\partial/\partial p)\psi(p,q) = (\Phi[p,q], H(P,Q)\Psi) \ .$$

Here, as above, we assume the relevant domain conditions are satisfied so as to justify these expressions. Finally, if $H(P,Q)$ corresponds physically to a Hamiltonian, then the phase-space continuous representation form of Schrödinger's equation reads

$$i\hbar\partial\psi(p,q;t)/\partial t = H(-i\hbar\partial/\partial q, q+i\hbar\partial/\partial p)\psi(p,q;t) \ .$$

Coherent States

Of all the phase-space continuous representations one is singled out by the requirement that $(Q+iP)\Phi_o = 0$ and this leads to the well-known coherent states.[5] If we set $a \equiv (Q+iP)/\sqrt{2\hbar}$ and $z \equiv (q+ip)/\sqrt{2\hbar}$, then it follows that

$$\begin{aligned} 0 = a\Phi_o &= U[p,q]a\Phi_o \\ &= U[p,q]aU^{\dagger}[p,q]U[p,q]\Phi_o \\ &= (a-z)\Phi[p,q] \ . \end{aligned}$$

Consequently we learn that

$$a\Phi[p,q] = z\Phi[p,q] \ ;$$

namely, for every q and p, $\Phi[p,q]$ is an eigenvector of a with complex eigenvalue z. These are not mutually orthogonal eigenvectors by any means. Up to a phase they correspond to the familiar coherent states. In particular

$$\begin{aligned} e^{i\frac{1}{2}pq/\hbar}\Phi[p,q] &= e^{i(pQ-qP)/\hbar}\Phi_o \\ &= e^{(a^{\dagger}z-z^*a)}\Phi_o \\ &= e^{-\frac{1}{2}|z|^2}e^{a^{\dagger}z}e^{-z^*a}\Phi_o \\ &= e^{-\frac{1}{2}|z|^2}e^{a^{\dagger}z}\Phi_o \\ &\equiv \Phi[z] \ , \end{aligned}$$

which is just the coherent state more commonly written as $|z\rangle$.

The resolution of unity with coherent states reads simply as

$$I = \pi^{-1}\int \Phi[z]\Phi^{\dagger}[z]d^2z$$

where $d^2z \equiv d(\mathrm{Re}\ z)d(\mathrm{Im}\ z)$. This fundamental relation is more commonly encountered[2,5] in the form

$$I = \pi^{-1}\int |z\rangle\langle z| d^2z \ .$$

One important property of coherent states regards temporal evolution when the Hamiltonian $H = \omega\hbar a^{\dagger}a$. Since

$$e^{-i\omega t a^{\dagger}a}a^{\dagger}e^{i\omega t a^{\dagger}a} = e^{-i\omega t}a^{\dagger} \ ,$$

it follows that

$$\begin{aligned} e^{-itH/\hbar}\Phi[z] &= e^{-i\omega t a^{\dagger}a}e^{-\frac{1}{2}|z|^2}e^{za^{\dagger}}\Phi_0 \\ &= e^{-\frac{1}{2}|z|^2}\exp(ze^{-i\omega t}a^{\dagger})\Phi_0 \\ &\equiv \Phi[e^{-i\omega t}z] \ . \end{aligned}$$

In other words, the evolution of a coherent state induced by the harmonic oscillator Hamiltonian is still a coherent state.

Apart from the normalization factor $\exp(-\frac{1}{2}|z|^2)$, $\Phi[z]$ is an entire function of the complex variable z. In consequence the coherent-state vector representative

$$\psi(z) \equiv (\Phi[z],\Psi) = e^{-\frac{1}{2}|z|^2}g(z^*) \ ,$$

where $g(z^*)$ is an entire function of the complex variable z^*. In particular it is clear that

$$\phi_0(z) = (\Phi[z],\Phi_0) = e^{-\frac{1}{2}|z|^2} \ ,$$

which establishes the fact that the reproducing kernel never vanishes if an harmonic oscillator ground state is taken as the fiducial vector.

The eigenproperty of annihilation operators makes the determination of operator matrix elements especially simple. Let $W(a^{\dagger},a)$ denote an operator function of $a^{\dagger}$ and a and $:W(a^{\dagger},a):$ denote the normal ordered form (all $a^{\dagger}$ to the left of all a). Then it follows that

$$(\Phi[z],:W(a^{\dagger},a):\Phi[z']) = W(z^*,z')(\Phi[z],\Phi[z']) \ ,$$

and the diagonal matrix elements become

$$(\Phi[z],:\mathcal{W}(a^\dagger,a):\Phi[z]) = \mathcal{W}(z^*,z) \ .$$

Moreover since the function $\mathcal{W}(z^*,z)$ uniquely determines the two-variable function $\mathcal{W}(z^*,z')$, it is clear that the diagonal matrix elements suffice to determine an operator. Further properties of coherent states may be found in Ref. 5.

Generalizations

It is important to appreciate that the foregoing discussion is just one chapter in a larger and more general story. There are many ways to single out an overcomplete family of states other than with the Heisenberg canonical group. The vectors that make up such a basic set may be denoted by $\Phi[\ell]$, where $\ell = (\ell^1,\dots,\ell^L)$ is a real point in an L-dimensional label space. These vectors are assumed to be continuous in the variables ℓ, and they may always be represented as $\Phi[\ell] = V[\ell]\Phi_o$ for some unitary operators $V[\ell]$ and fiducial vector Φ_o. If the operators $V[\ell]$ form an irreducible group representation, then the operator

$$\int\Phi[\ell]\Phi^\dagger[\ell]\delta\ell \ ,$$

where $\delta\ell$ denotes the (left-) invariant group measure, is guaranteed to be a multiple of the unit operator by Schur's Lemma. Normalization is afforded by the integral

$$\int|(\Phi_o,\Phi[\ell])|^2\delta\ell$$

provided that it converges. And only when the normalization integral converges can one form a valid resolution of unity and a representation space composed of continuous functions $\psi(\ell) \equiv (\Phi[\ell],\Psi)$ with inner product given as

$$(\chi,\psi)_c \equiv \int\chi^*(\ell)\psi(\ell)\delta\ell$$

where the measure is now assumed normalized. The physical significance of these representations clearly depends on the set of vectors $\{\Phi[\ell]\}$ and on the specific labelling chosen.

Further discussion of different overcomplete families of states and their associated continuous representations may be found in Ref. 3 and elsewhere.[6] Frequently, the vectors that make up the overcomplete family of states in such cases are referred to as generalized coherent states.

REINTERPRETED ACTION PRINCIPLE

In the framework of the abstract Hilbert space $\mathcal{H}$ the fundamental equation of quantum theory, the Schrödinger equation, reads

$$i\hbar\dot{\Psi} = H\Psi \ ,$$

where H denotes the Hamiltonian, and the solution to this equation is abstractly given by

$$\Psi(t) = e^{-itH/\hbar}\Psi(0) \ .$$

The Schrödinger equation can be derived from an action principle based on the quantum action functional

$$I = \int(\Psi, i\hbar\dot{\Psi} - H\Psi)dt \ .$$

In particular, Schrödinger's equation arises from stationarity of this action functional under unrestricted variations of the vector $\Psi(t)$. But stationarity under general variations of the unit vector $\Psi(t)$--unrestricted save for $||\Psi(t)|| \equiv 1$ and excluding pure phase variations of the form $\Psi(t) = \exp[i\alpha(t)]\Psi$--also leads to Schrödinger's equation. It is natural to inquire what are the consequences of stationarity of the quantum action under variations limited to a fairly restricted set of unit vectors such as those which constitute an overcomplete family of states.

Restricted Quantum Action Functional

The implications of such an inquiry are most easily seen if we assume as an example that

$$\Psi(t) \equiv \Phi[p(t),q(t)]$$

where $\Phi[p,q]$ are the vectors of a phase-space overcomplete family of states. For such a restricted set of vectors the quantum action functional becomes

$$I = \int\{i\hbar(\Phi[p,q],\dot{\Phi}[p,q]) - (\Phi[p,q],H\Phi[p,q])\}dt \ .$$

Let us examine these two terms separately. First suppose that $H = H(P,Q)$. Then as we already noted

$$\begin{aligned} H(p,q) &\equiv (\Phi[p,q],H(P,Q)\Phi[p,q]) \\ &= (\Phi_0,H(P+p,Q+q)\Phi_0) \\ &= H(p,q) + O(p,q;\hbar) \ . \end{aligned}$$

To compute the other term, first observe that

$$i\hbar\dot{\Phi}[p,q] = i\hbar \frac{d}{dt} e^{-iqP/\hbar} e^{ipQ/\hbar}\Phi_o$$
$$= e^{-iqP/\hbar}[\dot{q}P-\dot{p}Q]e^{ipQ/\hbar}\Phi_o \ .$$

Consequently

$$(\Phi[p,q],i\hbar\dot{\Phi}[p,q]) = (\Phi_o,e^{-ipQ/\hbar}[\dot{q}P-\dot{p}Q]e^{ipQ/\hbar}\Phi_o)$$
$$= (\Phi_o,[\dot{q}(P+p)-\dot{p}Q]\Phi_o)$$
$$= p\dot{q}$$

based on our assumption that the mean position and momentum in the state Φ_o both vanish. When we combine these terms it follows that the restricted form of the quantum action functional reads

$$I = \int[p\dot{q}-H(p,q)]dt \ .$$

Several remarks about this expression are in order.

Approximate Quantum Solution

Extremal variation of I with respect to arbitrary variations of p and q yields two first-order equations, namely

$$\dot{q} - \partial H(p,q)/\partial p = 0 \ ,$$

$$\dot{p} + \partial H(p,q)/\partial q = 0 \ ,$$

which have solutions we denote by $p_{c\ell}(t)$ and $q_{c\ell}(t)$ subject (say) to the initial conditions $p_{c\ell}(0) = p'$ and $q_{c\ell}(0) = q'$ which may be specified arbitrarily. From the way the restricted action functional was derived, it follows that

$$e^{-itH/\hbar}\Phi[p',q'] \approx \Phi[p_{c\ell}(t),q_{c\ell}(t)] \ ,$$

which represents an approximation to the true quantum evolution of the state $\Phi[p',q']$ that is optimal in some sense given the restriction on the class of allowed vectors. The approximate evolution of a general vector Ψ follows from linearity according to

$$e^{-itH/\hbar}\Psi = e^{-itH/\hbar}\int\Phi[p',q']\psi(p',q')(dp'dq'/2\pi\hbar)$$
$$\approx \int\Phi[p_{c\ell}(t),q_{c\ell}(t)]\psi(p',q')(dp'dq'/2\pi\hbar) \ .$$

Alternatively, one may use the approximate expression

$$\begin{aligned}\psi(p',q';t) &\equiv (\Phi[p',q'],e^{-itH/\hbar}\Psi) \\ &\approx (\Phi[p_{c\ell}(-t),q_{c\ell}(-t)],\Psi) \\ &= \psi(p_{c\ell}(-t),q_{c\ell}(-t)) \ .\end{aligned}$$

The quality of these approximations may be improved by an optimal choice of the fiducial vector.

For some special cases it may happen that the true evolution of any state in the set $\{\Phi[p',q']\}$ always remains in that set. In that case we have equality,

$$e^{-itH/\hbar}\Phi[p',q'] = \Phi[p_{c\ell}(t),q_{c\ell}(t)] \ ,$$

and the approximate evolutionary forms are actually exact. In particular, it follows that

$$e^{-itH/\hbar}\Psi = \int\Phi[p_{c\ell}(t),q_{c\ell}(t)]\psi(p',q')(dp'dq'/2\pi\hbar)$$

and

$$\psi(p',q';t) = \psi(p_{c\ell}(-t),q_{c\ell}(-t)) \ .$$

Such systems are called exact systems, but they are the exception rather than the rule. We note that harmonic oscillator Hamiltonians and coherent states form an exact system, a fact which was established in the preceding section.

Relation to Classical Mechanics

A second remark to make about the restricted form of the quantum action functional is its basic identity-in-form with the classical action functional. There are two reasons why it does not strictly equal the classical action functional. For one thing $H(p,q)$ generally contains the parameter $\hbar$, and for another the physical significance of p and q is different in the two cases. Classically, p and q refer to the momentum and coordinate of a particle which may in principle be measured; quantum mechanically, p and q here refer to the mean momentum and coordinate of the basic vector $\Phi[p,q]$. A system in the state $\Phi[p,q]$ corresponds to a particle localized approximately in a phase-space cell of order $\hbar$ (depending on Φ_o) centered at the point p and q. Of course, this interpretation depends critically on the particular parameterization by the so-called canonical coordinates. After a classical canonical transformation such a special parameterization no longer holds. More generally then, and in the absence of further information, p and q just represent labels for the vectors $\Phi[p,q]$.

There are no real conceptual roadblocks in adopting $H(p,q)$ rather than $\mathcal{H}(p,q)$ as the classical Hamiltonian. The possible appearance of an extra parameter ($\hbar$) causes no conceptual difficulty. One could even arrange to reorder in some manner the expression $\mathcal{H}(P,Q)$ so that $H(p,q)$ is independent of $\hbar$ and has a prescribed functional form. For example, normal ordering in the case of coherent states leads to these properties quite directly, and this concept may be generalized to other cases. Thus we are led to adopt the expression

$$H(p,q) = (\Phi[p,q],\mathcal{H}\Phi[p,q])$$

as the classical Hamiltonian corresponding to the quantum Hamiltonian $\mathcal{H}$. This identification is called the weak correspondence principle.[7] And with this identification it follows that the restricted quantum action functional

$$I = \int[p\dot{q}-H(p,q)]dt$$

has the form of the classical action functional.

The Meaning of Quantization

The importance of this interpretation of the restricted quantum action functional lies in the process of quantization. Given a classical action principle, such as that above, it can, merely by a reinterpretation of the parameters p and q as labels for Hilbert-space vectors, immediately be interpreted as a restricted evaluation of the quantum action functional. This argument is termed the reinterpreted action principle. The act of reinterpretation of the variables already constitutes, in this language, the first step in quantization. The parameter $\hbar$ has not suddenly been introduced; it was always there even if it did not appear explicitly. For an exact system such a reinterpretation is basically sufficient to yield a quantization since the classical solutions $p_{c\ell}(t)$ and $q_{c\ell}(t)$, i.e. the solutions of the classical Hamiltonian equations of motion, suffice to characterize correctly the time evolution of an arbitrary state Ψ, as we have noted above. But most systems are not exact systems, and the classical equations provide only an approximate solution. To do better a second step in quantization becomes necessary, namely the enlargement in one way or another of the domain of the quantum action functional beyond the original restricted set of unit vectors. What is entailed in this enlargement? In one sense it is first necessary to deduce from the canonical kinematical form

$$i\hbar(\Phi[p,q],d\Phi[p,q])$$

the basic vectors themselves or what is the same thing an expression for the general overlap

$$(\Phi[p,q],\Phi[p',q']) \;.$$

Evidently the simplest way to proceed is to presume the Heisenberg algebra and canonical coordinates and choose a convenient fiducial vector Φ_0, which leads directly to an expression for the general overlap. Actually, in a certain sense, this consequence can be deduced given some basic physical significance of the parameters p and q; this is shown in Ref. 3. Next, it is necessary to deduce from the classical Hamiltonian

$$(\Phi[p,q],\mathcal{H}\Phi[p,q])$$

an expression for the quantum Hamiltonian $\mathcal{H}$ itself or what is the same thing the general matrix elements

$$(\Phi[p,q],\mathcal{H}\Phi[p',q']) \;.$$

Some remarks pertaining to this problem appeared in the preceding section.

These quantities are the basic ingredients: The general overlap is just the reproducing kernel, and linear sums determine the vector representatives. Schrödinger's equation can then be formed using the general Hamiltonian matrix elements as an integral kernel. Quantum mechanics is fully characterized.

Discussion

Although we have described this second step of quantization as one big jump it need not proceed this way. One can imagine that the set of allowed vectors is gradually increased from the original overcomplete family of states to all possible unit vectors. After the initial step of reinterpretation, such a gradual increase describes a continuous procedure of passing from a classical to a quantum theory--and continuously back again, if desired, simply by gradually decreasing the allowed vectors back to the original set. In this whole process, we note again, the parameter $\hbar$ remains unchanged.

An analogy of the foregoing discussion with classical thermodynamic processes should not go unnoticed by the reader. Thermodynamic equilibrium of a system holds for the extremal of the entropy subject to prevailing constraints.[8] If some diaphragms are freed, some pistons are unpinned, etc., then the available parameters for the entropy increase and the extremal state is generally improved relative to the constrained case. If we make an analogy between the entropy and the action and between constraints on diaphragms, pistons, etc. and constraints on vectors, then we may assert that "quantum mechanics is just classical mechanics with all stops removed"!

By way of summary we remark that the classical action functional is expressed in the form

$$I = \int\{i\hbar(\Phi[p,q],\dot{\Phi}[p,q])-(\Phi[p,q],\mathcal{H}\Phi[p,q])\}dt \; .$$

Further support for this point of view will come from our study of the path integral expressed in a continuous representation in the next section.

Generalizations

It is evident that the entire analysis of this section may be repeated for a general overcomplete family of states $\{\Phi[\ell]\}$ and that the restricted quantum action functional when $\Psi(t) \equiv \Phi[\ell(t)]$ is given by

$$I = \int\{i\hbar(\Phi[\ell],\dot{\Phi}[\ell])-(\Phi[\ell],\mathcal{H}\Phi[\ell])\}dt$$

which can be interpreted as a "classical" action functional describing a "classical dynamics" on the label space. In this interpretation

$$H(\ell) \equiv (\Phi[\ell],\mathcal{H}\Phi[\ell])$$

plays the role of "Hamiltonian" (time-translation generator), and

$$i\hbar(\Phi[\ell],d\Phi[\ell])$$

is the basic canonical kinematical form.

It is very instructive to analyze various examples in this general context, particularly those for which the labels overspecify the vectors, i.e., $\Phi[\ell] = \Phi[\ell']$ does not imply that $\ell = \ell'$, and those for which the solution of the "classical Hamiltonian equations" are not unique. Some of these topics are treated in Ref. 3.

PATH INTEGRALS WITH CONTINUOUS REPRESENTATIONS

In the preceding section we dealt with a functional quantization scheme based on the reinterpreted action principle and continuous representations. Another and closely related functional quantization approach to be developed in this section is based on path integrals and continuous representations. It is pedagogically useful, however, if we first briefly review the conventional derivation of the path integral, say in its phase-space formulation.

Conventional Path Integral

In essence, most derivations of path integrals from an operator formalism proceed along a common path. The evolution operator $\exp[-i(t''-t')H/\hbar]$ is first factored into (N+1) equal terms according to

$$e^{-i(t''-t')H/\hbar} = e^{-i\varepsilon H/\hbar}e^{-i\varepsilon H/\hbar}\cdots e^{-i\varepsilon H/\hbar}$$

where $\varepsilon \equiv (t''-t')/(N+1)$. Next an initial and final state are chosen, and a resolution of unity is inserted between each of the factors. For example, suppose we are interested in the propagator in the coordinate or q-representation expressed in the form of a phase-space path integral. For this purpose we first observe that

$$\begin{aligned}\langle q'',t''|q',t'\rangle &\equiv \langle q''|e^{-i(t''-t')H/\hbar}|q'\rangle \\ &= \int\cdots\int\langle q_{N+1}|e^{-i\varepsilon H/\hbar}|q_N\rangle\langle q_N|e^{-i\varepsilon H/\hbar}|q_{N-1}\rangle\cdots \\ &\quad\times\langle q_2|e^{-i\varepsilon H/\hbar}|q_1\rangle\langle q_1|e^{-i\varepsilon H/\hbar}|q_0\rangle \\ &\quad\times dq_N\cdots dq_1\ ,\end{aligned}$$

where we have set $q_{N+1} \equiv q''$ and $q_0 \equiv q'$. This expression holds as an identity for any ε, and thus also in the limit $\varepsilon \to 0$; namely

$$\langle q'',t''|q',t'\rangle = \lim_{\varepsilon\to 0}\int\cdots\int\prod_{k=0}^{N}\langle q_{k+1}|e^{-i\varepsilon H/\hbar}|q_k\rangle\prod_{k=1}^{N}dq_k\ .$$

For small ε we may approximate each matrix element as

$$\langle q_{k+1}|e^{-i\varepsilon H/\hbar}|q_k\rangle \approx \langle q_{k+1}|(1-i\varepsilon H/\hbar)|q_k\rangle\ ,$$

and even with our goal set on obtaining a phase-space formulation there are still various ways to proceed at this point (see below). One approach asserts that

$$\langle q_{k+1}|(1-i\varepsilon H/\hbar)|q_k\rangle = \int\langle q_{k+1}|p_{k+\frac{1}{2}}\rangle\langle p_{k+\frac{1}{2}}|(1-i\varepsilon H/\hbar)|q_k\rangle dp_{k+\frac{1}{2}}$$

after having inserted one extra momentum-space resolution of unity on the left side of $\mathcal{H}$. Provided the integral converges it then follows that

$$\langle q_{k+1}|e^{-i\varepsilon\mathcal{H}/\hbar}|q_k\rangle$$
$$\approx (2\pi\hbar)^{-1}\int e^{i[p_{k+\frac{1}{2}}(q_{k+1}-q_k)-\varepsilon H(p_{k+\frac{1}{2}},q_k)]/\hbar}\,dp_{k+\frac{1}{2}}$$

where

$$H(p_{k+\frac{1}{2}},q_k) \equiv \langle p_{k+\frac{1}{2}}|\mathcal{H}|q_k\rangle/\langle p_{k+\frac{1}{2}}|q_k\rangle \ ;$$

the interpretation of H is discussed below. The proposed approximation is certainly correct to lowest order in ε. Consequently it follows that

$$\langle q'',t''|q',t'\rangle = \lim_{\varepsilon\to 0}(2\pi\hbar)^{-(N+1)}\int\ldots\int\exp\{i\sum_{k=0}^{N}[p_{k+\frac{1}{2}}(q_{k+1}-q_k)$$
$$- \varepsilon H(p_{k+\frac{1}{2}},q_k)]/\hbar\}\prod_{k=0}^{N}dp_{k+\frac{1}{2}}\prod_{k=1}^{N}dq_k \ .$$

Strictly speaking this is the end of the story, but the temptation to take the limit $\varepsilon \to 0$ before the integrations are performed is irrestible, at least to most physicists. In such a case the integrand is no longer defined for sequences $p_{k+\frac{1}{2}}$ and q_k but rather for real functions p(t) and q(t), at least formally, and it is traditional to adopt for the integrand the expression that it assumes in the limit for continuous and differentiable functions p(t) and q(t). Following this custom one is thus led to the formal functional integral or path integral

$$\langle q'',t''|q',t'\rangle = \mathcal{N}\int e^{i\int[p\dot{q}-H(p,q)]dt/\hbar}\mathcal{D}p\mathcal{D}q \ ,$$

where $\mathcal{N}$ denotes a formal normalization factor and formally

$$\mathcal{D}p \equiv \Pi_t dp(t), \quad \mathcal{D}q \equiv \Pi_t dq(t) \ .$$

This is the traditional phase-space path integral expression for the propagator based on the classical action

$$I = \int[p\dot{q}-H(p,q)]dt \ .$$

Discussion

It should be stressed that no satisfactory definition of functional integrals for quantum theory has been proposed that is anything other than a linear functional, but that simply begs the basic existence questions. [This situation is completely different for diffusion theory (e.g., Wiener measure) or imaginary-time quantum mechanics, but that is a different matter altogether.] The description of path integrals in continuous representations discussed below does not measurably improve the situation. In the author's view genuine functional integrals for quantum mechanical path integrals simply do not exist. But after all it doesn't really matter since in principle all necessary integrals can be performed for $\varepsilon > 0$ $(N < \infty)$ and the limit $\varepsilon \to 0$ $(N \to \infty)$ taken subsequently. On the other hand, having just argued against any rigorous interpretation for the path integral we hasten to stress its very important conceptual and heuristic value.

At this point let us discuss the quantity

$$H(p,q) = \langle p|\mathcal{H}|q\rangle / \langle p|q\rangle$$

that plays the role of the classical Hamiltonian. If $\mathcal{H}(P,Q) = \mathcal{H}_L(P,Q)$, where $\mathcal{H}_L$ is ordered with all P to the left of all Q, then $H(p,q) = \mathcal{H}_L(p,q)$. Such an expression coincides with the proper classical Hamiltonian up to terms of order $\hbar$. One can readily do better and completely eliminate any such reordering terms with more elaborate treatments of the term $\langle q_{k+1}|(1-i\varepsilon\mathcal{H}/\hbar)|q_k\rangle$ than that illustrated. For instance, the extra momentum-space resolution of unity can be inserted in a different place in the expression, or more generally, several resolutions of unity may be inserted (distinctions in factor ordering are handled in this fashion). The reader is encouraged to investigate these points for him or herself.

For completeness we add that whenever $H(p,q) = \frac{1}{2}p^2 + V(q)$ a straightforward Gaussian integration over each $p_{k+\frac{1}{2}}$ variable followed by a passage to the continuum limit leads to the formal expression

$$\langle q'',t''|q',t'\rangle = \mathcal{N}\int e^{i\int[\frac{1}{2}\dot{q}^2 - V(q)]dt/\hbar}\mathcal{D}q$$

which is the original Feynman q-space path integral. Here $\mathcal{N}$ denotes a formal normalization factor, but different than before.

The generalization to time-dependent Hamiltonians is straightforward, and this, as well as many other topics, may be found for example in the text of Feynman and Hibbs.[9]

Continuous-Representation Path Integrals

To derive the path integral for a phase-space continuous representation we choose suitable initial and final states and insert the appropriate continuous-representation resolution of unity repeatedly into the factored form of the propagator. This leads to the expression

$$<p'',q'',t''|p',q',t'> \equiv (\Phi[p'',q''],e^{-i(t''-t')H/\hbar}\Phi[p',q'])$$

$$= \lim_{\varepsilon\to 0}\int\ldots\int \prod_{k=0}^{N} (\Phi[p_{k+1},q_{k+1}],e^{-i\varepsilon H/\hbar}\Phi[p_k,q_k])$$

$$\times \prod_{k=1}^{N} (dp_k dq_k/2\pi\hbar) ,$$

where we have set $p_{N+1},q_{N+1} \equiv p'',q''$ and $p_o,q_o \equiv p',q'$. We have already exploited the fact that the right-hand side is an identity for any ε, and thus also in the limit $\varepsilon \to 0$. For small ε we approximate each factor by

$$(\Phi^{k+1},e^{-i\varepsilon H/\hbar}\Phi^k) \approx (\Phi^{k+1},(1-i\varepsilon H/\hbar)\Phi^k)$$

$$= (\Phi^{k+1},\Phi^k)\left[1 - \frac{i\varepsilon}{\hbar}\frac{(\Phi^{k+1},H\Phi^k)}{(\Phi^{k+1},\Phi^k)}\right]$$

$$\approx (\Phi^{k+1},\Phi^k)\exp\left[-\frac{i\varepsilon}{\hbar}\frac{(\Phi^{k+1},H\Phi^k)}{(\Phi^{k+1},\Phi^k)}\right] ,$$

where we have used the abbreviation $\Phi^k \equiv \Phi[p_k,q_k]$. This approximation is certainly correct to lowest order in ε, and it follows that

$$<p'',q'',t''|p',q',t'> = \lim_{\varepsilon\to 0}\int\ldots\int \prod_{k=0}^{N} \{(\Phi[p_{k+1},q_{k+1}],\Phi[p_k,q_k])$$

$$\times \exp[-i\varepsilon H(p_{k+1},q_{k+1};p_k,q_k)/\hbar]\}$$

$$\times \prod_{k=1}^{N} (dp_k dq_k/2\pi\hbar) ,$$

provided the integrals converge, where

$$H(p_{k+1},q_{k+1};p_k,q_k) \equiv \frac{(\Phi[p_{k+1},q_{k+1}],H\Phi[p_k,q_k])}{(\Phi[p_{k+1},q_{k+1}],\Phi[p_k,q_k])} .$$

As in the conventional treatment this is essentially the end of the story, but we proceed further in a formal fashion by interchanging the order of integration and the limit $\varepsilon \to 0$. The discrete integration variables p_k and q_k are replaced by the real functions $p(t)$ and $q(t)$, and we adopt for the integrand the expression it assumes for continuous and differentiable functions. To identify that expression we first observe that

$$\prod_{k=0}^{N} (\Phi^{k+1},\Phi^{k}) = \prod_{k=0}^{N} [1-(\Phi^{k+1},\Phi^{k+1}-\Phi^{k})]$$

$$\approx \exp\left[-\sum_{k=0}^{N} (\Phi^{k+1},\Phi^{k+1}-\Phi^{k})\right]$$

valid to the first order in each difference vector $\Phi^{k+1}-\Phi^{k}$. Consequently, in the appropriate limit we are led to the formal functional integral given by

$$\langle p'',q'',t''|p',q',t'\rangle$$

$$= \mathcal{N}\int \exp\{i\textstyle\int[i\hbar(\Phi[p,q],\dot{\Phi}[p,q])-(\Phi[p,q],\mathcal{H}\Phi[p,q])]dt/\hbar\}$$

$$\times\ \mathcal{D}p\mathcal{D}q\ ,$$

where $\mathcal{N}$ is a formal normalization factor, and $\mathcal{D}p$ and $\mathcal{D}q$ are as before. Observe that in this expression the role of the classical action functional is played by the quantum action functional restricted to vectors of the form $\Phi[p(t),q(t)]$. Here is further justification for our interpretation of the restricted quantum action functional as the classical action functional which was introduced in the preceding section. Drawing on that analysis we immediately see that

$$\langle p'',q'',t''|p',q',t'\rangle = \mathcal{N}\int e^{i\int[p\dot{q}-H(p,q)]dt/\hbar}\mathcal{D}p\mathcal{D}q\ .$$

But this is the very same formal expression we derived for the conventional phase-space path integral, and yet clearly the two propagators themselves are unequal. The resolution of this paradox is simply that neither formal expression is correct and that the true definitions retain the limit $\varepsilon \to 0$ to the last step. Thus their unequal final results are no surprise. Conversely, this remark implies that starting with a classical action functional, quantization by path integrals is possible in (at least) two distinct quantum mechanical representations depending on the particular choice of the discretization of the classical action functional. Even more, the ultimate interpretation of $\mathcal{D}p$ and $\mathcal{D}q$ can be different in the two choices since for the conventional case there is always one more "dp" than "dq" while for continuous representations there is always the same number. This fact is mandated by the different form of the inner product in the two cases. Thus not only does the inherent ambiguity in formal path integrals conceal factor-ordering problems, it also conceals the precise quantum mechanical representation.

These issues of ambiguity may be rephrased in another way. The formal phase-space functional integrals are expressed as if continuous and differentiable paths were all that were involved, but that is simply untrue. Generally speaking, the required paths are not differentiable and frequently not even continuous, and some rule must be adopted to extend the classical action to such unruly paths. The various ways in which this extension is possible give rise to the ambiguity we are discussing. And distinct discretizations generally lead to different extensions of the classical action to unruly paths. It is worth noting at this point that the so-called Itô and Stratonovich stochastic integrals represent two distinct resolutions of ambiguities analogous to factor-ordering problems that arise in the theory of stochastic processes precisely because Brownian motion paths, although continuous, are nowhere differentiable.[10]

Coherent States

In any attempt to evaluate the integrals involved in the discrete form of the phase-space continuous-representation path integral some choice of fiducial vector must be made, and the choice that leads to coherent states is generally most convenient. Since

$$(\Phi[z_{k+1}],\Phi[z_k]) = e^{-\frac{1}{2}|z_{k+1}|^2+z^*_{k+1}z_k-\frac{1}{2}|z_k|^2} ,$$

it follows that

$$\begin{aligned}\langle z'',t''|z',t'\rangle &\equiv (\Phi[z''],e^{-i(t''-t')H/\hbar}\Phi[z'])\\ &= \lim_{\varepsilon\to 0}\int\ldots\int\exp\left\{\sum_{k=0}^{N}[\tfrac{1}{2}(z^*_{k+1}-z^*_k)z_k-\tfrac{1}{2}z^*_{k+1}(z_{k+1}-z_k)\right.\\ &\qquad\left.-\,i\varepsilon H(z^*_{k+1};z_k)/\hbar]\right\}\\ &\qquad\times\prod_{k=1}^{N}(dx_k dy_k/\pi)\end{aligned}$$

where $z \equiv x + iy$, x and y real, and

$$H(z^*_{k+1};z_k) \equiv \frac{(\Phi[z_{k+1}],\mathcal{H}\Phi[z_k])}{(\Phi[z_{k+1}],\Phi[z_k])} \quad .$$

Note that the normalization factors cancel out and H is analytic in the two variables indicated. If $\mathcal{H}(a^\dagger,a)$ is expressed in normal-ordered form, then $H(z^*;z') = \mathcal{H}(z^*,z')$. We shall return to this expression for the propagator in the next section.

The formal expression for the path integral in coherent states reads

$$<z'',t''|z',t'> = \mathcal{N}\int \exp\{\int[\tfrac{1}{2}(\dot{z}^*z-z^*\dot{z})-iH(z^*;z)/\hbar]dt\}$$
$$\times \mathcal{D}x\mathcal{D}y \ ,$$

which formally corresponds to a functional integral over $z(t) \equiv x(t) + iy(t)$. Note that here, in phase-space form, the classical action reads

$$I = \int[\tfrac{1}{2}(p\dot{q}-\dot{p}q)-H(p,q)]dt$$

because of the phase convention for coherent states. We shall also return to this expression for the propagator in the next section.

Generalizations

The formulation of path integrals for general continuous representations based on the set $\{\Phi[\ell]\}$ follows exactly as above. We quote only the results of this analysis. For one thing, we have

$$<\ell'',t''|\ell',t'> \equiv (\Phi[\ell''],e^{-i(t''-t')\mathcal{H}/\hbar}\Phi[\ell'])$$

$$= \lim_{\varepsilon\to 0}\int\ldots\int \prod_{k=0}^{N} \{(\Phi[\ell_{k+1}],\Phi[\ell_k])$$

$$\times \exp[-i\varepsilon H(\ell_{k+1};\ell_k)/\hbar]\}$$

$$\times \prod_{k=1}^{N} \delta\ell_k \ ,$$

where $\ell_{N+1} \equiv \ell''$ and $\ell_o \equiv \ell'$, and where

$$H(\ell_{k+1};\ell_k) \equiv \frac{(\Phi[\ell_{k+1}],\mathcal{H}\Phi[\ell_k])}{(\Phi[\ell_{k+1}],\Phi[\ell_k])} .$$

For another, we have the formal expression

$$\langle \ell'',t''|\ell',t'\rangle = \mathcal{N}\int \exp\{i\int[i\hbar(\Phi[\ell],\dot{\Phi}[\ell])-(\Phi[\ell],\mathcal{H}\Phi[\ell])]dt/\hbar\} \times \mathcal{D}\ell ,$$

which is an integral over real-valued, multi-dimensional paths, $\ell(t)$. Here too we note that the quantum action functional restricted to vectors of the form $\Phi[\ell(t)]$ plays the role of the "classical" action functional in this path integral.

STATIONARY-PHASE APPROXIMATIONS

Unfortunately, very few path integrals can be evaluated exactly so various approximate methods are invoked. One such approach uses the stationary-phase approximation also called the steepest-descent method.[11] In this approach an integral (with sufficiently analytic integrand) of the form

$$A = \int e^{if(x)}dx ,$$

and integrated along the real line, is first transformed into an integral along a contour in the complexified x-plane that is arranged to cross over a saddle point defined by a solution of $f'(x_o) = 0$ and oriented so the integrand falls off as rapidly as possible on either side. The integrand is approximated by a Gaussian as it crosses over the saddle, and thus

$$A \approx e^{if(x_o)}\int e^{i\frac{1}{2}(x-x_o)^2 f''(x_o)}dx$$

$$= \sqrt{2\pi i/f''(x_o)}\; e^{if(x_o)}$$

If there are several solutions to $f'(x_o) = 0$, then the approximation is given by a sum of similar terms. The validity of such approximations improves as the overall scale of f is increased, and for large f, $\exp[if(x_o)]$ is the dominant term.

For example, the integral $\int\exp(\alpha x-\frac{1}{2}\beta x^2)dx$, $\mathrm{Re}\beta > 0$, over the real line has one extremal at $x = \alpha/\beta$, which may lie anywhere in the complex plane; the saddle point method leads to the result $\sqrt{2\pi/\beta}\,\exp(\frac{1}{2}\alpha^2/\beta)$, which for this case happens to be exact.

Conventional Path Integrals

The extension of the foregoing algorithm to conventional path integrals is fairly straightforward. For $\varepsilon > 0$, the phase-space form developed in the last section involves a (2N+1)-dimensional integral each variable of which possibly becomes complexified in the search for saddle points. The initial and final values, q' and q", remain real and fixed. Next an extremal of the effective action

$$I_N \equiv \sum_{k=0}^{N} [p_{k+\frac{1}{2}}(q_{k+1}-q_k)-\varepsilon H(p_{k+\frac{1}{2}},q_k)]$$

is sought, which leads to first-order difference equations. Subject to the proper boundary conditions, the solutions of these difference equations $\{p_{k+\frac{1}{2}},q_k\}$ give the special coordinate and mementa values of the extremal, and the dominant term of the stationary-phase approximation (valid for small $\hbar$), is given by $\exp(iS_N/\hbar)$, where we denote by S_N the value of I_N for the extremal solution. Whenever H is real the extremal solutions are also real, and thus so also is S_N. By construction the solutions of the difference equations are such that $q_{k+1}-q_k = O(\varepsilon)$ and likewise for $p_{k+\frac{1}{2}}-p_{k-\frac{1}{2}}$. In the limit $\varepsilon \to 0$ these solutions become continuous and differentiable functions p(t) and q(t) that satisfy the first-order differential equations

$$\dot{q} - \partial H/\partial p = 0$$

$$\dot{p} + \partial H/\partial q = 0$$

subject to $q(t') \equiv q'$ and $q(t'') \equiv q''$. In addition, as $\varepsilon \to 0$, the dominant approximation to the path integral tends to $\exp(iS/\hbar)$, where S is the value of the classical action

$$I = \int[p\dot{q}-H(p,q)]dt$$

evaluated for the extremal solutions p(t) and q(t). For simplicity, we assume there is only one extremal, and we shall confine our remarks just to the dominant approximation. In fact, the easiest way to choose an acceptable normalization for approximate propagators is to impose the composition law

$$\langle q''',t'''|q',t'\rangle = \int\langle q''',t'''|q'',t''\rangle dq''\langle q'',t''|q',t'\rangle ,$$

up to a suitable degree of approximation.[12]

Discussion

Superficially, the result for the dominant approximation seems also to come from a stationary-phase approximation to the formal phase-space path integral. But strictly speaking this cannot be true for the paths that enter the path integral are, in any reasonable sense, nondifferentiable and frequently not continuous. Thus, the ultimate extremal path, i.e., the continuous and differentiable solution of the classical Hamiltonian equations, is not among the paths that enter the path integral. Note that our deduction of such a path was based on the limiting behavior of extremal solutions of finite-dimensional integrals and did not presume that such a path was included in the path integral. Hence, given the fact that stationary-phase approximations of finite-dimensional integrals over the reals may necessitate an excursion outside the reals into the complex plane to find the extremal, we now note, analogously, that stationary-phase approximations of functional integrals may necessitate an excursion outside the class of paths involved in the integral to find the relevant extremal path.

This is an important point and one that is easily demonstrated. Let $\mu(s)$ be the measure on infinite sequences $\{s_k\}$ of real variables for which

$$\int \exp(\alpha\Sigma h_k s_k)d\mu(s) = \exp(\tfrac{1}{2}\alpha^2\Sigma h_k^2)$$

for any set $\{h_k\}$ such that $\Sigma h_k^2 < \infty$. It is known[13] that the measure $\mu(s)$ is concentrated on <u>non</u>square-summable sequences $\{s_k\}$ such that

$$\lim_{N\to\infty} N^{-1} \sum_{k=1}^{N} s_k^2 = 1, \quad \mu \text{ a.e.} \quad .$$

On the other hand, representation through a sequence of finite-dimensional integrals and stationary-phase approximations leads to the extremal conditions $s_k = \alpha h_k$ for all k, and thus to a square-summable sequence $\{s_k\}$ that lies outside the class on which μ is concentrated.

Continuous-Representation Path Integrals

In the last section both discrete and formally continuous path integral expressions were developed for the propagator in phase-space continuous representations. However, in any attempt to apply stationary-phase approximations to those integral expressions there appears to be an insurmountable obstacle since the very same classical Hamiltonian

$$I = \int[p\dot{q}-H(p,q)]dt$$

enters the continuous-representation case as entered the conventional case. The two first-order equations of motion that arise are simply incompatible with the presently required initial and final boundary conditions, p',q' and p",q", which generally overspecify the solution. Unfortunately this argument is too simplified and has led to some ambiguity in the literature.[14] A proper analysis is presented below.

For $\varepsilon > 0$ the effective action in the general phase-space continuous representation reads

$$I_N = \sum_{k=0}^{N} \{-i\hbar\ell n(\Phi[p_{k+1},q_{k+1}],\Phi[p_k,q_k])-\varepsilon H(p_{k+1},q_{k+1};p_k,q_k)\} .$$

Here $p_{N+1},q_{N+1} \equiv p'',q''$ and $p_0,q_0 \equiv p',q'$, while the remaining 2N variables are real-valued integration variables. In the stationary-phase approximation the boundary conditions remain real and fixed, while each of the 2N integration variables possibly becomes complexified in the search for saddle points. Extremization of the effective action I_N determines the location of the saddle points.

At this point it is convenient to specialize to coherent states (with the extra phase factor included), but to retain the p,q phase-space notation. In that case

$$\sum_{k=0}^{N} \{-i\hbar\ell n(\Phi[p_{k+1},q_{k+1}],\Phi[p_k,q_k])\} = \sum_{k=0}^{N} \{\tfrac{1}{2}(q_{k+1}p_k-p_{k+1}q_k)$$

$$+ \tfrac{1}{4}i[(p_{k+1}-p_k)^2+(q_{k+1}-q_k)^2]\} .$$

Assume for the moment that H = 0 and that this expression alone represented the effective action. The extremal relations then read

$$\tfrac{1}{2}(q_{k+1}-q_{k-1}) = \tfrac{1}{2}i(p_{k+1}-2p_k+p_{k-1}) ,$$

$$\tfrac{1}{2}(p_{k+1}-p_{k-1}) = -\tfrac{1}{2}i(q_{k+1}-2q_k+q_{k-1}) ,$$

which hold for $1 \leq k \leq N$. But these particular equations are degenerate and the only solution is $q_k \equiv q_{k-1}$, $p_k \equiv p_{k-1}$ which conflicts with the boundary conditions since in general $q'' \neq q'$, $p'' \neq p'$. To overcome this difficulty we first replace the Hamiltonian-less effective action by the expression

$$\sum_{k=0}^{N}\{\tfrac{1}{2}(q_{k+1}p_k-p_{k+1}q_k)+\tfrac{1}{4}iD[(p_{k+1}-p_k)^2+(q_{k+1}-q_k)^2]\}\ ,$$

where the variable $D > 1$. The limit $D \to 1$ is then reserved to the end of the computation. The modified extremal equations read

$$\tfrac{1}{2}(q_{k+1}-q_{k-1}) = \tfrac{1}{2}iD(p_{k+1}-2p_k+p_{k-1})\ ,$$

$$\tfrac{1}{2}(p_{k+1}-p_{k-1}) = -\tfrac{1}{2}iD(q_{k+1}-2q_k+q_{k-1})\ ,$$

which again hold for $1 \leq k \leq N$. These equations are nondegenerate and the general solution valid for large N is given by

$$q_k = \bar{q} + (q''-\bar{q})E^{(k-N-1)} + (q'-\bar{q})E^{-k}\ ,$$

$$p_k = \bar{p} + (p''-\bar{p})E^{(k-N-1)} + (p'-\bar{p})E^{-k}\ ,$$

where

$$E \equiv (D+1)/(D-1)$$

and

$$\bar{q} \equiv \tfrac{1}{2}(q''+q') - \tfrac{1}{2}i(p''-p')\ ,$$

$$\bar{p} \equiv \tfrac{1}{2}(p''+p') + \tfrac{1}{2}i(q''-q')\ .$$

Here, so long as $1 < E < \infty$, we have a solution that properly interpolates between arbitrary initial and final values, p',q' and p'',q''. Observe that these solutions rapidly vary (in just a few steps) from their boundary values to the complex values $\bar{q}$ and $\bar{p}$. In fact in the limit $\varepsilon \to 0$ the solutions are constant functions save at each end point where a finite jump occurs. Note also that the limiting solution is completely independent of E (or D) and so is unchanged as $D \to 1$. The extremal action evaluated for these solutions is the quantity that enters the stationary-phase approximation, and we shall study this quantity below in the context of the time-dependent problem.

When a Hamiltonian term is present in the effective action it leads to additional terms in the extremal equations of motion in such a way that for large N the solutions are given in the form

$$q_k \equiv \bar{q}_k + (q''-\bar{q}'')E^{(k-N-1)} + (q'-\bar{q}')E^{-k}\ ,$$

$$p_k \equiv \bar{p}_k + (p''-\bar{p}'')E^{(k-N-1)} + (p'-\bar{p}')E^{-k}\ ,$$

where $\bar{q}_k$ and $\bar{p}_k$ represent the discrete form of complex solutions to the Hamiltonian equations of motion (as was true in the case $H = 0$) subject to certain boundary conditions. Here $\bar{q}' \equiv \bar{q}_o$, $\bar{p}' \equiv \bar{p}_o$, $\bar{q}'' \equiv \bar{q}_{N+1}$, and $\bar{p}'' \equiv \bar{p}_{N+1}$, and this solution satisfies the boundary conditions $q' + ip' = \bar{q}' + i\bar{p}'$ and $q'' - ip'' = \bar{q}'' - i\bar{p}''$. To pick the proper solution, it is only necessary to choose the initial conditions as $\bar{q}' \equiv q' + w$ and $\bar{p}' \equiv p' + iw$ and adjust the complex variable w so that the time evolved solutions $\bar{q}''$ and $\bar{p}''$ satisfy the relation $\bar{q}'' - i\bar{p}'' = q'' - ip''$. Evidently the solutions $\bar{q}_k$ and $\bar{p}_k$ that satisfy these conditions are slowly varying, i.e., $\bar{q}_{k+1} - \bar{q}_k = O(\varepsilon)$, etc.

The evaluation of the extremal action involves terms that either are exclusively slowly varying or include at least one rapidly varying term. The former type are easily handled in the limit $\varepsilon \to 0$ in the usual integral fashion [and are independent of E (or D)], and so we concentrate on the latter. In evaluating the contribution of such terms to the effective action there arise nonvanishing quantities such as

$$\sum_{k=0}^{N} (q'-\bar{q}')^2(E^{-k}-E^{-k-1})^2 = D^{-1}(q'-\bar{q}')^2$$

and other quantities such as

$$\sum_{k=0}^{N} [\bar{p}_k(q'-\bar{q}')E^{-k-1}-\bar{p}_{k+1}(q'-\bar{q}')E^{-k}] = -\bar{p}'(q'-\bar{q}')$$

both of which have been evaluated in the limit $N \to \infty$. Combining all such expressions together, we find, in the limit $N \to \infty$, that the contribution to the effective action from expressions with rapidly varying ingredients is given by

$$\tfrac{1}{2}(\bar{p}''q''-\bar{q}''p''+\bar{q}'p'-\bar{p}'q')$$
$$+ \tfrac{1}{4}iD^{-1}[(p''-\bar{p}'')^2+(q''-\bar{q}'')^2+(p'-\bar{p}')^2+(q'-\bar{q}')^2] .$$

And when use is made of the boundary conditions this expression reduces just to the first term,

$$\tfrac{1}{2}(\bar{p}''q''-\bar{q}''p''+\bar{q}'p'-\bar{p}'q')$$

which is completely independent of D, and thus unchanged as $D \to 1$. By itself this expression leads to the correct answer in the case that $H \equiv 0$ for which $\bar{p}'' = \bar{p}' \equiv \bar{p}$ and $\bar{q}'' = \bar{q}' \equiv \bar{q}$, with $\bar{q}$ and $\bar{p}$ as defined before.

Summary

We are now ready to complete the expression for the extremal action in the limit $N \to \infty$. For that purpose we recognize that the complex solution we have termed $\bar{q}_k$ passes in the limit $\varepsilon \to 0$ to a complex solution we may more suggestively call $q_{c\ell}(t)$ since it is a solution of the classical equations of motion. With this terminology then $q''_{c\ell} \equiv q_{c\ell}(t'') \equiv \bar{q}''$, etc. Finally, after making this notational change and adding the purely continuum contributions we find that the dominant approximation for the propagator $\langle p'',q'',t''|p',q',t'\rangle$ is given by $\exp(iS/\hbar)$, where

$$S = \tfrac{1}{2}(p''_{c\ell}q''-q''_{c\ell}p''+q'_{c\ell}p'-p'_{c\ell}q')$$

$$+ \int[\tfrac{1}{2}(p_{c\ell}\dot{q}_{c\ell}-\dot{p}_{c\ell}q_{c\ell})-H(p_{c\ell},q_{c\ell})]dt \ .$$

Here the classical Hamiltonian is given by the diagonal coherent-state matrix elements of the quantum Hamiltonian. The action functional is expressed in terms of the complex solutions $q_{c\ell}$ and $p_{c\ell}$ of the classical equations of motion $\dot{q} - \partial H/\partial p = 0$ and $\dot{p} + \partial H/\partial q = 0$ subject to the boundary conditions $q'_{c\ell} + ip'_{c\ell} = q' + ip'$ and $q''_{c\ell} - ip''_{c\ell} = q'' - ip''$. As noted above such a solution is found by choosing the complex variable w, where $q'_{c\ell} \equiv q' + w$, $p'_{c\ell} \equiv p' + iw$, so that the time-evolved solution satisfies $q''_{c\ell} - ip''_{c\ell} = q'' - ip''$. Within this general solution a specialization to diagonal matrix elements $q'' = q'$ and $p'' = p'$ is entirely possible (e.g., so that trace expressions may be formed), but generally speaking, this will still imply that $q''_{c\ell} \neq q'_{c\ell}$ and $p''_{c\ell} \neq p'_{c\ell}$. We emphasize that in the limit $\varepsilon \to 0$ the extremal path is continuous and differentiable save at each end point where a finite jump occurs, and strictly speaking it is not one of the paths that enter the formal path integral.

Continuum Approach

Having now recognized the particular nature of the extremal path it is important to remark that it can be singled out by continuum methods. In particular one can choose the action functional

$$I_\varepsilon = \int[\tfrac{1}{2}(p\dot{q}-\dot{p}q)+\tfrac{1}{4}i\varepsilon(\dot{p}^2+\dot{q}^2)-H(p,q)]dt$$

for small but finite ε to determine the extremal solutions, and take the limit as $\varepsilon \to 0$ of this action evaluated for the extremal solutions to use in the dominant stationary-phase approximation. For $\varepsilon > 0$ the equations of motion derived from this action functional read

$$\dot{q} - \partial H/\partial p = \tfrac{1}{2}i\varepsilon\ddot{p} ,$$
$$\dot{p} + \partial H/\partial q = - \tfrac{1}{2}i\varepsilon\ddot{q} ,$$

and solutions to these equations valid for small ε are given by

$$q(t) = q_{c\ell}(t) + (q''-q''_{c\ell})e^{-2(t''-t)/\varepsilon} + (q'-q'_{c\ell})e^{-2(t-t')/\varepsilon} ,$$

$$p(t) = p_{c\ell}(t) + (p''-p''_{c\ell})e^{-2(t''-t)/\varepsilon} + (p'-p'_{c\ell})e^{-2(t-t')/\varepsilon} .$$

In these expressions the parameters have the same significance as before, and these solutions agree with the solutions of the difference equations when $t - t' \equiv k\varepsilon$ apart from the specification that $E = e^2 \approx 7.389$ or that $D \approx 1.313$. Since the principal results are independent of D this is of no consequence. In the limit $\varepsilon \to 0$ the extremal action is given by

$$S = \tfrac{1}{2}(p''_{c\ell}q''-q''_{c\ell}p''+q'_{c\ell}p'-p'_{c\ell}q')$$
$$+ \int[\tfrac{1}{2}(p_{c\ell}\dot{q}_{c\ell}-\dot{p}_{c\ell}q_{c\ell})-H(p_{c\ell},q_{c\ell})]dt$$

exactly as before. The validity of the approximate continuum solution derives from the boundary conditions and from the fact that for any E, $1 < E < \infty$,

$$\sum_{k=0}^{\infty} (E^{-k}-E^{-k-1}) = 1 = (2/\varepsilon)\int_0^{\infty} e^{-2t/\varepsilon}dt .$$

Unfortunately, the applicability of such a solution can only be justified a posteriori.

Generalization

The extension of the foregoing treatment of the stationary-phase approximation to phase-space continuous representations (other than coherent states) and to more general continuous representations is straightforward, at least in principle. For simplicity we treat these cases together by examining the situation for a general set of such vectors $\{\Phi[\ell]\}$, which may also be a phase-space set. Clearly the effective action for this situation is given by

$$I_N = \sum_{k=0}^{N} \{-i\hbar\ln(\Phi[\ell_{k+1}],\Phi[\ell_k])-\varepsilon H(\ell_{k+1};\ell_k)\}$$

where $\ell_0 \equiv \ell'$ and $\ell_{N+1} \equiv \ell''$. The extremal equations then are $\partial I_N/\partial \ell_k = 0$ for $1 \leq k \leq N$, and the solutions to these equations generally entail complex, multi-component variables, $\ell_1, \ldots, \ell_N$. Just what these equations are depends, of course, on the functional form of the terms in I_N, and in particular on the specific set of vectors $\{\Phi[\ell]\}$. In any case, it is clear that the extremal equations have the general form of a three-term recursion relation, $F(\ell_{k+1}, \ell_k, \ell_{k-1}) = 0$ for $1 \leq k \leq N$, and thus are compatible with general initial and final values, ℓ' and ℓ''. This conclusion assumes that F is nondegenerate; if it is degenerate then we may introduce an extra variable D to make F nondegenerate as was required in the case of coherent states. From analysis of the coherent state case it is also clear that in the limit $\varepsilon \to 0$ the solution in the general case tends to a continuous and differentiable, complex, multi-component function $\ell(t)$ save at each end point where a finite jump occurs. At the same time the extremal action S_N passes to a value S that enters the dominant stationary-phase approximation for the propagator $\langle \ell'', t'' | \ell', t' \rangle$.

Although the details of such an analysis cannot be worked out without further information, some insight into the limiting extremal function can be obtained by continuum methods. Analogous to the expression studied for coherent states, we now consider the action

$$I_\varepsilon = \int \{ i\hbar(\Phi[\ell], \dot{\Phi}[\ell]) - \tfrac{1}{2} i\hbar\varepsilon(\dot{\Phi}[\ell], (1 - \Phi[\ell]\Phi^\dagger[\ell])\dot{\Phi}[\ell])$$

$$- (\Phi[\ell], H\Phi[\ell]) \} dt \; .$$

The form of this expression follows directly from an expansion of the logarithm in the effective action to second order, and for coherent states this expression just reduces to the one presented earlier. The extremal equations here are second-order differential equations, and the solutions for small ε are, generally speaking, rapidly varying both initially and finally. In the limit $\varepsilon \to 0$ the solution obtained in this way is the correct limiting solution, but unfortunately it is not a priori clear that the limiting value of the action S_ε for the extremal solution leads to the correct value S. The calculation of S is of course correctly determined by the limiting behavior of the discrete formulation.

It is noteworthy in this expression for I_ε that besides the canonical kinematical form[3]

$$i\hbar(\Phi[\ell], d\Phi[\ell]) \equiv y_a(\ell) d\ell^a \; ,$$

with a summation over the L coordinates implied, we see the emergence of a second basic kinematical form,

$$\hbar(d\Phi[\ell],(1-\Phi[\ell]\Phi^{\dagger}[\ell])d\Phi[\ell]) \equiv g_{ab}(\ell)d\ell^a d\ell^b ,$$

which evidently governs the intrinsic geometry of the L-dimensional curved metric space formed as a submanifold of the Hilbert space H by the set of unit vectors $\{\Phi[\ell]\}$. For an arbitrary phase-space continuous representation expressed in canonical coordinates it is readily seen that this second form becomes

$$\hbar^{-1}<(\Delta Q)^2>dp^2+\hbar^{-1}<(\Delta P)^2>dq^2 + \hbar^{-1} <(\Delta P\Delta Q+\Delta Q\Delta P)>dpdq$$

where $\Delta Q \equiv Q - <Q>$ and $<\cdot> \equiv (\Phi_0,\cdot\Phi_0)$, etc. Thus in this special case we deal with a flat two-dimensional space, a property that is preserved under coordinate transformations such as those generated by a classical canonical transformation.

Conclusion

The approximations developed in this section certainly deserve further consideration. In particular, although we have developed the coherent-state approach most fully it is important to analyze correspondingly other phase-space approaches since for certain problems different choices of fiducial vectors may lead to improved stationary-phase approximations.

Finally, we note that the use of complex "classical" solutions in the form of a stationary-phase approximation developed in this section generally respects the unitarity condition

$$<\ell',t'|\ell'',t''> = <\ell'',t''|\ell',t'>^* .$$

This approximation is then preferable to the one based on real-valued "classical" solutions as they were derived and employed in the restricted quantum action functional since this approximation generally fails to satisfy the unitarity condition. Of course, the stationary-phase approximation of the conventional path integral discussed earlier also obeys its own form of unitarity condition, and it remains to be seen which of the two stationary-phase approximations to path integrals--conventional or continuous representation--leads to the better approximation in cases of specific interest. It is also worth pointing out that the continuous representation results may be converted into potentially different conventional results using, for example, the expression

$$\int <p'',q'',t''|p',q',t'>e^{\frac{1}{2}i(p''q''-p'q')/\hbar}dp''dp'$$
$$= 4(\pi\hbar)^{3/2}<q'',t''|q',t'>$$

which holds for coherent states. The converse connection is of course even more familiar.

REFERENCES

1. H. Weyl, The Theory of Groups and Quantum Mechanics (Dover, New York, 1931) §14; E. Wigner, Phys. Rev. 40, 749 (1932); J. E. Moyal, Proc. Cambr. Phil. Soc. 45, 99 (1949).
2. J. R. Klauder, Ann. of Physics 11, 123 (1960).
3. J. R. Klauder, J. Math. Phys. 4, 1055, 1058 (1963); ibid. 5, 177 (1964). In addition see J. McKenna and J. R. Klauder, J. Math. Phys. 5, 878 (1964).
4. See, e.g., P. M. Cohn, Lie Groups (Cambridge University Press, London, 1961), p. 110.
5. J. R. Klauder and E. C. G. Sudarshan, Fundamentals of Quantum Optics (Benjamin, New York, 1968), Chap. 7.
6. See, e.g., E. W. Aslaksen and J. R. Klauder, J. Math. Phys. 10, 2267 (1969); A. O. Barut and L. Girardello, Commun. Math. Phys. 21, 41 (1971); A. M. Perelomov, Commun. Math. Phys. 26, 222 (1972); F. A. Berezin, Commun. Math. Phys. 40, 153 (1975).
7. J. R. Klauder, J. Math. Phys. 8, 2392 (1967).
8. See, e.g., H. B. Callen, Thermodynamics (Wiley, New York, 1961), p. 24.
9. R. P. Feynman and A. R. Hibbs, Quantum Mechanic and Path Integrals (McGraw-Hill, New York, 1965).
10. See, e.g., K. Itô, Proc. Imperial Acad., Tokyo 20, 519 (1944); H. P. McKean, Jr., Stochastic Integrals (Academic Press, New York, 1969); R. L. Stratonovich, Conditional Markov Processes and Their Application to Optimal Control (Elsevier, New York, 1968); E. J. McShane, Stochastic Calculus and Stochastic Models (Academic Press, New York, 1974).
11. See, e.g., P. M. Morse and H. Feshbach, Methods of Theoretical Physics (McGraw-Hill, New York, 1953), Part I, p. 437.
12. See, e.g., W. H. Miller, J. Chem. Phys. 53, 1949 (1970), Appendix A.
13. T. Hida, Theory of Prob. Appl. (Moscow) 15, 119 (1970).
14. L. D. Faddeev, in Les Houches 1975, Proceedings, Methods in Field Theory, Edited by R. Balian and J. Zinn-Justin (North Holland, Amsterdam, 1970), p. 1.

PATH INTEGRAL ASSOCIATED WITH THE FOKKER-PLANCK EQUATION

B. Mühlschlegel

Institut für Theoretische Physik
Universität zu Köln
5000 Köln 41, Germany

SYNOPSIS

Path integrals are discussed from a general view point for classical processes and for quantum processes. The one-dimensional Fokker-Planck equation is then studied in detail. With reference to recent literature its Lagrangian is obtained. Besides explicite construction of the path integral, emphasis is put on the structure of the equation of motion and its relation to a self-adjoint problem.

1. MOTIVATION

Path-integral methods are used in quantum field theory and in statistical physics, and create there a fruitful relationship between the quantum system of interest and its classical background. This is not only for aesthetics, rather it allows to treat important practical problems. Here we mention only the possibility 1. of handling symmetries and their partial breaking, 2. of generating n-point functions by adding a source to the Lagrangian, 3. of building up perturbation theory and finally, 4. of studying the renormalizability of a field theory using the functional integral representation. We know, however, that historically the path integral in the first place was not used to relate quantum mechanical evolution to classical action (Feynman), but instead was introduced to describe the classical diffusion process (Wiener). Over the years, applications to quantum theory developed into an important and large domain. Though classical processes had been successfully extended from the simple diffusion process towards strongly non-linear phenomena for quite a while, the desire for more general path-integral formulations of such processes came into the center of interest with remarkable retardation, and only recently. Here the Fokker-Planck equation /1/ and its path integral has seen a number of contributions in the last couple of years, from which we solely mention the work of Dekker /2,3/, of Graham /4/, of Haken /5/ and of Horsthemke and Bach /6/. These publications, together with the work of Kubo /7/ have added to the education of the author and, subsequently, have stimulated him to prepare this article.

There is noticable enthusiasm about the implications of finding a "super-classical" Lagrangian contained in a path integral for classical processes /3,4/. The hope is expressed that this may be the clue for a rational formulation of non-equilibrium thermodynamics. It may be so for a general Liouville operator describing the classical (continuous Markov-) process. But it appears that just here the construction of a proper path integral is probably utopic since the operator ordering necessary /4/ is possible only in principle and not in practice. Short-time expansions of the propagator of the type

$$<x_2|e^{-\tau\mathcal{H}}|x_1> \approx <x_2|1-\tau\mathcal{H}|x_1> \qquad (1)$$

are, of course, always possible, and the "path sums" made this way for finite time serve a useful purpose as Kubo's work of the master equation shows /7/. But they are not unique /5/. Moreover, for the Fokker-Planck equation a $\tau^{1/\nu}$-expansion rather than a τ-expansion has to be used in order to arrive at the proper path integral; ν equals two due to second-order derivatives in the Liouvillean for that case. Treating the master equation with infinitely high derivatives the same way, one expects an essential singularity to appear at $\tau = 0$ which clearly invalidates a perturbative treatment.

Concerning the Fokker-Planck equation there is of course no more information in the corresponding path integral than there is in the differential equation itself; one realizes also that approximation methods for the latter are better developed and more powerful than for the former. The problematics here is, at first glance, quite reversed compared with the one in quantum field theory. Classical Lagrangians for gauge theories allow for an impressive modeling, and the path integral, as indicated above, is the great mean to quantize the theory /8/. For classical processes there is no such problem as calculating higher order corrections to scattering amplitudes. What one needs is the solution of a well behaved differential equation (or integral equation in the general case). To learn something about, let's say, the relaxation spectrum, one would not rely upon a path integral but instead would search for a suitable self-adjoint eigenvalue problem. In spite of all this, the question of the "Lagrangian" hidden behind the general diffusion process has been posed, and it deserves a unique answer, according to the authors mentioned above. One may hope that the Lagrangian of the Fokker-Planck equation might play a similar model-building role for classical processes as does, for example, the Lagrangian of the electromagnetic field for non-linear gauge theories of weak and electromagnetic interactions, but this is pure speculation, and no formalism seems to be at hand in order to find relevant solutions in the near future.

After these general remarks, let us briefly describe what is done on the following pages. We consider here the one-dimensional Fokker-Planck equation. Our treatment is thoroughly elementary and could be included in every course on quantum mechanics. We will contribute nothing to the mathematical rigor of the path-integral formulation; at the appropriate place x_2-x_1 is replaced by $\dot{x}dt$ as in previous works. In Section 2 we collect the well known properties of the Fokker-Planck equation with special emphasis on an imaginary gauge transformation. It appeared useful to us to obtain the Lagrangian associated with the process directly by contemplation about the equation of motion and without reference to the path-integral representation. The path integral is then considered at some length in Section 3. We agree with the Lagrangian of Ref. /6,4/ but have to reject the form offered in Ref. /3/. Graham's $\sqrt{\tau}$-expansion is viewed as some sort of WKB-approximation, and it is confirmed that this leads to a unique construction of the path integral. Finally in Section 4 we discuss questions already mentioned above in connection with the short-time expansion.

We regret that only the one-dimensional case is treated here. Several things become different for the Fokker-Planck equation in more than one dimension, and one has to consider conditions of integrability /1/. The $\sqrt{\tau}$-expansion is certainly independent of dimension /4/. There is a strong connection with problems of the Schrödinger equation of a particle in curved space. Covariant Schrödinger equations have recently been discussed again by Dowker /9/, and we may refer to this work also as a source of relevant literature on such problems.

2. GENERAL PROPERTIES

Consider a so called continuous Markov process x. The distribution function, or probability density w(x,t) of the variable x then obeys the Fokker-Planck equation /1/

$$-\frac{1}{2}\frac{\partial^2}{\partial x^2}\left[K_2(x)w\right]+\frac{\partial}{\partial x}\left[K_1(x)w\right]=-\frac{\partial w}{\partial t} \qquad (1)$$

where $K_2(x)>0$, $K_1(x)$ are the given diffusion and drift function.

Since (1) has the form

$$\frac{\partial G}{\partial x} + \frac{\partial w}{\partial t} = 0 \qquad (2)$$

the conservation law of probability is evident. Moreover, the stationary distribution $w_o(x)$ follows from

$$G_o = -\frac{1}{2}(K_2 w_o)' + K_1 w_o = \text{const} = 0 \qquad (3)$$

by means of a quadrature:

$$w_o(x) = e^{2\int \frac{K_1}{K_2} dx} \Big/ K_2(x) \qquad (4)$$

assuming that the K allow for natural boundary conditions at $x = \pm\infty$.

It is useful to view (1) as a Schrödinger equation with imaginary time and a non-hermitian Hamiltonian

$$(\mathcal{H} + \frac{\partial}{\partial t})\, w = 0, \quad \mathcal{H} = \frac{1}{2} p^2 K_2 + ipK_1 . \qquad (5)$$

The propagator, or Green function

$$G(2,1) = \langle x_2| e^{-(t_2-t_1)} |x_1\rangle\, \Theta(t_2-t_1) \qquad (6)$$

with $1 = (x_1,t_1)$, $2 = (x_2,t_2)$ satisfies

$$(\mathcal{H}_2 + \frac{\partial}{\partial t_2})\, G(2,1) = \delta(x_2-x_1)\, \delta(t_2-t_1) \qquad (7)$$

and governs the evolution of the distribution from time t_1 to a later time t_2:

$$w(2) = \int G(2,1) w(1)\, dx_1 . \qquad (8)$$

The adjoint equation to (1),(5) is

$$(\mathcal{H}^+ - \frac{\partial}{\partial t})\, v = 0, \qquad \mathcal{H}^+ = \frac{1}{2} K_2 p^2 - iK_1 p .$$

This "backward equation" is fulfilled by G(2,1) with respect to $1 = (x_1,t_1)$

$$(\mathcal{H}_1^+ - \frac{\partial}{\partial t_1})\, G(2,1) = \delta(x_2-x_1)\, \delta(t_2-t_1) .$$

But since G is supposed to be real:

$$G(2,1)=\langle x_2|e^{-\tau\mathcal{H}}|x_1\rangle\ \Theta(\tau) = \langle x_1|e^{-\tau\mathcal{H}^+}|x_2\rangle\Theta(\tau)\ ,\ \tau = t_2-t_1$$

we may drop the star on G .

Let λ_n, $|u_n\rangle$ be the eigenvalues and eigenstates of $\mathcal{H}$. We then assume that a general distribution function can be expanded in these states

$$w(x,\tau) = \sum_{n=0} c_n\langle x|u_n\rangle\ e^{-\lambda_n\tau}\ . \tag{9}$$

Here $\langle x|u_0\rangle = w_0(x)$, $\lambda_0 = 0$ represent the stationary distribution. The conservation law of probability and the relaxation towards equilibrium require

$$c_0 = 1,\ \int_{-\infty}^{\infty}\langle x|u_n\rangle\ dx = 0,\quad \lambda_n > 0,\quad n = 1, 2, 3\ldots \tag{10}$$

We will come back to this a few lines below.

Before considering the general case we confine our attention to constant diffusion. With

$$K_2(x) = 1,\quad K_1(x) = c(x) \tag{11}$$

it follows

$$\begin{aligned}\mathcal{H} = \frac{p^2}{2} + ipc &= \frac{p^2}{2} + \frac{i}{2}(pc + cp) + \frac{1}{2}c' \\ &= \frac{1}{2}(p + ic)^2 + \frac{1}{2}c' + \frac{1}{2}c^2\ .\end{aligned} \tag{12}$$

This looks like the Hamiltonian of a particle in an imaginary magnetic field. The field can of course be eliminated by means of an imaginary gauge transformation

$$\begin{aligned} e^{-\int^x c dx}\ \mathcal{H} e^{\int^x c dx} = e^{\int^x c dx}\ \mathcal{H}^+ e^{-\int^x c dx} &= H = H^+ \\ &= \frac{p^2}{2} + \frac{1}{2}(c' + c^2)\ .\end{aligned} \tag{13}$$

The transformation shifts the eigenvalue equation $\mathcal{H}|u_n\rangle = \lambda_n|u_n\rangle$ to a self-adjoint problem

$$H|n\rangle = \left(\frac{p^2}{2} + V\right)|n\rangle = \lambda_n|n\rangle,\ |n\rangle = e^{-\int^x c dx}|u_n\rangle \tag{14}$$

of a particle in the potential $V = 1/2(c' + c^2)$. With (4) we have $|n\rangle = w_0^{-1/2}\ |u_n\rangle$, and

$$\langle u_{n'}|w_o^{-1}|u_n\rangle = \langle n'|n\rangle = \delta_{nn'} \; . \tag{15}$$

Further, applying the inequality

$$\langle f|f\rangle + \langle g|g\rangle \geqslant 2|\mathrm{Im} \langle f|g\rangle| \tag{16}$$

to (13) it follows

$$2\langle\varphi|H|\varphi\rangle = \langle\varphi|p^2|\varphi\rangle + \langle\varphi|c^2|\varphi\rangle + \langle\varphi|i[p,c]|\varphi\rangle$$

$$= \langle p\varphi|p\varphi\rangle + \langle c\varphi|c\varphi\rangle - 2\,\mathrm{Im}\langle p\varphi|c\varphi\rangle \geqslant 0 \; . \tag{17}$$

Therefore all eigenvalues of H and consequently of $\mathcal{H}$ are in fact positive except $\lambda_o = 0$. With (15) the expansion coefficients in (9) become

$$c_n = \langle u_n|w_o^{-1}|w(\tau=0)\rangle = \int \frac{\langle u_n|x'\rangle w(x',0)}{w_o(x')} dx' \tag{18}$$

giving $c_o = 1$ as required before. We insert this in (9) and obtain

$$G(2,1) = \sum_n \frac{u_n(x_2)u_n(x_1)}{w_o(x_1)} e^{-\lambda_n\tau}\Theta(\tau) = w_o(x_2)^{1/2} G_H(2,1) w_o(x_1)^{-1/2} \tag{19}$$

where $G_H(2,1)$ is the propagator belonging to H given by (13). G_H is symmetric in x_2, x_1. The symmetry leads to

$$G(1,2) = \frac{w_o(x_1)}{w_o(x_2)} G(2,1) \tag{20}$$

with the understanding that $\tau = t_2 - t_1$ is not reversed.

Let us have a look at the "classical mechanics" associated with the Fokker-Planck equation (12). The Hamiltonian $H = p^2/2 + V$ leads to the velocity $i\dot{x} = \partial H/\partial p = p$, and the Lagrangian becomes

$$L_o(x,\dot{x}) = i\dot{x}p - H = -\frac{1}{2}\dot{x} - V(x) . \tag{21}$$

Note that we have used here the imaginary time in the same way as in the Schrödinger equation (5). "Pure" mechanics would have the time t', which is obtained from our time t by t = it'.

The gauge transformation which changes H into $\mathcal{H}$ transforms $L_o(x,\dot{x})$ into

$$L(x,\dot{x}) = L_o(x,\dot{x}) - (i\dot{x})(ic) = -\frac{1}{2}\dot{x}^2 + xc - V = -\frac{1}{2}(\dot{x}-c)^2 - \frac{1}{2}c' \tag{22}$$

The Lagrangian connected with the backward Hamiltonian $\mathcal{H}^+$ is obtained from this by reversing the velocity.

It should be emphasized that the above "classicalization" is unique as long as we insist upon the link between $\mathcal{H}$ and the hermitian H by means of the gauge transformation. When we disregard this property an ambiguity appears. Clearly,

$$\mathcal{H} = \frac{p^2}{2} + ipc \equiv \frac{p^2}{2} + i(1-\alpha)pc + i\alpha cp + \alpha c' \tag{23}$$

leads with classical quantities to

$$L^{(\alpha)}(x,\dot{x}) = \frac{1}{2}(\dot{x}-c)^2 - \alpha c' \tag{24}$$

where α is an arbitrary parameter.

We now turn to arbitrary diffusion $K_2(x) > 0$. Unlike the case of constant diffusion the p^2-part of the Hamiltonian $\mathcal{H} = p^2/2\ K_2 + ipK_1$ is here non-hermitian, too. As before we wish to collect p^2 and p into a square. For this to become possible it is inevitable to distribute K_2 symmetrically over the momenta:

$$\frac{p^2}{2} K_2 = \frac{1}{2} p\sqrt{K_2}\, p\sqrt{K_2} - i\,\frac{1}{4} p\sqrt{K_2}\, \frac{K'_2}{\sqrt{K_2}} . \tag{25a}$$

Shifting the $\sqrt{K_2}$ further to the left will produce two other -1/4 - terms and additional commutators, with the result

$$\frac{p^2}{2} K_2 = \frac{1}{2} \sqrt{K_2}p\sqrt{K_2}p - i\,\frac{3}{4} \sqrt{K_2}p\, \frac{K'_2}{\sqrt{K_2}} - \frac{3}{8}\, \frac{K'^2_2}{K_2} + \frac{1}{4} K''_2 . \tag{25b}$$

These relations will lead to two versions of the same Hamiltonian

$$\mathcal{H} = \mathcal{H}^{(1)} = \frac{1}{2}\left(p\sqrt{K_2} + i\,\frac{c^{(1)}}{\sqrt{K_2}}\right)^2 + \frac{1}{2}\sqrt{K_2}\,\frac{d}{dx}\left(\frac{c^{(1)}}{\sqrt{K_2}}\right) + \frac{1}{2}\left(\frac{c^{(1)}}{\sqrt{K_2}}\right)^2 \tag{26a}$$

with $c^{(1)} = K_1 - 1/4\ K'_2$, and

$$\mathcal{H} = \mathcal{H}^{(0)} = \frac{1}{2}\left(\sqrt{K_2}p + i\frac{c^{(0)}}{\sqrt{K_2}}\right)^2 + \frac{1}{2}\sqrt{K_2}\,\frac{d}{dx}\left(\frac{c^{(0)}}{\sqrt{K_2}}\right) + \frac{1}{2}\left(\frac{c^{(0)}}{\sqrt{K_2}}\right)^2 + \frac{1}{2}\left(\frac{K'_2}{K_2}\, c^{(0)} + \frac{1}{2} K''_2\right) \tag{26b}$$

with
$$c^{(o)} = K_1 - 3/4\ K_2'.$$

Both forms coincide, of course, for constant diffusion K_2 and reduce to (12), when K_2 is maintained there. The $p\sqrt{K_2}$ - version has actually the same structure as (12). This makes it possible to transform the general problem $\mathcal{H}|u\rangle = \lambda|u\rangle$ into one with constant diffusion. We see this at once by writing the eigenvalue equation in x-representation

$$\frac{1}{2}\frac{d}{dx}\sqrt{K_2}\frac{d}{dx}(\sqrt{K_2}u(x)) + \frac{d}{dx}\left(\frac{K_1 - \frac{1}{4}K_2'}{\sqrt{K_2}}\sqrt{K_2}u(x)\right) = \lambda u(x). \qquad (27)$$

The new variable y is introduced by

$$y = y(x) = \int^x \frac{dx}{\sqrt{K_2}} \quad . \qquad (28)$$

Multiplying (27) by K one obtains

$$-\frac{1}{2}\frac{d^2}{dy^2}\varphi + \frac{d}{dy}\bar{c}\varphi = \lambda\varphi \qquad (29)$$

where

$$\varphi = \sqrt{K_2}u, \qquad \bar{c} = \frac{K_1 - \frac{1}{4}\frac{dK_2}{dx}}{\sqrt{K_2}} \qquad (30)$$

have to be considered as functions of y through (28), assuming that K_2 allows the inversion $x = x(y)$. By the use of (13) we thus see that the problem of solving the general Fokker-Planck equation is reduced to solving the Schrödinger equation of a particle in the potential $V(y) = 1/2\ (\bar{c}^2 + d\bar{c}/dy)$. Or in mathematical terms: The general problem $\mathcal{H}|u\rangle = \lambda|u\rangle$ which is not self-adjoint, has to be written in the from $\mathcal{H} = \mathcal{H}^{(1)}$ of Eq. (26a). Then the point transformation (28) followed by a gauge transformation will produce the associated self-adjoint problem $H|n\rangle = \lambda|n\rangle$. We note that (28) is just the transformation which is also employed in the work of Horsthemke and Bach /6/.

Clearly, the orthogonality relation (15) and the symmetry of the propagator (20), proven for constant diffusion, remain true also for the general case. Explicitely:

$$\delta_{nn'} = \langle\varphi_{n'}|\bar{w}_o^{-1}|\varphi_n\rangle = \int dy\ \varphi_{n'}(y)\varphi_n(y)\ e^{-2\int\bar{c}\ dy}$$

$$= \int \frac{dx}{\sqrt{K_2}} \frac{u_{n'}(x)u_n(x)}{K_2-1} e^{-2\int dx \frac{K_1}{K_2} + \frac{1}{2}\log K_2} = \langle u_{n'}|w_o^{-1}|u_n\rangle \quad (31)$$

with $w_o(x)$ given by (4).

As before we wish to explore the "classical mechanics" behind the Fokker-Planck equation, now for arbitrary diffusion. When p,x are c-numbers, both expressions for $\mathcal{H}$ in (26) differ from one and another. We treat both cases simultaneously by introducing $\mathcal{H}^{(s)} = s\mathcal{H}^{(1)} + (1-s)\,\mathcal{H}^{(o)}$ with s=1,0. The velocity becomes

$$i\dot{x} = \frac{\partial^{(s)}}{\partial p} = \sqrt{K_2}\left[\sqrt{K_2}p + i\frac{c^{(s)}}{\sqrt{K_2}}\right] \quad (32)$$

and the elimination of p leads to the Lagrangian

$$L^{(s)}(x,\dot{x}) = i\dot{x}p - \mathcal{H}^{(s)}$$

$$= -\frac{1}{2K_2}(\dot{x}-c^{(s)})^2 - \frac{1}{2}\sqrt{K_2}\frac{d}{dx}\frac{c^{(s)}}{\sqrt{K_2}} - \frac{1}{2}(1-s)\left[\frac{K_2'}{K_2}c^{(o)} + \frac{1}{2}K_2''\right] \quad (33)$$

with

$$c^{(s)} = K_1 - \frac{3-2s}{4}K_2' \, . \quad (34)$$

With respect to the above discussion on the relation between $\mathcal{H}$ and a hermitian problem, L with s = 1 is certainly the strong favorite for the classical description. However, within the frame of formal considerations, also L with s = 0 may be possible. We may even change s between zero and one continuously. s = 1/2 would correspond to a Lagrangian which stems from a quantum Hamiltonian which is at least hermitian in its p^2-part. It is useful to make the difference between $L^{(s)}$ and $L^{(1)}$ more apparent. This is easily done by putting $c^{(s)} = c^{(1)} - 1/2(1-s)K_2'$ into (33) with the result

$$L^{(s)} = L^{(1)} - \frac{1}{2}(1-s)\frac{K_2'}{K_2}\dot{x} \, . \quad (35)$$

When one disregards the need to break K_2 into two square root factors one would face additional ambiguities of the type indicated in Eq. (24). We will come back to this question in Section 4 in connection with the discussion of short-time propagators.

3. PATH INTEGRAL

Feynman's "third way" to quantum theory /10/ is contained in his path integral representation

$$G(b,a) = \langle x_b | e^{-\frac{i}{\hbar} H(t_b - t_a)} | x_a \rangle \Theta(t_b - t_a)$$

$$= \int_{x_a}^{x_b} Dx \, e^{\frac{i}{\hbar} \int_{t_a}^{t_b} dt \, L} \Theta(t_b - t_a) \tag{1}$$

which derives the quantum mechanical propagator, or Green function of the Schrödinger equation from the classical action calculated and summed over all paths x between a and b in the indicated manner. With imaginary time, $\hbar = 1$, and $t_b \geq t_a$,

$$G(b,a) = \langle x_b | e^{-(t_b - t_a)H} | x_a \rangle = \int_{x_a}^{x_b} Dx(t) e^{\int_{t_a}^{t_b} dt \, L(x,\dot{x})} \tag{2}$$

takes the place of (1).

It is instructive to approach the problem of obtaining the path integral of the Fokker-Planck equation from several sides, and on different levels of sophistication. Here, the lowest level would be, simply to believe in the path integral, to let the problem with the measure function Dx(t) untouched and to look up straightaway at the Lagrangian in the exponent. Since we have Lagrangians at hand from our considerations in Section 2 about classical processes associated with the Fokker-Planck equation we may put (2.22) or (2.33) into (2).

For constant diffusion one is convinced by the literature /6,4,2/ that (2.22) is the right Lagrangian to use in (2). For general diffusion $K_2(x)$ the s = 1 - Lagrangian of Eq. (2.33)

$$L^{(1)} = -\frac{1}{2K_2}(\dot{x} - c^{(1)})^2 - \frac{1}{2}\sqrt{K_2}\frac{d}{dx}\frac{c^{(1)}}{\sqrt{K_2}}, \quad c^{(1)} = K_1 - \frac{1}{4}K_2' \tag{3}$$

agrees with the result of Horsthemke and Bach /6/ and of Graham /4/. The work of Dekker /3/ even offers a Lagrangian depending on a parameter $0 \leq s \leq 1$ which, at first glance, has quite a different origin than ours. Essential parts of this Lagrangian are already contained in (2.33) since inspection shows:

$$L^{(s)}_{Dekker} = L^{(s)} + \frac{1}{2}\, s(1-s)\, \sqrt{K_2}(\sqrt{K_2}^{-1})' + \frac{1}{32}\frac{K_2'^2}{K_2} . \qquad (4)$$

For $s = (0,1)$ both L are identical apart from the 1/32-term. This term depends on the discretization procedure employed by Dekker in the calculation of short-time propagators, which has subsequently been questioned by Graham /4/.

Encouraged by the easy success of the simple discussion which led to the Lagrangian $L^{(s)}$ in the previous section we now tend towards a somewhat higher level of understanding. Let us agree not to worry about some simple manipulations on functional integrals which are familiar to us from ordinary analysis but which may be doubtful here. The starting point which is the path integral

$$G_H(b,a) = \int_{y_a}^{y_b} Dy(t)\, e^{-\int_{t_a}^{t_b} dt\, [\frac{1}{2}\dot{y}^2 + V(y)]} \qquad (5)$$

for a particle in the potential V is, however, on safe mathematical ground. It represents the famous Wiener-Feynman-Kac formula

$$G_H(b,a) = E_a\{e^{-\int_{t_a}^{t_b} dt\, V(y(t))}\, \delta(y(t_b)-y_b)\} \qquad (6)$$

with $E_a\{\ldots\}$ being the expectation on the Brownian-motion process $y(t)$, $y(t_a) = y_a$.

We apply an imaginary gauge transformation

$$G(2,1) = e^{\int_{y_1}^{y_2} c(y)dy}\, G_H(2,1) \qquad (7)$$

which, of course, fits in nicely in the general convolution property of the Green functions:

$$G(3,1) = \int G(3,2)\, G(2,1)\, dy_2 \qquad (8)$$

Therefore, in

$$G_H(b,a) = \lim_{\substack{N\to\infty \\ \varepsilon\to 0}} G_H(b,N)G_H(N,N-1)\ldots G_H(2,1)G_H(1,a) \qquad (9)$$

(with the integration convention for equal coordinates y_i, $i=1\ldots N$, and $\varepsilon = \max(t_{i+1}-t_1)$) we may distribute the gauge factor over the product:

$$G(b,a) = e^{\int_{y_a}^{y_b} c(y)dy} G_H(b,a)$$

$$= \lim_{N\to\infty,\varepsilon\to 0} e^{\int_{y_n}^{y_b} cdy} G_H(b,N)e^{\int_{y_{n-1}}^{y_n} cdy} \ldots e^{\int_{y_a}^{y} cdy} G_H(1,a) \; . \quad (10)$$

Without changing the measure function, this leads to

$$G(b,a) = \int Dy(t)\, e^{-\int dt\,[\frac{1}{2}\dot{y}^2 + V(y) - c(y)\dot{y}]} \quad (11)$$

and is the Fokker-Planck propagator for constant diffusion when $V(y) = 1/2(c(y) + c'(y))$, as we know from Eq. (2.13). Therefore $L = -1/2(\dot{y}-c)^2 - 1/2c'$.

According to the discussion in Section 2, the case of arbitrary diffusion is obtained from the one with constant diffusion by means of the additional transformation $y = y(x)$ of Eq. (2.28). It acts upon a member of the convolution (10) as follows:

$$\int dy_i \; G_H(i+1,i)\, e^{\int_{y_{i-1}}^{y_i} cdy} G_H(i,i-1)$$

$$= \int \frac{dx_i}{\sqrt{K_2}(x_i)} \bar{G}_H(i+1,i) e^{\int_{x_{i-1}}^{x_i} \frac{cdx}{\sqrt{K_2}(x_i)}} \bar{G}_H(i,i-1) \quad (12)$$

where $\bar{G}_H(x_2t_2,x_1t_1) = G_H(y(x_2),t_2,y(x_1),t_1)$. Eq. (11) is then transformed into

$$G(b,a) = \int Dx(t) e^{-\int_{t_a}^{t_b} dt \left[\frac{1}{2K_2(x)}\dot{x}^2 + V(y(x)) - \frac{c(y(x))\dot{x}}{\sqrt{K_2}(x)}\right]} \quad (13)$$

and

$$L(x,\dot{x}) = -\frac{1}{2K_2(x)}(\dot{x} - \sqrt{K_2}(x)\; c)^2 - V + \frac{1}{2}c^2. \quad (14)$$

Following the arguments of Section 2 in the reverse direction, c has to be specified at the value of Eq. (2.30) in order to arrive at the general diffusion situation. This specification brings the Lagrangian (14) into the form of Eq. (13), as to be expected.

Further, we learn something from Eq. (12) about the transformation of the measure function:

$$Dy(t) = \lim_{\substack{N\to\infty \\ \varepsilon\to 0}} \frac{1}{\sqrt{2\pi(t_b-t_N)}} \prod_{i=1}^{N} \frac{dy_i}{\sqrt{2\pi\tau_i}}$$

$$= \sqrt{K_2(x_b)} \lim \frac{1}{\sqrt{2\pi(t_b-t_N)K_2(x_b)}} \prod_{i=1}^{N} \frac{dx_i}{\sqrt{2\pi K_2(x_i)\tau_i}} = \sqrt{K_2(x_b)}\, Dx(t) \tag{15}$$

where $\tau_i = t_i - t_{i-1}$, $t_0 = t_a$. The product (10) begins at its right end, after the $y \stackrel{\circ}{=} y(x)$-transformation has been performed, with

$$G(1,a) = \frac{1}{\sqrt{K_2(x_1)}}\, e^{\int_{x_a}^{x_1} \frac{c}{\sqrt{K_2}} dx}\; \bar{G}_H(1,a) \tag{16}$$

and is followed by similar factors by proceeding to the left. Calling i the postpoint and i-1 the prepoint in G(i,i-1), we may refer to (10), (12) as the postpoint representation of G(b,a) because G(i,i-1) contains the factor $K_2(x_i)^{-1/2}$. The pre-postpoint notion plays some role in the literature /2,3/. It is clear that the transformation method discussed here always corresponds to the postpoint-representation s = 1 in $L^{(s)}$. This is, of course, also seen in the transformation of the distribution function $\varphi = \sqrt{K_2}u$, $\varphi(y,t)dy = u(x,t)dx$:

$$\varphi(b) = \int G^y(b,a)\,\varphi(a)\,dy_a \rightarrow$$

$$u(b) = \int \frac{1}{\sqrt{K_2(x_b)}} G^{y(x)}(b,a)u(a)dx_a \equiv \int G^x(b,a)u(a)dx_a \,. \tag{17}$$

Superscripts are attached to the G in order to explain the respective dependence.

The third level of understanding the path integral is devoted to a discussion of Graham's $\sqrt{\tau}$-method /4/. Before we go into the details here, let us first have a look at the quantum effects (with $\hbar$) contained in a general matrix element of the type

$$M(2,1) = \langle x_2|e^{-F}|x_1\rangle \tag{18}$$

where the operator $F = F(p,x)$ is given as a function of quantum mechanical momentum and position. Imagine that in every term of the series expansion of $\exp(-F)$ the momentum parts are completely commuted through to the left of the position parts with the result

$$e^{-F} = S(p,x,\hbar) = \sum_{\nu=0}^{\infty} S_\nu(p,x)\hbar^\nu . \tag{19}$$

It follows then immediately

$$M(2,1) = \int dp \, \langle x_2|p\rangle\langle p|e^{-F}|x_1\rangle = \int\frac{dp}{2\pi\hbar}\, e^{\frac{ip}{\hbar}(x_2-x_1)} S(p,x_1,\hbar) \tag{20}$$

with p,x becoming c-numbers in passing from (19) to (20). Alternatively, one may push all p-parts to the right, and find an ordered function $T(p,x,\hbar)$ for the exponential which leads to

$$M(2,1) = \int dp\langle x_2|e^{-F}|p\rangle\langle p|x_1\rangle = \int\frac{dp}{2\pi\hbar}\, e^{\frac{ip}{\hbar}(x_2-x_1)} T(p,x_2,\hbar). \tag{21}$$

When F is hermitean, one has $S^\star = T$. With $F = F^+ + \hbar G$ the prepoint representation (20) can be related to the postpoint representation (21) in a non-hermitean case.

As an example of this general WKB-expansion consider $F = \beta H$ and the corresponding partition sum which is formally transformed into a partition integral by the use of (20):

$$Z = \Sigma e^{-\beta E_n} = \int dx\langle x|e^{-\beta H}|x\rangle = \int\frac{dpdx}{2\pi\hbar}\, e^{\log S(p,x,\hbar)} . \tag{22}$$

The power-series expansion (19) then leads to quantum corrections on classical properties which in principle (but of course not in practice) can be obtained to any desired order in $\hbar$.

The propagator of interest to us does not contain $\hbar$. However, the definition of a path integral for any process rests upon

$$G(b,a) = \lim_{\substack{N\to\infty \\ \varepsilon\to 0}} G(b,N)G(N,N-1) \ldots G(2,1)G(1,a) \tag{23}$$

(remember the integration convention) and therefore needs the propagator G(2,1) as representative factor in this product only in the limes $\tau = t_2 - t_1 \to 0^+$. Consider the Fokker-Planck propagator

$$G(2,1) = \langle x_2 | e^{-\tau[\frac{1}{2} p^2 K_2 + ipK_1]} | x_1 \rangle \tag{24}$$

which in the limit of vanishing time becomes $\delta(x_2 - x_1)$. For K_2 = const., $K_1 = 0$,

$$G_o(2,1) = \frac{1}{\sqrt{2\pi\tau K_2}} e^{-\frac{1}{2K_2}\left[\frac{x_2 - x_1}{\sqrt{\tau}}\right]^2} \tag{25}$$

shows how this delta-function peak is developed. Since the $\tau^{-1/2}$-singularity is expected also to be present in (24) it is advisable to factorize it from the rest. This, according to Graham /4/, is easily achieved by the replacement

$$\sqrt{\tau}\, p \to p \tag{26}$$

because one obtains with (20):

$$G(2,1) = \langle x_2 | e^{-[\frac{p^2}{2} K_2 + i\sqrt{\tau}\, pK_1]} | x_1 \rangle = \int \frac{dk}{2\pi\sqrt{\tau}} e^{ik\frac{x_2 - x_1}{\sqrt{\tau}}} S(k, x_1, \sqrt{\tau}) \tag{27}$$

where k is the c-number corresponding to the momentum $p = -i\sqrt{\tau}\, \partial/\partial x$. Trivially, $S_o = \exp(-1/2K_2 k^2)$ independent of x and $\sqrt{\tau}$ belongs to (25).

$\sqrt{\tau}$ takes the place of $\hbar$, and we wish to compute the first terms in the power-series expansion (19). More precisely, we want $S = S_o + S_1\sqrt{\tau} + S_2\tau + O(\tau^{3/2})$. It is to be expected that hereby all subtle effects due to non-commutativity are picked up, which in turn are brought into the product (23) by G(2,1) in the limit of vanishing τ.

For the following it is useful to note that the main problem of computing S is contained in the diffusion term rather than in the $i\sqrt{\tau}pK_1$-part because the latter can easily be handled by perturbation theory. For the sake of simplicity we put this term equal to zero and confine our attention to

$$F = \tau\mathcal{H} = \frac{1}{2} p^2 K_2. \tag{28}$$

Consider then F^n in the expansion (19) of exp(-F). One obtains

$$F^n = \left(\frac{p^2}{2}\right)^n K_2^n(x) + \left(\frac{p^2}{2}\right)^{n-1} [\, i\sqrt{\tau} p f_n(x) - \tau g_n(x) + O(\tau^{3/2})] \qquad (29)$$

where the function f_n and g_n can be found by means of recursion:

$$f_n = \frac{n(n-1)}{2} K_2^{n-1} K_2' , \qquad (30a)$$

$$g_n = \frac{n(n-1)}{2} \left[\frac{2-3n+n^2}{2} (K_2')^2 K_2^{n-2} + \frac{4n-5}{6} K_2'' K_2^{n-1}\right] . \qquad (30b)$$

These results follow by repeated use of the commutation relations

$$[\phi(x),p] = i\sqrt{\tau}\phi'(x), \; [\phi(x), \frac{p^2}{2}] = i\sqrt{\tau} p\phi'(x) \frac{\tau}{2} \phi''(x). \qquad (31)$$

Having in (29) all p-parts to the left of all functions of x, multiplication with $(-1)^n/n!$ and summation over n will lead to be desired ordered function S. Replacing p by the c-number k gives

$$S(k,x_1,\sqrt{\tau}) = e^{-\bar{F}} \Bigg[1 + \bar{F} \{ i\frac{\sqrt{\tau}}{2} kK_2' - \frac{\tau}{4} \Big(\frac{(K_2')^2}{K_2} (\bar{F}^2-2\bar{F}) + \frac{1}{3} K_2''(3-4\bar{F})\Big)\} + O(\tau^{3/2}) \Bigg] \qquad (32)$$

with $\bar{F} = 1/2K_2(x_1)k^2$.

The essential part of the calculation is by now settled. What still has to be done is the integration (27) which is gaussian due to the factor exp $(-\bar{F})$, and leads to

$$G(2,1) = \frac{e^{-\frac{1}{2K_2}\left(\frac{x_2-x_1}{\sqrt{\tau}}\right)^2}}{\sqrt{2\pi K_2 \tau}} \left[1 - \frac{3}{4} \frac{K_2'}{K_2} \left(\frac{x_2-x_1}{\sqrt{\tau}}\right) \sqrt{\tau} + \left(\frac{1}{8} K_2'' - \frac{3}{32} \frac{(K_2')^2}{K_2}\right) \tau + O(\tau^{3/2}) \right] \qquad (33)$$

with $K_2 = K_2(x_1)$. It is, of course, very tempting to replace x_2-x_1 by $\dot{x}\tau$, and to lift the bracket in the exponent with the (surprising?) result:

$$G(2,1) = \frac{e^{\tau L^{(0)}(\dot{x},x_1)}}{\sqrt{2\pi K_2(x_1)\tau}} \qquad (34)$$

where $L^{(0)}$ is given by (2.33) putting s = 0 and K_1 = 0 (the drift term can, however, easily be incorporated in (34), as mentioned above).

To complete the exercise in calculating "quantum corrections" of second order, we turn to the ordered function T(21), which corresponds to the situation where all p-parts are brought to the right of the x-parts. Here it is not necessary to repeat the whole calculation, rather for $F = 1/2p^2K_2$ we may employ the relation

$$F = F^+ - i\sqrt{\tau}\, K_2'p - \frac{\tau}{2} K_2'' \tag{35}$$

and treat in exp (-F) the terms besides F^+ by means of perturbation theory up to order τ. For $\exp(-F^+)$ the above results can be used with $i \to -i$ and x_1 replaced by x_2. We then find

$$G(2,1) = \frac{e^{\tau L^{(1)}(\dot{x},x_2)}}{\sqrt{2\pi K_2(x_2)\tau}} \tag{36}$$

where $L^{(1)}$ is the s = 1-Lagrangian of Eq. (2.33) or Eq.(3), respectively.

What have we achieved by the two formulas (34), (36) for G(2,1)? From the quantum-mechanical viewpoint, $\hbar$-expansion gives the WKB-approximation, therefore we will call (34), (36) the WKB-propagator of the Fokker-Planck equation. For constant mass $\sim K_2^{-1}$ the WKB-approximation is of course unique. For x-dependent mass, however, it matters whether one pins the mass on the prepoint 1 or on the postpoint 2, and we are not surprised to find two different WKB-expressions of the propagator. It is nice to see that the simple considerations of section 2 produce just precisely the right classical quantities needed in both cases, thus furnishing us with a deeper understanding of $L^{(s)}$. Concerning the path-integral construction (23) we use the postpoint WKB-propagator (36) with reference to the previous discussion, and obtain with (15) again

$$G(b,a) = \int_{x_a}^{x_b} Dx(t)\; e^{\int_{t_a}^{t_b} dt\, L^{(1)}(x,\dot{x})} \tag{37}$$

Someone might insist upon using instead of (36) the prepoint version (34) of the propagator in building up the path integral by means of the infinite product (23). Does this then give a different representation with $L^{(o)}$ proving that the form of the path integral is not unique? The answer is no. G(2,1) given by (34) must, as a member of the convolution (23) add in the right way to the measure function (15) and therefore needs $K_2(x_2)$ under the square root:

$$G(2,1) = \frac{e^{\tau L^{(o)}(\dot{x},x_1) + \frac{1}{2}\log\frac{K_2(x_2)}{K_2(x_1)}}}{\sqrt{2\pi K_2(x_2)\tau}} . \tag{38}$$

Expanding the log and replacing x_2-x_1 by $\dot{x}\tau$ as before, leads to $L^{(o)}(\dot{x},x_1) + 1/2\,[K_2'(x_1)/K_2(x_1)]\,\dot{x}$. One has now to remember that this precisely equals $L^{(1)}(\dot{x},x_1)$, according to Eq. (2.35). Within the WKB-approximation x_1 can be replaced by x_2 in L. Therefore we are back to Eq. (36), as consistency requires, and no other form of the path integral can be produced by the use of prepoint propagators.

4. SHORT-TIME PROPAGATORS

We look once more at the basic equation (2.8)

$$w(2) = \langle x_2|e^{-(t_2-t_1)\mathcal{H}}|w_{t_1}\rangle = \int\langle x_2|e^{-\tau\mathcal{H}}|x_1\rangle w(1)dx_1 = \int G(2,1)w(1)dx_1 \tag{1}$$

which gives the time evolution of the distribution function w. $\bar{G}(2,1)$ is called a short-time propagator when it approximates the correct G(2,1) under the integral in (1) in such a way that, for small τ, the distribution function $\bar{w}(2) = \bar{w}(x_2,t_1+\tau)$ produced by $\bar{G}$ instead of G, solves the equation of motion, i.e. here the Fokker-Planck equation in first order in τ.

At first glance, one is convinced that the so defined $\bar{G}$ is the right object to use as a factor in the convolution (3.23) which gives the propagation in finite time t_b-t_a, and defines the path integral unambiguously in the limit $\varepsilon \equiv \max\,(t_{i+1} - t_i) \to 0$. This is in fact true as long as $\mathcal{H} = 1/2p^2 + V(x)$, but it becomes wrong when $\mathcal{H}$ contains products between momentum and position functions as in $\mathcal{H} = 1/2p^2K_2 + ipK_1$. With reference to the previous section we want to make here the statement:
"Short-time propagators are no good WKB-propagators, WKB- propagators are no good short-time propagators".

We verify this statement with two examples. Clearly, a short-time propagator of the Fokker-Planck equation is

$$\bar{G}(2,1)=\langle x_2|1-\tau[\frac{p^2}{2}K_2+ipK_1]|x_1\rangle=\int\frac{dq}{2\pi}\,e^{iq(x_2-x_1)-\tau[\frac{q^2}{2}K_2(x_1)+iqK_1(x_1)]}$$

$$=\frac{e^{-\frac{\tau}{2K_2(x_1)}[\frac{x_2-x_1}{\tau}-K_1(x_1)]^2}}{\sqrt{2\pi K_2(x_1)\tau}} \qquad (2)$$

This is in general different from Eq. (3.34) since it is obtained by a τ-expansion rather than by a $\sqrt{\tau}$-expansion; it would lead to an incorrect Lagrangian $L=-\frac{1}{2K_2}(\dot{x}-K_1)^2+\frac{1}{2}(K_2'/K_2)\dot{x}$. To see the other direction, consider the WKB-propagator (3.34) for constant diffusion $K_2=1$, $K_1=c$, $c'\neq 0$

$$G(2,1)=\frac{e^{-\frac{\tau}{2}[(\frac{x_2-x_1}{\tau}-c(x_1))^2+c'(x_1)]}}{\sqrt{2\pi\tau}} \qquad (3)$$

which can be written as $G(2,1)=\bar{G}(2,1)(1-\frac{\tau}{2}c'(x_1))$. Inserted in (1) it gives

$$w(x_2,t_1+\tau)=(1-\tau\,{}_2-\frac{\tau}{2}c'(x_2))\,w(x_2,t_1)+O(\tau^2) \qquad (4)$$

with $\mathcal{H}_2=-\frac{1}{2}\frac{\partial^2}{\partial x_2^2}+\frac{\partial}{\partial x_2}c(x_2)$. w must solve the Fokker Planck equation for small τ which is the case only when the c'-term would be absent in (4).

It is obvious from this simple consideration that the $\sqrt{\tau}$-expansion is generally not superior to the τ-expansion. It is the necessary thing to do for constructing the path integral as the discussion in Section 3 has shown. But it fails when one single WKB-propagator is expected to drive the distribution from t_1 to $t_1+\tau$. We need, so to say, infinitely many WKB-propagators, contained in the path integral

$$G(2,1)=\int_{x_1}^{x_2}Dx\;e^{\int_{t_1}^{t_1+\tau}L\,dt} \qquad (5)$$

A short-time propagator $\bar{G}$ in the above sense would then be an approximation to (5) by means of a discretization procedure.

At this point, it is useful to remember that in spite of the negative statement made above, there is a version of G(2,1) which fulfills both requirements of being a short-time propagator and a good factor in the product which makes the path integral. This is the short-time approximation of Eq. (2.19) which also appears in (3.10),

$$G(2,1) = \frac{e^{-\frac{1}{2\tau}(x_2-x_1)^2 - \tau V(x_1) + \int_{x_1}^{x_2} c\,dx}}{\sqrt{2\pi\tau}} \tag{6}$$

We immediately see how the gauge transformation does the work: Taking from the integral only $c(x_1)(x_2-x_1)$ will together with $V = 1/2(c^2 + c')$ create the WKB-progagator (3). But including also the next term, $1/2c'(x_1)(x_2^2-x_1^2)$ will turn G into a good short-time propagator $\bar{G}$ as expansion with respect to $\xi = x_1 - x_2$ and Gaussian integration in (1) show. It should be further noted that G, as it stands in (6), fulfulls the general symmetry requirement (2.20) up to order τ. This is not so for both the WKB-propagator and $\bar{G}$ but they have the symmetry at least under the integral which is sufficient.

$\bar{G}$ is not unique; different forms will give the same result when inserted in the integral (1) as was noted by Haken /5/. Cause and extent of the ambiguity for the Fokker-Planck case are easily seen by incorporating all possible commutations in $\mathcal{H}$:

$$\mathcal{H} = \frac{1}{2}\, p^2 K_2 + ipK_1 = sK_2 \frac{p^2}{2} + (1-s)\frac{p^2}{2} K_2 +$$

$$+ i\,[(\alpha K_1 - s\bar{\alpha}K_2')p + p((1-\alpha)K_1 - s(1-\bar{\alpha})K_2')] + \alpha K_1' - \frac{s}{2}(2\bar{\alpha}-1)K_2'' \,. \tag{7}$$

We calculate $\bar{G}(2,1) = \langle x_2 | 1 - \tau\mathcal{H} | x_1 \rangle$ by inserting momentum eigenstates, and obtain similar to (2):

$$\bar{G}(2,1) = \frac{1}{\sqrt{2\pi K\tau}}\, e^{-\frac{\tau}{2K}\left[\frac{x_2-x_1}{\tau} - C\right]^2 - \tau D} \tag{8a}$$

with

$$K = sK_2(x_2) + (1-s)K_2(x_1)$$

$$C = \alpha K_1(x_2) + (1-\alpha)K_1(x_1) - s(\bar{\alpha}K_2'(x_2) + (1-\bar{\alpha})K_2'(x_1))$$

$$D = \alpha K_1'(x_1) - \frac{s}{2}(2\bar{\alpha} - 1)K_2''(x_1). \tag{8b}$$

Associated with p^2 and p is a threefold ambiguity in $\bar{G}$ described by the parameters α, $\bar{\alpha}$ and s. α survives for constant diffusion /5/. It is the same parameter we had in (2.24).

Short-time propagators can be calculated without difficulty also for far more complex Hamiltonians. We refer here especially to the work of Kubo et al. /7/ where

$$\mathcal{H} = \varepsilon^{-1} \int dr \, [1 - e^{-i\varepsilon r p}] \, W(x,r). \tag{9}$$

$(\mathcal{H} + \frac{\partial}{\partial t})\, w = 0$ is the master equation, scaled with ε, the reciprocal size of the system which possesses the stochastic variable x, and W is the basic transition probability. A truncation of (9) at the p^2-part leads back to the Fokker-Planck equation. One obtains with a τ-expansion

$$\bar{G}(2,1) = \int \frac{dq}{2\pi\varepsilon} \, e^{\frac{\tau}{\varepsilon}\left\{i \frac{(x_2 - x_1)}{\tau} q - \int dr \, [1 - e^{-irq}] W(x_1, r)\right\}} \tag{10}$$

For a large system (small ε) $\bar{G}$ can be further evaluated by a steepest-descents approximation.

A product of short-time propagators

$$\hat{G}(b,a) = \bar{G}(b,N)\bar{G}(N,N-1)\ldots\bar{G}(2,1)\bar{G}(1,a) \tag{11}$$

(with integration convention) should in the limit $N\to\infty$, $\tau_i \to 0$ serve as an approximation for the correct finite-time propagator under the integral. Such an expression, occasionally called "path sum" /2/, could then be useful although it does not approach the proper path integral in general. For example, it is easy to construct with (10) the path sum of the master equation which gives a possibility to discuss asymptotic solutions /7/. The construction of a path integral of the master equation by means of the methods of Section 3 is, however, impossible due to infinitely high derivatives contained in (9).

Can one calculate functional integrals numerically? There is some vague thinking about numerical functional integration /11/. One could imagine that for more complicated problems the path sums perhaps might play a positive role from a computational point of view, because of their easy accessibility.

I thank P. Entel and S. Krebs for help with the manuscript.

REFERENCES

/1/ R.L. Stratonovich, Topics in the Theory of Random Noise, Gordon and Breach, New York (1963)

/2/ H. Dekker, Physica 85A, 363 (1976)

/3/ H. Dekker, Physica 85A, 598 (1976)

/4/ R. Graham, Phys. Rev. Lett. 38, 51 (1977)
Z. Physik B26, 281 (1977)

/5/ H. Haken, Z. Physik B24, 321 (1976)

/6/ W. Horsthemke and A. Bach, Z. Physik, B22, 189 (1975)

/7/ R. Kubo in: Synergetics, H. Haken ed. B.G. Teubner, Stuttgart (1973)
R. Kubo, K. Matsuo and K. Kitahara, J. Statistical Phys. 9, 51 (1973)

/8/ E.S. Abers and B.W. Lee, Phys. Reports 9C, 1 (1973)

/9/ J.S. Dowker in: Functional integration and its applications, ed. A.M. Arthurs, Clarendon Press, Oxford (1975)

/10/ R.P. Feynman, Rev. Mod. Phys. 20, 267 (1948)

/11/ J.M. Hammersley, in: Functional integration and its applications, ed. A.M. Arthurs, Clarendon Press, Oxford (1975).

SOME ASPECTS OF FUNCTIONAL INTEGRALS AND MANY BODY THEORY

David Sherrington

Physics Department, Imperial College
Prince Consort Road
London SW7 2BZ, U.K.

1. LAGRANGIAN FORMULATION OF THE QUANTUM MANY BODY PROBLEM

Conventionally the quantum many body problem has been formulated in terms of a Hamiltonian language. In second quantization this involves the use of field operators which obey commutation (for bosons) or anticommutation (for fermions) relations for equal times;

$$\text{bosons:} \quad \Psi(\underline{x},t)\,\Psi^{+}(\underline{x}',t) - \Psi^{+}(\underline{x}',t)\,\Psi(\underline{x},t) = \delta(\underline{x}-\underline{x}') \qquad (1a)$$

$$\text{fermions:} \quad \Psi(\underline{x},t)\,\Psi^{+}(\underline{x}',t) + \Psi^{+}(\underline{x}',t)\,\Psi(\underline{x},t) = \delta(\underline{x}-\underline{x}') \qquad (1b)$$

Here, and below, we use $\underline{x}$ to denote symbolically all the relevant coordinates other than time; i.e. space, spin, etc. The δ-function is to be interpreted as Kronecker or Dirac according to whether the corresponding coordinate element is discrete or continuous.

An alternative formulation is however available employing a functional integral representation over c-number fields within a Lagrangian framework [1-4].
In this formulation the partition function is given by

$$Z = N^{-1}\int \delta\Psi \, \exp(iA) \qquad (2)$$

where A is the action, given in terms of the Lagrangian density $L\{\Psi\}$ by

$$A = \int d\underline{x} \int_{0}^{-i\beta} dt \, L\{\Psi\} \quad , \qquad (3)$$

and N is a normalization constant. In the expressions (2) and (3) the $\Psi(\underline{x},t)$ are complex c-number fields. For bosons they are within an ordinary (commuting) algebra;

$$\Psi_1^{(\star)}\ \Psi_2^{(\star)} = \Psi_2^{(\star)}\ \Psi_1^{(\star)}\ , \tag{4a}$$

and for fermions within a Grassman (anticommuting) algebra:

$$\Psi_1^{(\star)}\ \Psi_2^{(\star)} = -\ \Psi_2^{(\star)}\ \Psi_1^{(\star)}. \tag{4b}$$

t is as usual taken as pure imaginary and may be restricted to 0 to $-i\beta$ since the Ψ are periodic in t with period $i\beta$. The integration volume of the functional integral (2) is the full Hilbert space and $\delta\Psi$ is to be understood as $\delta(\mathrm{Re}\Psi)\ \delta(\mathrm{Im}\Psi)$ or more explicitly $\prod_{\underline{x},t} d(\mathrm{Re}\Psi\ (x,t)\ )\ d(\mathrm{Im}\Psi\ (x,t)\ )$. The same expressions may be taken over for the grand canonical ensemble if L is replaced by $L+\mu\Psi^{\star}\Psi$ where μ is the chemical potential. We shall normally employ the grand ensemble and subsume $\mu\Psi^{\star}\Psi$ into L.

Just as in quantum field theory the vacuum-to-vacuum amplitude provides a generator for the Green functions so in thermal physics the partition function in the presence of external or fictitious fields coupling to $\Psi,\Psi^{\star}$, or their products or derivatives, provides a generator for thermal Green functions. For example, if we add to Lagrangian a term

$$L_{ext} = -\int dq(J^{\star}(q)\ \Psi(q) + J(q)\ \Psi^{\star}(q)\ ), \tag{5}$$

where q denotes $(\underline{x},t)$ then the Green function $G(q,q')$ is given by

$$G(q,q')_J = -\ i\ \frac{1}{Z_J}\ \frac{\delta}{\delta J^{\star}(q)}\ \frac{\delta}{\delta J(q')}\ Z_J \tag{6}$$

$$= \frac{-\ i\int\delta\Psi\ \Psi(q)\Psi^{\star}(q')\ \exp(iA_J)}{\int\delta\Psi\ \exp(iA_J)} \tag{7}$$

where Z_J, A_J are the partition function and action in the presence of L_{ext}. One is normally interested in the limit $J \to 0$ so that the subscript may be dropped in (7). In conventional notation $G(q,q')$ takes the form

$$G(q,q') = -i\langle T_\tau \Psi(q)\ \Psi^{+}(q')\rangle \tag{8}$$

where T_τ is the imaginary-time ordering operator[5]. Generally the relation between Hamiltonian and Lagrangrian definitions of Green functions is

$$\langle T_\tau \tilde{B}\tilde{C} \ldots \tilde{D}\rangle$$

$$= N_\Psi^{-1} \int \delta\Psi\ \tilde{B}\tilde{C} \ldots \tilde{D} \exp(iA) \qquad (9)$$

where $\tilde{B}$, $\tilde{C}$ ... $\tilde{D}$ are field operators and

$$N_\Psi = \int \delta\Psi \exp(iA). \qquad (10)$$

For systems without interparticle interactions A has a quadratic form so that the functional integrals are Gaussian and trivially performed; e.g. for a system of free particles the action has the form

$$A_o = \int d\underline{r} \int dt\{(2i)^{-1}[\dot{\Psi}^\star(\underline{r},t)\ \Psi(\underline{r},t) - \Psi^\star(\underline{r},t)\ \dot{\Psi}(\underline{r},t)]$$

$$- (2m)^{-1}\underline{\nabla}\Psi^\star(\underline{r},t).\underline{\nabla}\Psi(\underline{r},t) + \mu\Psi^\star(\underline{r},t)\ \Psi(\underline{r},t)\} \qquad (11)$$

For simplicity we shall drop all labels other than those of space and time. Further, in this section we shall restrict explicit expressions to a single type of particle-field and to translationally-invariant systems. The extension is straight-forward.

The calculations are clearest if transformed to momentum-energy space using

$$\Psi(\underline{k},i\omega_n) = (-2\pi\Omega i\beta)^{-\frac{1}{2}} \int_\Omega d\underline{r} \int_0^{-i\beta} dt\ \Psi(\underline{r},t) \exp(-i\underline{k}.\underline{r} - \omega_n t) \qquad (12)$$

$$G(\underline{k},i\omega_n) = \int_\Omega d\underline{r} \int_0^{-i\beta} dt\ G(\underline{r},t) \exp(-i\underline{k}.\underline{r} - \omega_n t) \qquad (13)$$

where

$$\omega_n = (2n + 1)\pi\beta^{-1} \quad \text{for fermions} \qquad n = \text{integer} . \qquad (14)$$
$$\qquad 2n\pi\beta^{-1} \quad \text{for bosons}$$

A_o and G^o then take the forms

$$A_o = \sum_{\underline{k}} \sum_n (i\omega_n - k^2/2m + \mu)\ \Psi^\star(\underline{k},i\omega_n)\ \Psi(\underline{k},i\omega_n) \qquad (15)$$

and

$$G^{\circ}(\underline{k},i\omega_n) = -\,i \int \Psi(\underline{k},i\omega_n)\, \Psi^{\star}(\underline{k},i\omega_n)\, e^{iA} \times$$

$$\times \prod_{k',n'} d(\mathrm{Re}\Psi(\underline{k}',i\omega_{n'}))\, d(\mathrm{Im}\Psi(\underline{k}',i\omega_n)) \times$$

$$\times \left[\int e^{iA} \prod_{\underline{k}'n'} d(\mathrm{Re}\Psi(\underline{k}',i\omega_{n'}))\, d(\mathrm{Im}\Psi(\underline{k}',i\omega_{n'}))\right]^{-1} \qquad (16)$$

Because A is a simple sum the integrations over all $\Psi(\underline{k}',i\omega_n)$ reduce to products of simple integrals. Only those with $(\underline{k}',n') = (\underline{k},n)$ do not cancel in the numerator and denominator of (16) and these lead immediately to

$$G^{(0)}(\underline{k},i\omega_n) = (i\omega_n - k^2/2m + \mu)^{-1}\ , \qquad (17)$$

as is well known from Hamiltonian formulation.

More generally, whenever the measure function,

$$P = \exp(i\,A) \qquad (18)$$

takes the form

$$P = \exp\{-\int \Psi^{\star}(a) f^{-1}(a,b)\, \Psi(b)\}\ , \qquad (19)$$

then the Green function is

$$-\,iN^{-1} \int \Psi(a)\Psi^{\star}(b) p\ \delta\Psi = (-i)\ f(a,b). \qquad (20)$$

Conventional perturbation theory follows very directly from the Lagrangian functional integral formulation. The action of an interacting system may be written

$$A = A_o + A_1\ , \qquad (21)$$

where A_o is the action of the non-interacting system,

$$A_o = -\,i \int \Psi^{\star}(a)\ G^{(0)-1}(a,b)\ \Psi(b), \qquad (22)$$

where $G^{(0)}$ is the non-interacting one-body Green function (see eqn (20)). The true one-body Green function of the interacting system is given by

$$G = -\,i \int \Psi\Psi^{\star} P\delta\Psi \ /\ \int P\delta\Psi\ , \qquad (23)$$

where P is given by eqn (18). Perturbation theory results from expanding ,

$$P = \exp(iA_o)\,(1 + iA_1 + (iA_1)^2/2 + \dots), \tag{24}$$

and substituting into (23) to obtain an expansion for G in terms of Hermite integrals, which are trivially evaluated using the analogue of Wick's theorem; i.e. if

$$I^{(n)}(a,b,\dots z) \equiv \frac{\int \Psi(a)\Psi^{\star}(b)\dots\Psi^{\star}(z)\,P_o\,\delta\Psi}{\int P_o\,\delta\Psi} \tag{25}$$

where there are n each of $\Psi, \Psi^{\star}$, and P_o is given by (19), then

$$I^{(n)}(a,b;\dots z) = \mathrm{Perm}\,f(a,b)\dots f(y,z), \tag{26}$$

where Perm means that we add all n! permutations of a Ψ and a $\Psi^{\star}$ and further associate a factor (-1) for fermions if the rearrangement involves an odd number of interchanges. Hermite polynomials with an unequal number of $\Psi, \Psi^{\star}$ give zero integrals. The standard linked cluster expansion results immediately. To illustrate this let us consider a diagrammatic representation. Denoting a bare Green function by $\rightarrow$ and an interacting Green function by $\Rightarrow$, and taking as an example a system with a two-body interaction represented diagrammatically by

then equations (23) and (24) give

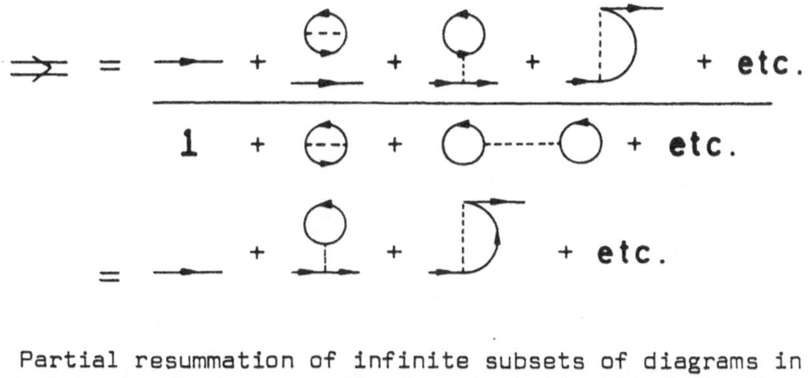

(27)

(28)

Partial resummation of infinite subsets of diagrams in (28) yields the standard renormalized perturbation theory. The lagrangian formulation does, however, also allow whole ranges of other possible perturbation expansions. One such range was discussed by Edwards and the author some years ago (2). In these short lectures there is insufficient time to discuss this approach in detail, so only an outline of the philosophy and general procedure will be given here. The idea is to expand the physical measure P about a maximally random Gaussian measure P_o

$$P_o = \exp\{ i \sum_{\underline{k},n} \Psi^{\star}(\underline{k},i\omega_n) G^{-1}(\underline{k},i\omega_n)\, \Psi(\underline{k},i\omega_n)\} \tag{29}$$

which has the property that it gives G exactly;

$$G(\underline{k},i\omega_n) = -\, i\int \Psi(\underline{k},i\omega_n)\, \Psi^{\star}(\underline{k},i\omega_n) P_o \delta\Psi \;/\; P_o\delta\Psi \quad , \tag{30}$$

so that

$$P = P_o + P_1 + P_2 \quad \ldots \tag{31}$$

with the $P_i; i > 0$ giving no extra contribution to G. To obtain the expansion we note that P satisfies the equation

$$QP = 0$$

where

$$Q = \sum_{\underline{k},n} \{ \frac{\delta}{\delta\Psi(\underline{k},i\omega_n)} [\, i \frac{\delta}{\delta\Psi^{\star}(\underline{k},i\omega_n)} + (\frac{\delta A}{\delta\Psi^{\star}(\underline{k},i\omega_n)}] + \text{conj.} \} \; ; \tag{32}$$

P_o satisfies

$$Q_o\, P_o = 0 \tag{33}$$

where

$$Q_o = \sum_{\underline{k},n} \frac{\delta}{\delta\Psi(\underline{k},i\omega_n)} [\, iD(\underline{k},i\omega_n) \frac{\delta}{\delta\Psi^{\star}(\underline{k},i\omega_n)} + W(\underline{k},i\omega_n)\Psi(\underline{k},i\omega_n)] + \text{conj.} \} \tag{34}$$

with $G(\underline{k},i\omega_n) = D(\underline{k},i\omega_n) \,/\, W(\underline{k},i\omega_n)$; (35)

another useful operator Q_2 is defined as Q_o but with D, W replaced by - R, - S given by

$$D(\underline{k},i\omega_n) = 1 + R(\underline{k},i\omega_n) \tag{36}$$

$$W(\underline{k},i\omega_n) = i\omega_n - k^2/2m + \mu + S(\underline{k},i\omega_n) \tag{37}$$

R and S are thus residue and self-energy functions related to the conventional self-energy $\Sigma(\underline{k},i\omega_n)$ by

$$\Sigma(\underline{k},i\omega_n) = - S(\underline{k},i\omega_n) + R(\underline{k},i\omega_n)G^{-1}(\underline{k},i\omega_n) \quad ; \tag{38}$$

and finally the operator Q_1 is defined by

$$Q = Q_o + Q_1 + Q_2 \quad .$$

The expression "conj" in the above refers to a related term with $\Psi \leftrightarrow \Psi^*$; in general it is a little different from a simple complex conjugate but the reader is refered to Ref. 2 for details.

A perturbation expansion for R, S follows from the solution to the sequence of equations for P_i;

$$\begin{aligned} &Q_o P_o = 0 \\ &Q_o P_1 = - Q_1 P_o - Q_2^{(1)} P_o \\ &Q_o P_2 = - Q_1 P_1 - Q_2^{(1)} P_1 - Q_2^{(2)} P_o \\ &\text{etc.} \end{aligned} \tag{39}$$

Q_o is a Hermite operator. The sequence (39) thus inverts readily to give P_i. The condition that P_i ; $i > 0$ gives no contribution to G - i.e. contains no Hermite functions of the form $(\Psi\Psi^* - iG)P_o$ - suffices to obtain straight-forwardly the perturbation expansion for R, S . A convenient diagrammatic representation is available. For details the reader is referred to the original papers (2,6). Let us note here, however, a freedom in the procedure in that D, W are not unique - nor are R, S - but need only satisfy (35). On the other hand the individual terms in the perturbation expansion depend upon the choice of the separation of D, W. A judicious choice can have advantages. This may be illustrated by considering the nominal second order perturbation term for R, S for a two-body interaction $\Psi^*\Psi^* V\Psi\Psi$;

$$S^{(2)}(q) - R^{(2)}(q)\, G^{-1}(q) = \sum_{lmn} \delta(q+l-m-n)\, V(q-n)\, [\, V(q-n) \mp V(l-n)]$$

$$\times \frac{[G(l)G(m) + G(l)G(n) - G(m)G(n) - G^{-1}(q)G(l)G(m)G(n)]}{[W(q) + W(l) - W(m) - W(n)]} \quad (40)$$

where the upper sign is for fermions, the lower for bosons. One choice is to take

$$W(q) = G^{-1}(q) \quad (41)$$

This produces conventional renormalized perturbation theory;

$$\Sigma^{(2)}(q) = - S^{(2)}(q) + R^{(2)}(q)\, G^{-1}(q)$$

$$= \sum_{lmn} \delta(q+l-m-n)\, V(q-n)[\, V(q-n) \mp V(l-n)\,]\, G(l)G(m)G(n) \quad (42)$$

Another choice, appropriate to the investigation of quasi-particles, is to choose S to be frequency independent. In this case the denominator in (40) becomes frequency-independent and an obvious separation of R, S is suggested;

$$S^{(2)}(q) = \sum_{lmn} \delta(q+l-m-n)\, V(q-n)\, [\, V(q-n) \mp V(l-n)]$$

$$\times \frac{[G(l)G(m) + G(l)G(n) - G(m)G(n)]}{[W(q) + W(l) - W(m) - W(n)]}, \quad (43)$$

$$R^{(2)}(q) = \sum_{lmn} \delta(q+l-m-n)V(q-n)[\, V(q-n) \mp V(L-n)]\, \frac{G(l)G(m)G(n)}{[W(q)+W(l)-W(m)-W(n)]} \quad (44)$$

S is thus self-consistently frequency-independent and the residue function R has no pole at the quasi-particle energy but describes the background continuum of excitations. Other special choices may be appropriate to other situations.

2. AUXILIARY FIELDS, LINEAR RESPONSE, AND LANDAU THEORY

In many problems the action can be written in the form

$$A = A_o + A_2 \quad (1)$$

where A_o corresponds to a non-interacting system and A_2 has the form

$$A_2 = - \int dq dq' \lambda^{*}(q)\, K(q,q')\, \lambda(q') \quad (2)$$

where $\lambda(q)$ is a linear combination of products of field variables Ψ,Ψ^*. A_2 describes interaction between the particles of the system; we shall refer to this as self-interaction.

The auxiliary field technique replaces the self-interaction by a suitably averaged interaction with a fictitious auxiliary field. In his lectures Prof. Mühlschlegel [7] is describing this procedure within a Hamiltonian formulation. In the Lagrangian formulation [4] it is carried out simply by employing the functional analogue of completing the square [8,9];

$$\exp\{\frac{1}{4}\int didj\, a(i)b(i,j)a(j)\}$$

$$= \frac{\int \delta x \exp\{-\int didj\, x(i)b^{-1}(i,j)\,x(j) - \int di\, a(i)x(i)\}}{\int \delta x \exp\{-\int didj\, x(i)b^{-1}(i,j)x(j)\}}\ ; \quad (3)$$

this is the analogue of the Stratonovich procedure [10,11] used in the Hamiltonian formulation. (3) leads to

$$\exp(iA_2) = N_\phi^{-1}\int \delta\phi \exp\{-\frac{i}{4}\int dqdq'\phi(q)K^{-1}(q,q')\phi(q') - \int dq\lambda(q)\phi(q)\} \quad (4)$$

where

$$N_\phi = \int \delta\phi \exp\{-\frac{i}{4}\int dqdq'\phi(q)K^{-1}(q,q')\phi(q')\}. \quad (5)$$

For simplicity we have taken λ real, but the extension is straightforward. With this substitution any Green function can be written in the form

$$\langle T\,\tilde{B}(1)\ \ldots\ \tilde{C}(n)\rangle = \frac{\int \tilde{B}(1)\ \ldots\ \tilde{C}(n)P'\delta\Psi\delta\phi}{\int P'\delta\Psi\delta\phi}, \quad (6)$$

where $\quad P' = \exp(iA_o' + iA_1'),$ (7)

with $\quad A_o' = A_o - \frac{1}{4}\int dqdq'\phi(q)K^{-1}(q,q')\phi(q')$ (8)

and $\quad A_1' = i\int dq\lambda(q)\phi(q)\ .$ (9)

$\tilde{B}, \ldots, \tilde{C}$, are again Ψ fields.

In the Hamiltonian formulation [7] ϕ is normally interpreted as an external time-dependent field which must be averaged over with the Gaussian measure $\exp\{-(i/4)\int dqdq'\phi(q)K^{-1}(q,q')\phi(q')\}$. In the Lagrangian formulation, however, Ψ,ϕ may be interpreted as interacting fields more nearly on equal terms with one another; c.f. the corresponding problem of the interaction of real fields, such as the electron-phonon problem. Indeed, just as we can define propagators for the physical fields by means of equation (6), so we can similarly introduce propagators for the auxiliary fields; for example, the analogue of the one-body Green function is

$$\Delta(q,q') = \frac{\int \phi(q)\phi(q')P'\delta\Psi\delta\phi}{\int P'\delta\Psi\delta\phi}\ ; \tag{10}$$

this function can be given a simple and useful interpretation. To see this we observe that a substitution

$$\phi'(q) = \phi(q) + 2i \int K(q,q')\lambda(q')dq' \tag{11}$$

leads to

$$A_o' + A_1' = A - \frac{1}{4}\int dqdq'\phi'(q)K^{-1}(q,q')\phi'(q'). \tag{12}$$

Thus

$$\Delta(q,q') = \Delta^{\circ}(q,q') - 4\int d\bar{q}d\bar{q}'K(q,\bar{q})\left\{ \frac{\int \lambda(\bar{q})\lambda^{*}(\bar{q}')e^{iA}\delta\Psi}{\int e^{iA}\delta\Psi} \right\} K(\bar{q}',q') \tag{13}$$

where Δ° is as Δ evaluated with

$$P' \rightarrow P_o' = \exp(iA_o); \tag{14}$$

i.e.

$$\Delta^{(o)}(q,q') = -2iK(q,q') \tag{15}$$

The expression in the curly brackets of (13) is the Green function, or propagator, $\langle T\lambda\lambda^{+}\rangle$, which is of great importance in characterising the linear response of the system to a probe which couples to λ, as well as fluctuations and dissipation in the system. We see, therefore, that, but for multiplicative factors, $(\Delta-\Delta^{\circ})$ is the propagator for λ, a higher order Green function*.

All the perturbative, diagrammatic and other techniques which have been developed for real field problems may be applied to the auxiliary field problem, but will not be discussed in detail here. Let us note explicitly only that Bose condensation of the auxiliary field corresponds to a phase transition in the real system to a state with $\langle\lambda\rangle \neq 0$.

In certain circumstances where one is primarily interested in ϕ or λ response it may be convenient to eliminate completely the

---

* For λ quadratic in $\Psi^{(*)}$ its propagator is a two-body Green function

physical particle field Ψ leading to an effective ϕ-action $A(\phi)$, the leading (quadratic - or "non interacting") part of which is

$$A_o(\phi) = -i/2 \int \phi(q) \Delta_{RPA}^{-1} (q,q') \phi(q') \tag{16}$$

where Δ_{RPA} is the random phase approximation to Δ. An approximation which has been of interest is the so-called "static approximation (12,7) in which all time-dependence of the ϕ is ignored. The further approximation in which all integrals are assumed to be dominated by the maxima of $(iA(\phi))_{static}$ is the Landau approximation; for example

$$F_{Landau} = - kT' \min (-iA(\phi)_{static}) \tag{17}$$

Much of the modern theory of critical phenomena[13] uses as its starting point the static approximation with $A(\phi)_{static}$ replaced by its continuous gradient expansion about the Landau extrema. The corresponding effective classical Hamiltonian is known as the Ginzburg-Landau-Wilson Hamiltonian.

Transformation from field to field has an appealing simplicity (at least at a qualitative level) within the Lagrangian formulation - for example, the introduction of an auxiliary field as above is essentially the inverse of the procedure used in the theory of superconductivity to eliminate the phonon field in favour of a (non-local retarded) electron-electron interaction. If, further, one's interest is in correlations of superconducting order parameter fluctuations, then one might wish to further transform the electron-electron interaction into an interaction of an auxiliary order-parameter field as obtained by taking $\lambda = \Psi\Psi$.

Most of the classic examples of the use of auxiliary fields are being discussed by Prof. Mühlschlegel (7) and so I shall not discuss them here. Instead, in the next section, I shall employ the technique in a discussion of magnetic alloys.

As stated earlier, auxiliary fields can be envisaged as effective external fields which must be averaged over with a suitable Gaussian measure, as in (4). It is interesting to consider how true spatial randomness fits into the general scheme. Suppose that the Lagrangian, or the Hamiltonian, depends locally upon some set of parameters $\{Y(x)\}$. Then any physical function f will be a functional of Y. Suppose now that $\{Y\}$ is not uniform but randomly distributed with some probability $P\{Y\}$ due to some physical disorder. Two extreme kinds of disorder can be distinguished, (i) quenched disorder in which the particular distribution $\{Y(x)\}$ is frozen and represents a constraint on the statistical mechanics, (ii) annealed disorder in which the disorder is free to take all possible values with overall probability $P\{Y\}$; in this case, the Y-distribution is summed over in the partition function just as are the spins or the spatial locations of the particles, etc. Quenched disorder is of greater physical importance although the consequences of annealed disorder may be easier to evaluate. In the case of annealed disorder the physical partition function is given, within a multiplicative temperature-independent constant, by the average partition function

$$\bar{Z} = \int \delta Y\ P\{Y\} Z\{Y\} \tag{18}$$

Thermodynamic functions are obtained from $\bar{Z}$ in the usual way; e.g.

$$F = -kT \mathrm{Ln} \bar{Z} = -kT \mathrm{Ln} \int \delta Y\ P\{Y\} Z\{Y\} \tag{19}$$

On the other hand, for the more physically relevant case of quenched disorder, averages must be taken over physical observables, such as the free energy itself, and in the thermodynamic limit a physical observable O will be equivalent to the average $\bar{O}$ given by

$$\bar{O} = \int \delta Y\ P\{Y\} O\{Y\}. \tag{20}$$

In this case the physical energy function is thus given by

$$F = -kT \int \delta Y\ P\{Y\} \mathrm{Ln} Z\{Y\}. \tag{21}$$

For example Y might be a one-body potential V(r) distributed in a Gaussian manner with zero mean but with the auto-correlation function

$$\langle V(\underline{r})\ V(\underline{r}')\rangle = W(\underline{r}-\underline{r}')$$

Then

$$P\{V\} = N_V^{-1} \exp\{-\tfrac{1}{2}\int d\underline{r} d\underline{r}' V(\underline{r}) W^{-1}(\underline{r}-\underline{r}') V(\underline{r}')\} \tag{22}$$

where

$$N_V = \int \delta V \exp\{-\tfrac{1}{2}\int dr d\underline{r}' V(\underline{r}) W^{-1}(\underline{r}-\underline{r}') V(\underline{r}')\} \tag{23}$$

Analogy with (4) is apparent. In particular if (4) is substituted into the expression for Z and the integration over the physical fields performed, one obtains direct analogy with (18).

It is much less convenient to average Ln Z than Z itself and a mathematical artifice has been proposed to put (21) into a form closer to (18) [14]. This involves noticing that

$$\mathrm{Ln} X = \lim_{n\to 0} \frac{1}{n}(X^n - 1) \tag{24}$$

for any quantity X, in particular for X = Z. If n is integral Z^n can be further expressed as $\Pi^n_{\alpha=1} Z_\alpha$, where α is a dummy label. This is equivalent to giving every field entering Z an extra dummy quantum number α, which can take on n possible values. Let us call the resulting partition function Z_n. Then

$$\int \delta Y P\{Y\} \; \mathrm{Ln} Z\{Y\} = \lim_{n\to 0} \frac{1}{n}\{\int \delta Y P\{Y\} Z_n\{Y\} - 1\} \tag{25}$$

provided the analytic continuation from discrete to continuous n can be justified. No rigorous justification is yet available, but the recent use of (25) by Edwards and Anderson [15] in demonstrating the possibility of a new phase, the spin glass phase, in strongly and competively disordered magnetic alloys has stimulated much study of both the mathematical approximation and the physical phase. We shall return to this problem in the next section.

3. MAGNETIC ALLOYS

There are many interesting problems associated with magnetic alloys. One, which excited a great many physicists for a number of years, concerns the properties of a single magnetic impurity atom in a non-magnetic matrix - the Kondo or spin fluctuation problem. We shall not discuss it here. Instead we shall consider some of the problems associated with concentrated alloys of magnetic materials and shall address two main problems, (i) the formation of magnetic moments with particular regard to the effect of statistical clustering in the case of weak moment impurities and (ii) the cooperative ordering of randomly located magnetic moments, particularly with regard to the so-called spin glass phase. For a theoretical understanding of the effects, it is imperative that the spatial disorder not be averaged out at too early a stage in the analysis or in too naive a fashion. For this reason we shall concentrate on treating as completely as possible the effects of spatial randomness and pay less attention here to treating the magnetism at the sophisticated levels now current for considering pure materials - indeed we shall use mean field theory for the most part.

Moment Formation in Transition Metal Alloys [16]

Let us consider first a simple crude model for an alloy of a magnetic transition metal, which we shall call A, and a non-magnetic (but possibly exchange-enhanced) transition metal, which we shall call B; examples of A are Ni, Fe, Co, of B are Pd, Pt, Ni. The model we shall consider is characterized by a Hamiltonian

$$H = \sum_{ij\sigma} t_{ij} \Psi^+_{i\sigma} \Psi_{j\sigma} + \sum_{i\sigma} V_i n_{i\sigma} + \sum_i V_i n_{i+} n_{i-} \tag{1}$$

where

$$n_{i\sigma} = \Psi^{+}_{i\sigma}\,\Psi_{i\sigma} \tag{2}$$

and we allow, schematically, only for 2-fold spin degeneracy, $\sigma = +$ denoting spin up, $\sigma = -$ denoting spin down. The field variables are expressed in a discrete Wannier representation and simulate the d-electrons of a true transition metal. In general the parameters t, U, V depend upon which sites are occupied by A and which by B atoms. s-p charge transfer effects will be assumed subsumed into V. Our interest in these lectures will be more in spin fluctuations of the d-electrons than in their density fluctuations and it is thus convenient to re-write the only many body term in (1), that invloving U, in such a way as to separate them; viz

$$\sum_i U_i n_{i+} n_{i-} = \frac{1}{4} \sum_i U_i \left[(n_{i+} + n_{i-})^2 - (n_{i+} - n_{i-})^2 \right] \tag{3}$$

Auxiliary fields can be introduced as in § 2, conjugate to the density (or charge) fluctuations $(n_{i+} + n_{i-})$ and the spin fluctuations $(n_{i+} - n_{i-})$. For simplicity however let us treat the charge fluctuations from the outset in a Hartree Fock approximation,

$$\sum_i (n_{i+} + n_{i-})^2 \rightarrow 2 \sum_{i\sigma} n_i n_{i\sigma} \tag{4}$$

where

$$n_i = \sum_\sigma \langle n_{i\sigma} \rangle , \tag{5}$$

and further make the following three simplifying (but inessential) restrictions on V, t, U;

(A) $V_i + U_i n_i/2 = \text{constant}$
(B) t_{ij} independent of the distribution of A and B atoms over the lattice sites
(C) $U_i = U_A$ or U_B depending upon whether the atom at i is of type A or B.

The application of these constraints results in an effective Hamiltonian with only magnetic disorder;

$$H = \sum_{\underline{k}\sigma} \varepsilon_{\underline{k}}\, \Psi^{+}_{\underline{k}\sigma} \Psi_{\underline{k}\sigma} - \frac{1}{4} \sum_i U_i (n_{i+} - n_{i-})^2 , \tag{6}$$

where the first term has been written in Bloch representation to emphasise its lattice invariance. All the spatial randomness now lies in the last (spin fluctuation) term. Treating this term within the auxiliary field technique there results the expression for the partition function

$$Z = Z_o \int \prod_i \delta\left[\left(\frac{\beta U_2}{4\pi}\right)^{1/2} m_i\right] \exp\{-\beta F([m])\}, \tag{7}$$

where

$$\begin{aligned} F([m]) = {} & \frac{1}{4} \sum_n \sum_i U_i m_i(i\omega_n) m_i(-i\omega_n) \\ & - \frac{1}{4} \sum_n \sum_{ij} U_i m_i(i\omega_n) \chi^{(o)}_{ij}(i\omega_n) U_j m_j(-i\omega_n) \\ & - \frac{1}{32} \sum_{nn'n''} \sum_{ijkL} \Lambda^{(o)}_{ijkL}(i\omega_n, i\omega_{n'}, i\omega_{n''}, -i\omega_n, -i\omega_{n'}, -i\omega_{n''}) \\ & U_i m_i(i\omega_n) U_j m_j(i\omega_{n'}) U_k m_k(i\omega_{n''}) \\ & \qquad U_L m_L(-i\omega_n - i\omega_{n'} - i\omega_{n''}), \end{aligned} \tag{8}$$

where $\chi^{(o)}$ is the space and frequency-dependent succeptibility and $\Lambda^{(o)}$ the corresponding four-point function associated with the bare band structure ε_k. The frequency summation is boson-like (see §1 eqn (14)). In the static approximation only the $\omega_n = 0$ terms are retained. Further taking the minimum of the static $F([m])$ provides the Landau (mean field) equation

$$m_i - \sum_j \chi^{(o)}_{ij} U_j m_j - \frac{1}{4} \sum_{jkL} \Lambda^{(o)}_{ijkL} U_j m_j U_k m_k U_L m_L + \ldots \tag{9}$$

Here m_i denotes the local moment (or magnetization) on site i, to be determined self-consistently. $\chi^{(o)}$ and $\Lambda^{(o)}$ are now the appropriate static functions and are independent of the AB distribution; all the spatial randomness effects are contained in the U distribution. In principle (9) will yield all the effects of spatial disorder within a general mean field theory. A complete solution will not be attempted here but it is instructive to consider some special cases

(i) All U_i equal. If $\chi^{(o)}(q)(= \sum_{ij} \chi^{(o)}_{ij} \exp(iq.R_{ij}))$ has its maximum at q = 0 all the m_i are equal and given by

$$m(1 - U\chi^{(o)}(0)) = B(Um)^3 + \ldots , \tag{10}$$

$$\text{where } B = \frac{1}{4} \sum_{jkL} \Lambda^{(o)}_{ijkL} \tag{11}$$

B is negative so that

$$m = 0 \text{ if} \quad (1 - U\chi^{(o)}(0)) > 0, \tag{12a}$$

$$m \neq 0 \text{ if} \quad (1 - U\chi^{(o)}(0)) < 0. \tag{12b}$$

This is the Stoner-Wohlfarth criterion for magnetism. If $\chi^{(o)}(q)$ has its maximum at $\underset{\sim}{q} = \underset{\sim}{Q} \neq 0$ there results a spin density wave structure with period $\underset{\sim}{Q}$, if $(1 - U^{(o)}(Q)) < 0$. For the A, B alloy under consideration we are assuming

$$1 - U_A\chi^{(o)}(0) < 0, \tag{13a}$$

$$1 - U_B\chi^{(o)}(0) > 0. \tag{13b}$$

(ii) All $U_i = 0$ except at $i = 0$. m_o is then given by

$$m_o(1 - U_o\chi^{(o)}_{oo}) = CU_o^3\, m_o^3 + \ldots \tag{14}$$

where $C = \frac{1}{4}\Lambda^{(o)}_{oooo}$ is again negative. This leads to the Friedel-Anderson-Wolff condition for a local moment;

$$1 - U\chi^{(o)}_{oo} > 0 \quad \rightarrow \quad m_o = 0 \tag{15a}$$

$$1 - U\chi^{(o)}_{oo} < 0 \quad \rightarrow \quad m_o \neq 0 \tag{15b}$$

An extension of this case is that with all U_i equal to U (say) except at site 0 where $U_o = U + \delta U$. In this case the condition for a local moment becomes

$$(1 - \delta U\tilde{\chi}_{oo}) < 0, \tag{16}$$

where

$$\tilde{\chi} = \chi^{(o)} \;/\; (1 - U\chi^{(o)}). \tag{17}$$

(iii) Particular small cluster distributions can be studied explicitly by an extension of (ii) but it is interesting to relate the problem of cluster moment formation in concentrated alloys to that of electron localization in random potentials. An analogy is found by redefining as variable

$$M_i = U_i m_i \tag{18}$$

so that (9) becomes

$$U_i^{-1} M_i - \sum_j \chi_{ij}^{(o)} M_j - \frac{1}{4} \sum_{jkL} \Lambda_{ijkL}^{(o)} M_j M_k M_L + \ldots = 0 \qquad (19)$$

so that all randomness is now in the first term, χ, Λ etc being AB-independent band-structure parameters. This can be compared with the Schrödinger equation for an electron in a disordered potential distribution [17]

$$E_i \phi_i + \sum_j t_{ij} \phi_j - E\phi_i = 0 \qquad (20)$$

In the most studied (simplest) model E_i is taken to vary from site to site but t_{ij} is assumed constant. Taking

$$E_i = U_i^{-1} \, , \, t_{ij} = - \chi_{ij}^{(o)}(0) \qquad (21)$$

and noting that

$$\sum_{jkL} \Lambda_{ijkL}^{(o)} (0) < 0 \qquad (22)$$

we see that local solutions to (19) corresponds essentially to negative energy solutions to (20). The solutions to (20) may however be either extended or localized depending upon whether their energy E is greater or less than the mobility edge energy E_c. E_c may be greater than or less than zero - it is of order

$\{\overline{U^{-1}} - \chi^{(o)}(0)\}$ where

$$\bar{f} = c_A f_A + c_B f_B, \qquad (23)$$

c_A, c_B being the concentration of A, B atoms. Thus, crudely speaking, if $E_c < 0$ ferromagnetic solutions to (19) seem likely, while if $E_c > 0$ local magnetic moments are likely in a Friedel-Anderson sense associated with randomly occurring clusters of higher-than average A concentration, leading to a picture of Curie-like regions embedded in a Pauli soup. In the latter case the magnetic clusters may be either uncorrelated from one to another or may be ordered in a ferromagnetic or a spin-glass type of sense (We shall return to this subject below). For all E_c, however, we note that (19), or the extensions without simplifications (A), (B), (C), makes allowance for statistical clustering in a manner lost in more conventional effective uniform approaches, such as the CPA.

Some further analysis of (9) may be found in earlier work of the author [16], as also a brief consideration of the analogous situation for a model alloy of a magnetic transition metal and a simple or noble metal, but relatively little application of the above ideas has been made to date.

Spin Glasses

Even if the ions of one constituent of an alloy carry well-defined local moments the problem of their possible cooperative ordering due to exchange interactions presents an interesting challenge particularly if the individual interactions have competing alignment tendencies. The study of this problem has been increasingly active in recent years.

The classical experimental manifestation of the problem is found in alloys of transition and noble metals such as CuMn or AuFe. At high concentrations of the magnetic constituent these alloys exhibit the order expected of the pure material but at lower, but still finite concentrations (typically up to order 15%), a new type of order is found in which the magnetic moments appear frozen but there is no average long-range order or magnetization. This new phase has been given the name "spin glass". Mathematically, the above manifestions may be expressed as follows

(i) frozen moment; $\overline{|\langle S_i \rangle|^n} \neq 0$ (24)

where S_i is the spin operator on site i, <> refers to a thermodynamic average and the bar to an average over magnetic sites. n is arbitrary.

(ii) no average long range order;

$$\overline{|\langle \underline{S}_i \rangle . \langle \underline{S}_j \rangle|} \to 0 \text{ as } |\underline{R}_i - \underline{R}_j| \to \infty, \qquad (25)$$

where the bar is now an average over equivalent pairs of sites.

Note that for a conventional phase, ferromagnetism, antiferromagnetism, etc. the limit in (25) would be to a periodic function of $(\underline{R}_i - \underline{R}_j)$.

The onset of the order is accompanied by a cusp in the susceptibility at some temperature T_g, as may readily be demonstrated would occur if such a freeze out of moment occured at some critical phase transition temperature T_g with $\overline{|\langle S \rangle|}$ growing continuously below T_g. On the other hand, the specific heat shows a broad rounded maximum at a temperature higher than T_g with no discernible singularity at T_g (at the time of writing)- this suggests that much entropy freezes out in effectively isolated clusters above T_g and that if there is a true phase transition the specific heat singularity at T_g is very weak.

Another feature observed experimentally is that of slow time-dependence of reactions to various stimuli (such as the application or removal of a magnetic field). This suggest that the system is very glassy-like, having a number of metastable low-lying energy states with slow relaxation between them and to the true lowest energy state or states.

A Hamiltonian appropriate to the magnetic ordering in these alloys is

$$H = - \sum_{i,j} J(\underline{R}_i - \underline{R}_j)S_i.S_j \tag{26}$$

where i,j run over the sites occupied by the magnetic ions and $J(\underline{R}_i - \underline{R}_j)$ has the Ruderman-Kittel-Kasuya-Yosida (RKKY) form, oscillating with $(\underline{R}_i - \underline{R}_j)$ with a period of k_F^{-1}, where k_F is the Fermi wave vector. The origin of J lies in interaction via polarization of the conduction electrons and can be derived from more fundamental models such as discussed in the previous section. Since the magnetic ions are distributed randomly on the underlying lattice and the oscillations in $J(\underline{R})$ are incommensurate with the lattice, a pair of magnetic ions chosen randomly may have either a ferromagnetic or antiferromagnetic direct interaction. Thus, at low concentrations of the magnetic constituent the total ordering information carried by all exchange routes, direct or indirect (via intermediate non-magnetic ions), will be randomly ferromagnetic or antiferromagnetic for different pairs of ions even at equivalent separations. The standard deviation of the total interaction strength will scale more slowly than the number of contributing paths (in fact as $n^{1/2}$ where n is the number of the paths). In the high concentration limit, on the other hand, equivalent pairs will receive predominantly equivalent information of a mean strength which will scale directly with the number of contributing paths. The random competition occurring for the low concentration case has been given the name "frustration" (20) and is believed responsible for the occurrence of a spin glass phase*.

---

*A quantitative measure of the degree of competition can be given in terms of a determination function D defined as

$$D = \frac{\sum_{\text{pairs}} \left| \sum_{\text{paths}} \left(\prod_{\text{pathlinks}} J_{pL} / |J_{pL}| \right) W_{\text{path}} \right| W_{\text{pair}}}{\sum_{\text{pairs}} \sum_{\text{paths}} W_{\text{path}} W_{\text{pair}}}$$

where "pair" refers to a pair of magnetic ions, "path" to a path between such a pair via an arbitrary number of intermediate magnetic ions, J_{pL} is an exchange interaction associated with the corresponding pathlink, W_{path} is a suitable weighting factor associated with the strength of the J_{pL} along the path, and W_{pair} is a pair weighting factor. For a pure ferromagnet, or a simple antiferromagnet in which all the bonds can be simultaneously satisfied, D = 1, but if D = 0 no simple conventional order is possible. A measure of frustration is given by $f = (D^{-1} - 1)$.

Let us demonstrate heuristically within a crude mean field theory [21] how the above-mentioned phase structures arise. For simplicity we shall use the Ising analogue of (26). If P(H) is the distribution of local effective fields, then mean field theory gives

$$m = \overline{\langle S_i \rangle} = \int dH\, P(H) \tanh(\beta H), \tag{27}$$

$$q = \overline{\langle S_i \rangle^2} = \int dH\, P(H) \tanh^2(\beta H), \tag{28}$$

The simplest approximation for P(H) is to assume a mean field

$$\bar{H} = m\, \bar{J} = m\, \overline{\sum_j J(\underline{R}_i - \underline{R}_j)} \tag{29}$$

and a variance

$$(\Delta H)^2 = \overline{H^2} - (\bar{H})^2 = q(\Delta J)^2 = q\,\{\overline{\sum_j J^2(\underline{R}_i - \underline{R}_j)} - \overline{(\sum_j J(\underline{R}_i - \underline{R}_j)\,)^2}\} \tag{30}$$

We shall ignore all other moments. Substituting (29) and (30) into (27) and (28) leads to a prediction of the possible phases (i) paramagnet $q = m = 0$ (ii) ferromagnet $q \neq 0$, $m \neq 0$ (iii) spin glass $q \neq 0$, $m = 0$. The critical temperatures for transitions between paramagnetic and ordered phases are given by the non-trivial solutions to the linearized equations for q, m. They are

$$\text{para to ferro:} \quad T_c = \bar{J}/k \tag{31}$$

$$\text{para to spin glass:} \quad T_g = \Delta J/k \tag{32}$$

The phase transition with the higher critical temperature is that which will occur. At high concentrations of the magnetic constituent $\bar{J} > \Delta J$ and the transition is to ferromagnetism, but at low concentrations $\Delta J > \bar{J}$ and spin glass character results.

Edwards and Anderson introduced a more subtle mean field theory for this problem based on equations (24) and (25) of §2. The simplest model within which to discuss the procedure is characterized by

$$H = -\frac{1}{2} \sum J_{ij} \sigma_i \sigma_j \; ; \quad \sigma_i = \pm 1 \tag{33}$$

where there is an Ising spin σ on each lattice site i, and the bonds run over nearest neighbours and are distributed randomly and independently of one another with a symmetric distribution

$$P(J_{ij}) = P(-J_{ij}), \tag{34}$$

which we shall take as Gaussian of standard deviation $\tilde{J}$:

$$P\{J_{ij}\} = \prod_{(ij)} \left((2\pi)^{1/2}\, \tilde{J}\right)^{-1} \exp(-J_{ij}^2/2(\tilde{J})^2); \tag{35}$$

where (ij) refers to a nearest neighbour bond. Applying (25) yields

$$\bar{F} = -kT \lim_{n\to 0} \{\mathrm{Tr}\, \exp(-\frac{\beta^2}{2} \sum_{ij} \sum_{\alpha,\beta=1}^{n} (\tilde{J})^2 \sigma_i^\alpha \sigma_j^\alpha \sigma_i^\beta \sigma_j^\beta\,) - 1\} \tag{36}$$

Here, α, β are dummy indices, each taking n values; they may be considered to label replicas which do not interact in the unaveraged system but which become effectively interacting when the J_{ij} average is performed. Edwards-Anderson mean field theory corresponds to introducing the variational parameter

$$q = \langle\sigma_i^\alpha\, \sigma_i^\beta\rangle \quad ; \qquad \alpha \neq \beta \tag{37}$$

making the substitution* in (36)

$$\frac{1}{2} \sum_{ij} \sigma_i^\alpha \sigma_i^\alpha \sigma_i^\beta \sigma_i^\beta \rightarrow zq\sum_i (\sigma^\alpha\sigma^\beta - q/2) \; ; \qquad \alpha \neq \beta, \tag{38}$$

where z is the coordination number, and finding the physical F and q from the extremal condition

$$\delta F/\delta q = 0 \; . \tag{39}$$

The physical interpretation of q is as in (28); i.e. $q = \overline{\langle S_i\rangle^2}$, and the above procedure for this special case (35) yields the same self-consistency equation for q as given by (28) with P(H) Gaussian with the variance ΔJ equal to $\sqrt{z}\tilde{J}$ and with zero mean. If in place of (35) a shifted Gaussian P(J) with mean J_o is employed and the additional variational mean field parameter $m = \langle\sigma^\alpha\rangle$ introduced,

---

*This is the analogue of the approximation made for a pure ferromagnet:

$$\frac{1}{2} \sum_{ij} \sigma_i\sigma_j \rightarrow zm \sum_i (\sigma_i - m/2) \quad \text{where } m = \langle\sigma_i\rangle$$

there results precisely the pair of equations (27), (28) with Gaussian P(H) shifted to have mean $\bar{J} = zJ_o$. Details of the self-consistent solutions may be found in Refs. 19 and 23; briefly, for the interesting case of $\Delta J \gg \bar{J}$, q is found to be zero for $T > T_g$ and to grow continuously to 1 as T is lowered from T_g to zero, and the susceptibility and the specific heat are predicted to have cusp-like maxima at T_g.

Both of the above mean field theories are wrong in their details and could not be expected to give either the dynamic glassy relaxation behaviour found experimentally or the cluster entropic freeze-out effects seen in the experimental specific heat. The EA mean field theory further appears to have thermodynamically unacceptable consequences (such as negative entropy as $T \to 0$ even in the Ising model), but there seems no reason to believe either that the simple model (33) is inadequate or that the procedure of §2 equation (25) (or this section, equation (36)) is unacceptable, although clearly its combination with (38) is. Much work is currently in progress on understanding spin glasses such as modelled by (36), but will not be discussed further here.

Poor Moment Spin Glasses

In the previous sub-section were discussed only good moment spin glasses - those in which the magnetic constituent carried a good local moment. I would like to conclude by discussing briefly the possibility of spin glasses in poor moment systems such as discussed in the first sub-section of this chapter. As discussed there, even in an AB alloy in which an isolated A atom does not carry a good local moment, statistically occuring regions of greater-than-average A-atom density can stabilize a moment in a generalized Friedel-Anderson sense. Interactions between such clusters, mediated by the non-magnetic matrix and A atoms in less dense regions, can lead to cooperative ordering, provided they are strong enough to overcome the destabilizing dynamic Kondo-like effects which have been ignored in our Hartree-like treatment. If the intercluster interactions are sufficiently frustrated the ordering might be expected to be of generalized spin-glass character; the concentration region in which this might be expected is an intermediate one, sufficiently concentrated to overcome the Kondo-like destabilizing dynamic processes but not sufficient to reduce the frustration to a point where ferromagnetic order takes over. RhCo shows the anticipated sequence of phases with increasing concentration [22]; Pauli-paramagnet, spin-glass, ferromagnet.

A starting point to analyse such systems would be the model of equation (6) and its associated partition function (7), employed in an analysis based on §2 equation (25), analogous to that discussed in the previous sub-section.

References

1. J.S. Bell, 1962, 'Lectures on the Many Body Problem (Naples Spring School)' Ed. E.R. Caianiello (New York, London: Academic Press), pp 81-9
2. S.F. Edwards and D. Sherrington, 1967, Proc. Phys. Soc. 90, 3-22
3. D. Sherrington, 1967, Proc. Phys. Soc. 90, 583-4
4. D. Sherrington, 1971, J. Phys. C4, 401-416
5. L.P. Kadanoff and G. Baym, 1962, 'Quantum Statistical Mechanics' (New York; Benjamin)
6. D. Sherrington, 1966, Ph. D. thesis (University of Manchester) unpublished
7. B. Mühlschlegel, 1977, these lectures
8. S.F. Edwards, 1955, Proc. Roy. Soc. A 232, 371-6
9. I.M. Gel'fand and A.M. Yaglom, 1960, J. Math. Phys. 1, 48-69
10. R.L. Stratonovich, 1957, Dokl. Akad. Nauk SSSR 115, 1097-1100 (Sov. Phys. Dokl. 2, 416-9)
11. J. Hubbard, 1959, Phys. Rev. Lett. 3, 77-8
12. B. Mühlschlegel, unpublished notes University of Pennsylvania, referenced by Wang et al, 1969, Phys. Rev. Lett. 23, 92-5
13. See for example S.K. Ma, 1976, 'Modern Theory of Critical Phenomena' (New York: Benjamin) or C. Domb and M.S. Green (ed), 1976, 'Phase Transitions and Critical Phenomena (New York: Academic Press)
14. S.F. Edwards, 1970, in '4th Int. Conf. on Amorphous Materials' (ed. R.W. Douglas and W. Ellis, New York: Wiley)
15. S.F. Edwards and P.W. Anderson, 1975, J. Phys. F5, 965-74
16. D. Sherrington and K. Mihill, 1974, Proc. Int. Conf. Mag. (Moscow 1973) Vol. 1 (1), 283-87 ; J. de Phys. 35, C4, 199-201.
17. P.W. Anderson, 1958, Phys. Rev. 109, 1492-1505
18. See for example the review by K. Fisher, Int. Conf. on Magnetism (Amsterdam 1975)
19. D. Sherrington, 1975, AIP Conf. Proc. 29, 224-228
20. G. Toulouse, 1977, Comm. Phys. 2 , 115-119
21. B.W. Southern, 1976, J. Phys. C9, 4011-4020
22. B.R. Coles, A. Tari and H.C. Jamieson, 1974, Proc. L.T. XIII, 414
23. D. Sherrington and S. Kirkpatrick, 1975, Phys. Rev. Lett. 35, 1792-96

PATH INTEGRALS IN QUANTUM AND STATISTICAL PHYSICS

G.J. Papadopoulos

Department of Physics, University of Athens

Panepistimiopolis, Athens 621, Greece

I. HISTORICAL SURVEY

Once upon a time, mathematics was a product of everyday experience. Later, it's main source became the study of the physical world. This, on account of the unique capability of mathematics to describe the physical law, and furthermore, its ability to extract information from the mathematical form of the law. These two facts kept physicist and mathematician close for almost throughout the history of science. However, in recent years, the professional pure mathematician turned his mathematics into a self-propelling discipline for its own sake, thus progressively alienating himself from the needs of the physicist. This fall off in partnership made the physicist switch from mainly being the best customer to a manufacturer of mathematics as well. It would not then be unfair if we labelled the mathematics expounded in the present course as mathematics made in physical community.

The purpose of the present lectures is to provide an introduction to path (or functional) integrals using, as a vehicle for presentation, certain applications from which the subject originated. We shall in particular begin with examples from Brownian motion.

To facilitate the discussion in the survey it will suffice to imagine a functional to be a function of infinitely many variables which are labelled by a continuous index.

The systematic development of the functional calculus began with Volterra (1) at the start of the century. In his work various limiting processes of the analysis of functions, such as continuity and differentiability were appropriately transcribed to the functional regime. Perhaps the most significant contribution made by Volterra was a general method for handling functional operations. This consists in approximating a functional with a function of N variables; thus initially reducing the problem of infinitely many variables to one involving functions of N variables. The results, thus obtained, depend on N which in the end is made to go to infinity. An example of such a procedure is the well known technique for solving Fredholm's integral equation.

Among the first who established integration of functionals is Daniell (2) (1919) in connection with the mean value of a functional. Functional integration as a means for solving partial differential equations of stochastic nature was produced by Wiener (3) in the early 20's. Wiener succeeded in obtaining the fundamental solution (propagator) of the diffusion equation.

In general integration of a functional over all its variables, by analogy to the integration of multi-variable functions, leads to divergences. However, this is not a case for concern, for in the cases of physical interest there is always a cure dictated by the circumstances of the problem. The remedy lies in the introduction of a weighting factor per variable of integration, or a joint weighting function. The joint weight for all variables constitutes the measure of integration, which for many purposes is a probability, but that is not always necessary. Appropriate choice of the measure of integration of special functionals leads to the fundamental solution of certain partial differential equations of mathematical physics. In this way Wiener obtained the propagator of the diffusion equation, using as a measure Einstein's expression for the joint probability of finding a brownian particle in a succession of space intervals during a corresponding succession of time intervals.

Chandrasekhar (4) (1943) treated the theory of Brownian motion definitely by way of functional integration, though nowhere did he explicitly state it.

In the year 1932 Dirac (5) with his paper on the role of Lagrangian mechanics in quantum theory, laid the foundation stone of what was destined to become in the hands of Feynman, a new formulation of quantum mechanics. Indeed it was Feynman (1943) who began the building of the theory, through his thesis on space-time approach to non-relativistic quantum mechanics. With his subsequent work (6) on quantum electrodynamics, the statistical treatment of liquid helium, and other areas, he raised the subject to the rank of a new discipline. As with all great developments it takes some time before they become common property of the theoretical physics community. However, in recent years, we have been witnessing further and deeper penetration of the Feynman flavour in all sorts of areas. In years to come the growth will be greater and hopefully outstanding, yet unresolved, problems will draw nearer to a solution.

Before we go any further, for the record of the early history, we should mention that Kirkwood (7) already speculated in a short conference address in 1933, that integration of special functionals in a Wiener sense, could be applied to quantum physics in connection with the evaluation of the statistical sum.

With the quantum theory of fields (8) the names of Schwinger, Bogoliubov, Edwards, Salam, Matthews, Friedrichs, Shapiro, Segal, Symanzik, and others are associated. Edwards and Peierls (9) in the early 50's achieved approximate integration of functional differential equations in this connection.

The first functional formalism of hydrodynamics is due to Hopf (10) (1952). He obtained a characteristic functional embracing all hydrodynamic information and thus he hoped to treat the problem of turbulence. However, Hopf succeeded in extracting rather limited information from his characteristic functional. Further progress was made (11) by Edwards and Rosen in the 60's. Brittin and Chappell (12) (1966) worked out a formalism for MHD.

The classical many-particle theory was tackled by Bogoliubov (13) (1946) who developed a generating functional containing all reduced 1,2,...,-particle distributions. This functional obeys a linear equation of motion, first order in time. Hosokawa (14) (1967) obtained a functional integral representation for Bogoliubov's generating functional. The formalism can also be transcribed to the quantum regime.

The most important feature of the many-particle and hydrodynamic functional theories, is their linearity. This is on account of the capability of functionals to embrace an infinite system of linear equations (relating to the various reduced distributions) into which the original nonlinear equation of motion can be decomposed. However, gain from linearity is dissipated in the increasing complexity of functional manipulations.

In recent years Edwards (15) has applied path integral methods in polymerized matter. Of the application of path integrals in solid state involving nonequilibrium statistical physics, we mention here the Thornber-Feynman theory (16) for conductivity calculations.

This discussion clearly shows that there is practically no branch of theoretical physics where one cannot proceed via a functional formalism. The question arises whether functional procedures can reach results inaccessible to other methods. In certain cases this is so; others are more suited to nonfunctional techniques. But, if one's taste is functional at the present stage of the calculus the store of algorithms, although growing, is not sufficient for one to expect full benefit from the functional treatment. However, in the case of the polaron problem (17) the path integral treatment has prevailed over the other methods. The various perturbation expansions in functional integrals are independent of the expansion parameter size. Of the disadvantages at the moment, there appears to be the difficulty of performing in a closed form evaluation in curvilinear co-ordinates (18). The path integral evaluation of the propagator in spherical polar co-ordinates for a Coulomb potential still remains to be done. However, the curvilinear path integral has gone through several stages of construction (19) and its understanding has reached a satisfactory state of refinement.

In recent years some progress has been made towards inclusion of the spin in path integral evaluations relating to the Dirac equation (20). This will constitute Professor Devreese's lectures.

One thing we have not yet spoken about is how

statistics gets into the functional integrals. In the Feynman path integral, appropriate for systems described by Schrödinger hamiltonian, it is attained by permuting the particles in the final positions of the paths in all possible ways (21). In the case of Bose statistics the resulting propagators are summed as they are while for Fermi statistics the propagators associated with old permutations carry a factor of -1 in the sum.

Now, how are things done when the statistics is inherent in the hamiltonians themselves, the so called second quantized hamiltonians? Functional integrals for Bose hamiltonians have been developed by Matthews and Salam (1955) and Klauder (1960) originally and followed by others (22). These functional integrals were essentially path integrals over complex paths represented by c-numbers, but as Matthews and Salam pointed out, they could be extended to cover the Fermi case with use of quaternions. Klauder in his 1960 paper where he introduced the states that later on became known as coherent, was the first to introduce an algorithm for Fermi states using just complex numbers. Later (23) (1976) we introduced a more convenient as we believe, form of complex number labelled states for Fermi systems which can derive from an appropriate form of the coherent states expressed through Bose operators, by mere replacement of the Bose operators by Fermi operators. The spin (1/2) statistics can also be derived through replacement of the Bose operators by an appropriate hybrid of Bose-Fermi operators. On this we shall talk in the present course. This about ends the present state of the calculus of path integrals, but a lot remains to be done by way of evaluations.

The presentation of these lectures will be rather heuristic and leading directly to the heart of the subject. We shall make no claim to rigour, but really you do not always require rigour to create confidence in your derivations. The development of concepts will be by way of example, for we prefer to combine presentation with comfort.

Now I would like to apologise for obviously having left out important references. This is not only due to the inevitable impossibility of including these in the space of a short survey, but also to the fact that I am not aware of everything that has to do with path integrals.

Next I will enter into the main part of the lectures.

II. NOTION OF THE FUNCTIONAL BY EXAMPLE

We shall introduce the notion of the functional using a particular example from brownian motion, interweaving in this way physics with mathematics.

When you have particles in a liquid environment they suffer continuously collisions from the molecules of the surrounding medium due to the thermal agitation of the latter. As a result of the thermal kicks a particle of approximately colloidal size and below develops an irregular random walk and this is Einstein's (24) conception of brownian motion. Langevin (25) took this picture and gave it a mathematical dress, and in this way he produced his dynamical equation for the motion of a brownian particle. The Langevin model simulates the many-particle medium interaction by a hydrodynamic dissipative force and a random force.

For simplicity we write down the one dimensional Langevin equation for the free brownian particle :

$$m \frac{d^2x}{dt^2} = - \frac{m}{t_o} \frac{dx}{dt} + f(t) \qquad \text{(II.1)}$$

where t_o is the particle relaxation time, $-(m/t_o)\dot{x}$ is the frictional force, and $f(t)$ is a random force with statistical properties deriving from the thermal state of the medium. The force $-(m/t_o)\dot{x} + f(t)$ takes care of the total medium interaction on the brownian particle.

Let us now obtain the solution of (II.1) for the velocity $V(t)$ which satisfies the initial condition $\dot{X}(t') = V(t') = v'$:

$$V(t) = \exp\left(\frac{t'-t}{t_o}\right)v' + \int_{t'}^{t} \exp\left(\frac{\tau-t}{t_o}\right)m^{-1}\, f(\tau)\, d\tau \qquad \text{(II.2)}$$

For the velocity at time t complete knowledge of the force $f(\tau)$ from 0 tot t is required. By changing $f(\tau)|t' \leqslant \tau \leqslant t$ we get another $V(t)$. $V(t)$ depends on the whole range of values of the force f from t' to t. With each function $f(\tau)\|[t',t]$ a value of $V(t)$ is associated according to (II.2). We have a functional and indicate this by the notation $V_{t'}^{t}[f(\tau)]$.

Actually the Langevin equation is an equation of motion which is in common for each member of an ensemble of brownian particles each of which in general experiences a different succession of random forces $f(\tau)$ during the time interval $[t',t]$. Expression (II.2) gives the velocity of the brownian particle at time t under the influence of a particular random force $f(\tau)$ over the interval $[t',t]$. For another brownian particle we have a different random force, say $f'(\tau)$ during the same interval. For finding the velocity associated with the random force $f'(\tau)$ (same initial condition) we just remove the force $f(\tau)$ from (II.2) and plug in the force $f'(\tau)$. Here, the ensemble of particles during the time interval t' to t generates a collection of forces $f(\tau)$ with each of which a particle velocity is paired, and this pairing makes the velocity functional.

In general, a functional is a mathematical device which pairs each function f, out of a given collection of functions with a value $V[f]$.

As it was pointed out in the survey, a functional can be thought of as a function of infinitely many variables labelled by a continuous index. In our example of the functional (II.2) from brownian motion the $f(\tau)$'s are the variables and τ forms the continuous label. To make the statement more visible we draw a diagram with a few random forces.

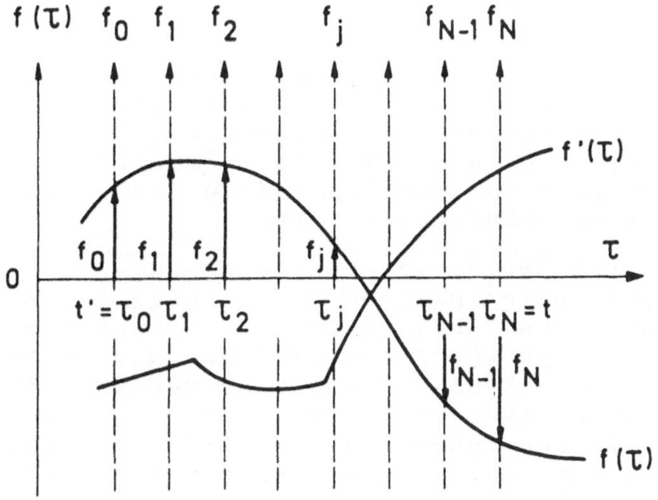

Fig. (II.1)

The various $f(\tau_j) = (f_j)$ are used as variables and are able to supply any function $f(\tau)$ over the interval $[t',t]$ in a specimen form.

If we make use of a fine partition P_N with points $t' = \tau_0 < \tau_1 \ldots < \tau_{N-1} < \tau_N = t$ of the interval $[t',t]$ we can write our functional $V[f]$ for a given argument function $f(\tau)$ approximately as :

$$V[f] \approx \exp\left(\frac{t'-t}{t_0}\right)v' + m^{-1}\sum_{j=0}^{N-1} \exp\left(\frac{\tau_j - t}{t_0}\right) f(\tau_j)\Delta\tau_j$$

$$= V_{P_N}(f_0, f_1, \ldots, f_j, \ldots, f_{N-1}) \qquad \text{(II.3)}$$

If we take another force f what we have to do to find the corresponding velocity $V(t)$ is to change the various f_j's so that they sample the new force f.

By the sampling function procedure we can approach a given functional for any of its argument functions. Now if we wish to increase the accuracy of the approximation what we have to do is to make use of finer and finer partitions which tend to cover the interval $[t',t]$. Given a functional $\Phi[f]$ the net result of using a sequence of such finer and finer partitions, $\{P_N\}$, is to obtain a corresponding sequence of functions, $\{\Phi_{P_N}\}$ with an ever increasing number of variables which in the limit as $N \to \infty$ goes to the given functional. We have :

$$\lim_{N\to\infty} \Phi_{P_N} = \Phi\,[f] \qquad \text{(II.4)}$$

As stated earlier, the partitions employed should cover the interval $[t',t]$ as $N\to\infty$. This condition is ensured using a sequence of partitions $\{P_N\}$ for which : $\max.\Delta\tau_j$ (of P_N) $\to 0$ as $N \to \infty$. The isomeric partition (all $\Delta\tau_j$ equal) does this job and we shall subsequently adopt it, unless otherwise stated.

Clearly the preceding discussion establishes an intimate association between functionals and many-variable functions. This is very important for it is the window through which one will be able to transfer the well known analysis of the many-variable functions to the regime of the functionals and thus without much effort acquire the functional calculus. But most important, we can define a functional (and this will prove more significant and more general in the good sense of generality) using a sequence of many-variable functions associated with a sequence of partitions tending to cover a given interval of indices. We shall meet this situation frequently in the construction of path integrals for obtaining the propagator of the Schrödinger equation.

Next we proceed to familiarize ourselves with the functional integral.

III. A FUNCTIONAL INTEGRAL

The propagators of several important differential equations of mathematical physics, such as the Schrödinger equation, the Fokker-Planck equation and other equations of stochastic nature can be expressed in terms of functional integrals. In developing the present section we shall be guided by applications from the theory of brownian motion. However, initially, we shall introduce the notion of our functional integral without making explicit reference to applications.

Suppose $\Phi_{P_N} = \Phi_{P_N}(f_0, f_1, \ldots, f_{N-1})$, (III.1)

a function of N variables, is an approximation of the functional $\Phi[f]$ associated with the partition P_N of the interval $[t', t]$

If we have a function of N variables we know what its multiple integral over a certain region $\mathcal{R}$ is :

$$J_{P_N} = \int \ldots_{\mathcal{R}} \int \Phi_{P_N}(f_0, f_1, \ldots, f_{N-1})\, df_0\, df_1 \ldots df_{N-1} \qquad \text{(III.2)}$$

With finer and finer partitions, which tend to cover the interval $[t', t]$ the number of variables in (III.2) grows larger and larger and we hope to find a limit for the sequence J_{P_N} which will be our functional integral. But, in general, the unlimited increase in the number of variables leads to divergences. However, this should not be a case for concern, for in the cases of physical interest there is always cure, which is dictated by the problem under study. The remedy lies in the introduction of a weighting factor per variable, or a joint weighting function for all variables.

The weighting function forms the measure of integration, which for many purposes is a probability, but that is not necessary to be always so. Suppose we have for the measure :

$$W_{P_N} \prod_{j=0}^{N-1} df_j = W_{P_N}(f_0, f_1, \ldots, f_{N-1}) \prod_{j=0}^{N-1} df_j \qquad \text{(III.3)}$$

We form now a new sequence of multiple integrals :

$$I_{P_N} = \int_{-\infty}^{+\infty} \ldots \int \Phi_{P_N}(f_0, f_1, \ldots, f_{N-1}) W_{P_N}(f_0, f_1, \ldots, f_{N-1}) \prod_{j=0}^{N-1} df_j$$

(III.4)

With physical weighting functions we have good mathematics, i.e. finite limits. We now define : If there exists

$$\lim_{N\to\infty} I_{P_N} = I[f] \qquad \text{(III.5)}$$

irrespective of the sequence of partitions $\{P_N\}$ provided of course (max. subinterval of P_N) $\to 0$, then $I[f]$ is the functional integral of the functional $\Phi[f]$ given through the sequence (III.1) with respect to the measure of integration supplied by the weighting sequence (III.3). Clearly the functional integral $I[f]$ is a new functional.

Although it would be pointless to look for the limit of the sequence W_{P_N} of (III.3) alone, independently of the multiple integration process (III.4), nevertheless we agree to denote the limiting process (III.5) giving the functional integral $I[f]$ by the following suggestive notation :

$$I[f] = \int \Phi_{t'}^{t}[f(\tau)] \; W_{t'}^{t}[f(\tau)] \prod_{t' \leqslant \tau < t} df(\tau) \qquad \text{(III.6)}$$

Furthermore, we shall make use of the term functional differential for what is denoted by the symbol $W_{t'}^{t}[f(\tau)] \prod_{t' \leqslant \tau < t} df(\tau)$ the precise meaning of which is obtained through the sequence (III.4). However one can give approximately a good meaning to the functional differential using an extremely fine partition of the interval $[t', t]$. The situation here is analogous to that of the generalized functions; indeed we are dealing here with a generalized functional, and the examples which follow will clarify the notion in question.

When the measure of integration, W_{P_N}, is a probability density then the functional integral, $I[f]$ is the mean value of the functional $\Phi[f]$, and in such a case we shall make use of the notation : $I[f] = \langle\Phi[f]\rangle$. Essentially, here we have a functional average, Such averages are useful in statistical physics and we shall display their utility with an example from the theory of brownian motion, to which the functional integrals owe their origin.

We follow Chandrasekhar's procedure (1943) (4). At the moment let us consider a brownian particle acted upon by a time prescribed external force, $\vec{F}(\tau)$. The equation of motion for such a particle is given by the following Langevin equation :

$$m\ddot{\vec{x}} = -\frac{m}{t_0}\dot{\vec{x}} + \vec{f}(\tau) + \vec{F}(\tau) \qquad \text{(III.7)}$$

It was pointed out in section (II) that the Langevin equation is an equation of motion in common for each member of and ensemble of brownian particles. The random element of the ensemble is the thermal force $\vec{f}(\tau)$ experienced by a brownian particle, as a result of the thermal agitation of the surrounding medium. Now, the collision frequency for a typical brownian particle is of the order of $10^{20}\ s^{-1}$ which means that we can almost talk about the distribution of the random force at each moment of time, meaning of course within extremely short time intervals; many, many orders of magnitude smaller than the relaxation time t_0. These considerations, essentially, justify the use of a functional distribution for the thermal force.

The basic problem is, once given the functional distribution of the thermal force, what are the average values of quantities deriving from the stochastic equation (III.7)? These quantities are functionals of the thermal force, and their averages are obtained through functional integration. The functional distribution of the random force derives from the following hypotheses, usually made about the thermal force : (i) For an extremely short time interval $[\tau, \tau + \Delta\tau]$, $(\Delta\tau \ll t_0)$ the probability of finding the random force with values lying between $\vec{f}(\tau)$, and $\vec{f}(\tau) + d\vec{f}(\tau)$ (here τ just serves as a label for the variable $\vec{f}(\tau)$) is Gaussian :

$$W(\vec{f}(\tau))\, d\vec{f}(\tau) = \left(\frac{\Delta\tau}{2\pi C}\right)^{3/2} \exp\left(-\frac{1}{2C}\vec{f}^{\,2}(\tau)\Delta\tau\right) d\vec{f}(\tau) \qquad \text{(III.8)}$$

Clearly (III.8) is normalised.
(ii) The random forces at two different times are not correlated. So, the joint probability for the random

force over the subintervals of an extremely fine partition P_N of a finite interval [t',t] will be the product of the individual probabilities over the various subintervals.

$$W_{P_N}[\vec{f}_j] \prod_{j=0}^{N-1} d\vec{f}_j =$$

$$\exp\left(-\sum_{j=0}^{N-1} \frac{1}{2C} \vec{f}_j^{\,2} \Delta\tau_j\right) \prod_{j=0}^{N-1} \left(\frac{\Delta\tau_j}{2\pi C}\right)^{3/2} d\vec{f}_j \qquad \text{(III.9)}$$

Furthermore, C is determined from the random force autocorrelation so that the average kinetic energy of the free brownian particle coincides with (3/2)kT. In this way C is found to be :

$$C = 2\frac{m}{t_o} kT \qquad \text{(III.10)}$$

With the functional probability (III.9) we have all of what we need to do a few functional integrations. As a first example let us obtain the average velocity of our brownian particle.

The solution of (III.7) for the particle velocity $\vec{V}(t)$, satisfying the condition $\vec{V}(t') = \vec{v}'$, is given by the functional :

$$\begin{aligned}\vec{V}(t) &= \vec{V}^{t}_{t'}[\vec{f}(\tau)] \\ &= \exp\left(\frac{t'-t}{t_o}\right)\vec{v}' + \int_{t'}^{t} \exp\left(\frac{\tau-t}{t_o}\right) m^{-2} \vec{F}(\tau)\, d\tau \\ &\quad + \int_{t'}^{t} \exp\left(\frac{\tau-t}{t_o}\right) m^{-1} \vec{f}(\tau)\, d\tau\end{aligned} \qquad \text{(III.11)}$$

In order to obtain the functional average of (III.11) we construct a sequence $\{V_{P_N}\}$ of N-variable functions using a sequence of partitions P_N with $\max.\Delta\tau^{(N)} \to 0$. We have :

$$\vec{V}_{P_N} = \overline{\vec{V}(t)} + \sum_{j=0}^{N-1} \exp\left(\frac{\tau_j - t}{t_o}\right) m^{-1} \vec{f}_j \Delta\tau_j \qquad \text{(III.12)}$$

where :

$$\overline{\vec{V}(t)} = \exp\left(\frac{t'-t}{t_o}\right)\vec{v}' + \int_{t'}^{t} \exp\left(\frac{\tau-t}{t_o}\right) m^{-1} \vec{F}(\tau)\, d\tau \qquad \text{(III.13)}$$

is the particle velocity produced by the application of only the external force.

Next, we obtain the average value of V_{P_N} against the force distribution W_{P_N} given by (III.9). We have :

$$< \vec{V}_{P_N} > = \int \vec{V}_{P_N} W_{P_N} \prod_{j=0}^{N-1} d\vec{f}_j = \overline{\vec{V}(t)} \qquad \text{(III.14)}$$

Passing to the limit of continuality we obtain the average value of our velocity functional, i.e. as $N \to \infty$ we have :

$$< \vec{V}^{t}_{t'}[f]> = \overline{\vec{V}(t)} \qquad \text{(III.15)}$$

This completes the functional integration at hand.

For obtaining (III.14) we have made use of the fact that W_{P_N} is a normalised distribution, and furthermore it is a product of Gaussians each with zero mean. Let us now write down these two facts together with a further average, the random force autocorrelation :

$$\int \begin{cases} 1 \\ f_k^{(\alpha)} \\ f_k^{(\alpha)} \; f_\ell^{(\alpha)} \end{cases} \times W_{P_N} \prod_{j=0}^{N-1} d\vec{f}_j = \begin{cases} 1 \\ 0 \\ C \; \delta_{\alpha\beta} \dfrac{\delta k\ell}{\Delta\tau_k} \end{cases} \qquad \text{(III.16)}$$

The superscripts α,β stand for the components 1,2,3 of the force $\vec{f}$, while the indices k,ℓ tag the variables associated with the points τ_k, τ_ℓ of the partition P_N.

It is quite clear that for any two times τ and τ' we can always arrange so that they are contained in all of our partitions. With this in mind, we now pass to the limit of infinite refinement of our time interval partitioning, and so we obtain the following functional integrals :

$$\int \begin{cases} 1 \\ \vec{f}(\tau) \\ f^{(\alpha)}_{(\tau)} \; f^{(\beta)}_{(\tau')} \end{cases} \times W^{t}_{t'}[f(\tau)] \prod_{t' \leq \tau < t} d\vec{f}(\tau) = \begin{cases} 1 \\ 0 \\ C\delta_{\alpha\beta}\delta\,(\tau-\tau') \end{cases}$$

(III.17)

In other words we have, with respect to the functional distribution (III.9) for the random force, the averages :

$$< \vec{f}(\tau) > = 0, \quad < f^{(\alpha)}_{(\tau)} f^{(\beta)}_{(\tau')} > = C\delta_{\alpha\beta}\, \delta(\tau-\tau') \tag{III.18}$$

It should be noted here that in performing our functional averages we could have used a functional distribution over a wider time interval $[t_1, t_2]$ i.e. with $t_1 < t'$ and $t_2 > t$.

Let us now make use of the averages (III.18) for determining the value of the coefficient C in the thermal force autocorrelation, so that the average kinetic energy of the free brownian particle equals (3/2)kT, a fact dictated by osmotic pressure data. Indeed, the possibility of such an evaluation provides a statistical mechanical consistency test for the model.

Now, the velocity $\vec{V}_o(t)$, of an individual free brownian particle is obtained from (III.11) by setting the external force $\vec{F}$, equal to zero. The mean kinetic energy per particle is obtained by averaging over the random elements in the expression $(m/2)\, \vec{V}_o^2(t)$, which are the initial particle velocity, $\vec{v}'$, and the thermal force, $\vec{f}$. We have :

$$< \frac{m}{2} \vec{V}^2(t) > = \exp\left(2\, \frac{t'-t}{t_o}\right) < \frac{m}{2} \vec{v}'^{\,2} >$$

$$+ \int_{t'}^{t} \exp\left(\frac{t' + \tau - t}{t_o}\right) < \vec{f}(\tau) >< \vec{v}'> d\tau$$

$$+ \frac{1}{2m} \int_{t'}^{t} \int_{t'}^{t} \exp\left(\frac{\tau + \tau' - 2t}{t_o}\right) < \vec{f}(\tau).\vec{f}(\tau) > d\tau\, d\tau' \tag{III.19}$$

Next, inserting in (III.19) the averages (III.18) for the thermal force and demanding that the mean kinetic energy is at all times (3/2)kT, we establish the value : $C = 2(m/t_o)\, kT$, given earlier by (III.10).

IV. FURTHER APPLICATIONS FROM BROWNIAN NOTION

In the previous section we gave detailed explanations concerning the mechanics of a particular case of functional integration. We shall now proceed with further applications deriving from the theory of brownian motion.

To simplify matters we consider the Einstein approximation for the brownian particle. The approximation applies when interest is focused on times larger than the characteristic time t_0 and the relaxation process. This approximation is equivalent to dropping out the inertia term from the Langevin equation of motion. In the case in which the brownian particle is acted upon by an external force $\vec{F}$ the Langevin equation of motion is modified to :

$$\frac{m}{t_0}\dot{\vec{x}} = \vec{f}(\tau) + \vec{F}(\vec{x},\tau) \qquad \text{(IV.1)}$$

Suppose we are wanting certain moments of the position vector $\vec{X}(t)$ or more generally the average value of a function $\Phi(\vec{X}(t))$ of $\vec{X}(t)$, given that $\vec{X}(o) = \vec{x}$. Let the solution of the equation of motion (IV.1) with the given initial condition be :

$$\vec{X}(t) = \vec{X}\,(\vec{x}',\ {}_o^t[\vec{f}(\tau)]) \qquad \text{(IV.2)}$$

It is a functional of the thermal force $\vec{f}(\tau)$ and so will be the function $\Phi(\vec{X}(t))$. The average value of Φ over the random force is obtained through the functional integral :

$$<\Phi(\vec{X}(t))> = \int \Phi(\vec{X}(\vec{x}',{}_o^t[\vec{f}(\tau)]))\ D[\vec{f}] \qquad \text{(IV.3a)}$$

where for simplicity we have denoted by $D[\vec{f}]$ the functional differential

$$D[\vec{f}] = W\ {}_o^t[\vec{f}(\tau)] \prod_{0 \leqslant \tau < t} d\vec{f}(\tau) \qquad \text{(IV.3b)}$$

The question arises whether we can obtain the functional average (IV.3a) by ordinary integration, rather than functional integration, by using an appropriate (moment) generating function. The answer is yes, and is attained by the following construction. Notice that :

$$\begin{aligned}\Phi(\vec{X}(t))\ D[\vec{f}] &= \int(\int\Phi(x)\ \delta(\vec{x} - \vec{X}(t))\ d\vec{x})\ D[\vec{f}] \\ &= \int\Phi(\vec{x}) <\delta(\vec{x} - \vec{X}(t))>_{fn\ell} d\vec{x}\end{aligned} \qquad \text{(IV.4)}$$

From (IV.4) follows the generating function yielding the moments is :

$$G\,(\vec{x}\;t|\vec{x}'0) = <\delta\,(\vec{x} - \vec{X}\,(\vec{x}', {}_{0}^{t}[\vec{f}]))>_{fnl} \qquad \text{(IV.5)}$$

$\delta(\vec{x} - \vec{X}(\vec{x},[\vec{f}]))$ can be thought of as the deterministic conditional probability distribution of finding the brownian particle in the vicinity of $\vec{x}$ at time t, provided that at time t_o it occupied the position $\vec{x}'$. The (functionally) averaged deterministic distribution can be thought of as an ensemble average conditional distribution. So the distribution yielding the ensemble averages is the ensemble average of the deterministic distribution.

The equation governing the motion of the deterministic distribution $\delta(\vec{x} - \vec{X}(t)) = \Delta(\vec{x}t|\vec{x}'0)$ is easily obtained by taking its rate of change (which is equal to $[-\dot{\vec{X}}(t).\partial/\partial\vec{x}]$. Taking account of the equation of motion (IV.1) and the fact that one of the factors in the rate of change involves $\delta(\vec{x} - \vec{X}(t))$ we are led to a Licuville type equation for the motion of the deterministic distribution :

$$\{\frac{m}{t_o}\frac{\partial}{\partial t} + [\vec{F}(\vec{x},t) + \vec{f}(t)].\frac{\partial}{\partial\vec{x}}\}\;\Delta(\vec{x}t|\vec{x}'0) = 0 \qquad \text{(IV.6)}$$

$\Delta(\vec{x}t|\vec{x}'0)$ is of course, the propagator of (IV.6). Having the equation of motion for the deterministic distribution of the $\vec{x}$'s we would like to know the equation of motion for its ensemble average value G, but, we shall put this off for a little while. At the moment, we turn to our evaluation of the functional average of Δ, given by (IV.5).

Consider the particular case of a time prescribed external force $\vec{F} = \vec{F}(t)$. The solution of (IV.1) in this case under the condition $\vec{X}(0) = \vec{x}'$, is :

$$\vec{X}(t) = <\vec{x}> + \frac{t_o}{m}\int_0^t \vec{f}(\tau)\;d\tau \qquad \text{(IV.7a)}$$

where

$$<\vec{x}> = <\vec{X}(t)> = \vec{x}' + \frac{t_o}{m}\int_0^t \vec{F}(\tau)\;d\tau \qquad \text{(IV.7b)}$$

We have now everything we need to find the ensemble average conditional distribution. We have :

$$G(\vec{x}t|\vec{x}'0) = \int\delta(\vec{x} - <\vec{x}> - \frac{t_o}{m}\int_0^t \vec{f}(\tau)\;d\tau)\;D[\vec{f}] \qquad \text{(IV.8a)}$$

where $D[\vec{f}]$ is explained through (IV.3b) and (III.9).

To perform the functional integral (IV.8a) we write the δ-function in Fourier form and furthermore write the functional differential in its explicit form. We have :

$$G(\vec{x}t|\vec{x}'0) = \int \frac{1}{(2\pi)^3} \int d\vec{x} \exp [i\vec{k}.(\vec{x} - <\vec{x}>) \\ i \int_0^t \frac{t_0}{m} \vec{k}.\vec{f}(\tau)\, d\tau] \}$$

$$\times \exp [- \frac{1}{2C} \int_0^t \vec{f}^2(\tau)\, d\tau] \prod_{0\leqslant\tau<t} (\frac{d\tau}{2\pi C})^{3/2} d\vec{f}(\tau) \qquad \text{(IV.8b)}$$

For the evaluation of (IV.8b) we make use of a fine partition of the interval $[o,t]$, so that we convert the integrals into sums and one has to deal repeatedly with this integral :

$$(\frac{\Delta\tau_j}{2\pi C})^{3/2} \iiint_{-\infty}^{+\infty} \exp (- \frac{1}{2C} \vec{f}_j^2 \Delta\tau_j - i \frac{t_0}{m} \vec{k}.\vec{f}_j \Delta\tau_j) d\vec{f}_j$$

$$= \exp [- \frac{1}{2} k^2 C (\frac{t_0}{m})^2 \Delta\tau_j] \qquad \text{(IV.9)}$$

Integrating, in this way, over all $\vec{f}_j$'s and passing to the limit of finer and finer partitions, we obtain :

$$G (\vec{x}t|\vec{x}'0) = \frac{1}{(2\pi)^3} \int \exp [i\vec{k}.(\vec{x} - <\vec{x}>) - \frac{1}{2} (\frac{t_0}{m})^2 C\, k^2] d\vec{k}$$

$$= (4 \pi Dt)^{-3/2} \exp [- \frac{(\vec{x} - < \vec{x} >)^2}{4\, Dt}] \qquad \text{(IV.10)}$$

where $D = KT\, t_0/m$ is the diffusion coefficient.

So, we have managed to obtain the conditional distribution of diffusing particles under the influence of a time dependent external force by means of functional integration. After this we turn back to some general properties.

Now, $G(\vec{x}t|\vec{x}'0)$ is the distribution of a diffusing particle, at time t, which at time t = 0 started from position $\vec{x}'$. Suppose now that the density of the particles at time t = 0 where $\rho(\vec{x})$, then their distribution $\rho(\vec{x},t)$ at time t (irrespective of their initial positions) will be the average value of G with respect to the initial positions. We have :

$$\rho(\vec{x},t) = \int G(\vec{x}t|\vec{x}'0)\, \rho_0(\vec{x}')\, d\vec{x}' \qquad \text{(IV.11)}$$

It is clear that

$$G(\vec{x}t|\vec{x}'0) = \langle \delta(\vec{x} - \vec{X}(\vec{x}', {}_0^t[\vec{f}])) \rangle_{fn\ell} \to \delta(\vec{x} - \vec{x}') \qquad \text{(IV.12)}$$
$$\text{as } t \to 0$$

for this is the initial conditional distribution. The initial condition (IV.12) is reflected in the fact that the density $\rho(\vec{x},t)$ evolves from its initial state $\rho_0(\vec{x})$ even if at time $t = 0$ and onwards one abruptly switches on the particles new forces. We have:

$$\rho(\vec{x},t) \to \rho_0(\vec{x}), \quad \text{as } t \to 0$$

The linear transformation (IV.11) is essentially the propagation equation for the particle density; $G(\vec{x}t|\vec{x}'0)$ is the propagator so, from (IV.11) the density $\rho(\vec{x},t)$ is the mean propagator with the averaging being performed against the distribution of the initial conditions.

Further the conditional distribution obeys the composition law :

$$G(\vec{x}t|\vec{x}'t') = \int G(\vec{x}t|\vec{x}_1\tau_1)\, G(\vec{x}_1\tau_1|\vec{x}'t')\, d\vec{x}_1 \qquad \text{(IV.13)}$$

for every τ_1 of the interval $[t',t]$. Equation (IV.13) bears the names of Smoluchowski, Kolmogorov and Chapman. Properties (IV.11) and (IV.13) are the defining properties for a semigroup.

It would be of interest to show how the composition law (IV.13) derives from our functional considerations.

Since the equation of motion (IV.1) contains no memory it follows that :

$$\vec{X}(\vec{x}', {}_{t'}^{t}[\vec{f}]) = \vec{X}(\vec{X}(\tau_1), {}_{\tau_1}^{t}[\vec{f}]) \qquad \text{(IV.14a)}$$

where :

$$\vec{X}(\tau_1) = \vec{X}(\vec{x}', {}_{t'}^{\tau_1}[\vec{f}]) \qquad \text{(IV.14b)}$$

With the aid of (IV.14) we verify that

$$\delta(\vec{x} - \vec{X}(\vec{x}', {}_{t'}^{t}[\vec{f}])) = \int \delta(\vec{x} - \vec{X}(\vec{x}_1, {}_{\tau_1}^{t}[\vec{f}]))$$
$$\times\ \delta(\vec{x}_1 - \vec{X}(\vec{x}', {}_{t'}^{\tau_1}[\vec{f}]))\, d\vec{x}_1 \qquad \text{(IV.15)}$$

This is the composition law for the deterministic distribution. Now the thermal distribution (III.9) factorizes for any pair of disjoint subintervals covering [t',t], i.e. :

$$W_{t'}^{t}[\vec{f}] = W_{t'}^{\tau_1}[\vec{f}].W_{\tau_1}^{t}[\vec{f}] \qquad \text{(IV.16)}$$

We multiply now (IV.15) and (IV.16) by members and integrate both sides over all $\vec{f}(\tau)$, $\tau \in [t',t]$ and in this way using (IV.5) for the conditional distribution we obtain the composition law (IV.13). It should be noted here that if the Langevin equation contains memory, or the thermal distribution does not factorise then the composition law fails to apply.

The composition rule (IV.13) enables us to derive the differential equation obeyed by G or the density $\rho(\vec{x},t)$. We choose the two times in the propagator differing by a short time Δt. In this case the evaluation of $G(\vec{x},t+\Delta t|\vec{x}_1,t)$ is feasible, even when the external force is both space and time dependent, i.e. $\vec{F} = \vec{F}(\vec{x},t)$, but slowly varying with time and position. In this case we have :

$$\vec{X}(t+\Delta t) \simeq \vec{x}_1 + \frac{t_0}{m}\int_t^{t+\Delta t} \vec{F}(\vec{x}_1,\tau)d\tau + \frac{t_0}{m}\int_t^{t+\Delta t} \vec{f}(\tau)\, d\tau \qquad \text{(IV.17)}$$

Following now the previous procedure we have for the short-time propagator :

$$G(\vec{x},t+\Delta t|\vec{x}_1,t) =$$
$$= (4\pi D\,\Delta t)^{-3/2} \exp\left[-\frac{1}{4\,D\Delta t}\left(\vec{x}-\vec{x}_1-\frac{t_0}{m}\vec{F}(\vec{x}_1,t)\,\Delta t\right)^2\right] \qquad \text{(IV.18)}$$

(IV.18) can be interpreted as the transition probability density for a diffusing particle which at time t is at $\vec{x}_1$ to find itself in the vicinity of $\vec{x}$ after time Δt.

With the aid of the transition probability (IV.18) we can obtain from the composition law (IV.13), following a procedure analogous to the one for the derivation of the Fokker-Planck equation (see (4)), the differential equation for the motion of G or the particle density ρ. We have :

$$\left[\frac{m}{t_0}\frac{\partial}{\partial t} + \frac{\partial}{\partial\vec{x}}.\vec{F}(\vec{x},t) - \kappa T\frac{\partial^2}{\partial\vec{x}^2}\right]\rho(\vec{x},t) = 0 \qquad \text{(IV.19)}$$

This is the Smoluchowski equation for a forced brownian particle. It resembles the Schrödinger equation, but it is not quite that. For the situation in phase space the reader is referred to (4).

With the expression (IV.18) for the short-time propagator we have all we need for making a functional integral over particle positions (path integrals) giving the propagator of the Smoluchowski equation. We work as follows : We take a fine partition of the interval $[t',t]$ and apply repeatedly the composition rule for all of its internal points. We have :

$$G(\vec{x}t|\vec{x}'t') = \int G(\vec{x}t|\vec{x}_{N-1}\tau_{N-1})\, G(\vec{x}_{N-1}\tau_{N-1}|\vec{x}_{N-2}\tau_{N-2}) \ldots$$

$$\ldots\, G(\vec{x}_2\tau_2|\vec{x}_1\tau_1)\, G(\vec{x}_1\tau_1|\vec{x}'t') \prod_{j=1}^{N-1} d\vec{x}_j \tag{IV.20}$$

We simplify matters and restrict ourselves to isomeric partitions, i.e. for the Nth partition we have : $\Delta\tau = (t-t')/N$.

Utilising (IV.19) in (IV.20) and passing to the limit of $\Delta\tau \to 0$ we have the following path integral for the conditional distribution of our diffusing particle :

$$G(\vec{x}t|\vec{x}'t') = \lim_{\Delta\tau\to 0} \int \exp\{-\sum_{j=0}^{N-1} \frac{1}{4D}\, [\,(\frac{\vec{x}_{j+1} - \vec{x}_j}{\Delta\tau})^2$$

$$- 2\,\frac{t_o}{m}\,\vec{F}(\vec{x}_j,\tau_j).\frac{\vec{x}_{j+1} - \vec{x}_j}{\Delta\tau} + (\frac{t_o}{m})^2\, \vec{F}^2(\vec{x}_j,\tau_j)]\;\Delta\tau\}$$

$$(4\,\pi D\,\Delta\tau)^{-3/2} \prod_{j=1}^{N-1} (4\,\pi D\,\Delta\tau)^{-3/2}\, d\vec{x}_j \tag{IV.21}$$

with the end conditions : $\vec{x}_o = \vec{x}'$, $\vec{x}_N = \vec{x}$.

The path integral (IV.21) over the particle positions and the functional integral (IV.5) over the thermal forces give the same thing; the conditional probability distribution for our forced diffusing particle. We have here the result of a transformation in functional integrals; from thermal force histories to particle paths.

We note here that (IV.21) has similarities with the path integral for the propagator of the Schrödinger equation for a particle in a vector and a scalar potential.

V. APPLICATION INVOLVING QUANTUM BROWNIAN MOTION

As is well known, brownian motion is not restricted to particles of colloidal size, but may be eqyally well exhibited by particles of atomic or even eletronic dimensions. In such cases, particularly in the latter, one can no longer employ classical dynamics, but has to have recourse to the quantum regime. Furthermore, in the case of electrons at low temperature, it would be necessary to couple Fermi-Dirac statistics with quantum dynamics. Presently we shall deal with low energy electron fluctuations, which may have application in the measurement of low temperatures.

The classical theory of brownian motion, apart from explaining a fundamental phenomenon, also deals with mathematically tractable problems of nonequilibrium statistical physics. The theory is based on the Langevin model which simulates the many-particle medium interaction experienced by the brownian particle by two single-particle forces; a dissipative, and a random force. The model displays a great attraction in that it reduces an essentially many-particle problem to a single-particle one. Now, due to the fact that dissipative forces are not in general hamiltonian derivable, it would seem that a quantum-mechanical construction of a density matrix for the brownian particle, based on the Langevin model, would inevitably entail difficulties. A circumstantial discussion of the difficulties involved has been given by Kubo (1969) (26). However, the simplicity of the model is such that it would be worthwhile attempting to develop a quantum-mechanical counterpart. The transcription to the quantum regime requires knowledge of the random force autocorrelation, which, unlike the classical state of affairs, varies from case to case. The evaluation of this autocorrelation is easily accomplished within the framework of the Heisenberg picture by comparing the mean energy of the brownian particle in a state of thermodynamic equilibrium, obtained by equilibrium statistical mechanics, with that obtained from an appropriate nonequilibrium method.

In the case of classical dynamics the Langevin equation of motion, for the brownian particle in a potential field $U(\vec{x})$ and under the influence of a time prescribed force $\vec{F}(t)$, is :

$$m \frac{d^2\vec{x}}{dt^2} + \frac{\partial}{\partial \vec{x}} U(\vec{x}) - \vec{F}(t) = - \frac{m}{t_o} \frac{d\vec{x}}{dt} + \vec{f}(t) \qquad (V.1)$$

The forces on the right-hand side of (V.1) simulate the many-body medium interaction experienced by the brownian particle. As is well known,the particle constantly loses energy through the dissipative force $-(m/t_o)d\vec{x}/dt$ while through the force $\vec{f}(t)$, taken to be a random force, its energy is restored on average. t_o is the particle relaxation time.

Chandrasekhar (4) assumed the distribution of the random force within a short time interval to be gaussian, and the random forces at two distinct times to be uncorrelated. Now, the high collision rate (10^{20} collisions/s) enables the use of a functional probability for the random force.

$$W_o^t\,[\vec{f}(\tau)] \prod_{0\leq\tau<t} d\vec{f}(\tau)$$

$$= \exp\left(- \int_o^t \frac{1}{2C} \vec{f}^{\,2}(\tau)d\tau\right) \prod_{0\leq\tau<t} \left(\frac{d\tau}{2\pi C}\right)^{3/2} d\vec{f}(\tau) \qquad (V.2)$$

embodying the above assumptions.

The distribution given by (V.2) implies the averages :

$$< \vec{f}(\tau) > = 0 \qquad < \vec{f}_\alpha(\tau_1)\vec{f}_\beta(\tau_2) > = C\,\delta_{\alpha\beta}\,\delta(\tau_1 - \tau_2) \qquad (V.3)$$

The coefficient C in the random force autocorrelation given in (V.3) and appearing in the distribution (V.2) in the classical case is the same irrespective of the potential $U(\vec{x})$ and is given by :

$$C = 2\,\frac{m}{t_o}\,kT \qquad (V.4)$$

where T is the medium temperature and k Boltzmann's constant. However, in the case of quantum dynamics the uncertainty principle correlates the kinetic and potential energies, and the situation in which C is independent of the potential energy is no longer valid.

For the evaluation of the random force autocorrelation we shall rely on two different procedures for obtaining the average energy in a state of thermodynamic equilibrium; an equilibrium and a nonequilibrium procedure. Our system will consist of brownian particles considered independent of each other. The state of thermodynamic equilibrium is brought about by environmental and very slight interparticle interactions which are simulated by the frictional and the random force. Although these interactions are responsible for bringing about the state of equilibrium they do not finally appear in the equilibrium distribution of the system. Actually the distribution is fully determined by the system hamiltonian $H(\vec{p},\vec{x})$ alone. Since the medium interactions are mainly responsible for the eventual establishment of equilibrium, it should be possible to obtain the average energy per particle of the system of interest determined by its own forces and the environmental interactions. The necessary averages are to be taken over distributions pertaining to the system plus the environment.

More specifically, let $\vec{X}(t)$ and $\vec{P}(t)$ be the position and momentum operators of a brownian particle, at time t, obtained from its (operator) equation of motion, which involves the medium interactions. In the case of the Langevin model they will be functions of the initial values of the position and momentum operators and functionals of the random force.

In short they will look like :

$$\vec{X}(t) = \vec{X}(\vec{x}', \vec{p}', {}_0^t[\vec{f}(\tau)])$$

$$\vec{P}(t) = \vec{P}(\vec{x}',\vec{p}', {}_0^t[f(\tau)]) \qquad (V.5)$$

where $\vec{x}'$ and $\vec{p}'$ $(= -i\hbar\partial/\partial\vec{x}')$ are the initial operator conditions of the particle.

Introducing the expressions (V.5) into the hamiltonian for the brownian particle, we may interpret the result as the 'instantaneous energy operator' of a particular brownian particle. Now, in order to obtain the mean energy per particle we have to average both against the density matrix of the initial conditions, ρ_{in}, and the distribution of the random force. The result will be the double average :

$$<< H(\vec{P}(t), \vec{X}(t)) >>$$

$$= \int \left(\int [H(\vec{P}(t),\vec{X}(t))\rho_{in}(\vec{x}|\vec{x}')]_{\vec{x}'=\vec{x}} d\vec{x}\right) W_o^t[\vec{f}(\tau)] \prod_{0\leq\tau<1} d\vec{f}(\tau) \qquad (V.6)$$

In the absence of external forces, equilibrium statistical mechanics tells us that the average energy per particle is obtained via the equilibrium density matrix ρ_{eq} as follows :

$$< H(\vec{p},\vec{x}) > = \int [H(\vec{p},\vec{x})\rho_{eq}(\vec{x}|\vec{x}')]_{\vec{x}'=\vec{x}} d\vec{x} \qquad (V.7)$$

When the initial conditions in (V.6) relate to the state of equilibrium, described by the density matrix ρ_{eq} of (V.7) then the average energies obtained through the nonequilibrium procedure described by (V.6) and the equilibrium way of (V.7) should coincide. Equating (V.6) (with $\rho_{in} = \rho_{eq}$) to (V.7) we are led to an equation involving the coefficient C appearing in the random force autocorrelation (V.3) from which it can be evaluated.

$$<< H(\vec{P}(t), \vec{X}(t)) >> = < H(\vec{p},\vec{x}) > \; ; \; \rho_{in} = \rho_{eq} \qquad (V.8)$$

We now ask the question what happens when external forces are switched on our system of interest? Are they likely to engender changes in the random force autocorrelation? Certainly if they are capable of creating, eventually, a new state of equilibrium affairs they are bound to modify the random force autocorrelation accordingly. However, in the case of a time prescribed external force, in the cases of free and harmonically bound particles, such a force superimposes itself upon the fluctuating force, and under these circumstances its characteristics remain unaltered.

In what follows we shall exemplify the above procedure in the case of the free particle. The case of the harmonically bound brownian particle as well as the construction of the corresponding density matrices and a general equation of motion for the density matrix can be found in the last of ref. (4).

For the free particle the Langevin operator equation is :

$$\frac{d\vec{p}}{dt} = \frac{1}{t_o}\vec{p} + \vec{f}(t) \qquad (V.9)$$

The solution which satisfies the initial condition

$$\vec{P}(o) = \vec{p}' = \frac{\hbar}{i}\frac{\partial}{\partial \vec{x}'}$$

As :

$$\vec{P}(t) = \exp\left(-\frac{t}{t_o}\right)\vec{p}' + \int_0^t d\tau \exp\left(\frac{\tau - t}{t_o}\right)\vec{f}(\tau) \qquad (V.10)$$

The hamiltonian is $H = \vec{p}^2/2m$. Therefore (V.6) becomes :

$$\langle\langle \frac{1}{2m}\vec{P}^2(t) \rangle\rangle = \exp\left(-2\frac{t}{t_o}\right)\frac{1}{2m}\langle \vec{p}'^2 \rangle +$$

$$\frac{Ct_o}{2m}\left[1 - \exp\left(-2\frac{t}{t_o}\right)\right] \qquad (V.11)$$

where for the derivation (V.11) we have made use of the random force autocorrelation, given in (V.3).

Since the system of our brownian particles remains in the same macroscopic equilibrium, the average initial energy $\langle \vec{p}'^2/2m \rangle$, and the mean energy, $\langle \vec{p}^2/2m \rangle$, at any time t should be the same. This energy is obtained via the density matrix pertaining to the statistics of the system.

Making use of (V.8) we find C, in the case of free particles, to be given by :

$$C = 2\frac{m}{t_o}\frac{2}{3}\langle \frac{1}{2m}\vec{p}^2 \rangle \qquad (V.12)$$

In the case of Boltzmann statistics the equilibrium density matrix is :

$$\rho^B_{eq}(\vec{x}|\vec{x}') = \frac{1}{V}\exp\left[-\frac{m}{2\hbar^2\beta}(\vec{x} - \vec{x}')^2\right] \qquad (V.13a)$$

while the corresponding distribution for Fermi-Dirac statistics is :

$$\rho^{FD}_{eq}(\vec{x}|\vec{x}') = \frac{1}{N}\int\frac{dx}{(2\pi)^3}\frac{2}{\exp[\beta(\varepsilon_k - \zeta)] + 1}\exp[i\vec{k}.(\vec{x}-\vec{x}')] \qquad (V.13b)$$

In (V.13b) N is the number of brownian particles in the system $\varepsilon_k = \hbar^2 k^2/2m$) is the free particle energy associated with the wavevector $\vec{k}$, and ζ is the chemical potential of the system.

Evaluating now the mean equilibrium energy per particle in the two regimes of statistics, we obtain, utilizing (V.12), the values of the random force autocorrelation coefficient, C_B and C_F, for Boltzmann and Fermi-Dirac statistics as :

$$C_B = 2 \frac{m}{t_o} kT \qquad (V.14)$$

(in agreement with Chandrasekhar's result (4)).

$$C_F = 2 \frac{m}{t_o} kT \left(\frac{kT}{F}\right)^{3/2} \int_0^\infty du \frac{u^{3/2}}{\exp(u - \beta\zeta) + 1} \qquad (V.14b)$$

where ε_F is the Fermi energy given by :

$$\varepsilon_F = \frac{\hbar^2}{2m} \left(\frac{3 N\pi^2}{V}\right)^{2/3}$$

and u is a dimensionless variable ($u = \hbar^2 k^2/2m kT$) introduced in (V.13b) to facilitate the derivation of (V.14b).

The expression for C associated with Boltzmann statistics, in spite of the quantum dynamics used for its derivation, does not involve Planck's constant and therefore coincides with the corresponding classical result. This is so, since with Boltzmann statistics, in the limit of large volume, the free-particle classical and quantal statistical treatments converge to each other. However, with Fermi-Dirac statistics the quantum effects persist irrespective of the system size. As is well known, for large temperatures the quantum statistical result tends to the classical one. At low temperatures the difference becomes more pronounced. Thus, the low temperature asymptotic expression for C_F, obtained from the corresponding mean energy of free fermions, through (V.12) is given by :

$$C_F = \frac{4}{3} \frac{m}{t_o} \left[\frac{3}{5} \varepsilon_F + \frac{\pi^2}{4} \frac{(kT)^2}{\varepsilon_F}\right] \qquad (V.14c)$$

From this we see that at absolute zero $C_F = 4m\ \varepsilon_F/5\ t_o$, while its corresponding value with Boltzmann statistics is $C_B = 0$.

As a simple application let us now evaluate the voltage autocorrelation $\langle\langle V(t)\ V(t+\tau)\rangle\rangle$ of a resistor due to the brownian motion of its conduction electrons. The voltage operator for such a resistor of length l, resistance R, with N conduction electrons and which is taken along the x direction, will be :

$$V(t) = \frac{e}{l} R \sum_{j=1}^{N} \frac{P_x^{(j)}(t)}{m} \tag{V.15a}$$

where $P_x^{(j)}(t)$ is the x component of the momentum operator of the jth conduction electron at time t. Utilizing (V.15a) we write the thermal voltage autocorrelation as :

$$\langle\langle V(t)\ V(t+\tau)\rangle\rangle =$$

$$\frac{e^2}{m^2\ l^2} \sum_{j,s=1}^{N} \langle\langle P_x^{(j)}(t)\ P_x^{(s)}(t+\tau)\rangle\rangle \tag{V.15b}$$

where the double averaging refers to the averages taken against the density matrix of the initial conditions and the distribution of the random force.

Essentially, the evaluation of (V.15b) requires a two-body density matrix for the initial conditions, but here we are dealing with nearly independent particles and the momentum correlations left with the evaluation of a single-particle momentum autocorrelation. The result is :

$$\langle\langle V(t)\ V(t+\tau)\rangle\rangle = \frac{e^2}{m^2 l^2} RN \langle\langle P_x(t)\ P_x(t+\tau)\rangle\rangle$$

$$= R\ \frac{C_F}{2m} \exp\left(-\frac{|\tau|}{t_o}\right) \tag{V.15c}$$

where for the derivation of (V.15c) we have combined (V.10), (V.3) and (V.13b). Furthermore we have made use of the formula, $R = l^2 m/e^2\ N\ t_o$ expressing the resistance in terms of the electron relaxation time. C_F is given by (V.14b).

Fourier transforming the voltage fluctuation formula (V.15c) with respect to τ and specializing to the case of near absolute zero temperature, using (V.14c), we have :

$$\frac{1}{6\pi}\int_{-\infty}^{+\infty} \langle\langle V(t)\, V(t+\tau) \rangle\rangle \exp(i\omega\tau)\, d\tau$$

$$= \frac{2}{\pi} R \left[\frac{1}{5}\varepsilon_F + \frac{3}{4}\pi^2 \frac{(kT)^2}{\varepsilon_F}\right] \frac{1}{1+(\omega t_o)^2} \; ; \quad T \simeq 0\ K \tag{V.15d}$$

The result (V.15d), taking account of the proper electron statistics, may indicate a more appropriate scale for the calibration of the low temperature flux thermometers (Kamper 1967, Kamper and Zimmerman 1971, Giffard et al 1972 (27)) as compared with the linear scale in use.

Let us now work out the case when an externally applied time dependent force $\vec{F}(t)$ acts on the free brownian particle. We shall be interested in obtaining the averages for the momentum position, and energy.

In this case the operator equation is :

$$\dot{\vec{p}} = -\frac{1}{t_o}\vec{p} + \vec{F}(t) + \vec{f}(t) \tag{V.16}$$

The solution, for the momentum and position, satisfying the initial conditions :

$$\vec{P}(o) = \vec{p}\,' = \frac{\hbar}{i}\frac{\partial}{\partial \vec{x}\,'} \quad , \qquad \vec{X}(o) = \vec{x}\,'$$

is given by :

$$\vec{P}(t) = \exp\left(-\frac{t}{t_o}\right)\vec{p}\,' + \int_0^t d\tau \exp\left(\frac{\tau - t}{t_o}\right)(\vec{F}(\tau) + \vec{f}(\tau)) \tag{V.17a}$$

$$\vec{X}(t) = \vec{x}\,' + m^{-1} t_o \left[1 - \exp\left(-\frac{t}{t_o}\right)\right]\vec{p}\,'$$
$$+ m^{-1} t_o \int_0^t d\tau \left[1 - \exp\left(\frac{\tau - t}{t_o}\right)\right](\vec{F}(\tau) + \vec{f}(\tau)) \tag{V.17b}$$

The above expressions for the momentum and position involve as random elements the thermal force $\vec{f}(t)$ and the initial conditions $(\vec{x}\,', \vec{p}\,')$. The average value of any quantity which is a function of $\vec{P}(t)$ or $\vec{X}(t)$ is to be taken against the thermal force distribution (V.2) and the density matrix ρ_{eq} which, depending on the statistics, will be (V.13a) in the case of Boltzmann statistics or (V.13b) in the case of Fermi-Dirac statistics.

Proceeding in this way we obtain the average values of the momentum and position, the momentum auto-correlation and the kinetic energy.

$$<< \vec{P}(t) >> = \int_0^t d\tau \exp\left(\frac{\tau - t}{t_0}\right) \vec{F}(\tau) \qquad \text{(V.18a)}$$

$$<< \vec{X}(t) >> = m^{-1} t_0 \int_0^t d\tau \left[1 - \exp\left(\frac{\tau - t}{t_0}\right)\right] \vec{F}(\tau) \qquad \text{(V.18c)}$$

$$<< P_\alpha(t)\, P_\alpha(o) >> = < p_\alpha^2 >_{eq} \delta_{\alpha\beta} \exp\left(-\frac{t}{t_0}\right) \qquad \text{(V.18c)}$$

$$<< \frac{1}{2m} \vec{P}^2(t) >> =$$

$$= \begin{cases} \frac{3}{2} kT & \text{Bo statistics} \\ \frac{3}{2} kT \left(\frac{kT}{\varepsilon_F}\right)^{1/2} \int_0^t du \frac{u^{3/2}}{\exp(u - \beta\xi) + 1} & \text{FD statistics} \end{cases}$$

$$+ \frac{1}{2m} \left[\int_0^t d\tau \exp\left(\frac{\tau - t}{t_0}\right) \vec{F}(\tau)\right]^2 \qquad \text{(V.18d)}$$

The double averages emphasize the two averaging procedures relating to the thermal force on one hand and the initial conditions on the other. Single averages with the suffix 'eq' indicate that the averaging is performed against the equilibrium density matrix.

The first two averages (V.18a,b) express the response of the system of brownian particles, in terms of momentum and displacement, to the external force $\vec{F}(t)$. The momentum autocorrelation, given in (V.18c), shows the decay pattern of this quantity.

The final relation gives the average energy of our particle at time t in terms of the mean equilibrium energy and an additional energy picked up from the external field.

For more relating to Brownian motion with quantum dynamics see (4).

VI. THE FEYNMAN PATH INTEGRAL

We are now entering fully into the regime of quantum mechanics. In the present section, we shall construct a path integral giving the propagator of the Schrödinger equation in the case of a time-independent hamiltonian; we really like to be simple when introducing a method. We shall also confine ourselves to the case of rectilinear co-ordinates, where most of propagator evaluations have been done. The procedure we shall follow, will be deductive rather than postulative. Although the seeds for such an approach are found in Feynman's work (1948) (5), Feynman followed an essentially postulative presentation. The development of the present method, as far as we know, originated from Abe' (1954) (28).

As is well known, the propagator of the Schrödinger equation for a system contains all quantum mechanical information about the system. Thus, it would seem appropriate to begin with considerations to how the propagator arises and telling us at the same time what thing this propagator is.

Let us consider a system with hamiltonian :

$$H(\vec{x}) = -\frac{\hbar^2}{2m}\frac{\partial^2}{\partial\vec{x}^2} + U(\vec{x}) \qquad \text{(VI.1)}$$

The problem we are faced with here is to find the evolution in time of a wavefunction, which at a given time (say t = 0) is given through $\Psi_0(\vec{x})$.

The answer to the problem is obtained by solving Schrödinger's equation :

$$\left[i\hbar\frac{\partial}{\partial t} - H(\vec{x})\right]\Psi(\vec{x},t) = 0 \qquad \text{(VI.2a)}$$

with the initial condition :

$$\Psi(\vec{x},0) = \Psi_0(\vec{x}) \qquad \text{(VI.2b)}$$

In terms of the evolution operator, exp $(-\ iHt/\hbar)$, the solution is given by :

$$\Psi(\vec{x},t) = \exp\left[-\frac{i}{\hbar} H(\vec{x})\ t\right]\Psi_0(\vec{x}) \qquad \text{(VI.3)}$$

This clearly satisfies the Schrödinger equation, together with the appropriate initial condition.

In (VI.3) the evolution operator acts locally on the initial wavefunction, which it transforms in time, so that it obeys the Schrödinger equation. If we wish now to disengage the process of propagation from the particular content of the wavefunction we rewrite (VI.3) as follows :

$$\Psi(\vec{x},t) = \int \exp\left[-\frac{i}{\hbar} H(\vec{x})\ t\right]\ \delta(\vec{x}-\vec{x}\,')\ \Psi_0(\vec{x}\,')\ d\vec{x}\,'$$

$$= \int K(\vec{x}t|\vec{x}\,'0)\ \Psi_0(\vec{x}\,')\ d\vec{x}\,' \qquad \text{(VI.4a)}$$

The kernel :

$$K(\vec{x}t|\vec{x}\,'0) = \exp\left[-\frac{i}{\hbar} H(\vec{x})t\right]\ \delta(\vec{x}-\vec{x}\,') \qquad \text{(VI.4b)}$$

is the propagator in co-ordinate representation. It supplies the wavefunction at time t (not necessarily later) from the information carried by the wave-function at t = 0.

Since :

$$\delta(\vec{x}-\vec{x}\,') = \sum_{\vec{k}} <\vec{x}|\vec{k}><\vec{k}|\vec{x}\,'> = <\vec{x}|\vec{x}\,'> \qquad \text{(VI.5)}$$

where all $<\vec{x}|\vec{k}>$ form a complete set of wavefunctions, (VI.4b) can take the form :

$$K(\vec{x}t|\vec{x}\,'0) = \exp\left[-\frac{i}{\hbar} H(\vec{x})t\right]\ <\vec{x}|\vec{x}\,'>$$

$$= <\vec{x}|\exp\left[-\frac{i}{\hbar} H(\vec{x})t\right]\ |\vec{x}\,'> \qquad \text{(VI.6)}$$

Now, one way to obtain an explicit expression for the propagator is to operate as follows :

$$K = \left[1 - \frac{i}{\hbar} Ht + \frac{1}{2!}\left(-\frac{i}{\hbar}\right)^2 H^2t^2 + \ldots\right]\ \delta(\vec{x}-\vec{x}\,')$$

This is an additive procedure based on a series expansion of the evolution operator.

Another way to go about it is via a multiplicative process based on a product expansion of the evolution operator as follows :

$$K \simeq \underset{1}{(1 - \frac{i}{\hbar} H \frac{t}{N})}\underset{2}{(1 - \frac{i}{\hbar} H \frac{t}{N})} \underset{\cdots}{\cdots} \underset{N}{(1 - \frac{i}{\hbar} H \frac{t}{N})}\, \delta(\vec{x}-\vec{x}') \qquad \text{(VI.7)}$$

If we take the limit as $N \to \infty$ we have the required propagator, since :

$$\lim_{N\to\infty} (1 - \frac{i}{\hbar} H \frac{t}{N})^N = \exp(-\frac{i}{\hbar} Ht)$$

The multiplicative procedure is the one employed in path integral constructions.

Next, before we go any further, notice that by adding a term of $O((\Delta t)^2)$, ($\Delta t = t/N$) to any one of the factors $(1 - \frac{i}{\hbar} H\, \Delta t)$ in (VI.7) the limit of the product as $N \to \infty$ is not affected. So we can obtain our propagator as a limiting case of the matrix element :

$$< \vec{x}|(1 - \frac{i}{\hbar} H\Delta t)(1 - \frac{i}{\hbar} H\Delta t) \; \ldots \; (1 - \frac{i}{\hbar} H\Delta t)|\vec{x}' >$$

Inserting complete sets of wavefunctions between the various operators we have :

$$K_N\, (\vec{x}t|\vec{x}'0) =$$

$$\int<\vec{x}|1 - \frac{i}{\hbar} H(\vec{x})\Delta t|\vec{x}_{N-1}><x_{N-1}|1 - \frac{i}{\hbar} H(\vec{x}_{N-1})\Delta t|\vec{x}_{N-2}>$$

$$\ldots<\vec{x}_2|1 - \frac{i}{\hbar} H(\vec{x}_2)\Delta t|\vec{x}_1><\vec{x}_1|1 - \frac{i}{\hbar} H(\vec{x}_1)\Delta t|\vec{x}'>d\vec{x}_1\, d\vec{x}_2 \ldots$$

$$\ldots\, d\vec{x}_{N-1} \qquad \text{(VI.8)}$$

We have in (VI.8) N short-time propagators of the form :

$$< \vec{x}_{j+1}\; 1 - \frac{i}{\hbar} H(\vec{x}_{j+1})\Delta t|\vec{x}_j > = (1 - \frac{i}{\hbar} H(\vec{x}_{j+1})\Delta t)\delta(\vec{x}_{j+1}-\vec{x}_j) \qquad \text{(VI.9)}$$

but (N-1) 3D integrations.

To obtain the form of the Feynman path integral

we employ the plane-wave decomposition of the identity transformation, i.e. :

$$\delta(\vec{x}_{j+1} - \vec{x}_j) = \frac{1}{(2\pi)^3} \int d\vec{k} \exp [i\vec{k}.(\vec{x}_{j+1} - \vec{x}_j)] \tag{VI.10}$$

The decomposition is done using essentially the eigen-functions of the kinetic energy operator.

Putting the hamiltonian (VI.1) into(VI.9), we have with the aid of (VI.10) an approximate expression for the short-time propagator, as :

$$< \vec{x}_{j+1}| 1 - \frac{i}{\hbar} [- \frac{\hbar^2}{2m} \frac{\partial^2}{\partial \vec{x}^2_{j+1}} + U(x_{j+1}) \Delta t] | \vec{x}_j >$$

$$= \frac{1}{(2\pi)^3} \int d\vec{k} \{1 - [i \frac{\hbar k^2}{2m} + \frac{i}{\hbar} U(\vec{x}_j)] \Delta t\} \exp [i\vec{k}.(\vec{x}_{j+1}-\vec{x}_j)] \tag{VI.11}$$

Notice that on the r.h.s. of (VI.11) $\vec{x}_{j+1}$ in $U(\vec{x}_{j+1})$ was replaced by $\vec{x}_j$ on account of the δ-function accompanying it.

As pointed out earlier we can add to the short-time operators of (VI.11) any terms of order higher than Δt, without this affecting the limit, as long as our short-time propagators are correct to first order in Δt we are O.K. So, we can replace in (VI.11) the term in the angular brackets by an expontial, which has the same expansion to order Δt. i.e. $< \vec{x}_{j+1} | 1 - \frac{i}{\hbar} H(\vec{x}_{j+1}) \Delta t | \vec{x}_j >$ is replaced by :

$$\frac{1}{(2\pi)^3} \int d\vec{k} \exp [-i \frac{\hbar k^2}{2m} \Delta t - \frac{i}{\hbar} U(\vec{x}_j) \Delta t] \exp[i\vec{k}.(\vec{x}_{j+1}-\vec{x}_j)]$$

$$= (\frac{m}{2\pi i \hbar \Delta t})^{3/2} \exp \{\frac{i}{\hbar} [\frac{m}{2} (\frac{\vec{x}_{j+1} - \vec{x}_j}{\Delta t})^2 - U(\vec{x}_j)] \Delta t\} \tag{VI.12}$$

Making the replacements (VI.12) in (VI.8) we have for the approximate propagator the expression :

$$K_N'(\vec{x}t|\vec{x}'0) = \int \exp \{\frac{i}{\hbar} [\frac{m}{2} (\frac{\vec{x}_{j+1} - \vec{x}_j}{\Delta t})^2 - U(\vec{x}_j)] \Delta t\}$$

$$\times (\frac{m}{2\pi i \hbar \Delta t})^{3/2} \prod_{j=1}^{N-1} (\frac{m}{2\pi i \hbar \Delta t})^{3/2} d\vec{x}_j \tag{VI.13}$$

with : $\vec{x}_o = \vec{x}'$, $\vec{x}_N = \vec{x}$.

Notice that in the (VI.13) we have the Lagrangian (in discrete form) of our problem revealed.

The product accompanying the exponential on the r.h.s. of (VI.13) is the path differential $D[\vec{x}_1, \vec{x}_2, \ldots, \vec{x}_{N-1}]$. It contains the right normalizing factors (measure of integration) for obtaining the propagator, through the multiple process of integration in the limit of infinite sub-division of the interval [0,t]. In other words, in the limit as $N \to \infty$, the sequence $[K'_N]$ goes to the required propagator.

$$K'_N(\vec{x}t|\vec{x}'0) \to K(\vec{x}t|\vec{x}'0)$$

The above method of path integral construction is essentially hamiltonian based, and the steps (VI.7) - (VI.13) originated from Abe' (28).

In the limit of infinite refinement of the partitions of the time interval [0,t] we make use of a notation, naturally emanating from (VI.13), and indicating the multiple integrations (integration over paths). We have :

$$K(\vec{x}t|\vec{x}'0) = \int \exp\{\frac{i}{\hbar}\int_0^t [\frac{m}{2}\dot{\vec{x}}^2(\tau) - U(\vec{x}(\tau))]\,d\tau\}D[\vec{x}(\tau)] \tag{VI.14a}$$

with $\vec{x}(0) = \vec{x}'$, $\vec{x}(t) = \vec{x}$ and where

$$D[\vec{x}(\tau)] = (\frac{m}{2\pi i\hbar\,d\tau})^{3/2} \prod_{0<\tau<t} (\frac{m}{2\pi i\hbar\,d\tau})^{3/2}\,d\vec{x}(\tau) \tag{VI.14b}$$

However, a word of caution is more than in order here. Expression (VI.14) may mislead (especially when vector potentials are involved) to a sequential form, from which although it can be reproduced for smooth (differentiable) paths in the limit of infinite refinement of [0,t], it may fail as a functional integral to lead to the correct propagator. The difficulty arises from contribution made by the discontinuous paths.

In view of the above, we should like to draw a picture depicting the axes of our multidimensional space integration, appropriate for a 1-dimensional space per variable. See fig.(VI.1). The 1-dimensional motion analogue of (VI.13) is an (N-1)-fold integral for which the range of each of its variables, x_1, x_2, ..., x_{N-1} extends from $-\infty$ to $+\infty$. These variables take their values independently of each other. Let us then freeze one particular set of these values and see what picture we reach. We get a set of points each of which lies on a different axis of the set of axes x_0, x_1, ..., x_{N-1}, x_N, as shown in the figure. If we connect these points by a continuous curve we get an idea of the continuous paths entering the process of multiple integration. Actually, this is somewhat illusive, for the continuous paths are not by any means the only paths involved in the integration process; for imagine a sequence of denser and denser partitions of [0,t], which implies that any two consecutive axes are separated from each other by a smaller and smaller distance, tending to zero. Now, since the variables of integration take any values from $-\infty$ to $+\infty$ (which are represented by points on the various axes) this means that the representative points on two consecutive axes (which tend to coalesce) can be separated by an appreciable distance. This sort of situation in the limit of infinite subdivision of our interval produces a broken path. In fact our integration involves all imaginable paths which cross only once each of the axes $x(\tau)$.

It has now become evident that the propagator $K(\vec{x}t|\vec{x}'0)$, from time 0 to time t, can be found from knowledge of the propagators $K(\vec{x},\tau|\vec{x}'0)$ and $K(\vec{x}t|\vec{x}_\tau\tau)$, from time 0 to τ and from τ to t respectively by summing over all the intermediate positions $\vec{x}_\tau$ at time τ as in the formula :

$$K(\vec{x}t|\vec{x}'0) = \int K(\vec{x}t|\vec{x}_\tau\tau)\, K(\vec{x}_\tau\tau|\vec{x}'0)\, d\vec{x}_\tau \qquad \text{(VI.15)}$$

This is the markoffian property of the propagator, characteristic of the formation of quantal amplitudes through all possible intermediate stages linked by a given time.

In the above construction of the path integral the hamiltonian was taken time-independent. When the hamiltonian is time dependent, $H(\tau) = (H(\vec{x},\tau))$ at time τ does not commute with the hamiltonian $H(\tau')$ at another τ'. The non-commutativity of the hamiltonian produces a complication and one requires Dyson's time-ordering operator T for obtaining the evolution operator $U(t|t')$ from time t' to time t :

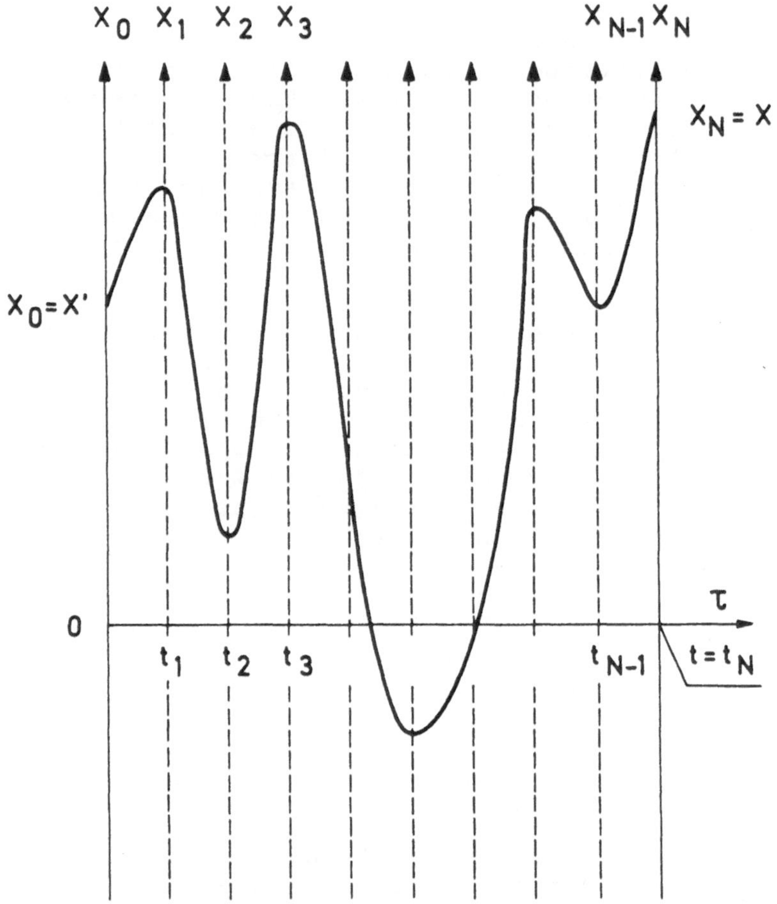

Fig. VI.1. Shows the multidimensional space of integration for one-dimensional motion.

$$U(t|t') = T \exp\left(-\frac{i}{\hbar}\int_{t'}^{t} H(\tau)\, d\tau\right) \qquad \text{(VI.16)}$$

The required propagator $K(\vec{x}t|\vec{x}'0)$ from the space-time point $(\vec{x}',t')$ to the spacetime point $(\vec{x},t)$ is given by :

$$K(\vec{x}t|\vec{x}'0) = \langle\, \vec{x}|U(t|t')|\vec{x}'\, \rangle \qquad \text{(VI.17)}$$

However, it is one thing writing down the evolution operator with formal use of the time-ordering operator and another thing obtaining, via a path integral, its configuration representation; the propagator. What we really need is an (approximate) form of the evolution operator suitable for path integral constructions. The way to proceed is via the infinitesimal generators of the evolution operator.

Let us now consider a fine partition, P_N, which for simplicity we take to be isomeric. Then we form the time ordered product, $U(t|t')$, of the various infinitesimal generators $(1 - iH(\vec{x}_j,\tau_j)\, \Delta t/\hbar)$ as follows :

$$U_N(t|t') =$$
$$\left(1 - \frac{i}{\hbar} H(\vec{x},t)\Delta t\right)\ldots\left(1 - \frac{i}{\hbar} H(\vec{x},\tau_2)\Delta t\right)\left(1 - \frac{i}{\hbar} H(\vec{x},\tau_1)\Delta t\right) \qquad \text{(VI.18)}$$

The evolution operator $U(t|t')$ associated with the time dependent hamiltonian $H(\vec{x},t)$ is obtained from $U_N(t|t')$ by passing to the limit of infinite refinement of the interval $[t',t]$ since :

(i) $U(t|t') \to 1$ as $t \to t'$ and (ii) $U(t|t')$ satisfies the (operator) Schrödinger equation.

(i) is easily seen, while (ii) can be obtained by considering the time interval $[t',t + \Delta t]$ and the associated sequence of ordered product, $U_{N+1}(t + \Delta t|t')$ approximating the evolution operator in the extended interval. We have :

$$U_{N+1}(t + \Delta t|t') = \left(1 - \frac{i}{\hbar} H(\vec{x},t+\Delta t)\, \Delta t\right) U_N(t|t') \qquad \text{(VI.19a)}$$

which can be written as follows :

$$i\hbar \frac{1}{\Delta t} [U_{N+1}(t + \Delta t|t') - U_N(t|t')] = H(\vec{x},t + \Delta t)\, U_N(t|t') \qquad \text{(VI.19b)}$$

We now take the limit as $N \to \infty$ (or $\Delta t = (t-t')/N \to 0$) and find :

$$i\hbar \frac{\partial}{\partial t} U(t|t') = H(\vec{x},t)\, U(t|t') \qquad \text{(VI.19c)}$$

which tells us that $U(t|t')$ produced from (VI.18) through the process of infinite refinement of the interval $[t',t]$ obeys the Schrödinger equation.

Furthermore, as before, it can be seen that by adding to the various infinitesimal generators of (VI.18) any quantity of $O((\Delta t)^2)$ these will not affect the limiting value of the operator $U(t|t')$ for they will only produce an additional operator with limiting value zero. This is very important since it also applies to the matrix elements of these operators, and the rest of the story regarding the construction of our path integral follows the lines of the time-independent case.

In fact our path integral for the propagator can be taken to be :

$$K(\vec{x}t|\vec{x}'t') = \lim_{\Delta t \to 0} \int [\prod_{r=0}^{N-1} < \vec{x}_{r+1}|1 - \frac{i}{\hbar} H(\tau_{r+1})\Delta t|\vec{x}_r >] \prod_{r=1}^{N-1} d\vec{x}_r \qquad \text{(VI.20)}$$

with the end conditions $\vec{x}_o = \vec{x}'$ and $\vec{x}_N = \vec{x}$.

Eventually (VI.20) in the case of a time dependent hamiltonian with a scalar potential leads again to formula (VI.13) for the approximate propagator, but this time with $U(\vec{x}_j)$ replaced by $U(\vec{x}_j,\tau_j)$. Also in the expression (VI.14a) for the propagator we have the replacement of $U(\vec{x}(\tau))$ by $U(\vec{x}(\tau),\tau)$ i.e. the lagrangian of the problem appears in the phase again.

VII. FURTHER ON PATH INTEGRALS

It is about time to deal with some evaluations of path integrals, and with what else other than the 3D-harmonic oscillator to start?

The Lagrangian is :

$$L = \frac{m}{2}\,\dot{\vec{x}}^2(\tau) - \frac{m}{2}\,\Omega^2\,\vec{x}^2(\tau) \qquad \text{(VII.1)}$$

The approximating sequence of propagators for isomeric partitions is :

$$K_N(\vec{x}t|\vec{x}'0) = \int \exp\left(\frac{i}{\hbar}\,S_N[\vec{x}_j]\right)\left(\frac{m}{2\pi i\hbar\Delta\tau}\right)^{3/2}\prod_{j=1}^{N-1}\left(\frac{m}{2\pi i\hbar\Delta\tau}\right)d\vec{x}_j \qquad \text{(VII.2a)}$$

where the discrete form of the action is :

$$S_N\,[\vec{x}_j] = \sum_{j=0}^{N-1}\frac{m}{2}\left[\left(\frac{\vec{x}_{j+1} - \vec{x}_j}{\Delta\tau}\right)^2 - \Omega^2\,\vec{x}_j\right]\Delta\tau \qquad \text{(VII.2b)}$$

with the end conditions : $\vec{x}_0 = \vec{x}$, $\vec{x}_N = \vec{x}$

We have got to perform the multiple integration over the various $\vec{x}_j$'s. Owing to the end conditions the kinetic energy introduces linear terms involving the variables $\vec{x}_1$ and $\vec{x}_{N-1}$ while the rest of the exponential phase is quadratic, but off-diagonal. We have got to complete the square.

The first thing we do is to make the phase a homogeneous expression in the variables of integration. We then introduce the transformation $\vec{x} \to \vec{y}$:

$$\vec{x}_j = \vec{X}_j + \vec{y}_j \quad (j = 0,1,2,\ldots,N) \qquad \text{(VII.3)}$$

Since $\vec{x}_0 = \vec{x}'$ and $\vec{x}_N = \vec{x}$, if we fix $\vec{X}_0$ and $\vec{X}_N$ by $\vec{x}'$ and $\vec{x}$ then $\vec{y}_0$ and $\vec{y}_N$ are fixed by the condition $\vec{y}_0 = \vec{y}_N = 0$.

Inserting the transformation (VII.3) into (VII.2b) we have by Taylor expansion about the $\vec{X}_j$'s :

$$S_N[\vec{X}_j + \vec{y}_j] = S_N\,[\vec{X}_j] + \sum_{j=1}^{N-1}\frac{\partial S_N\,[\vec{X}_j]}{\partial\vec{X}_j\Delta\tau}\,.\vec{y}_j\Delta\tau + S_N\,[\vec{y}_j] \qquad \text{(VII.4)}$$

$S_N\,[\vec{y}_j]$ is homogeneous in the new variables $\vec{y}_j$ ($j = 1, 2,\ldots,N-1$) on account of the end conditions on the $\vec{y}$'s. It seems however that with our transformation we have created plenty of other linear terms, the cross term involving the $\vec{X}$'s and $\vec{y}$'s. But really this is not a big problem, we just choose our $\vec{X}_j$'s so that :

$$\frac{\partial S_N[\vec{X}_j]}{\partial \vec{X}_k \Delta\tau} = 0 \qquad (k = 1,2,\ldots,N-1) \qquad \text{(VII.5)}$$

But, in the limit as $\Delta\tau \to 0$ equations (VII.5) are nothing else but the classical equation of motion

$$m \frac{d\vec{X}^2}{d\tau^2} = - m\, \Omega^2\, \vec{X} \qquad \text{(VII.6)}$$

which have to solve with the end conditions $\vec{X}(0) = \vec{x}'$, $\vec{X}(t) = \vec{x}$ in order to obtain the continuous form of the required path $\vec{X}_j$.

In terms of the $\vec{y}_j$ variables the multiple integral (VII.2) becomes :

$$K_N(\vec{x}t|\vec{x}'0) = \exp\left(\frac{i}{\hbar} S_N[\vec{X}_j]\right) K_N(\vec{0}t|\vec{0}0) \qquad \text{(VII.7a)}$$

where :

$$K_N(\vec{0}t|\vec{0}0) = \int_{\vec{y}_0=\vec{y}_N=\vec{0}} \exp\left(\frac{i}{\hbar} S_N[\vec{y}_j]\right)\left(\frac{m}{2\pi i\hbar\Delta\tau}\right)^{3/2} \prod_{j=1}^{N-1} \left(\frac{m}{2\pi i\hbar\Delta\tau}\right)^{3/2} d\vec{y}_j \qquad \text{(VII.7b)}$$

It should be noted here that the jacobian of the transformation from $\vec{x}$ to $\vec{y}$ is 1, and so we forget about it.

We have to evaluate (VII.7b) and follow Abé (1954) (28). The phase in (VII.7b) is now a homogeneous quadratic expression in the $\vec{y}_1$, $\vec{y}_2,\ldots,\vec{y}_{N-1}$-variables, and with some re-arrangement it takes the form :

$$S_N[\vec{y}_j] = \frac{m}{2\Delta\tau}\left\{\sum_{j=1}^{N-1}[2 - \Omega^2(\Delta\tau)^2]\vec{y}_j^{\,2} - 2\sum_{j=1}^{N-2}\vec{y}_j\cdot\vec{y}_{j+1}\right\} \qquad \text{(VII.8a)}$$

This expression can now be put in the form :

$$S_N[\vec{y}_j] = \frac{m}{2\Delta\tau}\sum_{j=1}^{N-1} A_j\,(\vec{y}_j - Q_j\,\vec{y}_{j+1})^2;\ \vec{y}_N = \vec{0} \qquad \text{(VII.8b)}$$

The A_j's and the Q_j's which make the two expressions identical are :

$$\left.\begin{aligned} A_1 &= 2 - \Omega^2(\Delta\tau)^2 \\ A_{j+1} + A_j Q_j &= 2 - \Omega^2(\Delta\tau)^2 \\ A_j Q_j &= 1 \end{aligned}\right\} \quad (j = 1,2,\ldots,N-2) \qquad \text{(VII.9a)}$$

We eliminate Q_j and obtain :

$$A_{j+1} = A_j^{-1} + 2 - \Omega^2(\Delta\tau)^2 \quad (j = 1,2,\ldots,N-2) \qquad \text{(VII.9b)}$$

The action form (VII.8b) clearly suggests the transformation :

$$\vec{\eta}_j = \vec{y}_j - Q_j \vec{y}_{j+1} \qquad (j = 1,2,\ldots,N-1) \qquad \text{(VII.10)}$$

which again has jacobian unity.

In terms of the $\vec{\eta}_j$ variables the multiple integral (VII.7b) becomes :

$$K_N(\vec{0}t|\vec{0}0) =$$

$$\int \exp\left(\frac{i}{\hbar}\frac{m}{2\Delta\tau}\sum_{j=1}^{N-1} A_j \vec{\eta}_j^{\,2}\right)\left(\frac{m}{2\pi i\hbar\Delta\tau}\right)^{3/2}\prod_{j=1}^{N-1}\left(\frac{m}{2\pi i\hbar\Delta\tau}\right)^{3/2} d\vec{\eta}_i \qquad \text{(VII.11)}$$

Performing the $\vec{\eta}_j$- integrations we have :

$$K_N(\vec{0}t|\vec{0}0) = \left(\frac{m}{2\pi i\hbar D_N(t)}\right)^{3/2} \qquad \text{(VII.12a)}$$

where :

$$D_N(t) = \Delta\tau\, A_1 A_2 \ldots A_{N-1} \qquad \text{(VII.12b)}$$

To complete the evaluation we require the value of D_N in the limit as $N \to \infty$ or what amounts to the same thing as $\Delta\tau \to 0$.

Let us then consider a given partition of the interval $[0,\tau]$ with k subdivision points $(\Delta\tau = \tau/k)$. We have $D_k(\tau) = \Delta\tau\, A_1 A_2 \ldots A_{k-1}$ and for the interval $[0,\tau + \Delta\tau]$ with $(k + 1)$ points :

$$D_{k+1}(\tau + \Delta\tau) = \Delta\tau\ A_1 A_2 \ldots A_{N-1} A_k = D_k(\tau) A_k \quad \text{(VII.13a)}$$

Utilizing (VII.9b) in (VII.13a) we find :

$$\frac{1}{(\Delta\tau)^2}\left[D_{k+1}(\tau+\Delta\tau) - 2\, D_k(\tau) + D_{k-1}(\tau-\Delta\tau)\right] = -\,\Omega^2\, D_k(\tau) \quad \text{(VII.13b)}$$

This is a difference equation of second order. We require for its solution two conditions e.g. $D_1 = 1.\Delta\tau$ and $D_2 = \Delta\tau\ A_1$ or :

$$D_1 = \Delta\tau\ , \quad \frac{D_2 - D_1}{\Delta\tau} = 1 - \Omega^2(\Delta\tau)^2 \quad \text{(VII.13c)}$$

But what we really want is the limiting value of $D_k(\tau)$ when $k \to \infty$ (or $\Delta\tau \to 0$). Under these circumstances the difference equation (VII.13b) and the conditions (VII.13c) go over to the differential equation :

$$\ddot{D}(\tau) = -\,\Omega^2 D(\tau) \quad \text{(VII.14a)}$$

and the (initial) conditions :

$$D(0) = 0, \quad \dot{D}(0) = 1 \quad \text{(VII.14b)}$$

The solution of (VII.14a,b) for $\tau = t$ leads to :

$$D(t) = \frac{\sin \Omega t}{\Omega} \quad \text{(VII.15)}$$

Taking the limit of (VII.7a) as $N \to \infty$ we have for the propagator of the harmonic oscillator :

$$K(\vec{x}t|\vec{x}'0) = \exp\left[\frac{i}{\hbar}\, S_c(\vec{x}t|\vec{x}'0)\right] K(\vec{0}t|\vec{0}0) \quad \text{(VII.16)}$$

where S_c is the classical action of the oscillator starting from $\vec{x}'$ to reach $\vec{x}$ in time t, $K(\vec{0}t|\vec{0}0)$ is the propagator of the harmonic oscillator with cyclic return to the origin in time t. The classical action can be either evaluated from the Hamilton-Jacobi equation of motion for the action or through the equation of motion (VII.6). We shall not enter the details of the classical action evaluation, but point out that Feynman and Hibbs (1965) (8) show how one can save himself from extra work. The evaluation of $K(\vec{0}t|\vec{0}0)$ comes from $K_N(\vec{0}t|\vec{0}0)$ in the limit as $N \to \infty$ which is obtained through (VII.15). The final result is:

$$K(\vec{x}0|\vec{x}'0) = \left(\frac{m\Omega}{2\pi i \quad \sin \Omega t}\right)^{3/2}$$

$$\times \exp \left\{\frac{i}{\hbar} \frac{m\Omega}{2 \sin \Omega t} [(\vec{x}^2 + \vec{x}'^2) \cos \Omega t - 2 \vec{x}.\vec{x}']\right\} \qquad \text{(VII.17)}$$

The above evaluation was given as an example of how path integral methods work in practice. For more about evaluations the reader may consult Feynman and Hibbs (6), Abé (1954) (28) and other references.

Closely allied to the method of evaluation of the harmonic oscillator propagator is the WKB approximation in path integrals. This approximation shows at the same time the proximity of the path integral method to classical mechanics. This is so for most of the contribution of the multiple integrations of the path integral to the propagator, comes when the variables $\vec{x}(\tau)$ lie in the vicinity of the classical path. Elsewhere there are rapid oscillations causing cancellations.

Let us now consider a case of a general lagrangian and let $\vec{X}(\tau)$ be the classical path of our problem. If we perform the transformation : $\vec{x}(\tau) = \vec{X}(\tau) + \vec{y}(\tau)$ the action functional of our system becomes :

$$S[\vec{x}(\tau)] = \int_0^t L(\vec{x}(\tau),\dot{\vec{x}}(\tau),\tau)d\tau$$

$$= S_c(\vec{x}t|\vec{x}'0) + \int_0^t \frac{\delta S}{\delta\vec{X}(\tau)}.\vec{y}(\tau)d\tau$$

$$+ \frac{1}{2}\int_0^t\int_0^t \vec{y}^T(\tau) \frac{\delta^2 S}{\delta\vec{X}(\tau)\ \delta\vec{X}(\tau')} \vec{y}(\tau')\ d\tau\ d\tau' + O(y^3) \qquad \text{(VII.18a)}$$

where $S_c(\vec{x}t|\vec{x}'0)$ is the action along the classical path. The first order term in $\vec{y}(\tau)$ is zero on account of the equations of motion. Now, if we approximate our functional to second order in $\vec{y}(\tau)$, we have the WKB approximation.

We write the WKB approximation for the propagator more explicitly , assuming $L = T - U$:

$$K_{WKB}(\vec{x}t|\vec{x}'0) = \exp[\frac{i}{\hbar} S_c(\vec{x}t|\vec{x}'0)]$$

$$\int \exp[\frac{i}{\hbar} \int_o^t (\frac{m}{2} \vec{y}^2(\tau) - \vec{y}^t(\tau) \frac{\delta^2 U(X(\tau))}{\delta \vec{X}^2(\tau)} \vec{y}(\tau))d\tau] D[\vec{y}]$$

$$\vec{y}(o) = \vec{y}(t) = \vec{0} \qquad \text{(VII.18b)}$$

where $D[\vec{y}]$ is the path differential (VI.14b).

For quadratic lagrangians the path integral in (VII.18b) does not depend on $\vec{x},\vec{x}'$ and is solely dependent on t. In this case the WKB expression for the propagator gives exact results; see the previous evaluation for the harmonic oscillator.

It can be shown that the propagator in the WKB approximation can be expressed fully in terms of the classical action as :

$$K_{WKB}(\vec{x}t|\vec{x}'0) =$$
$$[\det \frac{1}{2\pi i\hbar} \frac{\partial^2 S_c}{\partial \vec{x}\, \partial \vec{x}'}]^{1/2} \exp[\frac{i}{\hbar} S_c(\vec{x}t|\vec{x}'0)] \qquad \text{(VII.19)}$$

This result is due to Van Vleck (1928) (29) in connection with the correspondence principle. For more about this see De Witt (1957) (19). It should be mentioned here that if the lagrangian contains memory the above formula fails to apply. For an example of this nature see Papadopoulos (1974) (30).

Pauli (1952) (31) got involved a bit with path integrals and suggested the use of WKB short-time propagators in the composition law. These propagators being correct to order $\Delta\tau$ give exact results when the partitioning of the time interval is infinitely refined. However, owing to the fact that the WKB propagators are nearer to the exact propagator, one requires less refinement of the finite time interval for a good approximation of the propagator.

Next we present the propagator for a harmonically bound electron in a constant magnetic field B obtained by the path integral method. The lagrangian is :

$$L_1 = \frac{m}{2} \dot{\vec{x}}^2 - \frac{m}{2} \Omega^2 \vec{x}^2 + \frac{m}{2} \omega(x_1\dot{x}_2 - x_2\dot{x}_1) \qquad \text{(VII.20a)}$$

where $\omega = |e|B/m$ (= cyclotron frequency) and B has been taken in the z-direction.

We have a quadratic lagrangian and the Van Vleck formula can be also applied exactly. The propagator takes the explicit form :

$$K_1(\vec{x}t|\vec{x}'0) = \frac{m\Omega'}{2\pi i\hbar \sin\Omega' t}\left(\frac{m\Omega}{2\pi i\hbar \sin\Omega t}\right)^{1/2}$$

$$\times \exp\left\{\frac{i}{\hbar}\frac{m\Omega'}{2\sin\Omega' t}\left[(\vec{x}_\perp^2+\vec{x}_\perp'^2)\cos\Omega' t - 2\,\vec{x}_\perp\cdot\vec{x}_\perp'\cos\frac{\omega}{2}t + 2\,\vec{x}_\perp \wedge \vec{x}_\perp' \sin\frac{\omega}{2}t\right]\right\}$$

$$\times \exp\left\{\frac{i}{\hbar}\frac{m\Omega}{2\sin\Omega t}\left[(x_3^2 + x_3'^2)\cos\Omega t - 2x_3x_3\right]\right\} \quad \text{(VII.20b)}$$

where :

$$\Omega' = [\Omega^2 + (\omega/2)^2]^{1/2}$$

Let us now extract the spectrum of our charged particle. As per usual we find the partition function

$$Z(\beta) = \int d\vec{x}\, K_1(\vec{x}, -i\hbar\beta|\vec{x}0)$$

$$= \{2^3 \sinh[\tfrac{1}{2}\beta(\Omega'+\tfrac{\omega}{2})]\sinh[\tfrac{1}{2}\beta\hbar(\Omega'-\tfrac{\omega}{2})]\ \sinh(\tfrac{1}{2}\beta\,\Omega)\}^{-1}$$

(VII.21)

From the partition function we can extract the density of states $g(\varepsilon)$, since

$$Z(\beta) = \int_0^\infty g(\varepsilon)\exp(-\beta\varepsilon)d\varepsilon \quad \text{(VII.22a)}$$

So, the density of states is the inverse Laplace transform of $Z(\beta)$. Utilizing the formula for the 1-dimensional harmonic oscillator :

$$[2\sinh(\tfrac{1}{2}\beta\hbar\Omega)]^{-1} = \int_0^\infty \sum_{n=0}^\infty \delta[\varepsilon - (n+\tfrac{1}{2})\hbar\Omega]\exp(-\beta\varepsilon)\,d\varepsilon$$

(VII.22b)

we find for its density of states $g_3(\varepsilon)$:

$$g_3(\varepsilon) = \sum_{n=0}^\infty \delta[\varepsilon - (n+\tfrac{1}{2})\hbar\Omega] \quad \text{(VII.22c)}$$

If $g_1(\varepsilon)$, $g_2(\varepsilon)$ are the inverse Laplace transforms corresponding to the other two factors of the partition function (VII.21) associated with the x_1 and x_2 motions then by the convolution theorem we find for the density of states of our particle the expression :

$$g(\varepsilon) = \int g_1(\varepsilon-\varepsilon')\, g_2(\varepsilon'-\varepsilon'')\, g_3(\varepsilon'')\, d\varepsilon'\, d\varepsilon''$$

$$= \sum_{n_1,n_2,n_3=0}^{\infty} \delta(\varepsilon - \varepsilon_{n_1 n_2 n_3}) \qquad \text{(VII.22d)}$$

where :

$$\varepsilon_{n_1 n_2 n_3} = (n_1 + \tfrac{1}{2})\hbar \left(\sqrt{\Omega^2 + (\tfrac{\omega}{2})^2} + \tfrac{\omega}{2}\right)$$

$$+ (n_2 + \tfrac{1}{2})\hbar \left(\sqrt{\Omega^2 + (\tfrac{\omega}{2})^2} - \tfrac{\omega}{2}\right) + (n_3 + \tfrac{1}{2})\, \hbar\Omega \qquad \text{(VII.22e)}$$

This is the required spectrum, which represents the Zeemann effect for a harmonically bound electron.

The above density of states enables us to obtain the magnetization of harmonically bound electrons (whatever worth this might be) but nevertheless interesting for one can show the existence at low temperature pronounced oscillations of the succeptibility with the magnetic field resulting from the quantum statistics. For the details see our 1971 references (30).

VIII. FUNCTIONAL INTEGRALS FOR BOSE SYSTEMS

Functional integrals have been widely employed for the study of both the dynamical and the statistical properties of quantum systems. For systems described by Schrödinger hamiltonians we have the Feynman path integral (6). Functional integrals for second quantized hamiltonians have also been devised (Matthews and Salam, 1955 (8); Klauder, 1960 (22); Schweber, 1962; Casher, Lurie and Revzen, 1969 (22) and others). These works refer mainly for boson systems, but as Matthews and Salam (1955) pointed out, the functional integrals for Bose systems can be applied to Fermi systems with the use of quaternions (see applications to many-body expansions (Hubbard (1959); Edwards and Sherrington (1967); Sherrington (1967) (32).

In this section we shall develop functional integrals for second quantized hamiltonians with Bose operators. We move on lines parallel to Klauder's original work (22). The section contains part of a lecture given a couple of years ago at the Antwerp Summer School (22) in which I promised to return on the subject with more; here we are.

For the purpose of developing our functional integral, we shall simplify matters as per usual, and deal with the following simple hamiltonian

$$H = \Omega a^{+} a \qquad \text{(VIII.1)}$$

with the annihilation and creation operators a, a^{+} obeying the commutation relation

$$[a, a^{+}] = 1$$

In the occupation number representation, the effects of these operators on the state with n particles, $|n>$, as is well known are :

$$a|n> = n^{1/2}|n-1>, \quad a^{+}|n> = (n+1)^{1/2}|n+1>$$

from which we have the number operators $a^{+}a$, through

$$a^{+}a|n> = n|n>$$

Talking about occupation numbers, this is the natural representation system for our hamiltonian (VIII.1), for in this system H becomes diagonal, and so the required propagator can be most easily obtained, in this representation, as :

$$<m|\exp\left(-\frac{i}{\hbar}\hbar\Omega a^{+}a\ t\right)|n> = \exp(-i\Omega nt)\quad \delta_{nm}$$

but we shall pretend that we have no knowledge of this simple fact here, and proceed on purpose to make things a little more complicated.

We shall use the system of coherent states

$$|\alpha> = \frac{1}{\sqrt{\pi}} \sum_{n=0}^{\infty} \frac{\alpha^{n}}{\sqrt{n!}} \exp\left(-\frac{1}{2}|\alpha|^{2}\right)|n> \qquad \text{(VIII.2a)}$$

for our representation.

Where α labelling the state $|\alpha>$ is complex. Before we go any further it might be an idea to assemble together a few needed facts about these states. (More about coherent states and functional integrals is included in references (22) and (33).

The bra form of the coherent state $|\alpha>$ is given by :

$$<\alpha| = \frac{1}{\sqrt{\pi}} \sum_{m=0}^{\infty} \frac{\overline{\alpha}^m}{\sqrt{m!}} \exp\left(-\frac{1}{2}|\alpha|^2\right) < m | \quad \text{(VIII.2b)}$$

where $\overline{\alpha}$ denotes the complex conjugate of α.

The coherent state $|\alpha>$is an eigenfunction of the annihilation operator a with eigenvalue α :

$$a|\alpha> = a|\alpha> \quad \text{or} \quad <\alpha|a^+ = <\alpha|\overline{\alpha} \quad \text{(VIII.2c)}$$

The scalar product of two coherent states is given by :

$$<\alpha|\alpha'> = \frac{1}{\pi} \exp\left(-\frac{1}{2}|\alpha'|^2 + \overline{\alpha}\alpha' - \frac{1}{2}|\alpha|^2\right) \quad \text{(VIII.2d)}$$

which means that two different such states are not orthogonal.

The coherent states obey the completeness relation :

$$\int <\lambda|\alpha> <\alpha|\mu> d^2\alpha = <\lambda|\mu> \quad \text{(VIII.2e)}$$

where $d^2\alpha = d(\text{Re}\,\alpha)\, d(\text{Im}\,\alpha)$

Finally we need the following integral identity :

$$\int \exp\left(-\gamma|\alpha|^2 + \lambda\overline{\alpha} + \mu\alpha\right) \frac{1}{\pi} d^2\alpha = \frac{1}{\gamma} \exp\left(\frac{\lambda\alpha}{\gamma}\right) \quad \text{(VIII.2f)}$$

where $\text{Re}\,\gamma > 0$.

The propagator : We are looking for the propagator

$$<\alpha|\exp\left(-\frac{i}{\hbar} Hs\right)|\alpha'>$$

in terms of the coherent states. The parameter s could be the time t or $-i\hbar\beta$ to get matrix elements of the Boltzmann factor, $\exp(-\beta H)$, or could be a complex parameter. We shall, nevertheless, call it 'time'.

For our example we have, by analogy to (VI.6) :

$$< \alpha|\exp(-i\Omega a^+a s)|\alpha' > \simeq \int < \alpha|1 - i\Omega a^+a \Delta s|\alpha_{N-1} >$$

$$\times < \alpha_{N-1}|1 - i\Omega a^+a \Delta s|\alpha_{N-2} >\ldots< \alpha_2|1 - i\Omega a^+a \Delta s|\alpha_1 >$$

$$\times < \alpha_1|1 - i\Omega a^+a \Delta s|\alpha' > d^2\alpha_1 d^2\alpha_2 \ldots d^2\alpha_{N-1} \quad \text{(VIII.3)}$$

where $\Delta s = s/N$ of the line joining 0 with s.

In (VIII.3) we have N short-time propagators, but N-1 integrations.

With the aid of (VIII.2c) a typical short-time propagator takes the form :

$$< \alpha_{j+1}|1 - i\Omega a^+a \Delta s|\alpha_j > = (1 - i\Omega \bar{\alpha}_{j+1}\alpha_j \Delta s) <\alpha_{j+1}|\alpha_j>$$

$$= \frac{1}{\pi} \exp [- \frac{1}{2} |\alpha_j|^2 + (1 - i\Omega \Delta s) \bar{\alpha}_{j+1}\alpha_j - \frac{1}{2} |\alpha_{j+1}|^2]$$

$$+ O((\Delta s)^2) \quad \text{(VIII.4)}$$

We have added $O((\Delta s)^2)$ on the far r.h.s. of (VIII.4) in order to remove the excess terms of higher order in Δs (and thus maintain the equality with the l.h.s.) introduced through the replacement of $(1 - i\Omega \bar{\alpha}_{j+1}\alpha_j\Delta s)$ by the exponential $\exp(- i\Omega \bar{\alpha}_{j+1}\alpha_j \Delta s)$

As pointed out earlier on the typical short-time propagator, appearing in the (approximate) composition law (VIII.3), which is exact to first order in Δs can be replaced by another one again exact to first order in s without this affecting the limit as $\Delta s \to 0$, which gives the exact propagator :

$$< \alpha|\exp(- i\Omega a^+a s)|\alpha' >$$

With the above in mind let us replace the short-time propagators on the r.h.s. of (VIII;4). We have :

$$< \alpha|\exp(- i\Omega a^+a s)|\alpha' >$$

$$\simeq \int \exp [- \frac{1}{2} |\alpha'|^2 + (1 - i\Omega \Delta s) \bar{\alpha}_1\alpha' - |\alpha_1|^2$$

$$+ (1 - i\Omega \Delta s) \bar{\alpha}_2\alpha_1 - |\alpha_2|^2 + \ldots - |\alpha_{N-1}|^2$$

$$+ (1 - i\Omega \Delta s) \bar{\alpha}\alpha_{N-1} - \frac{1}{2} |\alpha|^2] \times \frac{1}{\pi} \prod_{j=1}^{N-1} \frac{1}{\pi} d^2\alpha_j \quad \text{(VIII.5a)}$$

The integrations over the α_i's can be performed one after the other with the aid of the integral identity (VIII.2f) and the result is :

$$< \alpha|\exp(-\, i\Omega\, a^{+}a\, s)\alpha' >$$
$$\approx \frac{1}{\pi} \exp\left[- \frac{1}{2} |\alpha|^2 + (1 - i\Omega \frac{s}{N})^N \bar{\alpha}\alpha' - \frac{1}{2} |\alpha'|^2\right] \tag{VIII.5b}$$

which in the limit as $N \to \infty$ leads to our propagator in terms of coherent states :

$$< \alpha|\exp(-\, i\Omega\, a^{+}a\, s)|\alpha' >$$
$$= \frac{1}{\pi} \exp\left[- \frac{1}{2} |\alpha|^2 + e^{-i\Omega s} \bar{\alpha}\alpha' - \frac{1}{2} |\alpha'|^2\right] \tag{VIII.5c}$$

Let us now work out a simple example by obtaining the average occupation number for a system of free phonons or photons. We just consider one frequency.

The density matrix ρ in terms of the coherent states is given by :

$$< \alpha|\rho|\alpha' >$$
$$= \left[\int <\alpha|\exp(-\, i\Omega\, a^{+}a)|\alpha > d^2\alpha\right]^{-1} < \alpha|\exp(-i\Omega\, a^{+}a\, s)\alpha' >$$
$$= \left[1 - \exp(-\, i\Omega s)\right] \frac{1}{\pi} \exp\left[- \frac{1}{2} |\alpha|^2 + e^{-i\Omega s} \bar{\alpha}\alpha' - \frac{1}{2} |\alpha'|^2\right] \tag{VIII.6}$$

with : $s = -\, i\hbar\beta = -\, i\hbar/kT$.

The average occupation number is obtained as :

$$< a^{+}a > = \mathrm{Tr}\, \rho\, a^{+}a$$
$$= \int < \alpha|\rho|\alpha' > < \alpha'|a^{+}a)\alpha > d^2\alpha' d^2\alpha \tag{VIII.7a}$$

which with the aid of (VIII.6) and $< \alpha'|a^{+}a|\alpha > = \bar{\alpha}'\alpha < \alpha'|\alpha >$ leads to :

$$< a^{+}a > = \frac{1}{\exp\left(\frac{\hbar\Omega}{kT}\right) - 1} \tag{VIII.7b}$$

the well known Bose function.

Before we deal with more general hamiltonians we wish to rearrange the terms in the exponent of (VIII.5a). We shall initially concentrate our attention on the expression deriving from the product of the scalar products of the form $< \alpha_{j+1} | \alpha_j >$ see (VIII.4).

We have the following three equivalent expressions :

$$<\alpha_N|\alpha_{N-1}><\alpha_{N-1}|\alpha_{N-2}>\ldots<\alpha_3|\alpha_2><\alpha_2|\alpha_1><\alpha_1|\alpha_0> \prod_{j=1}^{N-1} d^2\alpha_j$$

$$= \exp\{\frac{1}{2} \sum_{j=0}^{N-1} [\alpha_j(\bar{\alpha}_{j+1}-\alpha_j) - \bar{\alpha}_{j+1}(\alpha_{j+1}-\alpha_j)]\}\frac{1}{\pi} \prod_{j=1}^{N-1} \frac{1}{\pi} d^2\alpha_j \tag{VIII.8a}$$

$$= \exp[-\frac{1}{2}|\alpha_0|^2 - \sum_{j=0}^{N-1} \bar{\alpha}_{j+1}(\alpha_{j+1}-\alpha_j) + \frac{1}{2}|\alpha_N|^2]\frac{1}{\pi} \prod_{j=1}^{N-1} \frac{1}{\pi} d^2\alpha_j \tag{VIII.8b}$$

$$= \exp[\frac{1}{2}|\alpha_0|^2 + \sum_{j=0}^{N-1} \alpha_j(\bar{\alpha}_{j+1}-\alpha_j) - \frac{1}{2}|\alpha_N|^2] \frac{1}{\pi} \prod_{j=1}^{N-1} \frac{1}{\pi} d^2\alpha_j \tag{VIII.8c}$$

In the limit of infinite subdivision of the interval [0,t] (hope you have no objection to reverting back to straight time) the notation suggested by these expressions is respectively :

$$\exp\{\frac{1}{2} \int_0^t [\alpha(\tau) \frac{\partial\bar{\alpha}(\tau)}{\partial\tau} - \bar{\alpha}(\tau) \frac{\partial\alpha(\tau)}{\partial\tau}] d\tau\}\frac{1}{\pi} \prod_{0<\tau<t} \frac{1}{\pi} d^2\alpha(\tau) \tag{VIII.9a}$$

$$\exp [- \frac{1}{2}|\alpha(0)|^2 - \int_0^t \bar{\alpha}(\tau) \frac{\partial\alpha(\tau)}{\partial\tau} d\tau + \frac{1}{2}|\alpha(t)|^2]\frac{1}{\pi} \prod_{0<\tau<t} \frac{1}{\pi}d^2\alpha(\tau) \tag{VIII.9b}$$

$$= \exp[\frac{1}{2}|\alpha(0)|^2 + \int_0^t \alpha(\tau) \frac{\partial\bar{\alpha}(\tau)}{\partial\tau}d\tau - \frac{1}{2}|\alpha(t)|^2]\frac{1}{\pi} \prod_{0<\tau<t} \frac{1}{\pi}d^2\alpha(\tau) \tag{VIII.9c}$$

Although integration by parts transforms the expressions (VIII.9a,b,c) one from the other, it would be dangerous to begin with them for the game of functional integration. The trouble comes when writing back discrete forms from the 'limiting' expressions, for these forms could be different from those given by (VIII.8a,b,c). Consider for example (VIII.9b) :

The integral $\int_0^t \bar{\alpha}(\partial\alpha/\partial\tau)d\tau$ can be interpreted in a different way from the one in (VIII.8b) namely :

$\Sigma\, \bar{\alpha}_j(\alpha_{j+1} - \alpha_j)$.

This will certainly lead to erroneous evaluations. The moral is that really, as far as the process of multiple integration is concerned, we have no other choice than using (VIII.8a,b,c) for interpreting (VIII.9a,b,c). The reason for having to adhere rigidly to the discrete forms given by (VIII.8a,b,c) lies in the fact that these expressions constitute the various δ-functions of our problem and which we are not allowed to spoil. However, we have some freedom in the choice of the discrete forms which derive from the hamiltonian of the problem, but in this case we have to adjust the normalizing factors accordingly.

Consider now a more general hamiltonian $H(a^+,a)$ which is expressed in its normal form (the creation operators preceeding the annihilation operators) we have :

$$< \alpha_{j+1}|H(a^+,a)|\alpha_j > = H(\bar{\alpha}_{j+1},\alpha_j) < \alpha_{j+1}|\alpha_j > \tag{VIII.10}$$

The propagator associated with such a hamiltonian, as a functional integral, can take the form :

$$< a|T \exp [- \frac{i}{\hbar} \int_0^t H(a^+,a)d\tau]\alpha' >$$

$$= \exp \{- \frac{1}{2}|\alpha'|^2 - \int_0^t [\bar{\alpha}(\tau) \frac{\partial\alpha(\tau)}{\partial\tau} + \frac{i}{\hbar} H(\bar{\alpha}(\tau),\alpha(\tau))] d\tau + \frac{1}{2}|\alpha|\} \frac{1}{\pi} \prod_{0<\tau<t} \frac{1}{\pi} d^2\alpha(\tau) \tag{VIII.11}$$

with : $\alpha(0) = \alpha'$, $\alpha(t) = \alpha$

Again in writing the integral $\int_0^t \bar{\alpha}\,\dot{\alpha}\,d\tau$ in (VIII.11) we have in mind the discrete form (VIII.8b). The discrete form of the integral $\int_0^t H(\bar{\alpha}(\tau),\alpha(\tau))d\tau$ is obtained by use of (VIII.10) and is

$$\sum_{j=0}^{N-1} H(\bar{\alpha}_{j+1},\alpha_j)\,\Delta\tau$$

Quadratic hamiltonian with linear terms : Let us now proceed to evaluating the propagator associated with the following hamiltonian :

$$H_1 = \hbar\Omega a^+ a + \bar{\phi}(t)\, a^+ + \phi(t) a \qquad \text{(VIII.12)}$$

The solution to this problem can be used for handling the vibrational degrees of freedom in the polaron hamiltonian.

To obtain the propagator associated with the hamiltonian H_1 given in (VIII.12) we make use of the functional integral (VIII.11) with H replaced by H_1.

With this replacement the integral appearing in (VIII.11) will be :

$$\int_0^t [\bar{\alpha}(\tau)\,\frac{\partial\alpha(\tau)}{\partial\tau} + \frac{i}{\hbar}\,H_1(\bar{\alpha}(\tau),\alpha(\tau))]\,d\tau$$
$$= \int_0^t \{\bar{\alpha}(\tau)\,\frac{\partial\alpha(\tau)}{\partial\tau} + i\Omega\bar{\alpha}(\tau)d(\tau)$$
$$+ \frac{i}{\hbar}\,[\,\bar{\phi}(\tau)\,\bar{\alpha}(\tau) + \phi(\tau)\,\alpha(\tau)]\,\}d\tau \qquad \text{(VIII.13)}$$

This differs from the corresponding integral of the previous evaluation, where the hamiltonian were $\hbar\Omega a^+ a$, by the linear functional in $\bar{\alpha}$ and α in the phase.

A linear transformation which reduces the present evaluation to the previous one can be formed using a particular solution of the Euler-Lagrange equations, obtained, as per usual, by setting the first variational derivative of (VIII.13) w.r.t. $\bar{\alpha}$ equal to zero, i.e. :

$$\frac{\delta}{\delta\bar{\alpha}(\tau)}\int_0^t [\,\bar{\alpha}(\tau)\,\frac{\partial\alpha(\tau')}{\partial\tau'} + \frac{i}{\hbar}\,H_1(\bar{\alpha}(\tau'),\alpha(\tau'))]\,d\tau' = 0 \qquad \text{(VIII.14a)}$$

This leads to the following (classical) equation of motion :

$$\dot{\alpha}(\tau) + i\Omega\alpha(\tau) + \frac{i}{\hbar}\,\overline{\phi}(\tau) = 0 \qquad \text{(VIII.14b)}$$

Or equivalently its complex conjugate

$$\dot{\overline{\alpha}}(\tau) - i\Omega\overline{\alpha}(\tau) - \frac{i}{\hbar}\,\phi(\tau) = 0 \qquad \text{(VIII.14c)}$$

We now pick up a particular solution of (VIII.14b), say,

$$A(\tau) = -\frac{i}{\hbar}\int_0^{\tau} \exp\,[-\,i\Omega(\tau-\tau')]\;\overline{\phi}(\tau')d\tau' \qquad \text{(VIII.14d)}$$

and form the linear transformation :

$$\alpha(\tau) = A(\tau) + \eta(\tau) \qquad \text{(VIII.15a)}$$

In order to preserve the end conditions on the $\alpha(\tau)$ variables ($\alpha(0) = \alpha'$, $\alpha(t) = \alpha$) the $\eta(\tau)$ variables must obey the end conditions :

$$\eta(0) = \alpha', \qquad \eta(t) = \alpha - A(t) \qquad \text{(VIII.15b)}$$

If we put now the transformation of (VIII.15,a,b) into (VIII.13) and use the resulting expression in (VIII.11) the functional integral giving the required propagator takes (in terms of the new variables $\eta(\tau)$) the form :

$$\langle\,\alpha|T\,\exp\,[-\,\frac{i}{\hbar}\int_0^t H_1(a^+,a)d\tau]\,|\alpha'\,\rangle$$

$$= \exp[-\,\tfrac{1}{2}|\alpha'|^2 - \overline{A}(t)\eta(t) - \frac{i}{\hbar}\int_0^t \phi(\tau)\;A(\tau)d\tau + \tfrac{1}{2}|\alpha|^2]$$

$$\times\;\exp\,\{-\int_0^t[\,\overline{\eta}(\tau)\dot{\eta}(\tau) + i\Omega\overline{\eta}(\tau)\eta(\tau)]\,d\tau\}\frac{1}{\pi}\prod_{0<\tau<t}\frac{1}{\pi}\,d^2\eta(\tau)$$

$$\text{(VIII.16)}$$

For the derivation of (VIII.16) we have made use of the equation of motion given by (VIII.14b) and its equivalent complex conjugate form (VIII.14c), satisfied by A and $\overline{A}$. Furthermore, we did not bother ourselves with the Jacobian of the transformation (VIII.15) for this is just unity.

The procedure followed for obtaining (VIII.16) has essential analogies with the one presented in Feynman and Hibbs (6) for the treatment of the forced harmonic oscillator. In fact here the mathematical manipulations are somewhat simpler, for the classical equations of motion are only of first order w.r.t. the time.

To complete the evaluation of our propagator we need to perform the functional integration on the r.h.s. of (VIII.16), which by now is something we really know how to do. We recall first of all that the integral $\int_0^t \bar{\eta}\, \dot{\eta}\, d\tau$ appearing in the functional integral of (VIII.16) corresponds to the discrete form given by the sum over j in (VIII.8b) (the role of the α_j is now taken by the η_j). Furthermore, remembering that $i\Omega\bar{\eta}\eta$ comes from the hamiltonian $\hbar\Omega a^+ a$ then by adding the fixed quantity $[-\frac{1}{2}|\eta(o)|^2 + \frac{1}{2}|\eta(t)|^2]$ to the exponent in our functional integral, we make, according to (VIII.11) the propagator for the hamiltonian $\hbar\Omega a^+ a$ known from (VIII.5c). We do not want to alter our functional integral and so we must subtract from the exponent what we have added to it, and in this way we have established for ourselves the formula :

$$\exp\{-\int_0^t [\bar{\eta}(\tau)\, \dot{\eta}(\tau) + i\Omega\bar{\eta}(\tau)\, \eta(\tau)]\, d\tau\} \prod_{0<\tau<t} \frac{1}{\pi}\, d^2\eta(\tau)$$

$$= \exp\, [\bar{\eta}(t)\, \eta(o)\, \exp\, (-\, i\Omega t) - |\eta(t)|^2] \qquad \text{(VIII.17)}$$

It is now a matter of routine substitutions to be made in (VIII.16) by use of (VIII.17) and the formulae for A(t), and $\eta(o)$ $\eta(t)$, from (VIII.15d) and (VIII.15b) for obtaining the propagator associated with the hamiltonian H_1 given by (VIII.12). We have :

$$< \alpha|T \exp\, [-\frac{i}{\hbar} \int_0^t H_1(a^+,a)\, d\tau]\, |\alpha' >$$

$$= \frac{1}{\pi} \exp\, \{-\frac{1}{2}|\alpha'|^2 + \bar{\alpha}\alpha' \exp\, (-\, i\Omega t) - \frac{1}{2}|\alpha|^2$$

$$- \frac{i}{\hbar}\, \bar{\alpha} \int_0^t \exp\, [-\, i\Omega(t-\tau)]\, \bar{\phi}(\tau)\, d\tau - \frac{i}{\hbar}\, \alpha' \int_0^t \exp(-i\Omega\tau)\, \phi(\tau)\, d\tau$$

$$- \frac{1}{\hbar^2} \int_0^t \int_0^\tau \exp[-i\Omega(\tau-\tau')] \phi(\tau) \bar{\phi}(\tau') d\tau d\tau'\}$$

(VIII.18)

This completes the lecture. More about these wonderful integrals on another occasion.

That is how the lecture ended at that time. Now we go a little further. Since our purpose has been to give a little from a wide variety of areas on functional integration it is about time that we should do something on perturbation theory.

Let us consider a many-boson system with hamiltonian :

$$H = \Sigma \varepsilon_k a_k^+ a_k + \Sigma V_{kk'k''k'''} a_k^+ a_{k'}^+ a_{k''} a_{k'''}$$

(VIII.19)

We are interested in obtaining the grand function of the system :

$$Z = T_r \exp[-\beta(H - \mu N)] \qquad \text{(VIII.20)}$$

where μ is the chemical potential and $N = \Sigma a_k^+ a_k$ the number operator.

The target of finding the partition function exactly cannot be achieved and we have to recourse to perturbation procedures. Let us then introduce :

$$H' = H_o' + \Sigma V_{kk'k''k'''} a_k^+ a_{k'}^+ a_{k''} a_{k'''}$$

(VIII.21a)

where :

$$H_o' = \Sigma \varepsilon_k' a_k^+ a_k \ ; \quad \varepsilon_k' = \varepsilon_k - \mu \qquad \text{(VIII.21b)}$$

So we deal with the propagator of the hamiltonian H' made out of the hamiltonian H by replacing ε_k with ε_k'.

With reference to the coherence state system the partition function takes the form of the functional integral :

$$Z = \int <\{\alpha_k\}|\exp(-\beta H')|\{\alpha_k\}> \prod_k d^2\alpha_k$$

$$= \exp\{-\int_0^\beta [\Sigma(\bar{\alpha}_k(s)\frac{\partial\alpha_k(s)}{\partial s} + \varepsilon_k'\,\bar{\alpha}_k(s)\,\alpha_k(s))$$

$$+ \Sigma\, V_{kk'k''k'''}\,\bar{\alpha}_k(s)\,\bar{\alpha}_{k'}(s)\,\alpha_{k''}(s)\,\alpha_{k'''}(s)]\,ds\,\}$$

$$\times \prod_{\substack{k\\ 0<s\leqslant\beta}} \frac{1}{\pi}\,d^2\alpha_k(s) \qquad \text{(VIII.22)}$$

with $\alpha_k(o) = \alpha_k(\beta)$.

The explanation for the meaning of (VIII.22) is the one given for (VIII.11). Evidently, on account of the end condition $\alpha_k(o) = \alpha_k(\beta)$ the terms $-\frac{1}{2}|\alpha_k(o)|^2$ and $+\frac{1}{2}|\alpha_k(\beta)|^2$ which appear in the exponential argument in the propagator case, now cancel out. Furthermore, we have switched from time to inverse temperature via the transformation $\tau \to -i\hbar s$.

We expand now (VIII.22) in a power series of V.

$$Z = \exp[-\int_0^\beta \Sigma\,(\bar{\alpha}_k\,\frac{\partial\alpha_k}{\partial s} + \varepsilon_k'\,\bar{\alpha}_k\,\alpha_k)ds]$$

$$\times [1 - \int_0^\beta \Sigma\, V_{kk'k''k'''}\,\bar{\alpha}_k\,\bar{\alpha}_{k'}\,\alpha_{k''}\,\alpha_{k'''}ds + O(V^2)] \qquad \text{(VIII.23)}$$

We can generate the series through functional differentiations of an appropriate linear exponential functional. Explicitly we have :

$$Z_o\,[\bar{\phi},\phi] = \exp[-\int_0^\beta \Sigma\,(\bar{\alpha}_k\,\frac{\partial\alpha_k}{\partial s} + \varepsilon_k'\,\bar{\alpha}_k\,\alpha_k$$

$$+ \bar{\phi}_k\,\bar{\alpha}_k + \phi_k\,\alpha_k)ds] \times \prod_{\substack{k\\ 0<s\leqslant\beta}} \frac{1}{\pi}\,d^2\alpha_k(s) \qquad \text{(VIII.24)}$$

with : $\alpha_k(o) = \alpha_k(\beta)$.

The partition function, Z, is obtained as follows :

$$Z = \exp\left(-\int_0^\beta ds \sum_k V_{kk'k''k''} \frac{\delta^4}{\delta\bar{\phi}_k(s)\delta\bar{\phi}_{k'}(s)\delta\phi_{k''}(s)\delta\phi_{k'''}(s)}\right.$$

$$Z_o\,[\bar{\phi},\phi] \qquad \text{(VIII.25)}$$

The functional $Z_o\,[\bar{\phi},\phi]$ is evaluated explicitly through our formula (VIII.18) (giving the propagator for one mode) and taking its trace. We have :

$$Z_o\,[\bar{\phi},\phi] = Z_o \exp\{-\sum_k \int_0^\beta \int_0^\beta [\frac{1}{e^{\beta\varepsilon'_k}-1} + \theta(s_1 - s_2)]$$

$$\times \exp[-\varepsilon'_k.(s_1-s_2)]\ \phi_k(s_1)\ \bar{\phi}_k(s_2)\ ds_1\ ds_2\} \qquad \text{(VIII.26a)}$$

where :

$$Z_o = \prod_k \frac{1}{e^{\beta\varepsilon'_k}-1} \qquad \text{(VIII.26b)}$$

Looks like the partition function for the non-interacting system, but not quite so, for the chemical potential relates to that of the interacting system.

It is now a matter of routine functional differentiations to obtain through (VIII.25) any term of the perturbation series (VIII.23).

We evaluate the first order term in V :

$$Z^{(1)} = -\int_0^\beta ds\ \Sigma\ V_{kk'k''k'''} \frac{\delta^4}{\delta\bar{\phi}_k(s)\delta\bar{\phi}_{k'}(s)\delta\phi_{k''}(s)\delta\phi_{k'''}(s)}$$

$$\times Z_o\,[\bar{\phi},\phi] = Z_o \sum_{k,k'} \frac{(V_{kk'kk'} + V_{kk'k'k})}{(e^{\beta\varepsilon'_k}-1)\ (e^{\beta\varepsilon'_{k'}}-1)} \qquad \text{(VIII.27)}$$

Next we shall develop functional integrals for Fermi-systems working as closely as possible along the lines adopted in the Bose case.

IX. FUNCTIONAL INTEGRALS FOR FERMI SYSTEMS

In this section we shall deal with functional integrals for Fermi systems utilizing a complete system of states, closely related to the coherent states, and which incorporate the necessary algebra of fermion states.

As per usual when we wish to produce a functional integral giving the propagator associated with a given hamiltonian, H, in terms of a complete set of states $|\vec{\gamma}>$ we make use of the formula developed in section (VI). We have :

$$< \vec{\gamma}|T \exp [- \frac{i}{\hbar} \int_{t'}^{t} H(\tau)d\tau] |\vec{\gamma}' >$$

$$= \lim \int [\prod_{r=0}^{N-1} <\vec{\gamma}^{(r+1)}|1 - \frac{i}{\hbar} H(\tau_r)\Delta\tau_r + O((\Delta\tau)^2)|\vec{\gamma}^{(r)}>] \prod_{r=1}^{N-1} d\vec{\gamma}^{(r)}$$

$$\max.(\Delta\tau_r) \to 0 \qquad (IX.1)$$

with : $\vec{\gamma}^{(o)} = \vec{\gamma}', \ \vec{\gamma}^{(N)} = \vec{\gamma}$.

The time interval $[t',t]$ is partitioned, as usual by points τ_r with $\tau_o = t'$, $\tau_N = t$; $\Delta\tau_r = \tau_{r+1} - \tau_r$. T signifies the time ordering operator $d\vec{\gamma}$ is the product of differentials of the components of $\vec{\gamma}$ in the case of rectilinear variables, and this is the case we are dealing with.

The r.h.s. of (IX.1) is essentially our functional integral. It is quite general and accomodates first and second quantized hamiltonians, irrespective of the statistics obeyed by the system under study. While with first quantized hamiltonians we can obtain from (IX.1) the Feynman path integral which involves an exponential phase equal to $(i/\hbar)$ times the action functional of the classical equations of motion, the functional integrals associated with the second quantized boson hamiltonians lead to a phase functional relating to the variational principle yielding the Schrödinger equation (see Edwards and Freed 1970) (15). We have not given this particular form of functional integral here which can also be found in references of the previous section. The phase functional of section (VIII) is associated with the variational principle (VIII.14a) leading to the classical equations of motion of the system. Well now, according to Matthews and Salam (1955) (8), if we replace the complex variables in the boson functional integral with autocommuting complex variables (quaternions) we produce the corresponding functional integral for the fermion propagator. Functional integrals with quaternions are good for obtaining relations, but not so easy for handling direct evaluations.

However, the system of complex variables as originally demonstrated by Klauder (1960) (22) is quite capable of handling the Fermi case, provided one constructs an appropriate system of states. Klauder did so and evaluated the noninteracting case. In his algorithm the states are labelled by the complex numbers of the unit circle. However, one finds it desirable to introduce states that form the analogue of the coherent states. This was done in our reference (23) and we proceed to explain below.

We consider a Fermi system for which its particles can go into n channels and introduce the following n-channel state :

$$|\gamma_n\ldots\gamma_2\gamma_1\rangle = \pi^{-n/2}\exp\left(-\frac{1}{2}\sum_{j=1}^{n}|\gamma_j|^2\right)$$

$$\times\left(1+\sum_{j=1}^{n}\gamma_j C_j^{+}+\sum_{j>\ell=1}^{n}\gamma_j\gamma_\ell C_j^{+}C_\ell^{+}+\sum_{j>\ell>k=1}^{n}\gamma_j\gamma_\ell\gamma_k C_j^{+}C_\ell^{+}C_k^{+}\right.$$

$$\left.+\ldots+\gamma_n\ldots\gamma_2\gamma_1 C_n^{+}\ldots C_2^{+}C_1^{+}\right)|0\rangle \qquad \text{(IX.2a)}$$

where the γ_j's are complex numbers (labelling the state), $|0\rangle$ is the vacuum state, and C_j^{+} is the fermion creation operator for the channel j.

The state (IX.2a) contains all possible independent 0,1,2,..., n fermion states; each of which once. Thus, the 2-fermion states for n=3 are taken to be : $C_2^{+}C_1^{+}|0\rangle$, $C_3^{+}C_1^{+}|0\rangle$, $C_3^{+}C_2^{+}|0\rangle$. Any permutation of their operators will generate dependent states. As is well known, these states are orthogonal between themselves and all states with the lower or higher number of particles. It should be noted here that the ordering of the particle creation is not important, but if one keeps a certain convention this will facilitate the evaluations.

The bra-form of the state (IX.2a) is written in terms of the annihilation operators C_j and the complex conjugates $\overline{\gamma}_j$ of γ_j as :

$$< \gamma_1\gamma_2 \cdots \gamma_2| = \pi^{-n/2} \exp \left(-\frac{1}{2}\sum_{j=1}^{n}|\gamma_j|^2\right)$$

$$\times < 0|(1 + \sum_{j=1}^{n} \overline{\gamma}_j C_j + \sum_{j>\ell=1}^{n} \overline{\gamma}_\ell \overline{\gamma}_j C_\ell C_j + \sum_{j>\ell>k=1}^{n} \overline{\gamma}_k \overline{\gamma}_\ell \overline{\gamma}_j C_k C_\ell C_j$$

$$+ \overline{\gamma}_1\overline{\gamma}_2 \cdots \overline{\gamma}_n C_1 C_2 \cdots C_n) \qquad \text{(IX.2b)}$$

The totality of the states $|\gamma_n \cdots \gamma_2\gamma_1 >$ forms a complete set in the sense :

$$\int |\gamma_n \cdots \gamma_2\gamma_1 >< \gamma_1\gamma_2 \cdots \gamma_n| \prod_{j=1}^{n} d^2\gamma_j = 1 \qquad \text{(IX.3)}$$

$d^2\gamma$ means : $d(\mathrm{Re}\,\gamma)d(\mathrm{Im}\gamma)$.

For establishing (IX.3) one needs repeated use of integral

$$\int \pi^{-1} \exp(-|\gamma|^2) \begin{pmatrix} 1 \\ |\gamma|^2 \\ \gamma, \overline{\gamma} \end{pmatrix} d^2\gamma = \begin{pmatrix} 1 \\ 1 \\ 0,0 \end{pmatrix} \qquad \text{(IX.4)}$$

and the completeness of the various independent 0,1,2, ...,n fermion states produced through the action of the various combinations of products of creation operators on the vacuum state, as in (IX.2a).

The essentials of functional integration with our states can be shown in their entirety even in the simplest case with a hamiltonian of the form $H = \sum \varepsilon_j C_j^+ C_j$. In fact, from the technical point of view the multi-channel case with interaction adds nothing other than further manipulations of the same nature. However, we prefer to begin the discussion using a hamiltonian involving interaction.

Consider the hamiltonian :

$$H_1 = \sum_{j=1}^{3} \varepsilon_j C_j^+ C_j + \sum_{j>\ell=1}^{3} V_{j\ell} C_j^+ C_\ell^+ C_\ell C_j \qquad (IX.5)$$

We wish to evaluate the propagator :

$$< \gamma_1\gamma_2\gamma_3 | \exp(- i H_1 t) | \gamma_3'\gamma_2'\gamma_1' >$$

$$= \lim_{\Delta t\to 0} \int [\prod_{r=0}^{N-1} <\gamma_1^{(r+1)}\gamma_2^{(r+1)}\gamma_3^{(r+1)} | 1 - i H_1\Delta t | \gamma_3^{(r)}\gamma_2^{(r)}\gamma_1^{(r)}>]$$

$$\times \prod_{r=1}^{N-1} d^2\gamma_1^{(r)} \, d^2\gamma_2^{(r)} \, d^2\gamma_3^{(r)} \qquad (IX.6)$$

with : $\vec{\gamma}^{(o)} = \vec{\gamma}'$, $\vec{\gamma}^{(N)} = \vec{\gamma}$.

In (IX.6) the superscript, r, labels the complex variables associated with the point of partition $t_r = r\,\Delta t$ ($\Delta t = t/N$). Furthermore, for simplicity, we have absorbed in the t Planck's constant $\hbar$, i.e. our t is (time/$\hbar$).

To do the evaluation in (IX.6) we just need the matrix element :

$$< \gamma_1^{(1)} \gamma_2^{(1)} \gamma_3^{(1)} | 1 - i H_1 \Delta t | \gamma_3' \gamma_2' \gamma_1' >$$

$$= \pi^{-3} \exp[- \frac{1}{2} \sum_{j=1}^{3} (|\gamma_j^{(1)}|^2 + |\gamma_j'|^2)][1 + \sum_{j=1}^{3} A_j \bar{\gamma}_j^{(1)} \gamma_j'$$

$$+ \sum_{j>\ell=1}^{3} A_{j\ell} \bar{\gamma}_j^{(1)} \gamma_j' \bar{\gamma}_\ell^{(1)} \gamma_\ell' + A_{123} \bar{\gamma}_1^{(1)} \gamma' \bar{\gamma}_2^{(1)} \gamma_2' \bar{\gamma}_1^{(1)} \gamma_1']$$

(IX.7)

where :

$$A_j = 1 - i\varepsilon_j \Delta t, \quad A_{j\ell} = 1 - i(\varepsilon_j + \varepsilon_\ell + V_{j\ell}) \Delta t$$

and :

$$A_{123} = 1 - i(\varepsilon_1 + \varepsilon_2 + \varepsilon_3 + V_{12} + V_{13} + V_{23}) \Delta t$$

(IV.7a)

It is apparent from (IX.7) that in matrix elements the various creation and annihilation operators involved in the states $|\vec{\gamma}'>$ or $<\vec{\gamma}|$ disapear, and more important the ordering of the complex variables $\vec{\gamma}$ becomes of no consequence.

Next we perform the integral :

$$\int \langle \gamma_1^{(2)}\gamma_2^{(2)}\gamma_3^{(2)}|1 - i\, H_1 \Delta t|\gamma_3^{(1)}\gamma_2^{(1)}\gamma_1^{(1)}\rangle$$

$$\langle \gamma_1^{(1)}\gamma_2^{(1)}\gamma_3^{(1)}|1 - i\, H_1 \Delta t|\gamma_3'\ \gamma_2'\ \gamma_1'\rangle \times \prod_{j=1}^{3} d^2\gamma_j^{(1)}$$

$$= \pi^{-3} \exp\left[-\frac{1}{2}\sum_{j=1}^{3}(|\gamma_j^{(2)}|^2 + |\gamma_j'|^2)\right]$$

$$\times \left(1 + \sum_{j=1}^{3} A_j^2\ \bar{\gamma}_j^{(2)}\ \gamma_j' + \sum_{j>\ell=1}^{3} A_{j\ell}^2\ \bar{\gamma}_j^{(2)}\ \gamma_j'\ \bar{\gamma}_\ell^{(2)}\ \gamma_\ell'\right.$$

$$\left. + A_{123}^2\ \bar{\gamma}_1^{(2)}\ \gamma_1'\ \bar{\gamma}_2^{(2)}\ \gamma_2'\ \bar{\gamma}_3^{(2)}\ \gamma_3'\right) \qquad \text{(IX.8)}$$

The integrations are extremely simple and involve repeated use of formula (IX.4). It is clear from (IX.8) that by folding two consecutive short-time propagators the resulting (approximate) propagator, corresponding to time 2 Δt, has the same form, but with the quantities A_j, $A_{j\ell}$, A_{123} squared. The complete folding of the short-time propagators corresponding to finite time, t, leads to the same form, but with the various A's raised to the power of N.

In the limit as $N \to \infty$ (i.e. $\Delta t \to 0$) these quantities (see equation (IX.7a)) become :

$$A_j^N \to \exp(-i\varepsilon_j t),\quad A_{j\ell}^N \to \exp[-i(\varepsilon_j + \varepsilon_\ell + V_{j\ell})t]$$

$$A_{123}^N \to \exp[-i(\varepsilon_1 + \varepsilon_2 + \varepsilon_3 + V_{12} + V_{13} + V_{23})t]$$

So, the required propagator takes the form :

$$\langle \gamma_1\gamma_2\gamma_3|\exp(-iH_1 t)|\gamma_3'\ \gamma_2'\ \gamma_1'\rangle$$

$$= \pi^{-3} \exp\left[-\frac{1}{2}\sum_{j=1}^{3}(|\gamma_j|^2 + |\gamma_j'|^2)\right]\left\{1 + \sum_{j=1}^{3}\exp(-i\varepsilon_j t)\bar{\gamma}_j\gamma_j'\right.$$

$$+ \sum_{j>\ell=1}^{3} \exp[-i(\varepsilon_j + \varepsilon_\ell + V_{j\ell})t]\ \bar{\gamma}_j\gamma_j'\ \bar{\gamma}_\ell\gamma_\ell'$$

$$\left. + \exp[-i(\varepsilon_1 + \varepsilon_2 + \varepsilon_3 + V_{12} + V_{13} + V_{23})t]\right\} \qquad \text{(IX.9)}$$

The propagator (IX.9) associated with the second quantized hamiltonian (IX.5) is exact, and incorporates the Fermi-Dirac statistics. The ease with which such a propagator has been obtained is due to the highly convenient states employed. The various terms in (IX.9) are (separately) symmetric w.r.t. any exchange of a pair of indices. This enables one to remove restrictions on the indices in the sums and introduce appropriate factorials and δ's of Kronecker, although this might not be helpful in an actual evaluation. Nevertheless, it helps to shorten the analytic expressions for more general cases. As pointed out earlier on, the treatment of more general cases with our states is a matter of routine exercise, and here we shall be content with writing down certain results associated with the hamiltonian.

$$H = \sum_{\vec{k}} \varepsilon_{\vec{k}} C^{+}_{\vec{k}} C_{\vec{k}} + \frac{1}{2} \sum_{\vec{k},\vec{k}'} V_{\vec{k}\vec{k}'} C^{+}_{\vec{k}} C^{+}_{\vec{k}'} C_{\vec{k}'} C_{\vec{k}} \qquad \text{(IX.10)}$$

Inspection of (IX.9) enables one to obtain the matrix elements (in terms of the $\vec{\gamma}$-states) of the statistical operator $\exp[-i(H-\mu N)t]$. Denoting collectively, the state with all the γ_k's, by $|\{\gamma_k\}>$ we have :

$$< \{\gamma_k\}|\exp[-i(H-\mu N)t]|\{\gamma'_k\}> =$$

$$(\prod_{\vec{k}} \pi^{-1}) \exp[-\frac{1}{2}\sum_{\vec{k}} (|\gamma_{\vec{k}}|^2 + |\gamma'_{\vec{k}}|^2)]$$

$$\times \{1 + \sum_{\vec{k}} \exp[-i(\varepsilon_{\vec{k}}-\mu)t]\bar{\gamma}_{\vec{k}}\gamma'_{\vec{k}}$$

$$+ \frac{1}{2!} \sum_{\vec{k}_1,\vec{k}_2} \exp[-i(\sum_{j=1}^{2}(\varepsilon_{\vec{k}_j}-\mu) + \frac{1}{2}\sum_{j,\ell=1}^{2}{}' V_{\vec{k}_j\vec{k}_\ell})t]$$

$$\times (1-\delta_{\vec{k}_1\vec{k}_2})\bar{\gamma}_{\vec{k}_1}\gamma'_{\vec{k}_1}\bar{\gamma}_{\vec{k}_2}\gamma'_{\vec{k}_2} + \ldots$$

$$+ \frac{1}{n!} \sum_{\substack{\{\vec{k}_j\} \\ (i=1,2,\ldots,n}} \exp[-i(\sum_{j=1}^{n}(\varepsilon_{\vec{k}_j}-\mu) + \frac{1}{2}\sum_{j=\ell=1}^{n}{}' V_{\vec{k}_j\vec{k}_\ell})t]$$

$$\times [(\prod_{j>\ell=1}^{n}{}' (1-\delta_{\vec{k}_j\vec{k}_\ell})] \prod_{j=1}^{n} \bar{\gamma}_{\vec{k}_j}\gamma'_{\vec{k}_j} + \ldots \} \qquad \text{(IX.11)}$$

Σ' has the usual meaning of excluding in the sum the terms with $j = \ell$. Π' has the corresponding meaning in the product. The indicators $(1 - \delta_{\vec{k}\vec{k}'})$ do not allow counting a state more than once in each sum.

Finally we shall obtain from (IX.11) the thermodynamic potential $\Omega = - \kappa T \ln Z$ associated with the hamiltonian (IX.10). At this stage we switch from time to inverse temperature, via the substitution : $t \to i\beta$. With this replacement the partition function is obtained as the trace of our propagator (IX.11). Thanks to our systems of states, for the trace operation, we only have to use repeatedly (IX.4). Furthermore, we shall use as our channels $\vec{k}$ the free particle states with $\varepsilon_{\vec{k}} = \hbar^2 k^2/2m$ and this will allow convert the sums over the various $\vec{k}_j$'s into integrals. In this case the Kronecker δ's will become δ-functions times υ^{-2}, where $\upsilon = (2\pi)^{-3} \times$ (volume of the system).

We write down the final result :

$$\Omega = - \kappa T \ln \{1 + \frac{\upsilon}{1!} \int \exp [- \beta(\varepsilon_k - \mu)] d\vec{k}$$

$$+ \sum_{n=2}^{\infty} \frac{\upsilon^n}{n!} \int \exp [- \beta \sum_{j=1}^{n} (\varepsilon_{\vec{k}_j} - \mu) - \frac{\beta}{2} \sum_{j,\ell=1}^{n}{}' V_{\vec{k}_j \vec{k}_\ell}]$$

$$\times [\prod_{i>\ell=1}^{n}{}' (1 - \upsilon^{-1} \delta(\vec{k}_j - \vec{k}_\ell))] \prod_{j=1}^{n} d\vec{k}_j \qquad \text{(IX.12)}$$

The case of independent particles leads to the well known thermodynamic potential :

$$\Omega_o = - \kappa T \sum_{\vec{k}} \ln \{1 + \exp [- \beta(\varepsilon_{\vec{k}} - \mu)]\}$$

This can be deduced from the step prior to (IX.11) (as well as from (XI.12)) and it is quite simple to see that the trace of (IX.9) is just a product of

$$[1 + \exp (- \beta\varepsilon_j)] \qquad (j = 1,2,3)$$

In conclusion the present system of states seems promising and may be capable of reaching results not so accessible to other methods.

X. FUNCTIONAL INTEGRALS FOR SPIN SYSTEMS

Until now we have dealt with functional integrals for systems described by first and second quantized hamiltonians. One thing we have not yet done is some functional integration for spin hamiltonians. This can also be done provided we have at hand an appropriate and convenient (complete) set of states. As a first step in this direction it would be that of constructing a suitable functional integral giving the propagator associated with a spin (1/2) hamiltonian. Apparently the states of the previous section, with appropriate modification, can be used for the construction of functional integrals for spin (1/2) systems. This will be the object of the present section.

For the purpose of functional integration one needs a complete set of states labelled continuously. The state for an n-site system will be given by :

$$|\sigma_n \dots \sigma_2 \sigma_1 > = \pi^{-n/2} \exp(-\frac{1}{2}\sum_{j=1}^{n} |\sigma_j|^2) \prod_{j=1}^{n} (1+\sigma_j S_j^+)|0> \quad \text{(X.1a)}$$

where the σ_j's are complex numbers (labelling the state), $|0>$ is the spin vacuum state (all spins down), and S_j^+ is the raising operator for the spin on the jth site. The lowering operator will be simply denoted by S. The bra form of (1a) is :

$$< \sigma_1 \sigma_2 \dots \sigma_n | = \pi^{-n/2} \exp(-\frac{1}{2}\sum_{j=1}^{n} |\sigma_j|^2) < 0| \prod_{j=1}^{n} (1+\bar{\sigma}_j S_j) \quad \text{(X.1b)}$$

$\bar{\sigma}$ is the complex conjugate of σ.
The operators S_j^+, S_j obey for the same site fermion properties :

$$S_j^+ S_j + S_j S_j^+ = 1 , \qquad S_j^{+2} = S_j^2 = 0 \quad \text{(X.2a)}$$

while for different sites $(j \neq \ell)$ boson properties :

$$S_j S_\ell^+ - S_\ell^+ S_j = S_j^+ S_\ell^+ - S_\ell^+ S_j^+ = S_j S_\ell - S_\ell S_j = 0 \quad \text{(X.2b)}$$

The state (X.1a) can be put in the same form as the n-fermion state of section (IX) but here on account of properties (X.2b) of the spin operators we are free to commute the factors $(1 + \sigma_j S_j^+)$ without inducing a change in the state; this facility is lacking in the fermion case.

The totality of the n-site states forms a complete set in the sense :

$$\int |\sigma_n \ldots \sigma_2 \sigma_1 > < \sigma_1 \sigma_2 \ldots \sigma_n| \prod_{j=1}^{n} d^2\sigma_j = 1 \qquad (X.3)$$

where $d^2\sigma = d(\mathrm{Re}\,\sigma)\, d(\mathrm{Im}\,\sigma)$.

The derivation of (X.3) proceeds along the lines of (IX). Two different states are not orthogonal and they are normalized to 2^n which is the number of all possible mixtures of up and down spins.

The functional integral giving the propagator associated with an n-site spin hamiltonian in terms of the states $|\vec{\sigma}>$(= $|\sigma_n \ldots \sigma_2 \sigma_1>$) is given by :

$$< \vec{\sigma}|T \exp [- \frac{i}{\hbar} \int_o^t H(\tau)d\tau] |\vec{\sigma}' >$$

$$= \lim_{\max\Delta\tau_r \to o} \int [\prod_{r=1}^{N-1} < \vec{\sigma}^{(r+1)}|1 - \frac{i}{\hbar} H(\tau_r)\Delta\tau_r|\sigma^{(r)}>] \prod_{r=1}^{N-1} d^2\vec{\sigma}^{(r)}$$

$$\text{with : } \vec{\sigma}^{(o)} = \vec{\sigma}', \quad \vec{\sigma}^{(N)} = \vec{\sigma} . \qquad (X.4)$$

Further details have been given in (IX).

What one really needs for the construction of the functional integral (X.4) is just matrix elements of the operator $(1 - iH(\tau_r) \quad \Delta\tau_r/\hbar)$ in terms of the states $|\vec{\sigma}>$. These states prove to be very convenient for the purpose.

Before we go any further we recall that any spin hamiltonian is a function of the operators σ_j^x, σ_j^y, σ_j^z and these are expressed in terms of the operators S_j^+, S_j by :

$$\sigma_j^x = S_j + S_j^+ , \quad \sigma_j^y = i(S_j - S_j^+), \quad \sigma_j^z = 2\, S_j^+ S_j - 1 \qquad (X.5)$$

Through the relations (X.5) one expresses a spin hamiltonian in terms of the operators S_j^+, S_j alone.

Now, by way of illustration, let us deal with an Ising model involving a magnetic field B in the z-direction. The hamiltonian is :

$$H = -\mu B \sum_{j=1}^{n} \sigma_j^z - \sum_{j>\ell=1}^{n} J_{j\ell}\, \sigma_j^z\, \sigma_\ell^z \qquad (X.6)$$

which in terms of the operators S_j, S_j^+ takes the form :

$$H = \varepsilon_o + \sum_{j=1}^{n} \varepsilon_j S_j^+ S_j - 4 \sum_{j>\ell=1}^{n} J_{j\ell} S_j^+ S_j S_\ell^+ S_\ell \qquad (X.6a)$$

where :

$$\varepsilon_o = n\mu B - \sum_{j>\ell=1}^{n} J_{j\ell}, \quad \varepsilon_j = 2\,(\sum_{\ell\neq j=1}^{n} J_{\ell j} - \mu B) \qquad (X.6b)$$

Proceeding as in section (IX) one can now obtain the partition function associated with the hamiltonian (X.6) but this time when evaluating matrix elements one has to make use of the spin operator relations (X.2a,b) instead of the corresponding fermion relations. We quote the results :

$$Z = \int \langle \vec{\sigma} | \exp\,(-\beta H) | \vec{\sigma} \rangle\, d^2\vec{\sigma}$$

$$= \exp\,(-\beta\varepsilon_o) \Big\{ 1 + \sum_{j=1}^{n} \exp\,(-\beta\varepsilon_j) + \sum_{j>\ell=1}^{n} \exp\,[-\beta(\varepsilon_j + \varepsilon_\ell - 4J_{j\ell})]$$

$$+ \sum_{j>\ell>k=1}^{n} \exp\,\{-\beta[\varepsilon_j + \varepsilon_\ell + \varepsilon_k - 4\,(J_{i\ell} + J_{\ell k} + J_{jk})]\}$$

$$+ \ldots + \exp\,[-\beta(\sum_{j=1}^{n} \varepsilon_j - 4 \sum_{j>\ell=1}^{n} J_{j\ell})] \Big) \qquad (X.7)$$

known from the lattice gas theory. In (X.7) use has been made of the symmetry of the exchange energies $J_{j\ell}$.

We wish now to make some further use of the method (though without conclusion) to a mathematically less tractable problem.

We consider the Heisenberg hamiltonian :

$$H = -\mu B \sum_{j=1}^{n} \sigma_j^z - J \sum_{j=1}^{n-1} (\sigma_j^x\, \sigma_{j+1}^x + \sigma_j^y\, \sigma_{j+1}^y + \sigma_j^z\, \sigma_{j+1}^z) \qquad (X.8a)$$

and see how far our method can get us. In terms of the operators S^+, S our hamiltonian (X.8a) takes the form :

$$H = n\mu B + (1-n)J + 2\,(J-\mu B)(S_1^+ S_1 + S_n^+ S_n) + 2\,(2J-\mu B) \sum_{j=2}^{n-1} S_j^+ S_j - 2J \sum_{j=1}^{n-1} (S_j^+ S_{j+1} + S_{j+1}^+ S_j) - 4J \sum_{j=1}^{n-1} S_j^+ S_j S_{j+1}^+ S_{j+1} \qquad \text{(X.8b)}$$

One can clearly see the apparent similarity of the hamiltonian (X.8b) with the Hubbard hamiltonian and therefore, with minor modification, the possibility of the present method to deal with this problem to some extent.

The main facility provided through employing our states is in that the propagator is obtained in its irreducible form. The form can be found by a single folding of two short time propagators. Here the functional integration saves you from guessing. For the one dimensional Heisenberg model (X.8) the form of the propagator is :

$$< \vec{\sigma} | \exp\,(-\tfrac{i}{\hbar} Ht) | \vec{\sigma}\,'> = A(t) + \sum_{r=1}^{n} \sum_{j,k=1}^{\binom{n}{r}} A_{jk}^{(r)}(t)\, \bar{\eta}_j^{(r)}\, \eta_k'^{(r)} \qquad \text{(X.9)}$$

where the $\eta_j^{(r)}$ are given by :

$$\eta_j^{(1)} = \sigma_j ; \quad \eta_1^{(2)} = \sigma_1 \sigma_2$$

$$\eta_2^{(2)} = \sigma_1 \sigma_3 \;\ldots\; \eta_{\binom{n}{2}}^{(2)} = \sigma_{n-1}\,\sigma_n ; \quad \eta_1^{(3)} = \sigma_1 \sigma_2 \sigma_3 ,$$

$$\ldots\; \eta_1^{(n)} = \sigma_1 \sigma_2 \ldots \sigma_n$$

The integral of $\eta_j^{(r)}\, \bar{\eta}_k^{(q)}$ against $\pi^{-n} \exp\,(-|\vec{\sigma}|^2)$ is $\delta_{rq}\, \delta_{jk}$.
The partition function is obtained from the diagonal elements and is given by :

$$Z(\beta) = A(-i\hbar\beta) + \sum_{r=1}^{n-1} \sum_{j=1}^{\binom{n}{r}} A_{jj}^{(r)} (-i\hbar\beta) + A_{11}^{(n)} (-i\hbar\beta) \tag{X.10}$$

For obtaining the various matrix elements $A_{jk}^{(r)}$ (and in particular the diagonal elements) by means of functional integration we proceed as follows. We just relate the values of the various A's at time t + Δt to those at time t. In particular we exploit the relation :

$$< \vec{\sigma}|(1 - \frac{i}{\hbar} H\Delta t)^{N+1}|\vec{\sigma}' >$$

$$= \int < \vec{\sigma}|1 - \frac{i}{\hbar} H\Delta t|\vec{\sigma}" > < \vec{\sigma}"|(1 - \frac{i}{\hbar} H\Delta t)^{N}|\vec{\sigma}' > d^2\vec{\sigma}" \tag{X.11}$$

In the limit of $N \to \infty$ with $\Delta t = t/N$ we obtain the following linear differential equations :

$$i\hbar\dot{A} = H_o A , \quad i\hbar\dot{A}_{11}^{(n)} = H_{11}^{(n)} A_{11}^{(n)}$$

$$i\hbar\dot{A}_{jr}^{(r)} = \sum_{\ell=1}^{\binom{n}{r}} H_{j\ell}^{(r)} A_{\ell k}^{(r)} \quad (r = 1,2,\ldots,n-1) \tag{X.12}$$

where $H_{jk}^{(r)}$ is the coefficient of

$\pi^{-n} \exp(-\frac{1}{2}|\vec{\sigma}|^2 - \frac{1}{2}|\vec{\sigma}'|^2)\ \bar{\eta}_i^{(r)}\ \eta_k'^{(r)}$ of the matrix

element $< \vec{\sigma}|H|\vec{\sigma}' >$ and H_o is the non-operator constant

part of the hamiltonian [$H_o = n\mu B + (1-n)J$, $H_{11}^{(n)} =$

$- n\mu B + (1-n)$] . The rest of the coefficients $H_{jk}^{(r)}$ are made available from a first order calculation, always feasible.

The boundary conditions for the required coefficients $A_{jk}^{(r)}$ are obtained from the expression for the short time propagator between 0 and Δt and they are :

$$A(0) = 1 , \qquad A_{jk}^{(r)}(0) = \delta_{jk} \tag{X.12a}$$

The first two equations of (X.12) give :

$$A(t) = \exp\left[-\frac{i}{\hbar}(n B - nJ + J)t\right] ,$$

$$A_{11}^{(n)} = \exp\left[\frac{i}{\hbar}(n B - nJ - J)\, t\right] \tag{X.13}$$

The matrix element of the rth submatrix can be obtained via the technique used in normal mode analysis, without having to recourse to explicit diagonalization. We just compute the roots of the secular equation

$$\det\,(H_{jk}^{(r)} - \lambda\delta_{jk}) = 0 \tag{X.14}$$

If the roots are all different from each other the solution is :

$$A_{jk}^{(r)}(t) = \sum_{\alpha=1}^{\binom{n}{r}} C_{jk}^{(r)\alpha} \exp\left(-\frac{i}{\hbar}\lambda_{\alpha}^{(r)}\, t\right) \tag{X.15}$$

If there is multiplicity we have to replace the constant coefficient C (wherever the multiplicity occurs) by an appropriate polynomial of t. Since the procedure for obtaining the coefficients C is the same for both cases we treat the case without multiplicity. In fact we can pick up any of the matrix elements $A_{jk}^{(r)}$ and find its coefficients $C_{jk}^{(r)}$. For this reason we shall drop in our equations the indices j, k and r. We have :

$$\sum_{\alpha} C^{\alpha} = A_{jo}(o) = \delta_{jk}$$

$$\sum_{\alpha} \lambda\alpha\, C^{\alpha} = i\hbar \dot{A}_{jk}(o)$$

$$\sum_{\alpha} \lambda_{\alpha}^{n-1}\, C^{\alpha} = (i\hbar)^{n-1}\, A_{jk}^{(n-1)\cdot}(o) \tag{X.16}$$

The various derivatives at t = 0 are obtained from the equations of motion (X.12). The procedure is particularly helpful when we seek limited information from our propagator, e.g. when we require just the diagonal elements.

In general the form of the propagator is :

$$< \vec{\sigma} | \exp \left(- \frac{i}{\hbar} Ht\right) | \vec{\sigma}' > = A(t)$$

$$+ \sum_{r=1}^{n} \sum_{j,k,\alpha}^{\binom{n}{r}} C_{jk}^{(r)\alpha} \exp \left(- \frac{i}{\hbar} \rho_{\alpha}^{(r)} t\right) \bar{\eta}_j^{(r)} \eta_k'^{(r)} \qquad (X.17)$$

The solution of the problem has been reduced to that of solving the secular equation (X.14) and this requires computing in general.

The case for 3 sites can be treated analytically, and the partition function for such a model is :

$$Z_3(\beta) = 2 \cosh (3 \beta\mu B) \exp (2 \beta J)$$

$$+ 2 \cosh (\beta\mu B) [1 + \exp (2 \beta J) + \exp (- 4 \beta J)] \qquad (X.18)$$

In the present work emphasis has been focused on the states employed rather than on any particular applications. Applications have been used as a vehicle for conveying the method. Really, we cannot claim to have made any progress in bringing a particular problem to solution, but nevertheless we believe that the above states may be of some use for spin problems.
In case a false impression has been created the normal mode analysis is not functional integration. We just have been carried away.

REFERENCES

1. Volterra V., Theory of Functionals and of Integral and Integrodifferential Equations (Blackie and Son, London and Glasgow, 1930).

2. Daniell P.J., Ann. Math. 20 1 (1918).

3. Wiener N., J. Math. Phys. Mass. Inst. Techn. 2 131 (1923);
Wiener N., Proc. Lond. Math. Soc. 22 454 (1924).

4. Chandrasekhar S., Rev. Mod. Phys. 15 1 (1943)
Along similar lines :
Papadopoulos G.J., Lectures in Theoretical Physics X A 269 (Eds.: Barut A.O. and Brittin W.E., Gordon and Breach, N.Y., 1968).
Papadopoulos G.J., J. Phys. A 1 431 (1968).
Some things, but transcribed to the quantum regime :
Papadopoulos G.J., J. Phys. A 6 1479 (1973).

5. Dirac P.A.M., Selected Paper on Quantum Electrodynamics, p.312 (Ed.: Schwinger J., Dover Publi. Inc. N.Y., 1958).

6. Feynman R.P., Rev. Mod. Phys. 20 387 (1948).
Feynman R.P., Phys. Rev. 76 769 (1949).
Feynman R.P., Phys. Rev. 80 440 (1950).
Feynman R.P., Phys. Rev. 84 108 (1951).
Feynman R.P., Phys. Rev. 91 1301 (1953).
A collection of Feynman's works can be found in :
Feynman R.P. and Hibbs R.A., Quantum Mechanics and Path Integrals (Mc Graw-Hill, N.Y., 1965).

Feynman R.P., Statistical Mechanics (Addison-Wesley and Benjamin Inc., London, 1973).
Gelfand I.M. and Yaglom A.M., J. Math. Phys. 1 48 (1960).
Brush S.G., Rev. Mod. Phys. 33 79 (1961).

7. Kirkwood J., Phys. Rev. 44 31 (1933).

8. Schwinger J., Phys. Rev. 82 914 (1951).
Schwinger J., Phys. Rev. 91 713 (1953).
Schwinger J., Proc. Acad. Sci. (U.S.) 44 956 (1958).
Schwinger J., Phys. Rev. 115 721 (1959).
Bogoliubov N.N., Dok. 99, 225 (1954).
Edwards S.F., Proc. Roy. Soc. A 22424 (1954).
Edwards S.F., Analysis in Function Space, p.31 and p.167 (Eds.: Martin W.T. and Segal I., MIT Press Cambr. Mass. 1964).
Matthews P.T. and Salam A., Nuc. Cim. 2 120 (1955).
Zymanzik K., Z. Naturforsch 9 A 409 (1954).
Friedrichs K.O. and Shapiro H.N., Proc. Nat. Acad. Sci. (U.S.) 43 336 (1957).
Segal I.E., J. Math. Phys. 1 468 (1960).
For Functional integrals in quantum field theory with a variety of techniques see :
Tarski J., Lectures in Theoretical Physics X-A p.443 (Eds.: Barut A.O. and Brittin W.E., Gordon and Breach, N.Y., 1968).

9. Edwards S.F. and Peierls R.E., Proc. Roy. Soc. A 22424 (1954).

10. Hopf E., J. Mech. and Rat. Anal. 1 87 (1952).

11. Edwards S.F., J. Fluid. Mech. 18 239 (1964).
Rosen G., Phys. Fluids 10 2614 (1967).

12. Brittin W.E., Phys. Rev. 106 843 (1957).
Brittin W.E. and Chappell W.R., Lectures in Theoretical Physics VIII-A 101 (Eds.: Brittin, Univ. Cole Press, Boulder, 1965).

13. Bogoliubov N.N., Studies in Statistical Mechanics 1 1 (Eds.: de Boer J. and Uhlenbeck G.E., North-Holland Publ. Co., Amsterdam, 1962).

14. Hosokawa I., J. Math. Phys. 8 221 (1967).

15. Edwards S.F., Proc. Phys. Soc. 85 613 (1965).
Edwards S.F., Proc. Phys. Soc. 91 513 (1967).
Edwards S.F. and Freed K.F., J. Phys. C 3 739 (1970).
Edwards S.F. and Grant J.W., H. Phys. A 6 1169 (1973).
We mention here further works along Edwards' line:
Freed K.F.,J. Chem. Phys. 54 1453 (1971).
Papadopoulos G.J. and Thomchick J., J. Phys. A 10 No. 7 (1977).

16. Thornber K.K. and Feynman R.P., Phys. Rev. B 1 4099 (1970).
Thornber K.K., Phys. Rev. B3 1929 (1971).

Earlier related works :
Feynman R.P., Helwarth R.W., Iddings C.K., and Platzmann P.M., Phys. Rev. 127 1004 (1962).
Feynman R.P. and Vernon F.L., Ann. Phys. 24 118 (1963).

17. Feynman R.P., Phys. Rev. 97 660 (1955).

18. Early work with operational capability :
Edwards S.F. and Gulyaev Y.V., Proc. Roy. Soc. A 279 229 (1964).
There follow :
Peak D. and Inomata A., J. Math. Phys. 10 1042 (1969).
Khandekar D.C. and Lawande S.V., J. Phys. A 5 812 (1972).
Khandekar D.C., J. Phys. A 5 L 57 (1972).
For the hydrogen atom the treatment which exist proceeds via rectilinear c-ordinates :
Govaerts M.J. and Devreese J.T., J. Math. Phys. 13 1070 (1972) .

19. De Witt B.S., Rev. Mod. Phys. 29 377 (1975).
Arthurs A.M., Proc. Roy. Soc. A 313 445 (1969).
Arthurs A.M., Proc. Roy. Soc. A 318 523 (1970).
Dowker J.S. and Mayes I.W., Nucl. Phys. B29 259 (1971).

20. See Feynman (1950, 1951) in refs. (6)
Morette G., Phys. Rev. 81 448 (1951).
Schulman L.S., Phys. Rev. 176 1558 (1969).
Hamilton J.F. Jr. and Schulman L.S., J. Math. Phys. 12 160 (1971).
Papadopoulos G.J. and Devreese J.T., Phys. Rev. D13 2227 (1976).

21. Monttroll E.W. and Ward J.C., Phys. Fluids 1 55 (1958).
Ter Haar D., Lectures in Theoretical Physics VIII-

A 747 (Eds.: Brittin W.E., Univ. Cole Press, Boulder 1966).
See also books by Feynman and Hibbs and by Feynman in refs. (6).

22. See Matthews Salam (1955) in refs. (8).
Klauder J.R., Ann. Phys., N.Y. 11 123 (1960).
Schwaber S.S., J. Math. Phys. 3 831 (1962).
Revzen M., Phys. Rev. 185337 (1969).
Casher A., Lurié P., and Revzen M., J. Math. Phys. 9 1312 (1969).
Papadopoulos G.J., Proc. Antwerp Summer School, p.469 (Eds.: Devreese J.T. and Van Doren V., Plenum Press, London, 1976).

23. Papadopoulos G.J., Proc. Roy. Soc. A 350 547 (1976).

24. Einstein A., Investigations in the Theory of Brownian Movement p.1 (Dover, N.Y., 1956).

25. Langevin P., C.R. Acad. Sci. Paris 146 530 (1908).

26. Kubo R., J. Phys. Soc. Japan 26 Suppl.1 (1969).

27. Kamper R.A., Symposium on the Physics of Superconducting Devices, p.1 (ONR Report, Charlottesville, Virginia University, 1967).
Kamper R.A. and Zimmerman J.E., J. Appl. Phys. 42 132 (1971).
Giffard R.P., Webb R.A. and Wheatley J.G., J. Low Temp. Phys. 6 533 (1972).

28. Abé R., Busseiron Kenk-yu 79 101 (1954).

For evaluations and methods, see also earlier work of Montroll (1952) in refs. (6).

29. Van Vleck J.H., Proc. Natn. Acad. Sci. 14 178 (1928).

30. Papadopoulos G.J., J. Phys. A 7 183 (1974).
Papadopoulos G.J., J. Phys. A 4 773 (1971).
For further evaluations :
Papadopoulos G.J., Phys. Rev. D 11 2870 (1975).
Papadopoulos G.J. and Jones A.V., J. Phys. F 1 593 (1971).

31. Pauli W., Ausgewälte Kapitel der Feldquantisierung p.139 (ETH, Zürich, 1952).

32. Edwards S.F. and Sherrington D., Proc. Phys. Soc. 90 3 (1967).
Hubbard J., Phys. Rev. L 3 77 (1959).
Sherrington D., J. Phys. C 4 402 (1971).
Mühlschlegel B., Functional Integration and its Application, p.124 (Eds.: Arthurs A.M., Clarendon Press, Oxford, 1975).

33. Glauber R.J., Phys. Rev. 136 2766 (1963).
Louissell W.H., Radiation and Noise in Quantum Electronics (McGraw-Hill, N.Y., 1964).
Wiegel F.W., Phys. Letts 16 C No.2 (1975).

PATH INTEGRALS AND THE RELATION BETWEEN CLASSICAL AND QUANTUM MECHANICS

Martin C. Gutzwiller

IBM Thomas J. Watson Research Center

Yorktown Heights, New York 10598

0. INTRODUCTION

1. Purpose

Our perception of the world outside ourselves can best be described in the terms of classical physics. The phenomena on the atomic scale require the ideas of quantum mechanics for their understanding which remain rather formal and abstract in our mind. Hence the need to tie the two realms together. Many problems can be treated successfully, although only approximately, with the help of classical mechanics, e.g. transition rates in chemistry and self-bound states in quantum field theory (solitons and instantons).

The most common approach sofar used the wave function as the primary object to be determined. Green's function, however, seems more amenable to a general treatment which is independent of special assumptions about the mechanical system. The connection with the path integral formulation of quantum mechanics is straightforward. We are led to inquire about the nature of all classical trajectories between any two given endpoints, as a necessary prerequisite to the investigation of all possible paths between two given endpoints.

Such inquiries have been sorely neglected in the past, and physicists don't realize the complexity of classical behavior. Their intuition has been corrupted by the exclusive discussion of systems with one degree of freedom and of the Kepler problem. They have never seen an ergodic system displayed before them, and they don't realize the impossibility of constructing a quasi-classical wave function in that case.

Einstein[1] was the first to call attention to this difficulty, at a time when the idea of a wave function had not arisen as yet. His arguments form a particularly enlightening introduction to the problem ahead. Our main purpose is to deal with ergodic systems on the basis of path integrals in order to find an approximate energy spectrum for bound states. Explicit examples are emphasized because they are so difficult to grasp, even in the few situations where a complete analytical (rather than numerical) treatment can be carried out.

2. Einstein's Challenge

The quantization conditions of Bohr and Sommerfeld require that a canonical transformation be found into a set of mutually independent pairs of action and angle variables. Einstein asked in 1917 whether it is possible to formulate the quantization conditions without assuming such a canonical transformation. The background of this problem and Einstein's answer will be presented in this section.

A classical mechanical system has position coordinates q_i, momentum coordinates p_i $(i=1,\ldots,n)$, and a Hamiltonian $H(p,q)$. The letters p or q without indices refer to the collections $(p_i,\ldots,p_n)$ or $(q_1,\ldots,q_n)$. The equations of motion are

$$\frac{dp_i}{dt} = -\frac{\partial H}{\partial q_i}, \quad \frac{dq_i}{dt} = \frac{\partial H}{\partial p_i} \tag{1}$$

A complete integral J is a function of q and of a set α of n-1 parameters $\alpha_i,\ldots,\alpha_{n-1}$. The trajectories of (1) are given as the intersection of the manifolds

$$\frac{\partial J}{\partial \alpha_i} = \beta_i , \tag{2}$$

where $\beta = (\beta_1,\ldots,\beta_{n-1})$ is another set of parameters. The momentum p along the trajectory is given by

$$\frac{\partial J}{\partial q_i} = p_i \tag{3}$$

The function J satisfies Jacobi's partial differential equation

$$H\left(\frac{\partial J}{\partial q}, q\right) = \alpha_n , \tag{4}$$

where α_n is another fixed parameter, the constant value of the energy.

It is now necessary, according to Bohr and Sommerfeld, to find a canonical transformation such that J has the separated form

$$J = \sum_{i=1}^{n} J_i(q_i,\alpha), \tag{5}$$

where J_i depends only on position coordinate q_i and on the set of parameters α. If we have succeeded in finding (5), then the quantization conditions are simply

$$\oint p_i dq_i = 2\pi n_i h. \tag{6}$$

The integral is extended over one period, and h is Planck's quantum divided by 2π. Further detaiis can be found in the book by Max Born.[2]

Einstein notes first that the "action integrals" (6) are independent of the particular canonical transformation because the line integral

$$\oint \sum_1^n p_i dq_i \tag{7}$$

over some arbitrary closed loop in (p,q) space remains constant as the momenta p and the positions q develop according as the equation of motion (1). Second, the loop can be deformed almost at will without changing the value of the integral (7) as long as p and q are connected by the relation (3). In particular, the value of (7) vanishes if the loop can be shrunk to a point while maintaining the relations (3) with fixed α.

Einstein considers the necessary conditions for the existence of a relation (3) between the momenta and the positions along a particular trajectory. Two possibilities arise as the trajectory crosses a particular neighborhood of position space over and over again. Either the values of p repeat themselves, say after m traversals, or infinitely many different values of p occur. In the first case, J can be constructed, although the function will have m different branches. In the second case, there is no way of constructing J even with many branches, and the conditions (6) cannot be written down.

The movement of particle in a two-dimensional circularly symmetric potential provides a simple example where m=2. Since the angular momentum is a constant of motion, the azimuthal component of the momentum is fixed in each location, while the radial component can point either inward or outward. In general, the accessible region in position space is an annulus which depends on the values of the energy and the angular momentum. A two-valued function over an annulus can be considered as a single-valued function on a torus.

If a typical trajectory is followed along through phase space its points will fill a two-dimensional manifold in the shape of a torus. There are two types of closed paths which cannot be contracted into a point while remaining on the torus. As Einstein points out, the quantization conditions (6) are equivalent to setting the action integral (7) equal to $2\pi n_i h$ for one of these two types of closed paths. This formulation is obviously much more satisfactory and general. On the other hand, Einstein proposes no scheme in the case where the trajectory has infinitely many values for the momentum in the same neighborhood of position space. That is exactly the ergodic movement we are interested in.

3. Method

Ergodic behavior in a conservative system is characterized in phase space by the absence of invariant manifolds with dimension equal to the number of degrees of freedom, or lower. Einstein already hinted at the possibility that some parts of phase space may be lacking such invariant manifolds while other parts with the same energy might be covered by them. The quantization procedure of Bohr-Sommerfeld-Einstein can be applied to the latter, but not to the former. It is very unsatisfactory to let things remain in this incomplete state.

Although our attention will be focused on ergodic systems two examples will be discussed, one ergodic and the other not. These examples are non-trivial in that they are representative of their respective classes, but go beyond the ubiquitous harmonic oscillator and its modifications. Both arise as generalizations of Kepler's problem, and give us an intuitive feeling for the variety of classical paths.

Lunar theory provides an example with which the author is familiar. It is the area in physics where perturbation theory has been used most extensively, and it gives a neat demonstration of adiabatic perturbation. Its equations of motion are identical with those of a particle in a combined magnetic and electric field.

The anisotropic Kepler problem arises in the theory of donor impurities in semiconductors. It presents the best physical example of purely ergodic behavior which can be completely understood. The only other example which could conceivably be discussed in our context is the geodesic flow on a surface of constant negative curvature,[3] but it lacks the intuitive appeal a physicist needs.

The path integral formulation of quantum mechanics provides us with the tool which Einstein did not have in 1917. All we need are the classical trajectories and the first-order fluctuations around them. No effort is made to estimate the importance of the larger fluctuations, because their effect vanishes in the limit of small Planck's quantum.

The quantity of prime interest is the probability amplitude for a particle to end up with a given momentum p'' if it started out with the momentum p' and was moving with a total energy E. This so-called Green's function $F(p'p''E)$ is best suited to the investigation of bound states with a Coulomb like singularity in the potential. The classical approximation $\tilde{F}(p'p''E)$ to the quantum mechanical $F(p'p''E)$ gives both the correct energy eigenvalues and the correct wave function for the bound states of the ordinary Kepler problem. The more general case of a spherically symmetric potential shows very clearly how the constructive interference between classical trajectories leads to the singularities in $\tilde{F}$.

The last chapter is specifically devoted to Einsteins's challenge. The approximate spectrum is obtained directly from the approximate response function $\tilde{g}(E)$ where the momentum (or position) dependence of $\tilde{F}$ has been integrated out. In a crucial argument, only the (classical) periodic orbits will be shown to contribute to $\tilde{g}$ while the fluctuations around them appear only in the form of the stability exponents.

Ergodic systems are quantized by finding the singularities of $\tilde{g}(E)$, exactly as systems with invariant tori in phase space. Because of the instability of ergodic trajectories, however, the singularities in the density of states will appear as broadened peaks rather than δ-functions. That seems to be the prize to be paid for quantizing an ergodic system on the basis of its classical behavior.

I. TWO NON-TRIVIAL EXAMPLES FROM CLASSICAL MECHANICS

A system with n degrees of freedom has a phase space of 2n dimensions. If its Hamiltonian does not depend explicitly on time, every trajectory is restricted to the (2n-1) dimensional manifold $H(p,q) = E$ (the constant value the energy). The quantization of the system can proceed along the rules of Bohr-Sommerfeld-Einstein, provided the trajectories lie on tori of n dimensions. Einstein's challenge becomes effective when $n \geq 2$. It can be understood only in examples where $n \geq 2$. We shall consider two examples with $n=2$ which can easily be generalized to $n=3$, their natural setting.

The main problem of lunar theory in our version is a special case of the "restricted three-body problem".[4] Two heavy bodies (point masses) move around each other in a Kepler ellipse with fixed parameters (semi-major axis and excentricity) while the third, light one moves in the same plane and is attracted to both of them with the inverse square of the distance. If the Kepler ellipse degenerates to a circle, the Hamiltonian in the corotating coordinate system does not depend on the time and the movement of the third body conserves the energy in the rotating frame. There are various regions in phase space which are covered with tori of 2 dimensions.

The anisotropic Kepler problem originates from the motion of an electron in a semiconductor.[5] The kinetic energy is a quadratic form in the momenta, but the inverse mass tensor is anisotropic because the minimum of the conduction band is not at the center of the reciprocal space. A donor impurity generates an isotropic Coulomb field which is weakened by the large dielectric constant. In two dimensions the bound state trajectories are completely ergodic.

4. The Main Problem of Lunar Theory

The following notations will be used: The masses of the moon, earth, and sun will be called M, m, and μ. The center of mass of the earth and moon is located at the point Γ. Its point of closest approach to the sun is called Π. The point Γ moves around the sun in a Kepler ellipse of semimajor axis α and excentricity ε. The mean motion of Γ around the sun is given by the angle λ which is a linear function of time, $\lambda=\nu(t-t_0)$ and vanishes as Γ goes through Π. The true angle from Π to Γ as seen from the sun is given by $\lambda+\kappa$, and ρ is the distance from the sun to Γ. Both κ and ρ

are well known functions of λ with the period 2π. Kepler's third law says that $\nu^2\alpha^3=k^2(\mu+m+M)$ where k^2 is the gravitational constant.[6]

The moon's position relative to the earth is referred to a coordinate system (x,y) whose origin is located at the earth. Its xy-plane coincides with the plane of the Kepler orbit of Γ around the sun, and its negative x-axis goes through the mean position of the sun. The distance from the earth to the moon is called $r= (x^2+y^2)^{1/2}$, and s is the name for the cosine of the angle from the sun to the moon as seen from the earth, $rs = - x \cos \kappa - y \sin \kappa$. The equations of motion are

$$\begin{aligned} \ddot{x} - 2\nu\dot{y} &= \partial\Omega/\partial x, \\ \ddot{y} + 2\nu\dot{x} &= \partial\Omega/\partial y, \end{aligned} \tag{8}$$

with the potential function

$$\Omega = \frac{k^2(m+M)}{r} + \frac{\nu^2}{2}(x^2+y^2) + \frac{k^2\mu}{\rho}\sum_{n=2}^{\infty}\frac{m^{n-1}-(-M)^{n-1}}{(m+M)^{n-1}} \qquad (\frac{r}{\rho})^n P_n(s). \tag{9}$$

The expansion in Legendre polynomials starts with n=2, because the terms n=0 and n=1 are canceling out in this coordinate system.

The equations of motion (8) can be written in Hamiltonian form, if we introduce the momenta $u=\dot{x}-\nu y$ and $v= \dot{y}+\nu x$ with

$$H = \frac{1}{2}(u + \nu y)^2 + \frac{1}{2}(v - \nu x)^2 - \Omega \tag{10}$$

If $\varepsilon \neq 0$, then H depends explicitly on time through ρ and κ which appear in Ω and depend on λ. If $\varepsilon = 0$, however, we have $\rho = \alpha$ and $\kappa = 0$. The system in the rotating frame is conservative. Notice that the kinetic energy term in the Hamiltonian (10) corresponds exactly to the motion of a charged particle in a magnetic field B parallel to the z-axis. Its Larmor frequency is ν. The presence of the second and third terms in Ω corresponds to an additional electric field which makes the potential function Ω anisotropic, but not its kinetic energy.

Let us consider the case $\varepsilon = 0$ and fix the value of H at $E < 0$. There is exactly one simply closed orbit around the origin called the variation orbit, which intersects both the x-axis and the y-axis at a right angle. Its position coordinates (x_o,y_o) are periodic functions of time with a period T which is a monotonically increasing function of E. In terms of the variable $\tau = (n-\nu)(t-t_1)$ where $n-\nu= 2\pi/T$, we can write the Fourier expansion

$$x_o + iy_o = a \sum_{j=-\infty}^{j=+\infty} a_j\zeta^{j+1} \text{with} \quad \zeta = \exp(i\tau). \tag{11}$$

The factor a scales the length and corresponds intuitively to the semimajor axis of the moon's orbit provided we set $a_o=1$. At the time $t=t_1$ the sun and the moon are at opposite ends of the x-axis, and we have a full moon. The Fourier coefficients a_j

depend mainly on the ratio $\vartheta = \nu/(n-\nu)$ of the frequencies, in addition to the ratio of the lengths a/α, as well as the ratios of the masses M/m and $(M+m)/\mu$. The expansions of a_j in powers of these ratios have been worked out in great detail by the astronomers.

5. The Neighborhood of a Periodic Orbit

The moon does not move on the variation orbit, but it stays in its neighborhood. It was first recognized by G. W. Hill that the motion of the moon could be best understood as a relatively small deviation from an exactly periodic orbit of a simple shape which satisfies all the equations of motion including the perturbations of the sun. This approach leads to a much better convergent expansion than a series of canonical transformations which starts from ordinary Kepler motion and tries to eliminate, one after the other, all the perturbations in the Hamiltonian.

Hill's idea and his methods have become crucial to many problems. We shall use it to discuss the neighborhood of a periodic orbit in a modified form that is more appropriate to investigating the relations between classical and quantum mechanics. Let us go back to 3 degrees of freedom, i.e. allow the moon to move in the direction perpendicular to the ecliptic if a specific example is needed. The variation orbit stays the same as in the previous section.

At a fixed energy E, a particular periodic orbit is isolated in phase space. There are no other periodic orbits nearby with the same energy E, except in such special cases as the ordinary Kepler problem. As E varies continuously, however, we get a one-parameter family of periodic orbits in phase space. The resulting 2-dimensional manifold in phase space is transverse to each of the (2n-1)-dimensional manifolds of constant energy, i.e. their intersection defines the periodic orbit at a given energy.

Let us now consider the neighborhood of some point (p,q) on the periodic orbit of energy E. We choose local coordinates $(\delta\bar{p}, \delta\bar{q})$ through a linear canonical transformation of the original coordinates $(\delta p, \delta q)$ in the neighborhood of (p,q). Two arbitrary deviations $(\delta p', \delta q')$ and $(\delta p'', \delta q'')$ satisfy the relation

$$\delta\bar{p}' \ \delta\bar{q}'' - \delta\bar{q}' \ \delta\bar{p}'' = \delta p' \ \delta q'' - \delta q' \ \delta p'' \qquad (12)$$

where a product such as $\delta p \delta q$ stands for the scalar product $\delta p_1 \ \delta q_1 + \delta p_2 \ \delta q_2 + \delta p_3 \ \delta q_3$.

In the new local coordinates, $\delta\bar{p}_1$ describes the shift from the periodic orbit at energy E to the one at E + δE. If $(\Delta p, \Delta q)$ represents that shift in the original coordinates, then we get the relation

$$\frac{\partial H}{\partial p} \Delta p + \frac{\partial H}{\partial q} \Delta q = \delta E. \qquad (13)$$

The shift in phase space along the periodic orbit is described by $\delta\bar{q}_1$ which is given by

$\delta p = \dot{p}\delta t$ and $\delta q = \dot{q}\delta t$ in the original coordinates. In view of the equations of motion (1) we have

$$\Delta p\ \delta q - \Delta q\ \delta p = (\dot{q}\Delta p - \dot{p}\Delta q)\ \delta t = \delta E\ \delta t, \tag{14}$$

so that we can write $\delta\bar{p}_1 = \delta E$ and $\delta\bar{q}_1 = \delta t$

The remaining four-dimensional space of variations $(\delta p, \delta q)$ in the neighborhood of (p,q) can be chosen so as to satisfy the conditions

$$\delta p\ \Delta q - \delta q \Delta p = 0,$$

$$\delta p\ \dot{q} - \delta q\ \dot{p} = \delta p\ \frac{\partial H}{\partial p} + \delta q\ \frac{\partial H}{\partial q} = \delta H = 0. \tag{15}$$

The exact choice of $\delta\bar{p}_2$, $\delta\bar{q}_2$, $\delta\bar{p}_3$, and $\delta\bar{q}_3$ at each point (p,q) of the periodic orbit is largely arbitrary within the conditions (15). The linear transformation from $(\delta p, \delta q)$ to $(\delta\bar{p}, \delta\bar{q})$ is a function of the parameter t along the periodic orbit which will be assumed continuous, differentiable, and periodic of the same period T as the periodic orbit at energy E.

A trajectory in the neighborhood of the periodic orbit is now described by a sequence of points, one in each of the successive subspaces $(\delta\bar{p}_2, \delta\bar{p}_3, \delta\bar{q}_2, \delta\bar{q}_3)$. Therefore, these subspaces are mapped into one another, and these mappings are called symplectic because the bilinear from $\delta\bar{p}_2'\ \delta\bar{q}_2' + \delta\bar{p}_3'\ \delta\bar{q}_3'' - \delta\bar{q}_2'\ \delta\bar{p}_2'' - \delta\bar{q}_3'\ \delta\bar{p}_3''$ is preserved for any two variations $(\delta\bar{p}_2', \delta\bar{p}_3', \delta\bar{q}_2', \delta\bar{q}_3')$ and $(\delta\bar{p}_2'', \delta\bar{p}_3'', \delta\bar{q}_2'', \delta\bar{q}_3'')$. In a system with only two degrees of freedom, these consecutive subspaces have only two dimensions, and the conservation of the bilinear form coincides with the conservation of area.[7]

The action integral $S(q''q'E)$ can now be defined by

$$S(q''\ q'\ E) = \int_{q'}^{q''} p\,dq \tag{16}$$

for any two endpoints q' and q'' in this family of neighboring trajectories. Jacobi's partial differential equation requires that

$$H\left(-\frac{\partial S}{\partial q'}, q'\right) = E,\quad H\left(\frac{\partial S}{\partial q''}, q''\right) = E\ ; \tag{17}$$

the initial and final momenta are given by

$$p' = -\frac{\partial S(q''q'E)}{\partial q'}, \qquad p'' = \frac{\partial S(q''q'E)}{\partial q''}. \tag{18}$$

Let us now consider the special situation where the periodic orbit is followed around once from (p,q) back to (p,q). Assume further that the local coordinate system gives the velocity $\dot{q}$ no components in the 2 and 3 direction, i.e. $\dot{q}_2 = \dot{q}_3 = 0$. The initial position for neighboring trajectories is uniquely given by $\delta q_2'$ and $\delta q_3'$, while the final position is given by $\delta q_2''$ and $\delta q_3''$. The corresponding intial and final momenta are (i=2,3)

$$\delta p' = -\sum_{j=2}^{3} \frac{\partial^2 S}{\partial q_i' \partial q_j'} \delta q_j' - \sum_{j=2}^{3} \frac{\partial^2 S}{\partial q_i' \partial q_j''} \delta q_j''$$

$$\delta p'' = \sum_{j=2}^{3} \frac{\partial^2 S}{\partial q_i' \partial q_j'} \delta q_j' + \sum_{j=2}^{3} \frac{\partial^2 S}{\partial q_i' \partial q_j''} \delta q_j'' \tag{19}$$

because of (18). The deviations ($\delta p_2'$, $\delta p_3'$, $\delta q_2'$, $\delta q_3'$) and ($\delta p_2''$, $\delta p_3''$, $\delta q_2''$, $\delta q_3''$) from the periodic orbit now belong to the same four-dimensional subspace, and can be directly related to each other. There is no arbitrary symplectic transformation between the two subspaces.

With the 2 by 2 matrices

$$a = \frac{\partial^2 S}{\partial q' \partial q'}, \qquad b = \frac{\partial^2 S}{\partial q' \partial q''}, \qquad c = \frac{\partial^2 S}{\partial q'' \partial q''} \tag{20}$$

the equations (19) become

$$\begin{aligned} \delta p' &= -a\,\delta q' - b\,\delta q'' \\ \delta p'' &= b^+\,\delta q' + c\,\delta q'', \end{aligned} \tag{21}$$

where b^+ is the transposed of the matrix b. These equations can be solved for ($\delta p''$, $\delta q''$) in terms of ($\delta p'$,$\delta q'$) in the form

$$\begin{aligned} \delta q'' &= A\,\delta q' + B\,\delta p', \\ \delta p'' &= C\,\delta q' + D\,\delta p' \end{aligned} \tag{22}$$

By direct comparison we find that

$$\begin{aligned} &A = -b^{-1}a, B = -b^{-1}, \\ &C = b^+ - cb^{-1}a, D = cb^{-1}a, \end{aligned} \tag{23}$$

only the existence of b^{-1} is required, i.e. b is a regular matrix.

The eigenvalues of the mapping (22) are the zeros of the polynomial

$$F(\Lambda) = \det \begin{pmatrix} A - \Lambda I & B \\ C & D - \Lambda I \end{pmatrix} \tag{24}$$

This 4 by 4 determinant can be reduced to a 2 by 2 determinant by a series of trivial transformations. Thus, we get

$$F(\Lambda) = \det \mid b^{+} + \Lambda a + \Lambda c + \Lambda^2 b \mid \Big/ \det \mid b \mid , \tag{25}$$

whose zeros come in complex conjugate pairs, and, if Λ is a zero, then $1/\Lambda$ is also a zero. These properties are crucial in the further dicussion.

6. Invariant Tori and Adiabatic Perturbation

The general theory of the preceding section can be translated into more concrete terms if we use the example of lunar motion. The periodic orbit is now given explicitly by the variation orbit (11). Its momenta u_0 and v_0 are given by similar Fourier series as (11). If we want to discuss the full 3-dimensional motion we add its motion in the direction perpendicular to the ecliptic, $z_0=0$ and $w_0=0$, along with the corresponding terms in the Hamiltonian (10).

The deviations from the variation orbit are found by "linearizing" the equations of motion. The momenta (u,v) and the positions (x,y) are formally replaced by the expressions $(u_0+\delta u,\ v_0+\delta v)$ and $(x_0+\delta x,\ y_0+\delta y)$. The Hamiltonian (10) is expanded in powers of δu, δv, δx, δy to second order. The coefficients in this expansion are functions of u_0; v_0, x_0, y_0. They are known functions of t with the period T, which can be written as Fourier series of the type (11) provided the excentricity ε of the earth's orbit around the sun is assumed to vanish.

If $\varepsilon \neq 0$, however, the Hamiltonian (10) can be expanded in powers of ε. For our purposes it is enough to keep the terms which are linear in both ε and δu, δv, δx, δy. They arise from the potential function Ω because both ρ and κ in (9) are known functions of $\lambda=\nu(t-t_0)$.

$$\rho = \alpha[1 - \varepsilon \cos \lambda + \ldots], \quad \kappa = 2\varepsilon \sin\lambda + \ldots, \tag{26}$$

where the omitted terms are of second order and higher in ε. The coefficients of these bilinear (driving) terms in the Hamiltonian have a "double period", the short period T (=month) and the long period $2\pi/\nu$(=year).

The construction of the preceding section is relatively easy to carry out. The starting point for the variation orbit is chosen at its intersection with the positive x-axis. This intersection takes place at right angle so that $\dot{x}_0=0$, and $u_0=0$ because $y_0=0$. On the other hand $\dot{y}_0>0$, and $v_0>0$ because $x_0>0$. The local coordinates are $\delta p_2 = \delta u$ and $\delta q_2 = \delta x$ (with $\delta p_3 = \delta w$ and $\delta p_3 = \delta z$ if the motion perpendicular to the ecliptic is discussed).

The matrices A, B, C, D in (22) can be obtained directly from the numerical integration, if necessary. Since the Fourier series (11) converges well in the lunar problem, it is more satisfying to obtain the first few relevant terms analytically. In either case, the equations of motion for δu,... have to be discussed. These equations are linear with the inhomogeneous term directly proportional to ε.

The homoqeneous solution has the Hill-Floquet-Bloch property (HFB):[8] It is the linear combination of as many independent solutions as the number of equations (22) where each solution can be written in our local coordinates $(\delta\bar{p}_2,...,\delta\bar{q}_2,...)$ in the form $\exp(-\theta t)\cdot(\delta\tilde{p}_2,...,\delta\tilde{q}_2,...)$ and the functions $\delta\tilde{p}_2,...,\delta\tilde{q}_2,...$ have the period T. The value of θ is related to one of the zeros in the polynomial $F(\Lambda)$ of (24) through $\Lambda = \exp(\theta T)$.

In the lunar problem the motions in the plane of the ecliptic and perpendicular to it are independent in our approximation. The polynomial $F(\Lambda)$ decays into a product of two quadratic functions. This degeneracy results from the symmetry of the variation orbit, which in turn is due to the symmetry of the equations of motion with respect to a reflexion on the plane of the ecliptic.

The zeros of (25) fall into one of the following two categories when the matrices a, b, c, d are one-dimensional.

1) The two zeros are complex and of absolute value 1, i.e. they can be written as $\Lambda_1 = \exp(i\alpha)$ and $\Lambda_2 = \exp(-i\alpha)$.
2) The two zeros are real and reciprocal to each other, i.e. they can be written as $\Lambda_1 = \exp(\beta)$ and $\Lambda_2 = \exp(-\beta)$.

In the first case the corresponding periodic functions $\delta\tilde{p}_2$ and $\delta\tilde{q}_2$ are complex conjugate to each other. Such is the case in the lunar problem, and the deviation from the variation orbit becomes

$$\delta x + i\,\delta y = a\left[\xi\zeta^{\gamma}\sum_j \xi_j\zeta^{j+1} + \xi^*\zeta^{-\gamma}\sum_j \xi'_j\zeta^{j+1}\right] \tag{27}$$

with an arbitrary complex coefficient ξ and real constants ξ_j, ξ'_j which satisfy the normalization condition $\xi_0-\xi'_0=1$. If γ is an irrational number, $(n-\nu)\gamma=\alpha/T$ and t varies from $-\infty$ to $+\infty$, it is as if ξ and ζ varied independently over some appropriate circles in the complex plane. The resulting manifold in pł.ase space is obviously a 2-dimensional torus. Such a statement is not true when the zeros of (25) are real.

Let us now consider the situation when the excentricity ε does not vanish. The deviation from the variation orbit becomes

$$\delta x + i\,\delta y = a\left[\eta\,\zeta^{\vartheta}\sum_j \eta_j\,\zeta^{j+1} + \eta^*\zeta^{-\vartheta}\sum_j \eta'_j\,\zeta^{j+1}\right] \tag{28}$$

with real constants η_j, η'_j and a complex coefficient η such that $|\eta| = \varepsilon$. The number ϑ is small since it equals the ratio of the month to the year. Therefore, if (28) is added to the sum of (11) and (27), the torus in phase from (11) and (27) gets slowly modified by (28). The basic structure of the trajectories remains even in the presence of a time dependent perturbation, provided its frequency is low. That is just what is commonly called adiabatic perturbation.

7. The Anisotropic Kepler Problem

Many ergodic systems are known, particularly in various simplified versions of the three-body problem. Unfortunately, they have been explored only numerically. An interesting very special situation has been discovered by Sitnikov.[9] A complete enumeration is possible, but the motion is one-dimensional and time-dependent which does not suit our purpose of meeting Einstein's challenge. The anisotropic Kepler problem was first (and so far only) investigated by the author[10] The main results will be presented in this and the following section.

The Hamiltonian in two dimensions is given by

$$H = \frac{u^2}{2\mu} + \frac{v^2}{2\nu} - \frac{1}{\sqrt{x^2+y^2}} \qquad (29)$$

where $\mu \cdot \nu = 1$ and μ/ν has a typical value of ~5. The constant value of the energy $E<0$ can be normalized to -1/2, because both the momenta and the positions are easily scaled with E. In our natural units the action integral is given by

$$S = \frac{1}{\sqrt{-2E}} \int (u\,dx + v\,dy), \qquad (30)$$

which is the same as in the ordinary Kepler problem.

Every trajectory in position space is restricted to the circular disc $x^2 + y^2 = 4$ if $H = -1/2$. The (longitudinal, heavy) x-axis is crossed more often than the (transverse, light) y-axis. When the trajectory crosses the x-axis, i.e. when y=0, the coordinates x and u satisfy the inequalities

$$|x| \leq \frac{2}{1 + u^2/\mu}, \quad -\infty \leq u \leq +\infty. \qquad (31)$$

The area conserving transformation from (x,u) to (X,U)

$$X = x\left(1 + \frac{u^2}{\mu}\right), \quad U = \sqrt{\mu}\arctan\frac{u}{\sqrt{\mu}} \qquad (32)$$

reduces the inequalities (31) to the following simpler ones

$$|X| \leq 2, \quad |U| \leq \sqrt{\mu}\,\pi/2. \qquad (33)$$

Exactly one trajectory intersects the x-axis at the position $x = X\cos^2(U/\sqrt{\mu})$ with the forward momentum $u = \sqrt{\mu}\tan(U/\sqrt{\mu})$ going upward, $v>0$, at the energy $H = -1/2$.

Let us take this intersection as the starting point, and call it (U_0,X_0). The next intersection with the x-axis takes place at values (U_1,X_1) with $v<0$. As we continue the trajectory, we find successively the points (U_2,X_2), (U_3,X_3),... in the rectangle (33). This rectangle is mapped into itself in an area conserving manner each time the trajectory intersects the longitudianl axis. We shall call this the Poincare map. It is a straightforward generalization of the map (22) near a periodic orbit.

In a system where the trajectories in phase space restrict themselves to a torus, the consecutive points (U_0, X_0), (U_1,X_1),... are all located on a simple closed smooth curve. In this way an easy numerical check confirms whether or not the trajectories obey Einstein's restrictions. Very often, one finds smooth curves in some regions, and a wild scattering of points in others. The latter situation is typical of ergodic behavior[11].

In the two-dimensional anisotropic Kepler problem, the whole rectangle (33) is one ergodic region. There are no smooth curves of consecutive points. At most, one can find finite sequences of points which repeat themselves over and over again as the trajectory is pursued. Such an exceptional sequence is always associated with a periodic orbit. It represents a fix point for some power of the Poincare map. Such points form a set of measure zero which is dense in the rectangle (33).

8. Bernoulli Sequences

An ergodic system has random behavior. Its description must be related to that of a stochastic process. The most famous of these is the tossing of a coin, i.e. a sequence of 0 and 1, or -1 and +1, which may be infinitely long in both directions. Instead of simply two possible values in each position of the sequence, there may be 6 as in the throwing of a dice, or any other integer including ∞. Such sequences are called Bernoulli sequences.[12] They record the actual performance in a well defined sequence of events. There is no general rule how to associate a Bernoulli sequence with an ergodic trajectory. The following scheme works for the anisotropic Kepler problem.

Each intersection of the trajectory with the x-axis is described by a binary digit a. Put $a=1$, if $X>0$; put $a=-1$, if $X<0$. Thus a binary sequence ..., a_{-1}, a_0, a_1, a_2,... is associated with the sequence ...(U_{-1},X_{-1}), (U_0,X_0), (U_1,X_1), (U_2,X_2),... of points in the rectangle (33). The binary sequence tells us how the trajectory in (x,y) space switches back and forth between the right and the left half.

Any binary Bernoulli sequence can be mapped into a square if we define the real numbers ξ and η as

$$\xi = \sum_{i=1}^{\infty} a_i (1/2)^i, \quad \eta = \sum_{i=0}^{\infty} a_{-i}(1/2)^{i+1} \tag{34}$$

If the binary sequence varies freely, all the points in the square

$$-1 \leq \xi \leq +1, \quad -1 \leq \eta \leq +1 \tag{35}$$

will be covered. The relation between the points in the square and the binary sequences is not one-to-tone. E.g. if $\xi=1/2$, one can have either $a_1=a_2=1$ with $a_i=-1$ for $i>2$, or we can have $a_i=1$ and $a_2=-1$ with $a_i=+1$ for $i>2$.

The following assertion is central to the understanding of the anisotropic Kepler problem. There is a one-to-one correspondence between the points (U,X) of the rectangle (33) and the points (ξ,η) of the square (35). This correspondence is continuous, but not differentiable, except at the boundaries of (33) and at the center line $X=0$. The continuity is claimed with respect to the natural topology in the rectangle (33) and the square (35), although the mapping is defined with the help of binary Bernoulli sequences. The two sequences of the preceding paragraph correspond to the same trajectory which hits the origin of position space at its second crossing of the x-axis. Such an event is called a collision, henceforth. The topology of the square (35) is stronger than the one which is ordinarily associated with Bernoulli sequences.

The ergodic behavior of the trajectories shows up as follows: give any two points (U',X') and (U'',X'') in the rectangle (33) with some neighborhoods N' and N'' around them. Map both N' and N'' into the square (35) where they will define neighborhoods M' and M''. Each one of the latter contains a small square, Q' and Q'', whose vertices have simple coordinates; i.e. coordinates which are multiples of $(1/2)^n$ with a sufficiently large integer n. The points inside each square have Bernoulli sequences with well defined values of the binaries $a_{-n+1},\ldots, a_{-1}, a_0, a_1,\ldots, a_n$ which we assume to be given by a'_{-n+1}, a'_n and $a''_{-n+1},\ldots, a''_n$. A possible trajectory from the neighborhood N' to the neighborhood N'' in the rectangle (33) is given by a Bernoulli sequence which the special values $a'_{-n+1},\ldots, a'_n, a''_{-n+1},\ldots, a''_n$ in the positions from $-2n+1$ to $+2n$ and any arbitrary values before $-2n+1$ and after $2n$. The initial conditions for such a trajectory can be obtained if we map the corresponding small square (of dimension $(1/2)^{2n}$) from the square (35) back into the rectangle (33).

II. THE QUANTIZATION OF BOUND STATES

9. Path Integrals in Phase Space

Feymman introduced path integrals into quantum mechanics in terms of the initial and final position coordinates. Garrod[13] showed that Feymman's treatment can be significantly extended, and then put to good use by a relatively easy transformation.

Instead of a path in position space from q' to q'', we introduce the notion of a history in phase space which consists of an alternating sequence of momenta and positions, $q'=q_0, p_1, q_2, p_3,\ldots, p_{2n-1}, q_{2n} = q''$. Its average energy is

$$E_N = \frac{1}{2mN}\sum_1^N p_{2n-1}^2 + \frac{1}{2N}\left[\frac{1}{2}V(q_o) + \sum_1^{N-1} V(q_{2n}) + \frac{1}{2}V(q_{2N})\right]. \tag{36}$$

Green's function G(q''q'E) is obtained from the formula

$$\lim_{N\to\infty} (2\pi h)^{-3N} \int \prod_1^{N-1} d^3q_{2n} \int \prod_1^{N} d^3p_{2n-1} \exp [-\frac{i}{h}S_N]/(E-E_N), \tag{37}$$

where the action integral S_N in the exponent is

$$S_N = \sum_1^N p_{2n-1} (q_{2n} - q_{2n-2}). \tag{38}$$

Planck's quantum divided by 2π is represented by letter h.

The formula (37) for G can be reduced to Feymman's expression for the propagator K(q''q't) by the Fourier transform

$$K(q''q't) = \frac{i}{2\pi} \int G(q''q' E) \exp\left[-\frac{iEt}{h}\right] dE, \tag{39}$$

with the integral over E taken along a parallel to the real axis in the complex E-plane and a positive imaginary part. The expression (37) is inserted into (39); the order of the integrations is changed, so that the integration over E can be performed first, with the help of residues. Then the integrals over p_{2n-1} become standard Fresnel integrals, and we get

$$K(q''q't) = \lim_{N\to\infty} \left(\frac{mN}{2\pi iht}\right)^{3N/2} \int \prod_1^{N-1} d^3q_{2n} \exp\frac{i}{h} L_N \tag{40}$$

where the averaged Lagrangian L_N is given by the expression

$$\frac{m}{2N} \sum_1^N \left(\frac{q_{2n}-q_{2n-2}}{t/N}\right)^2 - \frac{1}{N}\left[\frac{1}{2}V(q_0) + \sum_1^{N-1} V(q_{2n}) + \frac{1}{2} V(q_{2N})\right] \tag{41}$$

Garrod's formula (37) is really more satisfactory than Feymman's formula (40), because the time step t/N does not explicitly appear in (37) and the normalization goes with Planck's constant, as one would expect it in phase space.

The Coulomb problem can be discussed more easily in momentum coordinates. Therefore, we introduce the propator

$$F(p''p'E) = (2\pi h)^{-3} \int d^3q'' \int d^3q' \, G(q''q'E) \exp [\,\frac{i}{h} (p'q' - p'q'')]. \tag{42}$$

The expression (37) is thereby converted into

$$\lim_{N\to\infty} (2\pi h)^{-3N-3} \int \prod_0^{N} d^3q_{2n} \int \prod_1^{N} d^3p_{2n-1} \exp [-\frac{i}{h}R_N]/(E-E_N), \tag{43}$$

with the virial integral

$$R_N = \sum_0^N q_{2n} (p_{2n+1} - p_{2n-1}), \tag{44}$$

and the momenta $p_{-1}=p'$ and $p_{2N+1}=p''$. The symmetry between the expressions (37) and (43) is striking. The same integral over histories in phase space gives the Green's function in both position and momentum coordinates.

These integrals were established in cartesian coordinates for both position and momentum space. Is it then possible to write similar expressions in some other system of canonically conjugate variables? The answer is yes, but with important modifications which have to be handled with great care.[14] This becomes already apparent if we go from the cartesian coordinates (u,v,w; x,y,z) to the polar coordinates (S,L,M; r,ϑ,φ) through the ordinary formulas

$$\begin{aligned} u &= \sin\vartheta \cos\varphi\, s + \cos\vartheta \cos\varphi \frac{L}{r} - \frac{\sin\varphi}{\sin\vartheta} \frac{M}{r} \\ v &= \sin\vartheta \sin\varphi\, s + \cos\vartheta \sin\varphi \frac{L}{r} + \frac{\cos\varphi}{\sin\vartheta} \frac{M}{r} \\ w &= \cos\vartheta\, s - \sin\vartheta \frac{L}{r} \end{aligned} \tag{45}$$

$$\begin{aligned} x &= r \sin\vartheta \cos\varphi, \\ y &= r \sin\vartheta \sin\varphi, \\ z &= r \cos\vartheta. \end{aligned} \tag{46}$$

There will now be an alternating sequence of momenta (S_{2n-1}, L_{2n-1}, M_{2n-1}) and positions (r_{2n}, ϑ_{2n}, φ_{2n}).

The expressions for the action integral (38) and the virial (44) transform, as one would expect, into

$$S_N = \sum_1^N [S_{2n-1} (r_{2n} - r_{2n-2}) + L_{2n-1} (\vartheta_{2n} - \vartheta_{2n-2}) + M_{2n-1} (\varphi_{2n} - \varphi_{2n-2})], \tag{47}$$

$$R_N = -\sum_0^N [r_{2n} (S_{2n+1} - S_{2n-1}) + \vartheta_{2n} (L_{2n+1} - L_{2n-1}) + \varphi_{2n} (M_{2n+1} - M_{2n-1})]. \tag{48}$$

The kinetic energy in (36), however, is obtained correctly provided p_{2n-1}^2 is replaced by the expression

$$S^2_{2n-1} + (r_{2n}r_{2n-2})^{-1}\,[L^2_{2n-1} - \frac{h^2}{4}] + (r_{2n}r_{2n-2}\sin\vartheta_{2n}\sin\vartheta_{2n-2})^{-1}\,[M^2_{2n-1} - \frac{h^2}{4}] \quad (49)$$

The symmetric occurence of (r_{2n}, ϑ_{2n})) and (r_{2n-2}, ϑ_{2n-2}) can be accepted rather easily. The correction $h^2/4$, however, although of higher order than L^2 and M^2 is important to obtain agreement of the classical quantization conditions with Schrodinger's equation for low quantum numbers. The derivation of (49) is both tricky and not very rigorous.

10. The Transition to Classical Mechanics

The integration over the intermediate momenta and positions in (37) and (42) will be referred to as the summation over all possible histories, in contrast to the integration over all possible paths which are restricted to position space.[15] There is no relation between the momentum p_{2n-1} and its adjacent positions, q_{2n-2} and q_{2n}.

The transition to classical mechanics is based on the choice of certain preferred histories (or paths) which dominate in the limit of small Planck's quantum. A particular history, $\bar{p}_1, \bar{q}_2, \bar{p}_3, \ldots, \bar{p}_{2N-1}$, is chosen because its length in terms of S_N or R_N is stationary with respect to small variations, $\delta p_1, \delta q_2, \delta p_3, \ldots, \delta p_{2N-1}$, of the intermediate momenta and positions. Let us, therefore, write $p_{2n-1} = \bar{p}_{2n-1} + \delta p_{n-1}$ and $q_{2n} = \bar{q}_{2n} + \delta q_{2n}$, and expand S_N or R_N with the subsidiary condition E_N = const.

If we write first the separate expansions

$$S_N = \bar{S}_N + \delta^1 S_N + \delta^2 S_N ,$$

$$E_N = \bar{E}_N + \delta^1 E_N + \delta^2 E_N + \ldots , \quad (50)$$

to quadratic terms in the variations, the subsidiary condition is enforced with the help of a Lagrange parameter τ through

$$\delta^1 S_N - \tau\, \delta^1 E_N = 0. \quad (51)$$

This condition is equivalent to the equations of motion

$$\bar{q}_{2n} - \bar{q}_{2n-2} = \frac{\tau}{mN}\,\bar{p}_{2n-1}\,,\; \bar{p}_{2n+1} - \bar{p}_{2n-1} = -\frac{\tau}{N}\frac{\partial V}{\partial q}\bigg|_{\bar{q}'} \quad (52)$$

where τ/N is the effective time step. In the limit $N \to \infty$, these equations become identical with the equations of motion (1).

The integrations in (37) or (42) are now carried out with S_N (or R_N) replaced by its expansion to quadratic terms in the variations

$$S_N \cong \bar{S}_N + [\delta^2 S_N - \tau\,\delta^2 E_N]_{\delta E = 0} \tag{53}$$

There are 6N-3 intermediate momenta and positions with one condition $\delta' E = 0$ among them. The resulting multiple Fresnel integral is straightforward. The determinant and the number of negative eigenvalues of the matrix in (53) has to be found. The same two quantities have to be known if the classical approximation to (41) or to (42) are to be evaluated.

The number of negative eigenvalues in the second variation of S_N, L_N, or R_N is crucial for the understanding of the variational principles by which the equations of motion are derived. The trajectory minimizes S_N, L_N, or R_N only if there are no such negative eigenvalues. How can one tell? This question is almost always left unanswered in the textbooks on classical mechanics. Yet, there exists a very satisfactory answer which was first discussed in detail by M. Morse.[16] It involves the notion of conjugate points along a trajectory, and is most easily understood for trajectories in position space, i.e in the variation of L_N.

Consider a family of classical trajectories which leave the fixed initial position q' with an initial momentum p' in a small neighborhood of momentum space with volume $d^3p' = d\Omega' dE$. Their endpoints q'' are defined by the time t over which we follow each trajectory and they come to lie in a small neighborhood of position space with volume $d^3q'' = d\Omega'' dt$. The linear mapping from $d\Omega'$ into $d\Omega''$ has the rank 2 except at isolated values of t, and these are called the conjugate points of q' along the trajectory. The order or multiplicity of a conjugate point is the reduction in rank. The Jacobian $\partial\Omega''/\partial\Omega'$ can be expressed in terms of the action S(q''q'E), and is found to be the reciprocal of a determinant D_S of second derivatives.

$$D_S = \begin{vmatrix} \dfrac{\partial^2 S}{\partial q' \partial q''} & \dfrac{\partial^2 S}{\partial q' \partial E} \\[2ex] \dfrac{\partial^2 S}{\partial E \partial q''} & \dfrac{\partial^2 S}{\partial E^2} \end{vmatrix} \tag{54}$$

The number of negative eigenvalues in the second variation of S_N, L_N, or R_N equals the number of conjugate points, each counted with its appropriate multiplicity.

The summation over all histories in phase space in the quadratic approximation (53) requires that one finds the determinant of the (big) matrix in (53). This seems difficult to do, in general, although it can be carried out in a system of spherical symmetry. Nevertheless, the result is hardly in doubt because it can be found directly, i.e. without the use of path integrals. Many authors have performed an expansion of G(q''q'E) in powers of h by inserting G into Schrodinger's equation. The approximate Green's function $\tilde{G}$(q''q'E) is then found to be

$$- \frac{1}{2\pi\hbar^2} \sum_{\text{classical tr.}} (|D_S|)^{1/2} \exp\left[\frac{i}{\hbar} S'(q''q'E) + \text{phases}\right]. \tag{55}$$

The "phases" are $-i\pi/2$ times the number of conjugate points with appropriate multiplicity. The determinant of the big matrix in (53) obviously equals $|D_s|$ except for some simple factors π and h.

The propagator $F(p''p'E)$ can be treated similarly, and its classical approximation $\tilde{F}(p''\,p'E)$ is given by a formula exactly like (55) where the action integrals S is replaced by the virial

$$R(p''p'\,E) = -\int_{p'}^{p''} q\,dp, \tag{56}$$

and the determinant of second derivatives of D_R appears instead of D_S. The direct derivation of this expression for $\tilde{F}$ by expanding Schrodinger's equation in powers of h is now quite delicate, because we are dealing with an integral equation.

11. The Wavefunctions of the Hydrogen Atom

Path integrals are of special interest with respect to bound states, because one has to deal with infinitely many classical paths whose constructive interference leads to resonances, alias bound states. These can be understood much better in momentum space on the basis of classical trajectories for the following reason. A particle of a given negative energy E has only a limited domain in position space available for its classical motion. The approximate wave functions are very poor at the boundary of this domain. On the other hand, all of momentum space is accessible classically, provided the attractive potential has a singularity of the Coulomb type. Much better wave functions can be expected in momentum space, and, indeed, the correct wave functions will be shown to arise in the case of the hydrogen atom.

The Kepler motion in momentum space (the hodograph of a planet around the sun) is much simpler than in position space.[17] If we restrict ourselves temporarily to two dimensions, the momentum space is a plane. A scale is determined in this plane by a circle of radius $\sqrt{-2mE}$ around the origin which we shall call the unit circlè. The trajectories are again circles which intersect the unit circle in diametrically opposite points. These orbits can be viewed more directly if the momentum plane is mapped onto the sphere over the unit circle by a sterographic projection. The orbits become the great circles on the sphere, and, remarkably, the ordinary distance between two points on this sphere equals the virial along the Kepler trajectory. Obviously, any two points on the sphere determine a great circle, unless they are diametrically opposite in which case there is an infinity of orbits.

The simple geometry of the Kepler orbits in momentum space allows us to write down an explicit expression for the virial $R(p''p'E)$ as a function of the end momenta p' and p''. The mass and the charge of the particle are m and e. With the abbreviations $\rho_0=\sqrt{-2mE}$, $\rho_1=|\,p'\,|$, $\rho_2=|\,p''\,|$, and

$$\tau = [(\rho_1^2 + \rho_0^2)(\rho_2^2 + \rho_0^2) - \rho_0^2 |p'' - p'|^2]^{1/2} / \rho_0 |p'' - p'| \tag{57}$$

we have the virial along the "shortest" great arc on the unit sphere

$$R(p''p'E) = 2 \frac{me^2}{\rho_0} \arctan\left(\frac{1}{\tau}\right) \tag{58}$$

The shortest arc from p' to p'' will be called the direct trajectory while the longest arc will be called the indirect trajectory. Its length is given by (58) with π-arctan$(1/\tau)$ replacing arctan $(1/\tau)$. τ vanishes when $p''=-\rho_0^2 p'/\rho_1^2$ i.e. when p' and p'' are diametrically opposite on the unit sphere, and τ becomes ∞ when p''-p' vanishes. The value of arctan $(1/\tau)$ is in the range $(0,\pi/2)$.

All the trajectories which start at p' go through the opposite momentum $P'=-\rho_0^2 p'/\rho_1^2$. In two dimensions, P' is the first conjugate point to p' and its multiplicity is 1. In three dimensions, however, a double infinity of trajectories starting in p' goes through P' so that P' is a conjugate point of order 2. If a trajectory is followed beyond P', it will go again through p' so that the starting point becomes the second conjugate point, again of multiplicity 2 in three dimensions. Therefore, in three dimensions all conjugate points of a Kepler orbit in momentum space are of order 2, whereas in position space the first and second conjugate points are simple. Only the third conjugate point is double, because it coincides with the initial point. This is an example of how the structure of conjugate points depends on the coordinates.

The determinant D_R corresponding to D_S can be worked out from (58) and (57). Notice that its value is the same for direct and indirect trajectories as well as for their arbitrary prolongations, because arctan $(1/\tau)$ in (58) gets only its sign changed or a multiple of π added. After some lengthy calculations one finds that

$$(|D_R|)^{1/2} = \frac{8m^3e^4\rho_0}{(\rho_0^2 + \rho_1^2)^2(\rho_0^2 + \rho_2^2)^2} \quad \frac{1+\tau^2}{2\tau} \tag{59}$$

The summation over the classical trajectories consists of two parts: the direct trajectories and their prolongations by a full circle in momentum space have phases which add up to a multiple of $-2\pi i$, the indirect trajectories and their prolongations have phases which add up to an odd multiple of $-i\pi$. The addition of exponentials in (55) then leads to

$$\sin\left(\frac{me^2}{h\rho_0} \cdot 2 \arctan \tau\right) \Big/ \sin\left(\frac{me^2}{h\rho_0}\pi\right). \tag{60}$$

The approximate Green's function $\widetilde{F}(p''p'E)$ equals $(-1/2\pi h^2)$ times the product of (59) and (60).

The poles occur where the sine in the denominator of (60) vanishes, i.e. $me^2/h\rho_0=n$ (some integer other than zero), and this gives us immediately the Bohr formula for the energy levels in the hydrogen atom. When $me^2/h\rho_0$ is an integer, the denominator in (60) becomes a rational function of τ. Together with (53) we see immediately that only even powers of τ occur, and that the denominator contains only the combinations $(1+\tau^2)$ and powers thereof. Therefore, the residues of $\widetilde{F}$ can be written as a sum of products $f(p')f(p'')$ where $f(p)$ is some polynomial in p divided by an appropriate power of $p^2+\rho_0^2$. A close examination reveals that these funcitons are exactly the hydrogen bound state wave functions, with the correct normalization.[12]

12. Potentials with Spherical Symmetry

All the important quantities can be calculated explicitly in the Coulomb potential, at least if one works in momentum space. Einstein discussed the quantization in a general spherical potential V(r) in order to explain how the concept of a torus in phase space comes about. It seems appropriate to treat this more general case so as to bring out the connection between the Green's function and the torus even if explicit algebraic formulas cannot be given. The coordinates in momentum space will not be used because the arguments are more familiar in position space.

The initial position q' and the final position q'' determine the plane of the trajectory. The action integral depends only on the absolute values $r'^2=|q'|$, $r''^2=|q''|$, and on the angular distance ϑ which is restricted by $0<\vartheta<\pi$. The determinant D_S of (54) now reduces to

$$D_S = \frac{S_\varphi}{r'^2 r''^2 \sin\varphi} \begin{vmatrix} S_{r'r''} & S_{r'\varphi''} & S_{r'E} \\ S_{\varphi'r''} & S_{\varphi'\varphi''} & S_{\varphi'E} \\ S_{Er''} & S_{E\varphi''} & S_{EE} \end{vmatrix} \tag{61}$$

where lower indices indicate partial derivatives. The angle $\varphi=\varphi''-\varphi'$ measures the total angular motion in the plane of the trajectory, and is not restricted in any way. If we work in a two-dimensional space, the expression (61) would have $1/r'r''$ in front of the determinant. The supplementary factor $S_\varphi/r'r''\sin\varphi$ in three dimensions gives the higher multiplicity of the eigenstates, and changes the shape of the angular dependence in the approximate eigenfunction.

Polar coordinates (r,φ) are used in the orbital plane. Their canonical conjugates are the radial component ρ of the momentum and the angular momentum M which is a constant of motion. The value of M for a particular trajectory determines a (set of generally more than one) interval $r_1\leq r\leq r_2$ which defines an anunlus in the orbital plane. The radial momentum ρ can have either one of the two values

$$\rho = \pm\,[2mE - 2m\,V(r) - M^2/r^2]^{1/2} \tag{62}$$

at each point in the annulus. An explicit representation of the torus for the trajectory is obtained if these two values are plotted along an axis which is perpendicular to the orbital plane.

Each endpoint of the trajectory is represented by two points on the torus, one above and one below the orbital plane, except when r equals r_1 or r_2 so that $\rho=0$. These two pairs can be connected in four different ways. Moreover, the angular separation between q' and q'' can be covered by the trajectory either by following the shorter path around the torus, i.e. by covering the angular separation ϑ, or by following the longer path, i.e. by covering the angle $2\pi-\vartheta$. These will be called the direct and the indirect trajectories as in the Coulomb problem, and we are free to add a multiple of 2π in either case. By convention, we shall associate a direct (indirect) trajectory with a positive (negative) value of the angular momentum M.

The enumeration of trajectories for given r', r'', (with r'<r'' for definiteness), and ϑ can be made in terms of the angles

$$(\alpha,\beta,\gamma) = \left(\int_{r_1}^{r'}, \int_{r''}^{r_2}, \int_{r_1}^{r_2}\right) \frac{dr}{r} \frac{M}{[2mr^2(E-V)-M^2]^{1/2}} \tag{63}$$

The angular momentum M has to be determined by finding two integers, $\kappa\geq 0$ and $\lambda\geq 0$, such that

$$\varphi'' - \varphi' = \varphi = 2\lambda\pi + \vartheta = (2\kappa+1)\gamma \pm \alpha \pm \beta . \tag{64}$$

The double signs in front of α and β are independent. They take care of the four ways of connecting the two pairs of endpoints on the torus. The values of α, β, and γ are positive or negative according as the sign of M. The integer λ is ≥ 0 for direct trajectories, and $\lambda<0$ for indirect ones.

The action integral S can be expressed as

$$S = (2\kappa + 1)\Theta \pm \sigma \pm \tau, \tag{65}$$

in terms of the partial actions

$$(\sigma,\tau,\Theta) = \left(\int_{r_1}^{r'}, \int_{r''}^{r_2}, \int_{r_1}^{r_2}\right) \frac{dr}{r} \frac{2mr^2(E-V)}{[2mr^2(E-V)-M^2]^{1/2}} \tag{66}$$

which corresponds to the angles (64). The determinant D_S becomes

$$\frac{M}{r'^2r''^2\sin\varphi} \left[2m(E-V) - \frac{M^2}{r'^2}\right]^{-1/2} \left[2m(E-V(r'')) - \frac{M^2}{r''^2}\right]^{-1/2} \frac{\partial\varphi}{\partial M} \tag{67}$$

The number of conjugate points depends on the sign of α and β in (64), as well as the signs of M and $\partial\varphi/\partial M$. E.g. for a direct trajectory (M>0) with outward radial momenta at the endpoints (upper signs in (64)) and positive $\partial\varphi/\partial M$, there are $2(\kappa+\lambda)$ conjugate points. The exponent in (55) is, therefore, given by

$$\frac{i}{h}[(2\kappa+1)\Theta-\sigma-\tau] - i\,(\kappa+\lambda)\pi - i\,\frac{\pi}{4} \tag{68}$$

We shall carry out the calculations in this case, and add the appropriate modifications for the other cases at the end.

The summation over all trajectories in (55) is a summation over all integer pairs (κ,λ) for which the conditon (64) can be fulfilled by finding the appropriate value of M. This complicated situation can be formulated as follows: integrate over M for a given κ and make sure that the quantity $(2\kappa+1)\gamma\pm\alpha\pm\beta-\vartheta)$ takes on only multiples of 2π. The last is accomplished if we add i.e. $(\vartheta\pm\alpha\pm\beta-(2\kappa+1)\gamma)$ to the exponent (68) in (55), and perform a summation over all integers $\ell\geq 0$. The integration over M requires a factor in the amplitude which will not be discussed. The exponent (68) can now be written as

$$(2\kappa+1)[\frac{\Theta}{h} - (\ell+\frac{1}{2})\gamma - \frac{\pi}{2}] \pm [\frac{\sigma}{h} - (\ell+\frac{1}{2})\alpha - \frac{\pi}{4}]$$

$$\pm [\frac{\tau}{h} - (\ell+\frac{1}{2})\beta - \frac{\pi}{4}] \pm [(\ell+\frac{1}{2})\vartheta\,\frac{\pi}{4}] \tag{69}$$

where a common factor i has been omitted. The double signs in front of the second and third term correspond to the double signs in (64), and the double sign in front of the last term has to be chosen accordingly as the sign of M. The conjugate points are properly accounted for in (69), provided the complications with $\partial\varphi/\partial M$ are neglected.

The constructive interference in (55) comes from the first term in (69) because it contains the arbitrarily large factor $(2\kappa+1)$. Therefore, the quantity $\frac{1}{h}\Theta-(\ell+\frac{1}{2})\gamma - \pi/2$ has to satisfy two requirements:

1) It has to be stationary with respect to the variation of M at a constant value of E. In terms of

$$\omega(E,M) = 2\int_{r_1}^{r_2} \frac{dr}{r}\,[2mr^2(E-V)-M^2]^{1/2} \tag{70}$$

we obtain $\Theta = \frac{1}{2}\omega + M\gamma$ from (63) and (66) as well as

$$\left(\frac{\partial\omega}{\partial M}\right)_E = -2\gamma(E,M) \tag{71}$$

Therefore, the stationarity condition becomes

$$\frac{\partial}{\partial M}\left[\frac{\Theta}{h} -(\ell + 1/2)\gamma - \frac{\pi}{2}\right]= \left(\frac{M}{h}- (\ell + 1/2)\right)\frac{\partial\gamma}{\partial M} =0. \tag{72}$$

Since we assume that $\partial\gamma/\partial M\neq 0$, we find the quantization

$$M = (\ell + \frac{1}{2})\ h. \tag{73}$$

2) The factor of $(2\kappa+1)$ in (69) has to be a multiple of π. After inserting the expression for Θ in terms of w and the condition (73) we get

$$\omega = (n + \frac{1}{2})\ 2\pi h. \tag{74}$$

Notice the half-integer quantization of the angular momentum which is characteristic of a problem in 3 dimensions and can be traced directly to the number $2(\kappa+\lambda)$ of conjugate points. In 2 dimensions there would be only 2κ conjugate points, and the angular momentum would be an integer multiple of h.

The constructive interference in the summation over κ and the integration over M in (55) with (69) leads to poles in the dependence of $\tilde{G}(q''q'E)$ on E. The residues can be read off (67) and (69). The angular dependence comes from the first factor in (67) and the last term (with a double sign) in (69). It is

$$\left[\frac{2\ell+1}{\sin\vartheta}\right]^{1/2}\cos[(\ell+1/2)\vartheta - \frac{\pi}{4}] \sim \left[\frac{2\ell+1}{4\pi}\right]^{1/2} P\ell(\cos\vartheta), \tag{75}$$

where the left-hand side is asymptotically (for large ℓ) equal to the Legendre polynomial on the right-hand side, including the correct normalization. Notice again the difference in 3 and 2 dimensions. In 2 dimensions we would have found the simpler expression $\cos[\ell\vartheta]$.

The radial parts of the wave function become exactly what the WKB approximation in one dimension with the potential $V(r) + M^2/2mr^2$ would yield, because with (73) the second and third terms in (69) become the radial action integrals starting at classical return point with the initial phase loss of $\pi/4$. With ρ as a function of r given by (62) with the upper sign we get

$$\frac{1}{r}\rho^{-1/4}\cos\left[\frac{1}{h}\int_{r_1}^{r}\rho dr - \frac{\pi}{4}\right],\ \frac{1}{r}\rho^{-1/4}\cos\left[\frac{1}{h}\int_{r}^{r_2}\rho dr - \frac{\pi}{4}\right]. \tag{76}$$

The two expressions are identical up to a factor $(-1)^n$ because of (74).

III. A METHOD FOR ARBITRARY SYSTEMS

13. The Importance of Periodic Orbits

Spherically symmetric potentials are representative of the situation where the classical trajectories in phase space lie on tori. There are no known algebraic or even analytic constants of motion in most cases, however, besides the Hamiltonian, e.g. in the restricted three body problem where two heavy masses move around each other in a fixed circle, and a third (light) mass is attracted to both of them. Such systems are often called non-separable because the separation of variables is not feasible with one simple canonical transformation.[19]

An efficient algorithm of successive canonical transformation may exist, however, as in the lunar problem, with convergence assured in parts of phase space. The author believes that such a procedure can be devised whenever the trajectories come to lie on tori of dimension no larger than the number of degrees of freedom. This would permit the same construction of bound states as in the preceding section.

The question then comes up whether approximate bound state energies can be found without recourse to the invariant tori. The construction of approximate wavefunctions would be sacrificed in return for approximate energies even in the absence of invariant tori, i.e. when the classical behavior of the system is ergodic. Such a method will be discussed in this chapter.

A response function g(E) is first defined by eliminating the coordinate dependence in Green's function G(q''q'E). The endpoints q' and q'', are identified and integrated over all space,

$$g(E) = \int d^3q\, G(q\,q\,E) = \sum_j \frac{1}{E-E_j} \tag{77}$$

The label j identifies the eigenstates of the quantum mechanical system, and E_j is the corresponding energy. The same procedure is now applied to $\tilde{G}(q''q'E)$ as given in (55) in order to define the approximate response function $\tilde{g}(E)$.

The starting position q' coincides with the final position q'' in each classical trajectory. The exponent of (55) varies at a rate which is given by the partial derivative

$$\frac{\partial S(qqE)}{\partial q} = \left(\frac{\partial S(q''q'E)}{\partial q''} + \frac{\partial S(q''q'E)}{\partial q'} \right)_{q''=q'} = p''-p', \tag{78}$$

where p' and p'' are the initial and the final momentum. The integral over q''=q'=q becomes very small due to destructive interference unless p''=p', i.e. the trajectory closes itself smoothly. The summation in (55) can, therefore, be limited to periodic orbits if we are calculating $\tilde{g}(E)$.[20]

Chose a coordinate system where q_i varies along the periodic orbit, and (δq_2, δq_3) are the variations perpendicular to it. If $\bar{q}$ is a point on the periodic orbit we can write

$$S(q\,q\,E) = S(\bar{q}\,\bar{q}\,E) + \frac{1}{2}\left[\frac{\partial^2 S}{\partial q'\partial q'} + 2\frac{\partial^2 S}{\partial q'\partial q''} + \frac{\partial^2 S}{\partial q''\partial q''}\right]_{q'=q''=\bar{q}} \delta q \delta q, \tag{79}$$

where the variations in the quadratic term contain only the components δq_2 and δq_3. The linear term vanishes because $\bar{p}'=\bar{p}''$. The integration of a term in (55) over q decays into an integration along the period orbit, i.e. over q_1, and an integration perpendicular to it, i.e. over (δq_2,δq_3) in the quadratic approximation (79). The latter leads to a double Fresnel integral so that (55) becomes

$$-\frac{1}{h}\sum_{\text{periodic orbits}} \oint d\bar{q} \bullet \exp\,[-\frac{i}{h}S(E)\text{-phases}]\,\mathbf{A}\,(\bar{q})\exp\,(\pm\frac{i\pi}{4}\pm\frac{i\pi}{4}) \tag{80}$$

The square of the amplitude $\mathbf{A}$ ($\bar{q}$) is given by the quotient of determinants

$$|D_S|/\det\left(\frac{\partial^2 S}{\partial q'\partial q'} + 2\frac{\partial^2 S}{\partial q'\partial q''} + \frac{\partial^2 S}{\partial q''\partial q''}\right)_{2,3} \tag{81}$$

and the double signs in (80) refer to the signs of the eigenvalues belonging to the quadratic term in (79). Both the action integral $S(\bar{q}\,\bar{q}\,E)=S(E)$ and the phases are independent of $\bar{q}_1$ and can be taken outside the integral in (80).

The determinant D_S can be evaluated in our special coordinates. If we differentiate the equations (17) with respect to E, or the first one with respect to q_j'', and the second one with respect to q_j', we get relations which always contain $\partial H/\partial p'$ or $\partial H/\partial p''$. From the equations of motion (1) we know that $\partial H/\partial p=\dot{q}$, and our coordinate system was chosen such that $\dot{q}=(|\dot{q}|,0,0)$. Therefore, we find that

$$\frac{\partial^2 S}{\partial E\partial q'} = -\frac{1}{|\dot{q}|}, \quad \frac{\partial^2 S}{\partial E\partial q''} = \frac{1}{|\dot{q}|} \tag{82}$$

and that $\partial^2 S/\partial q_j''\,\partial q_1'$ as well as $\partial^2 S/\partial q_j'\,\partial q_1''$ vanish. The determinant D_S of (54) simplifies to

$$D_S = \frac{1}{|\dot{q}|^2}\det\left(\frac{\partial^2 S}{\partial q'\partial q''}\right)_{2,3} \tag{83}$$

A comparison of (81) and (83) with (25) shows that the amplitude $\mathbf{A}$ ($\bar{q}$) is given by $|\dot{q}|^{-1} \bullet [F(1)]^{-1/2}$. The integral $\oint d\bar{q}$ in (80) over the periodic can now be performed because F(1) is independent of $\bar{q}$. The meaning of $\oint dq/|\dot{q}|$ is the time T it takes the system to go around the periodic orbit, i.e. the period. Thus, we have finally

$$\tilde{g}(E) = \frac{1}{h} \sum_{p.o.} T \exp\left[\frac{i}{h} S(E) - \text{phases} \pm i\frac{\pi}{4} \pm i\frac{\pi}{4} \right]/[F(1)]^{1/2} \tag{84}$$

The set of periodic orbits contains all the multiple traversals of a primitive periodic orbit (p.p.o.). From the origin of the factor T it is obvious that it always refers to the p.p.o., while the other terms are associated with the p.o. itself and include the effect of multiple traversal.

The resonances in $\tilde{g}(E)$ do not come from any particular primitive periodic orbit and its multiple traversals, but they originate from the whole pattern of periodic orbits.[21] This can be seen from the example of bound states in a spherically symmetric potential. Only the exponential in (84) is important in order to establish the constructive interference.

The condition for periodicity results directly from (64) in the form $2\lambda\pi = 2\kappa\gamma$ with the two positive integers, κ and λ. The angle γ is given by (63) as a function of E and M. If the trajectory closes itself smoothly, γ is a rational multiple λ/κ of π. The angular momentum M varies in a well defined interval which depends on the given energy E. In a typical screened atomic potential one has $\gamma=\pi$ for $M=0$ and $\partial\gamma/\partial M>0$, so that there is a one-to-one relation between γ and M. Let us consider E and γ, rather than E and M, as the independent parameters. The action integral Θ in (66) satisfies the relation

$$\left(\frac{\partial\Theta}{\partial\gamma} \right)_E = M(E,\gamma) \tag{85}$$

because of (71). The exponent in (84), including the effect of conjugate points, becomes

$$i \left(\frac{2\kappa\Theta}{h} - \kappa\pi - \lambda\pi \right) \tag{86}$$

which can be considered as a homogeneous function of first degree in κ and λ through $\gamma=2\pi/\kappa$.

Constructive interference occurs when the values of (86) differ by multiples of $2\pi i$, $2n\pi i$ or $2\ell\pi i$, as κ or λ increase by 1. The partial derivatives of (86) are good approximations for the differnces when both κ and λ are very large. Thus, we get the conditions

$$\frac{\partial(86)}{\partial\kappa} = \frac{2}{h}(\theta - M\gamma) - \pi = \frac{2\omega}{h} - \pi = 2n\pi,$$

$$\frac{\partial(86)}{\partial\lambda} = \frac{2\pi}{h} M - \pi = 2\ell\pi, \tag{87}$$

which are identical with the well known quantization conditions (73) and (74).

14. The Quantization of Stable Orbits

A periodic orbit in 3 dimensions without symmetry yields four eigenvalues Λ for the mapping (22). These come in pairs of complex conjugate numbers and pairs of mutually reciprocal numbers. They can be written in terms of two real numbers, α and β, as $\exp(i\alpha+\beta)$, $\exp(i\alpha-\beta)$, $\exp(-i\alpha-\beta)$, and $\exp(-i\alpha+\beta)$. This is the most general situation even in systems with more than 3 degrees of freedom. A neighboring trajectory winds around the period orbit if $\alpha\neq 0$, and it drifts towards or away from the periodic orbit if $\beta\neq 0$. The two occur simultaneously, in general.

If the periodic orbit has a simple symmetry such as the reflexion on the ecliptic in the lunar problem, the eigenvalues Λ come in simple pairs, either $\exp(i\alpha)$ and $\exp(-i\alpha)$ or $\exp(\beta)$ and $\exp(-\beta)$. A system with only two degrees of freedom like the lunar problem in the ecliptic or the anisotropic Kepler problem in two dimensions has the same two alternatives. The first alternative, linear stability, is realized for the variation orbit (11) in the lunar problem, while the second alternative, linear instability, holds for all periodic orbits of the 2-dimensional anisotropic Kepler problem.

The value of F(1) which appears in the denominator of (84) can be expressed in terms of α and β. The linear map (22) is transformed into a normal form for this purpose such that its eigenvalues appear in the diagonal, and zeros above the diagonal. In a stable system with two degrees of freedom like the motion of the moon in the ecliptic, the various matrices in (22) are one-dimensional and the value of F(1) is $4(\sin\alpha/2)^2$. In a stable system with three degrees of freedom like the motion of the moon both in and out of ecliptic, there are two independent stability indices, α_1 and α_2. The value of F(1) becomes $16\,(\sin\alpha_1/2)^2(\sin\alpha_2/2)^2$.

The further discussion requires an examination of the phases (originating at conjugate points) and the double signs in (84) which come from the eigenvalues of the quadratic term in (79). The formulas in the preceding section are written for 3 degrees of freedom. For two degrees of freedom, the matrix in (79) and the determinants in (81) and (83) are 1 by 1. The exponent in (84) has the term $+i\pi/4 \pm i\pi/4$ where the double sign agrees with the quadratic term in (79). These facts can be derived from the analog of (55) in two dimensions (Cf. III).

The sign changes in (83) as one follows the periodic orbit determine the number of conjugate points. The initial value of the matrix elements in (83) can be obtained from the approximate expression $S(q''q'E) \cong |q''-q'|\,[2m(E-V(q))]^{1/2}$ with $q = 1/2(q'+q'')$, which is valid for the direct trajectory from q' to a nearby point q''. The matrix in (83) is diagonal and negative when the distance $|q''-q'|$ is small. Because of (25) and $F(1)>0$ for stable orbits, the quadratic terms in (79) and the sign of D_S in (83) are not independent. A rather careful discussion is necessary to establish the various possibilities for the exponent in (84) as a function of the number of conjugate points.

The result can be stated rather simply for two degrees of freedom. The expression (84) simplifies to

$$\tilde{g}(E) = -\frac{1}{2h}\sum_{p.o.} T\frac{(\pm 1)}{|\sin\alpha/2|} \exp[\frac{i}{h}S(E)]. \tag{88}$$

with the upper sign for 0 or 1, 4 or 5 conjugate points, and the lower sign for 2 or 3, 6 or 7 conjugate points, and so on.

In order to interpret this strange result, we go back to the expression (27) for the deviation along the variation orbit in the lunar motion. Both the periodic (variation) orbit (11) and (27) have a common factor ζ which describes the overall approximate rotation of the moon around the earth. In both of the Fourier expansions, (11) and (27), the dominant terms are the ones with $j=0$. Therefore, apart from the overall rotation around the earth, the moon's trajectory is dominated by ζ^{γ} and $\zeta^{-\gamma}$ in (27). These terms describe how the trajectory winds around the periodic orbit (11). In one period T (month) there will be $2\pi\gamma$ turns since the variable ζ goes around the unit circle in that time, and the stability angle is given by $\alpha=2\pi\gamma$ in agreement with the discussion at the end of section 6.

The winding number $2\pi\gamma$ also determines the number of conjugate points as can be seen from (27). If $2\pi\gamma<\pi$, there will be none; for $\pi<2\pi\gamma<2\pi$ there is 1 conjugate point; for $2\pi<2\pi\gamma<3\pi$ there are 2; etc. If α is interpreted as $2\pi\gamma$ in (88), we can write $(\pm 1)|\sin\alpha 2| = \sin\alpha/2$. Obviously, what is important in (88) is the combination of stability exponent and phase loss due to conjugate points. Therefore, α will be interpreted henceforth as the winding angle $2\pi\gamma$,[22] and we can write

$$\tilde{g}(E) = \frac{1}{2h}\sum_{p.o.} \frac{T}{\sin\alpha/2} \exp[\frac{i}{h}S(E)]. \tag{89}$$

The value of α in the lunar motion is found to be $>2\pi$ for both the motion in the ecliptic and out of it. It takes the moon less than an ordinary (synodic) month to go from perigee to perigee or from node to node. On the other hand, the motion from perigee to perigee takes more than a sidereal month $(2\pi/n)$, and the motion from node to node takes less (regression of the nodes).

The last formula for $\tilde{g}(E)$, and a similar one for more than 2 degrees of freedom, is strikingly simple. Yet, the $\sin\alpha/2$ in the denomiator is not easy to understand. In order to make some progress insert the expansion

$$\frac{1}{2i\sin\alpha/2} = \sum_{\ell=0}^{\infty} e^{-i(\ell+1/2)\alpha} \tag{90}$$

into (89) and break the summation overall periodic orbits (p.o.) into a summation over the primitive periodic orbits (p.p.o.) and their multiple traversals. Then, (89) becomes

$$\tilde{g}(E) = -\sum_{p.p.o.} \frac{iT}{h} \sum_{k=1}^{\infty} \sum_{\ell=0}^{\infty} \exp k[\frac{i}{h}S(E) - i(\ell+1/2)\alpha]. \tag{91}$$

This formula is still formally correct, but it will be interpreted in a somewhat restrictive manner.

For a given p.p.o. and a fixed integer ℓ, the sum over k becomes singular when the following condition is satisfied

$$S(E) = [2n\pi + (\ell + 1/2)\alpha]h, \tag{92}$$

where n is integer. The first caveat against this formula has to do with the idea of associating a set of levels with each p.p.o. The example of a spherically symmetric potential in the preceding section shows to the contrary that the constructive interference originates in the set of all p.o. when adjacent ones have their action integrals differing by multiples of $2\pi h$. Nevertheless, the condition (92) looks very convincing at least for $\ell=0$, because the addition of $\alpha/2$ to $2n\pi$ brings in the phase losses at conjugate points in exactly the right amount. The question then arises how to explain the terms with $\ell>0$, i.e. the quantization condition with $\ell>0$.

Miller[22] introduced the idea of the stability frequency ω for this purpose. It corresponds exactly to our interpretation of the terms ζ^{γ} and $\zeta^{-\gamma}$ in (27), i.e. a rotatory or oscillatory motion around the periodic orbit with circular frequency $\omega=\alpha/T$. Since the essential approximation in going from quantal to classical mechanics consisted in treating all fluctuations around the classical trajectory only to second order, it is natural to quantize the transverse degrees of freedom around the periodic orbit like harmonic oscillators. If we recall that the period T is given by the derivative dS/dE, then we can write formula (92) in the form

$$S(E) - \sum_{j=1}^{N-1} (\ell_j+1/2)h\omega_j \frac{dS}{dE} + \ldots = 2n\pi h, \tag{93}$$

where we have assumed N degrees of freedom and N-1 stability frequencies $\alpha_1,\ldots,\alpha_{N-1}$ around the periodic orbit. The left-hand side looks like the Taylor expansion of S(E) when the insert E-ε for E and identify ε with the sum of oscillator energies $(\ell_j+1/2)h\omega_j$. The deviation from integer quantum numbers now appears to be due to the energy which goes into the transverse degrees of freedom.

This interpretation is very satisfying, indeed, although a second warning is called for. The higher levels of excitation $\ell>0$, are useful only when the transverse modes of oscillation are soft, i.e. $\alpha=\omega T$ does not compete with $2n\pi$ in (92). In the lunar problem α turns out to be of order 2π, and one would not call this stability soft. In such a case it seems more appropriate to study the set of periodic orbits more closely and try to find the ones which wind around each other not only in the linear approximation about a particular (simple) one. In this way we are thrown back to the situation in the sphercally symmetric potential where all periodic orbits act in concert to give resonances.

The quantization of stable orbits has been discussed recently by a number of French authors [23]. While Balian and Bloch take a rather intuitive approach, the work of Voros is close to the more abstract ideas of Duistermaat, Guillemin and Hormander which is also the inspiration for the applications to differential geometry by Chazarain, Colin de Verdiere and Weinstein.

Balian and Bloch study the density of modes $\rho(k)$ in a cavity as a function of the wave number k in the eigenvalue equation $(\Delta+k^2)\psi=0$ with the boundary conditions $\psi=0$. They argue on the basis of the formula (77) although they use a more restrictive form for Green's function than (55). Their hypothesis is equivalent to the existence of invariant tori in phase space because they assume the existence of a unique phase function $S(q''q'E)$ with $k^2=2mE/h^2$. Each closed classical path in the cavity is found to contribute an oscillatory term $\sin(kL/h)$ to $\rho(k)$ where L is the length of the path. The first few oscillating terms already lead to rather sharp peaks near the correct resonances.

These results are obviously very close to the ones which were discussed earlier in this section, and it is very satisfying to know that they are corroborated by some recent work in geometry. Both Chazarain and Colin de Verdiere study the eigenvalues λ_j with $j=0,1,\ldots$ and $0=\lambda_0\leq\lambda_1\leq\ldots$ of the Laplace operator on a compact Riemannian manifold. Chazarain investigates the function $f(t)=\sum\exp(i\surd\lambda_j t)$ which arises naturally in the propagation of waves, while Colin de Verdiere examines $h(z)=\sum\exp(-\lambda_j/z)$ which comes up in the diffusion of heat. The function f(t) is singular in a set of points {T} where T is the length of a closed geodesic line of the Riemannian manifold. The function h(z) has various oscillatory components when z becomes complex, $z=x+iy$, and $y>0$ goes to infinity, and each of them has the form $\exp(-zT^2/4)$.

Results of this type have been refined on the basis of Hormander's theory of Fourier integral operators by Duistermaat, Guillemin and Weinstein. They have established a direct connection between integrals of the type (7) over closed geodesics, and the asymptotic behavior of the eigenvalues λ_j.

15. Approximate Energy Eigenvalues for Ergodic Systems

This last section is devoted to the discussion of ergodic systems which was the original purpose of this whole review. Unfortunately, the results are far from complete, and only tentative conclusions can be stated. After a suggestion of Percival[24] there are two types of spectra, regular and irregular, which are associated essentially with non-ergodic and ergodic behavior of the classical trajectories in the corresponding parts of phase space. Such a distinction is quite speculative because our understanding of ergodic systems, let alone their quantization, is so poor at present. Some direct numerical checks were carried out by Pomphrey[25] who seemed to confirm the general idea in the case of two harmonic oscillators which are coupled by third order terms. The energy levels in the irregular spectrum seem randomly distributed, and their values are extremely sensitive to the strength of the coupling parameters. The last feature serves to distinguish the regular (stable) from the irregular (unstable) energy levels, and allows us to compare the "weight" of the irregular spectrum below some energy E with the total volume of ergodic trajectories in phase space below E.

According to Pomphrey the two quantities as a function of E are proportional to each other.

The only way to obtain approximate energy levels which is applicable to ergodic systems, is based on the formula (84) or on a scheme which was recently proposed by Miller.[26] The latter uses a more general type of generating function to perform the canonical transformation from ordinary momentum and position coordinates to occupation number and phase angle coordinates. This has not been tried out, and leads to calculations of a quite different nature than finding classical trajectories and, in particular, periodic orbits. Therefore, it does not fit into the framework of this conference.

It should be pointed out at once, that the summation over all periodic orbits of an ergodic system has only been used in a very limited manner as will be shown shortly. Contrary to the common belief, an ergodic system is full of periodic trajectories, just as the system of real numbers is full of rational numbers. They form a dense set of measure zero.[28] If the behavior of the whole system is smooth enough, then, according to an idea of Poincare, the periodic orbits ought to give us all the information we need. Poincare never stated what he hoped to get out of the study of periodic orbits, but formula (84) is a dramatic confirmation of his guess.

Let us go back to section 8 where the homeomorphism between classical trajectories and binary Bernoulli sequences with the fine topology was discussed for the anisotropic Kepler problem. There is exactly one periodic orbit for each periodic binary sequence of even length with the exception of the trivial sequences ...++++... and ...−−−−... The periodic binary sequences are easy to enumerate; there is a total of 2^N such sequences of length N. A cyclic permutation among the N binary digits of a particular sequence does not give rise to a new periodic orbit, but simply shifts the starting point from one intersection with x-axis to the next one. A reversal of the ordering yields a periodic orbit where the particle moves in the opposite sense. Such an orbit counts separately in the summation (84). There may be other symmetries in the periodic binary sequence which are directly reflected in the shape of the corresponding periodic orbit.

A periodic orbit in the anisotropic Kepler problem is necessarily unstable. A neighboring trajectory stays close as long as its binary sequence repeats the same pattern of digits, but it can drift arbitrarily far as soon as this pattern becomes different. It seems reasonable to assume that every periodic orbit in an ergodic system is linearly unstable. Therefore, we have to evaluate (84) for an unstable periodic orbit, and we shall discuss the case of two degrees of freedom. With the eigenvalues $\exp(\beta)$ and $\exp(-\beta)$ for the transformation (22), the determinant (24) for $\Lambda = 1$ becomes

$$F(1) = -4 \sinh^2(\beta/2) \tag{94}$$

The same discussion as in the preceding section is necessary to put together all the relations among the signs in (79) and (83) as well as the number of conjugate points. The resulting expression is

$$\tilde{g}(E) = -\frac{i}{2h} \sum_{p.o.} T \frac{(\pm 1)}{\sinh\beta/2} \exp[\frac{i}{h} S(E)] \tag{95}$$

with the upper sign for 0, 3 or 4, 7 or 8 conjugate points, and the lower sign for 1 or 2, 5 or 6 conjugate points, and so on.

The sign of β was assumed to be positive so that the denominator in (95) is always positive. The double sign in the numerator can again be incorporated into the exponent, but it does not acquire the usual form $-im\pi/2$ where m is the number of conjugate point. Rather we could write the formally correct but not very useful expression $-i\pi[1/2(m+1)]$ where [ ] indicates the integer nearest and below $1/2(m+1)$. The structure of conjugate points in an unstable periodic orbit is simpler than in a stable one. As one makes several passes around a primitive periodic orbit, the number of conjugate points perpass always remains an integer if the orbit is unstable, while that average number can have any real value for a stable orbit because the winding number α or the exponent γ in (27) is essentially arbitrary.

The situation in the anisotropic Kepler problem is particularly simple. There is exactly 1 conjugate point for each traversal of the x-axis, just like in the ordinary plane Kepler problem. There is, for the time being, only overwhelming numerical evidence for this proposition. The general proof is directly connected with the proof for the one-to-one relation between binary Bernoulli sequences and classical trajectories. The existence of at least one classical trajectory for each binary Bernoulli sequence can be demonstrated analytically, because a certain function (of a finite number of variables) can be shown to have at least one maximum in the interior of its domain. The uniqueness together with the proposition about the number of conjugate points could be proven, if there were no other extremum in the domain.

The qualitative consequences of (95) for the structure of the approximate density of states can be investigated by the same procedure of Miller as was discussed at the end of the preceding section. The expansion corresponding to (90) is now given by

$$\frac{1}{2\sinh\beta/2} = \sum_{\ell=0}^{\infty} \exp -(\ell + 1/2)\beta \tag{96}$$

and, instead of (91), we now have

$$\tilde{g}(E) = -\sum_{p.p.o.} \frac{iT}{h} \sum_{k=1}^{\infty} \sum_{\ell=0}^{\infty} \exp k\,[\frac{i}{h} S(E) - im\frac{\pi}{2} -(\ell+1/2)\beta], \tag{97}$$

where the contribution of the m conjugate points to the exponent is written in the slightly incorrect form $-im\pi/2$.

Complete constructive interference does not occur in (7) for any one primitive periodic orbit because of the real part $-(\ell+1/2)\beta$ in the exponent. The condition for resonance (93) now takes the complex form

$$S(E) + i\sum_{j=1}^{N-1}(\ell_j + 1/2)h\omega_j \frac{dS}{dE} = (n + m/4)2\pi h, \tag{98}$$

where we have introduced N-1 transverse degrees of freedom, all of them unstable, with unstability frequencies $\omega_j=\beta_j/T$. The resonance energy takes the complex form $E+i\varepsilon$ where the imaginary part ε comes from the unstable fluctuations. Again, this interpretation is very satisfactory, though frought with pitfalls. The density of states in the classical approximation consists of a series of broadened peaks, each centered around the usual condition $S(E)=(n+m/4)2\pi h$. The width of these peaks is given by $h\omega$ which is to be compared with the distance $\Delta E=2\pi h/T$ between them (recall that $T=dS/dE$). The sharpness of these classical resonances depends, therefore, on the value of $\beta/2\pi$. When $\beta/2\pi<<1$, one has sharp distinct peaks in the density of states. When $\beta/2\pi\gtrsim 1$, these peaks vanish into an averaged smooth density of states. The obvious warnings against such a simplistic view will not be recounted here.

The simplest periodic orbit in the two-dimensional anisotropic Kepler problem has the binary sequence ...+-+-... It is a simple loop around the origin in both position and momentum space. Its stability exponent for the mass ratio 5 is 1.16, and satisfies the criterion for peaks in the density of states. The stability exponent for other periodic orbits is roughly proportional to their binary length. Therefore, the longer ones will not satisfy the above criterion for a mass ratio 5, and it is not clear how this will affect the overall behavior of the classical density of bound states.

REFERENCES

1. A. Einstein, Verhandlungen der Deutschen Physikalischen Gesellschaft **19**, 82-92 (1917).

2. Max Born, The Mechanics of the Atom, republished by Frederick Ungar Publishing Co., New York, 1960. Neither Born nor Sommerfeld in Atombau und Spektrallinien (first published 1919) mention Einstein's article. It is hard to understand the extent to which this important contribution of Einstein to early quantum mechanics was ignored by his contemporaries. W. Pauli wrote two long review articles on the old quantum mechanics, "Quantentheorie" in Handbuch der Physik, H. Geiger and K. Scheel, eds., Vol. 23, Springer-Verlag 1926, pp. 1-278, and "Allgemeine Grundlagen der Quantentheorie des Atombaues", Chap. 29 in Muller-Pouillet's Lehrbuch, Vol. 2, part 2, F. Vieweg, Braunschweig, 1929, p.p. 1709-1842. He also wrote the article "Einstein's Beitrag zur Quantentheorie" for "Allbert Einstein als Philosoph und Naturforscher", P. A. Schilpp, ed., W. Kohlhammer Verlag, Stuttgart, 1955, pp. 74-83, which contains a list of Einstein's contributions to quantum theory. Yet, the article of reference 1 is mentioned nowhere.

3. There is a wide variety of articles in this area which started with the paper by J. Hadamard, J. Math. Pure Appl. **4**, 27-87 (1898). Further classic papers are by E. Artin, Abhandl. Math. Sem. Hamburg **3**, 170-175 (1924); G. A. Hedlund, Bull. Am. Math. Soc. **47**, 241-260 (1939); D. V. Anosov, Geodesic Flows on closed Riemamian manifolds with negative curvature, Proc. Steklov Inst. Math. **97**, translated by Am. Math. Soc., Providence, 1969; E. Hopf, Bull. Am. Math. Soc. **77**, 863-877 (1971).

4. The restricted three-body problem is the main topic in a monograph by V. Szebehely, Theory of Orbits, Academic Press, 1967. A lot of numerical exploration has been made since in order to establish various families of trajectories, e.g. a series of investigations by M. Henon, Ann. Astro. **28**, 499 and 992 (1965); Bull. Astro., Paris, **1**, fasc. **1**, p. 57 and fasc. **2**, p. 49 (1966); Astron. and Astrophys. **1**, 223 (1969) and **9**, 24 (1970).

5. The anisotropic Kepler problem has been investigated numerically in quantum mechanics by W. Kohn and J. M. Luttinger, Phys. Rev. **96**, 1488 (1954); R. A. Faulkner, Phys. Rev. **184**, 713 (1969).

6. The motion of the moon is the topic of E. W. Brown, An Introductory Treatise on Lunar Theory, Cambridge University Press 1896, republished by Dover Publications 1960; a less detailed account of lunar theory aud a very readable introduction to celestial mechanics is D. Brouwer and G. M. Clemence, Methods of Celestial Mechanics, Academic Press, 1961.

7. These general properties are derived in many textbooks on classical mechanics, e.g. L. A. Pars, A Treatise on Analytical Dynamics, Heinemann, London,

1965; C. L. Siegel and J. K. Moser, Lectures on Celestial Mechanics, Springer Verlay, 1971.

8. This important property was discovered by the astronomer G. W. Hill, Am. J. Math. **1**, 5-26, 129-147, 245-260 (1878); it was rediscovered by the mathematician Floquet, Ann. de l'Ecole norm. sup (2), XII, 43 (1883); and it entered physics through the work of F. Bloch, Zeits. f. Phys. **52**, 555 (1928).

9. A good survey of the mathematical accomplishments in this area is given by Jurgen Moser, Stable and Random Motions in Dynamical Systems, Princeton University Press, 1973.

10. M. C. Gutzwiller, J. Math. Phys. **14**, 139-152 (1973) contains the numerical investigations; while M. C. Gutzwiller, J. Math. Phys. **18**, 806-823 (1977) gives the analytical arguments.

11. A particularly instructive mechanical example is worked out by M. Henon and C. Heiles, Astron. J. **69**, 73 (1964). In contrast, a sequence of mathematical mappings to imitate the properties of a Poincare map is studied by M. Henon, Q. Appl. Math. **27**, 291 (1969).

12. Bernoulli sequences are discussed in ref. 9. Another very readable account from a mathematician viewpoint can be found in V. I. Arnold and A. Avez, Problemes ergodignes de la mecanique classique, Gauthier-Villars, 1967; translated into English at Benjamin, New York, 1968.

13. C. Garrod, Rev. Mod. Phys. **38**, 483 (1966).

14. Cf. S. F. Edwards and Y. V. Gulayev, Proc. Roy. Soc. (London) **279**, 229 (1964); A. M. Arthurs, Proc. Roy. Soc. (London) **318**, 523 (1970); also paper I in ref. 15.

15. The developments in the remainder are mostly taken from a series of papers to be designated as I, II, III, IV by the author, M. C. Gutzwiller, J. Math. Phys. **8**, 1979 (1967); **10**, 1004 (1969); **11**, 1791 (1971); **12**, 343 (1971).

16. M. Morse, The Calculus of Variations in the Large, Am. Math. Soc., Providence, Rhode Island, 1935. A more easily accessible account is given by N. Seifert and W. Threlfall, Variationsrechnung im Grossen (Theorie von Marston Morse) B. G. Teubner, 1928; republished by Chelsea, New York, 1951; J. Milnor, Morse Theory, Princeton University Press, 1962.

17. The simplicity of Kepler's problem in momentum space has been widely recognized in quantum mechanics since the treatment of V. Fock, Z. Phys. **98**, 145 (1935). It rests on the connection with the group of rotations 0(4) which was first used by W. Pauli, Z. Phys. **36**, 336 (1926).

18. This result was first derived by the author in I, but independently, by A.

Norcliffe and I. C. Percival, J. Phys. B. Ser. 2, **1**, 774, 784 (1968).

19. The classical quantization of non-separable mechanical systems with invariant tori has become a very active area of research, and only some early representative references can be given. R. A. Marcus, Faraday Discussions of the Chemical Society, **55**, 34 (1973); J. Chem. Phys. **61**, 4301 (1974) and **62**, 2119 (1975); M. V. Berry and M. Tabor, Proc. Roy. Soc. London A **349**, 101-123 (1976); K. S. Sorbie, Molecular Physics, **32**, 1577 and 1327 (1976); K. S. J. Nordholm and S. A. Rice, J. Chem. Phys. **61**, 203 and 768 (1974); **62**, 157 (1975).

20. This argument has been put in doubt by K. F. Freed, Faraday Discussions of the Chemical Society **55**, 68 (1973). He reasons that q' and q'' need only be inside the volume d^3q of integration which is assumed to be of linear extent δq, and the momenta may differ by an amount δp. The variation of S(q''q'E) is, therefore, of the order $\delta p\ \delta q$, and destructive interference does not occur as long as $\delta p \cdot \delta q < h$. Trajectories might be important when they close approximately, even if they cannot do it exactly. Our skill in treating mechanical systems, ergodic or not, does not seem sufficient at present to resolve this issue. For a start we shall assume the extreme position of excluding all except perfectly closed trajectories.

21. The authors' results have been misconstrued and misrepresented almost consistently in this respect. Cf. D. W. Noid and R. A. Marcus, J. Chem. Phys. **62**, 2119 (1975); M. V. Berry and M. Tabor, Proc. Roy. Soc. Lond. A. **349**, 101 (1976); and others in ref. 19 and 22. On the contrary, the first applications of (84) in III show that all periodic orbits are needed to give rise to resonance, not only a single primitive one. In IV, however, an example was investigated where only one periodic orbit was known at the time, and it was used to construct approximate eigenvalues for the lack of more information about the system, the anisotropic Kepler motion in two dimensions. With the many more periodic orbits which have been discovered since (cf. ref. 10), the approximate treatment of IV will have to be reexamined.

22. The idea that the stability angle α includes the effect of the conjugate points was clearly stated in IV. Yet, Miller, J. Chem. Phys. **63**, 996 (1975) states the contrary. Even more unfortunate is his assertion in J. Chem. Phys. **64**, 502 (1976) that the author's method based on (89) is just "an approximate version of Keller's formalism". The reference is to J. B. Keller, Ann. Phys. (N.Y.) **4**, 180 (1958); J. B. Keller and S. I. Rubinow, Ann. Phys. (N.Y.) **9**, 24 (1960) whose work is based entirely on the construction of wave functions for a classical mechanical system with invariant tori in phase space. Ergodic systems are not considered there.

23. R. Balian and C. Bloch, Am. Phys. (N.Y.) **60**, 401 (1970); **63**, 592 (1971); **64**, 271 (1971); **69**, 76 (1972); A. Voros, Am. Inst. H. Poincare **24**, 31 (1976); L. Hormander, Acta Math. **127**, 79 (1971); J. J. Duistermaat, Comm. Pure Appl. Math. **27**, 207 (1974); J. Chazarain, Inventiones Math. **24**, 65 (1974); Y.

Colin de Verdiere, Compositio math. **27**, 83 and 159 (1973); J. J. Duistermaat and V. W. Guillemin, Inventiones Math. **29**, 39 (1975); A. Weinstein, in Fourier Integral Operators, Lecture Notes in Mathematics, Springer, **459**, 341 (1975).

24. I. C. Percival, J. Phys. B **6** L, 229 (1973).

25. N. Pomphrey, J. Phys. B **7**, 1909 (1974).

26. W. H. Miller, J. Chem. Phys. **64**, 2880 (1976).

27. Cf. ref. 22. Also W. H. Miller, J. Chem. Phys. **56**, 38 (1972) and, in particular, J. Chem. Phys. **62**, 1899 (1975) are of interest in this connection. Path integral arguments are used in order to compute transition rates in quantum statistics. Periodic orbits now occur in the inverted potential and the stability exponents are indeed the frequencies for the transverse degrees of freedom. The expression (6) gives their occupation probability.

28. This fact has been known for some time, in particular since the geodesics on a surface of negative curvature can be constructed quite explicitly, cf. ref. 3. A detailed numerical study in a system where no such complete description is available, was started by G. H. Lunsford and J. Ford, J. Math. Phys. **13**, 700 (1972). They try to give a mathematical formulation (based on their empirical data) to the way in which periodic orbits occur in an ergodic system.

EXPLICIT FUNCTIONAL INTEGRATION METHOD FOR DETERMINING APPROXIMATE STATIONARY STATES IN QUANTUM FIELD THEORIES

Gerald Rosen

Department of Physics, Drexel University

Philadelphia, Pennsylvania 19104, USA

ABSTRACT

The Rayleigh-Ritz procedure for functionalities, an explicit functional integration method for determining approximate stationary states in quantum field theories, is described and illustrated here for certain models with simply chosen expressions for the trial state functionals. Approximate vacuum, one-particle and two-particle (noninteracting, bound or scattering) stationary states are derived for generic self-interacting real scalar theories without renormalization, for a nonrelativistic real scalar-Weyl spinor model with renormalization, and for a self-interacting complex scalar model with renormalization.

INTRODUCTION

The problem of calculating approximate stationary states in quantum field theory has evoked renewed interest, in view of the current importance of "extended objects" or quantized "particlelike" solutions in local essentially nonlinear field theories [1] for hadron physics. In atomic, molecular and nuclear quantum mechanics, stationary bound states can be determined to desired accuracy by means of Rayleigh-Ritz variational calculations based on the energy functional [2]. Ten years ago I worked to develop a field-theoretic analogue of the Rayleigh-Ritz procedure for approximating stationary bound states in quantum mechanics. My purpose here is to describe and illustrate this Rayleigh-Ritz functional integration method [3] for determining approximate stationary states in quantum field theories.

Let the Dirac ket $|.\rangle$ denote an abstract stationary state in the quantum field theory, so that $\underline{H}|.\rangle = E|.\rangle$ with $\underline{H}$ the abstract Hamiltonian and E the energy of the state. In the case of a boson field $\phi = \phi(x)$, one can evoke the coordinate-diagonal representation [4] and seek the Dirac representative state functional $\Psi[\phi] \equiv \langle\phi|.\rangle$. The abstract Hamiltonian has the coordinate-diagonal matrix elements $\langle\phi|\underline{H}|.\rangle = H\Psi[\phi]$, where H is the functional differential Hamiltonian operator in ϕ and $\delta/\delta\phi$. Generalizing the notion of a functional, a functionality is a rule that assigns a (real or complex) number to a (real or complex) functional. The energy functionality $E = E\{\Psi\} \equiv \langle.|\underline{H}|.\rangle/\langle.|.\rangle = \int\Psi^* H\Psi D(\phi)/\int|\Psi|^2 D(\phi)$ assigns the real-valued total field energy E to the complex-valued state functional, where $D(\phi) = D(\phi+\omega)$ is a displacement-invariant measure for the functional integrations over all fields $\phi = \phi(x)$. Both Ψ and $D(\phi)$, defined to within normalization factors independent of ϕ, are such that the numerator and denominator in $E\{\Psi\}$ exist as finite quantities. Generalizing the notion of functional differentiation, a functionality $F = F\{\Phi\}$ is differentiable at the functional Φ if

$$\left(\frac{\partial}{\partial z} F\{\Phi+z\Omega\}\right)_{z=0} \equiv \int (\mathcal{D}F/\mathcal{D}\Phi)\Omega D(\varphi)$$

exists as a linear functionality in Ω for a certain suitably dense class of functionals $\Omega = \Omega[\phi]$. With the energy functionality differentiable at the physical state functional Ψ, we have $D E/D\Psi = 0$ as a variational principle for the Schrödinger functional differential equation $H\Psi = E\Psi$. It follows that if $\Psi = \Psi[\phi;f]$ is prescribed in functional form but with a function f here as a "parameter", then the energy functionality can be evaluated as a functional $E = E[f]$, and f can be determined by solving the relatively simple Rayleigh-Ritz partial differential equation $\delta E[f]/\delta f = 0$, thus giving an approximate form for a physically admissible Ψ and a very good value for the associated E. The following "functional integration by parts" lemma is needed (and in fact usually sufficient) for the actual evaluation of E[f]: If $\delta\Omega[\phi]/\delta\phi(x)$ and the functional integral $\int\Omega[\phi]D(\phi)$ both exist, then $\int\{\delta\Omega[\phi]/\delta\phi(x)\}D(\phi) = 0$. The proof is based on the displacement invariance of the integration measure [4].

Hence, since $DE \equiv E\{\Psi+D\Psi\} - E\{\Psi\}$ vanishes for all variations $D\Psi$ if and only if Ψ is a stationary state functional, approximate $\Psi[\phi]$ are derivable by an immediate extension of the classical Rayleigh-Ritz procedure : (1) Assume a parameter-function dependent trial form for the state functional; (2) Evaluate the associated energy functionality by explicit functional integration; (3) Calculate the functional derivatives of the latter expression with respect to the parameter-functions and set these functional derivatives equal to zero; (4) Solve the resulting partial differential equations to determine the best form for the parameter-functions. This Rayleigh-Ritz procedure can be applied to determine approximate stationary states in quantum field theories of contemporary practical importance, and the method is illustrated here for model theories with simply chosen expressions for the trial state functionals. Because the trial functionals that appear in the following sections are obviously structured for ease of calculation rather than for substantial physical realism, the analysis and results herein are merely intended to serve as a primer for calculations of a more complicated nature.

1. SELF-INTERACTING REAL SCALAR THEORIES WITHOUT RENORMALIZATION

Let us first consider self-interacting real scalar local field theories based on Lorentz-invariant Lagrangians of the form

$$\mathcal{L} = -\frac{1}{2}\partial^{\mu}\phi\partial_{\mu}\phi - u(\phi) \tag{1.1}$$

where the generic self-interaction energy density

$$u(\phi) = \sum_{n=1}^{\infty} \alpha_{2n}\frac{\phi^{2n}}{(2n)!} \quad (\geqslant 0) \tag{1.2}$$

is a prescribed even entire analytic non-negative function of ϕ. The Hamiltonian operator

$$H = \int [-\frac{1}{2}\frac{\delta^2}{\delta\phi(\vec{x})^2} + \frac{1}{2}|\vec{\nabla}\phi(\vec{x})|^2 + u(\phi(\vec{x}))]d^3x \tag{1.3}$$

is associated with such theories. Our object is to determine physical suitability conditions for the self-interaction energy density (1.2) without renormalization of any of the constant α's, conditions on $u(\phi)$ which admit vacuum, one-quantum, and quasi-classical eigenstates for the Hamiltonian operator (1.3), and to derive the approximate stationary state functionals for the physically meaningful class of theories. A functional differential operator representation [4] of momentum density appears in (1.3), the ϕ field being diagonalized for all values of x at a fixed instant of time. We employ the Rayleigh-Ritz procedure for functionalities [3] as our principal analytical tool. Only so-called Class A theories, for which the self-interaction energy density divided by ϕ^2 is uniformly bounded for all values of the field, are shown to admit wholly acceptable one-quantum and quasiclassical stationary states without renormalization.

Vacuum state functionals for theories with a Hamiltonian operator (1.3) are assumed to take the approximate form

$$\langle\phi|vac\rangle = \exp[-\frac{1}{2}\iint\phi(\vec{x})f(\vec{x},\vec{y})\phi(\vec{y})d^3x d^3y] \tag{1.4}$$

where $f(\vec{x},\vec{y}) = f(\vec{y},\vec{x})$ is a real symmetric distribution to be determined by the Rayleigh-Ritz procedure. The energy functionality [3]

$$E_{vac} = E_{vac}[f] \equiv \langle vac|\underline{H}|vac\rangle/\langle vac|vac\rangle \tag{1.5}$$

is evaluated by explicit functional integration with a displacement-invariant measure or by evoking the functional integration by parts lemma [4]. We obtain¶

$$E_{vac} = \int [\frac{1}{4} f(\vec{x},\vec{x}) + \frac{1}{4} \vec{\nabla}_x . \vec{\nabla}_y \; g(\vec{x},\vec{y})|_{y=x} + U(g(\vec{x},\vec{x}))]d^3x \qquad (1.6)$$

in which $g(\vec{x},\vec{y}) = g(\vec{y},\vec{x})$ is the real symmetric distribution inverse to $f(\vec{x},\vec{y})$,

$$\int f(\vec{x},\vec{z})g(\vec{z},\vec{y})d^3z = \delta(\vec{x}-\vec{y}) \;, \qquad (1.7)$$

and the linear transform of (1.2) :

$$U(g) \equiv \sum_{n=1}^{\infty} \frac{\alpha_{2n}}{n!} (\frac{g}{4})^n = \pi^{-1/2} g^{-1/2} \int_{-\infty}^{\infty} e^{-\phi^2/g} u(\phi)d\phi, \qquad (1.8)$$

exists for all non-negative g less than a certain positive constant g_{cr} (which may be infinite) depending on the form of $u(\phi)$. The Rayleigh-Ritz equation $\delta E_{vac}/\delta f(\vec{x},\vec{y}) = 0$ is thus :

$$\int g(\vec{x},\vec{z}) \; [-\vec{\nabla}_z^2 + V(g(\vec{z},\vec{z}))]g(\vec{z},\vec{y})d^3z = \delta(\vec{x}-\vec{y}) \qquad (1.9)$$

¶ With the displacement-invariant measure $D(\phi) = D(\phi+\omega)$ normalized to give

$$<vac|vac> \equiv \int \{\exp\, [-\int\int \phi(\vec{x})f(\vec{x},\vec{y})\phi(\vec{y})d^3x\; d^3y]\} \; D(\phi) = 1$$

we have [3]

$$<vac|\phi(\vec{x})\phi(\vec{y})|vac> \equiv \int \phi(\vec{x})\phi(\vec{y})\{\exp[-\int\int \phi(\vec{x})f(\vec{x},\vec{y})\phi(\vec{y})d^3x\; d^3y]\}\; D(\phi) = \frac{1}{2}\, g(\vec{x},\vec{y})$$

and

$$<vac|\phi(\vec{x})^{2n}|vac> = \frac{(2n)!}{n!} (\frac{g(\vec{x},\vec{x})}{4})^n \;.$$

in which

$$V(g) \equiv 4\,U'(g) = \sum_{n=1}^{\infty} \frac{\alpha_{2n}}{(n-1)!} \left(\frac{g}{4}\right)^{n-1}$$

$$= 2\,\pi^{-1/2}\, g^{-3/2} \int_{-\infty}^{\infty} (2\,g^{-1}\phi^2 - 1)\, e^{-\phi^2/g}\, u(\phi)\, d\phi \qquad (1.10)$$

is usually a positive function on the domain of definition for (1.8), $0 \leqslant g < g_{cr}$.† We have $g(\vec{x},\vec{y})$ depending only on $|\vec{x}-\vec{y}|$ for the vacuum state, assumed to be spatially homogeneous and isotropic, and so the unique vacuum solution to (1.9) follows by evoking the convolution theorem,

---

† It is easy to verify the more direct relationship $V(g(\vec{x},\vec{x})) = \langle vac|u''(\phi(\vec{x}))|vac\rangle / \langle vac|vac\rangle$ and to show that certain even entire analytic non-negative $u(\phi)$ give rise to $V(g)$ that are not positive for all g on the domain of definition for the quantity (1.8). This is exemplified by the Class A theory $u(\phi) = a^2\phi^2 e^{-\phi^2}$ with $V(g) = a^2(2-g)(1+g)^{-5/2}$ negative for $2 < g < \infty$. Such theories are at best of marginal physical interest, with $u(\phi)$ decreasing as ϕ^2 increases over a significant range of values, $[\phi u'(\phi)]$ being negative for values of ϕ to the extent that a bounding relation of the form

$$[\phi u'(\phi)] \geqslant \int_0^{\infty} \phi(\sin\lambda\phi)\mu(\lambda)d\lambda$$

cannot hold for all values of ϕ with $\mu(\lambda)$ ($\geqslant 0$) a certain non-negative real distribution. Our classification scheme is complete if we require $[\phi u'(\phi)]$ to be bounded from below in this fashion, for then

$$V(g) = 4\pi^{-1/2} g^{-3/2} \int_0^{\infty} e^{-\phi^2/g} \phi u'(\phi) d\phi$$

$$\geqslant \int_0^{\infty} e^{-g\lambda^2/4} \lambda\mu(\lambda)d\lambda$$

is positive for all g on the domain of definition for the quantity (1.8).

$$g(\vec{x},\vec{y}) = \lim_{K\to+\infty} g_K(\vec{x},\vec{y}) \tag{1.11}$$

$$g_K(\vec{x},\vec{y}) \equiv \frac{1}{(2\pi)^3} \int_{|\vec{k}|\leq K} \frac{e^{i\vec{k}.(\vec{x}-\vec{y})}}{(|\vec{k}|^2 + c^2)^{1/2}} d^3k \; ; \tag{1.12}$$

where

$$V(g_K(\vec{z},\vec{z})) \equiv c^2 \quad [c \geqslant 0] \tag{1.13}$$

is a constant that depends on the positive cutoff parameter K. Evaluating (1.12) with $\vec{x} = \vec{y}$, we find

$$g_K(\vec{z},\vec{z}) = \frac{1}{4\pi^2} [K(K^2+c^2)^{1/2} - c^2 \sinh^{-1}(K/c)]$$

$$\cong \frac{K^2}{4\pi^2} \quad \text{for } \lim_{K\to+\infty} (c/K) = 0$$

$$\cong \frac{K^2}{4\pi^2} [(1 + \gamma^2)^{1/2} - \gamma^2 \sinh^{-1} \gamma^{-1}] \quad \text{for } \lim_{K\to+\infty} (c/K) = \gamma, \text{ a finite constant}$$

$$\cong \frac{K^2}{6\pi^2 c} \quad \text{for } \lim_{K\to+\infty} (c/K) = \infty, \tag{1.14}$$

and hence if (1.13) and (1.14) can be satisfied in a self-consistent fashion, then by virtue of (1.7), (1.11) and (1.12) the distribution in (1.4) is given by

$$f(\vec{x},\vec{y}) = \lim_{K\to+\infty} \frac{1}{(2\pi)^3} \int_{|\vec{k}|\leq K} (|\vec{k}|^2+c^2)^{1/2} e^{i\vec{k}.(\vec{x}-\vec{y})} d^3k \tag{1.15}$$

It follows immediately from (1.13) and (1.14) that the theories fall into four qualitatively distinct classes:

For Class A theories, $u(\phi)/\phi^2$ is uniformly bounded for all ϕ; then $V(g)$ is regular for all positive g with $\lim_{g\to+\infty} V(g) \equiv m^2$, a finite non-negative constant, and c is asymptotic to m for large K, the quantity $g(\vec{z},\vec{z})$ is infinite according to (1.11) and the first case in (1.14).

For Class B theories, $u(\phi)/\phi^4$ is uniformly bounded as $\phi^2 \to \infty$; then $V(g)$ is regular for all finite positive g with $\lim_{g\to\infty}[V(g)/g] = 4\pi^2\gamma^2 \, [(1+\gamma^2)^{1/2}-\gamma^2 \sinh^{-1}\gamma^{-1}]^{-1}$, a finite positive constant, and $c = \gamma K$; the quantity $g(\vec{z},\vec{z})$ is infinite according to (1.11) and the second case in (1.14).

For Class C theories, $u(\phi)/\phi^{2N}$ is uniformly bounded as $\phi^2 \to \infty$ for some positive value of $N > 2$; then $V(g)$ is regular for all finite positive g with $\lim_{g\to\infty}[V(g)/g^{N-1}]$ equal to a finite positive constant and c asymptotically proportional to $K^{3(N-1)/(N+1)}$; the quantity $g(\vec{z},\vec{z})$ is infinite according to (1.11) and the third case in (1.14).

For Class D theories, $u(\phi)/\phi^{2N}$ is unbounded as $\phi^2 \to \infty$ for all positive values of N; then $\lim_{g\to g_{cr}} V(g) = \infty$ for a finite positive g_{cr}; a self-consistent solution to (1.13) and (1.14) does not exist, except for certain special cases [with $g(\vec{z},\vec{z})$ finite and c asymptotically proportional to K^3] which are discussed in example 6 below.

The following are representative members from the classes of theories, each example expressed in natural physical units which leave $u(\phi)$ free of trivial constant parameters :

Example 1 : $u(\phi) = \frac{1}{2}\,[m^2\phi^2 + a^2(1 - e^{-\phi^2})]$, a Class A theory with $V(g) = m^2 + a^2(1+g)^{-3/2}$.

Example 2 : $u(\phi) = \frac{1}{2}\,[m^2\phi^2 + a^2(\sin\phi)^2]$, a Class A theory with $V(g) = m^2 + a^2\, e^{-g}$.

Example 3 : $u(\phi) = a^2\phi^4$, a Class B theory with $V(g) = 6\, a^2 g$ and γ the positive root of the equation

$$(2\pi\gamma^2/3)\;[(1+\gamma^2)^{1/2} - \gamma^2 \sinh^{-1}\gamma^{-1}]^{-1} = a^2$$

Example 4 : $u(\phi) = \frac{1}{2} a^2(\phi^2 - 2b\phi^4 + \phi^6)$ [$b \leqslant 1$];

a Class C theory with $V(g) = a^2[1 - 6bg + (45/4)g^2]$, $N = 3$, and

$$c = \frac{(5a^2)^{1/4}}{2\pi} K^{3/2} + O(K^{1/2})$$

Example 5 : $u(\phi) = a^2[\phi^{-2}(e^{\phi^2}-1) -1] = a^2 \sum_{n=1}^{\infty} \frac{\phi^{2n}}{(n+1)!}$

a typical Class D theory with

$$V(g) = 8a^2g^{-2}[(1 - \tfrac{1}{2}g)(1-g)^{-1/2} - 1]$$

$$= a^2 \sum_{n=1}^{\infty} \frac{(2n)!}{(n+1)!(n-1)!} (g/4)^{n-1}$$

and $g_{cr} = 1$; it is readily shown that there is no consistent solution to eq.(1.13) and the third case in (1.14) with

$$g_K(\vec{z},\vec{z}) = 1 - f(K), \lim_{K\to\infty} f(K) = 0$$

Example 6 : $u(\phi) = \frac{1}{2} a^2(e^{\phi^2}-1)$, an exceptional Class D theory with $V(g) = a^2(1-g)^{-3/2}$ and $g_{cr} = 1$, all other exceptional Class D theories having a $u(\phi)$ which is a linear combination of a function of this specific form and a $u(\phi)$ associated with a Class A, B, or C theory. For these exceptional Class D theories (1.13) and (1.14) are satisfied in a self-consistent manner with

$$c = \frac{K^3}{6\pi^2} + O(K^{-1}) ,$$

provided that $a = (\pi/6)(54/5)^{3/4}$ (in natural physical units); then

$$g_K(\vec{z},\vec{z}) = 1 - \frac{54\pi^4}{5\ K^4} + O(K^{-8})$$

and $g(\vec{z},\vec{z}) = 1$.

The energy for the vacuum state is obtained by substituting (1.15) and (1.11) into {1.6}; we have :

$$E_{vac} = \lim_{K \to \infty} \int \&_{vac} \; d^3x,$$

where $\&_{vac}$ is asymptotic to $K^4/16\pi^2$ for Class A theories, asymptotically proportional to K^4 for Class B theories, asymptotically proportional to $K^{6N/(N+1)}$ for Class C theories, and asymptotically proportional to K^6 for exceptional Class D theories.

One-quantum state functionals for theories with a Hamiltonian operator (1.3) are assumed to take the approximate form

$$<\phi|one> = \int \overline{\phi}(\vec{x})f(\vec{x},\vec{y})\phi(\vec{y}) \; d^3x \; d^3y \; <\phi|vac> \; ,$$

$$(1.16)$$

where $<\phi|vac>$ is given by (1.4). Both the real function $\overline{\phi}(\vec{x})$ and the real symmetric distribution $f(\vec{x},\vec{y}) = f(\vec{y},\vec{x})$ in (1.16) are to be determined by the Rayleigh-Ritz procedure. The energy functionality

$$E_{one} = E_{one} \; [\overline{\phi},f] \equiv \; <one|\underline{H}|one>/<one|one> \quad (1.17)$$

is evaluated as[††]

$$E_{one} = E_{vac} + \frac{1}{2\Gamma} \int [|\vec{\nabla}\overline{\phi}(\vec{x})|^2 + (\int f(\vec{x},\vec{y})\overline{\phi}(\vec{y})d^3y)^2$$

$$+ V(g(\vec{x},\vec{x}))\overline{\phi}(\vec{x})^2]d^3x \qquad (1.18)$$

[††]   With the vacuum state normalization for the measure*, we have

$$<one|one> = \frac{1}{2} \int\int \overline{\phi}(\vec{x})f(\vec{x},\vec{y})\overline{\phi}(\vec{y}) \; d^3x \; d^3y \; ,$$

$$<one| - \frac{\delta^2}{\delta\phi(\vec{x})^2}|one> = \frac{1}{2} f(\vec{x},\vec{x})<one|one>$$

$$+ \frac{1}{2} (\int f(\vec{x},\vec{y})\overline{\phi}(\vec{y})d^3y)^2 \; ,$$

$$<one||\vec{\nabla}\phi(\vec{x})|^2|one> = \frac{1}{2} \vec{\nabla}_x \cdot \vec{\nabla}_y g(\vec{x},\vec{y})|_{\vec{y}-\vec{x}}<one|one>$$

$$+ \frac{1}{2} |\vec{\nabla}\overline{\phi}(\vec{x})|^2 \; ,$$

and

$$<one|\phi(\vec{x})^{2n}|one> = \frac{(2n)!}{n!4^n}(g(\vec{x},\vec{x})^n<one|one>$$

$$+ ng(\vec{x},\vec{x})^{n-1} \; \overline{\phi}(\vec{x})^2) \; .$$

in which E_{vac} is given by (1.6), $V(g)$ by (1.10), $g(\vec{x},\vec{y})$ implicitly in terms of $f(\vec{x},\vec{y})$ by (1.7), and

$$\Gamma \equiv \iint \overline{\phi}(\vec{x}) f(\vec{x},\vec{y}) \overline{\phi}(\vec{y}) d^3x d^3y \tag{1.19}$$

From (1.18) we derive the Rayleigh-Ritz equations

$$\frac{\delta E_{one}}{\delta \overline{\phi}(\vec{x})} = \frac{1}{\Gamma} [- \nabla^2 \overline{\phi}(\vec{x}) + \iint f(\vec{x},\vec{y}) f(\vec{y},\vec{z}) \overline{\phi}(\vec{z}) d^3y d^3z$$
$$+ V(g(\vec{x},\vec{x})) \overline{\phi}(\vec{x}) - 2(E_{one}-E_{vac}) \int f(\vec{x},\vec{y}) \overline{\phi}(\vec{y}) d^3y] = 0, \tag{1.20}$$

$$\frac{\delta E_{one}}{\delta f(\vec{x},\vec{y})} = \frac{1}{4} \quad \delta(\vec{x}-\vec{y}) + \int g(\vec{x},\vec{z}) [\nabla_z^2 - V(g(\vec{z},\vec{z}))] g(\vec{z},\vec{y}) d^3z$$
$$+ \frac{1}{2\Gamma} [\int [\overline{\phi}(\vec{x}) f(\vec{y},\vec{z}) + \overline{\phi}(\vec{y}) f(\vec{x},\vec{z})] \overline{\phi}(\vec{z}) d^3z$$
$$- \int g(\vec{x},\vec{z}) V'(g(\vec{z},\vec{z}) \overline{\phi}(\vec{z})^2 g(\vec{z},\vec{y}) d^3z -$$
$$- 2(E_{one}-E_{vac}) \overline{\phi}(\vec{x}) \overline{\phi}(\vec{y}) \quad = 0 \tag{1.21}$$

We seek a solution of (1.20) and (1.21) compatible with the vacuum-state equation (1.9), expecting $f(\vec{x},\vec{y})$ to be altered insignificantly for the one-quantum state (1.16). By making use of the relation derived from eq.(1.9),

$$\int f(\vec{x},\vec{z}) f(\vec{z},\vec{y}) d^3z = [- \nabla_z^2 + V(g(\vec{x},\vec{x}))] \delta(\vec{x}-\vec{y}), \tag{1.22}$$

Eq.(1.20) reduces to

$$\int f(\vec{x},\vec{y}) \overline{\phi}(\vec{y}) d^3y - E \overline{\phi}(\vec{x}) = 0 , \tag{1.23}$$

or equivalently, by virtue of (1.15),

$$\lim_{K\to\infty} (- \nabla^2 + c^2)^{1/2} \; \overline{\phi}(\vec{x}) - E \overline{\phi}(\vec{x}) = 0 , \tag{1.24}$$

where $E \equiv E_{one} - E_{vac}$ is the observable energy of the field quantum, while (1.9) and (1.23) reduce (1.21) to the condition

$$\lim_{K\to\infty} [\frac{V'(g_K(\vec{z},\vec{z}))}{\Gamma_K}] = 0 , \tag{1.25}$$

where

$$\Gamma_K \equiv \int \overline{\phi}(\vec{x}) \, (-\nabla^2 + c^2)^{1/2} \, \overline{\phi}(\vec{x}) \, d^3x \qquad (1.26)$$

is such that $\lim_{K\to\infty} \Gamma_K = \Gamma$. The condition (1.25) is satisfied by Class A theories, Class B theories, and Class C theories with $N < 4$, but neither by Class C theories with $N \geqslant 4$ nor exceptional Class D theories. However, for Class B and C theories the relativistic wave equation (1.24) predicts an infinite mass for the field quanta in the limit of an infinite cutoff parameter, $\lim_{K\to\infty} c = \infty$. Thus, only Class A theories admit wholly acceptable one-quantum stationary states, and for the Class A theories (1.24) produces the proper relativistic wave equation for $\overline{\phi}(\vec{x})$:

$$(-\nabla^2 + m^2)^{1/2} \, \overline{\phi}(\vec{x}) = E\overline{\phi}(\vec{x}) \quad . \qquad (1.27)$$

Since the eigenfunction solutions of (1.27) are not square-integrable for all $\vec{x}$ with a definite eigenvalue E, it is obviously necessary to restrict the $\vec{x}$ integration to a large but finite spatial volume for a consistent evaluation of the integral terms in (1.18). Then the observable energy of the quantum $E = E_{one} - E_{vac}$ is verified to be free to assume all positive real values, $E \geqslant m \equiv \lim_{g\to\infty}[V(g)]^{1/2}$, the observable mass of the field quanta for Class A theories. Conversely, if (1.20) and (1.21) are supplemented with the condition $V'(g(\vec{x},\vec{x})) = 0$ for Class A theories, the former equations can be reduced by a straightforward calculation to the eqs.(1.27) and (1.9).

Quasiclassical state functionals for the theories under consideration take the approximate form

$$\langle\phi|q\text{-}c\rangle = \exp\Big[-\frac{1}{2}\iint [\phi(\vec{x}) - \overline{\phi}(\vec{x})] f(\vec{x},\vec{y}) \times [\phi(\vec{y}) - \overline{\phi}(\vec{y})] d^3x \, d^3y\Big] \quad , \qquad (1.28)$$

where the real function $\overline{\phi}(\vec{x})$ and the real symmetric distribution $f(\vec{x},\vec{y}) = f(\vec{y},\vec{x})$ are to be determined by the Rayleigh-Ritz procedure. The energy functionality

$$E_{q-c} = E_{q-c}\,[\bar{\phi},f] \equiv \langle q-c|H|q-c\rangle/\langle q-c|q-c\rangle \qquad (1.29)$$

is evaluated by functional integration as

$$E_{q-c} = \int[\tfrac{1}{4}\, f(\vec{x},\vec{x}) + \tfrac{1}{4}\,\vec{\nabla}_x\cdot\vec{\nabla}_y\, g(\vec{x},\vec{y})|_{y=x} + \tfrac{1}{2}|\vec{\nabla}\bar{\phi}(\vec{x})|^2 + W(g(\vec{x},\vec{x}),\bar{\phi}(\vec{x}))]d^3x \qquad (1.30)$$

in which $g(\vec{x},\vec{y}) = g(\vec{y},\vec{x})$ is the real symmetric distribution inverse to $f(\vec{x},\vec{y})$ and

$$W(g,\bar{\phi}) \equiv \sum_{n=1}^{\infty}\sum_{p=0}^{n} \frac{\alpha_{2n}}{(n-p)!(2p)!}\left(\frac{g}{4}\right)^{n-p}\bar{\phi}^{2p}$$

$$= \pi^{-1/2}\, g^{-1/2}\int_{-\infty}^{\infty}\{\exp[-(\phi-\bar{\phi})^2/g]\}u(\phi)d\phi \qquad (1.31)$$

is a linear transform of (1.2), existing for all non-negative values of g less than a certain positive constant g_{cr} (which may be infinite) that depends on the form of $u(\phi)$ but not on the value of $\bar{\phi}$. On its domain of definition $0 \leqslant g < g_{cr}$, the function (1.31) satisfies the diffusion equation

$$\frac{\partial W(g,\bar{\phi})}{\partial g} = \frac{1}{4}\,\frac{\partial^2 W(g,\bar{\phi})}{\partial\bar{\phi}^2} \qquad (1.32)$$

The Rayleigh-Ritz equations that follow from (1.30) are

$$\frac{\delta E_{q-c}}{\delta\bar{\phi}(\vec{x})} = -\nabla^2\bar{\phi}(\vec{x}) + \frac{\partial W}{\partial\bar{\phi}}(g(\vec{x},\vec{x}),\bar{\phi}(\vec{x})) = 0\ , \qquad (1.33)$$

$$\frac{\delta E_{q-c}}{\delta f(\vec{x},\vec{y})} = \frac{1}{4}\,[\delta(\vec{x}-\vec{y}) + \int g(\vec{x},\vec{z})[\nabla_z^2 - \frac{\partial^2 W}{\partial\bar{\phi}^2}(g(\vec{z},\vec{z}),\bar{\phi}(\vec{z}))] \times g(\vec{z},\vec{y})\, d^3z] = 0 \qquad (1.34)$$

with (1.32) being used in writing (1.34).

To facilitate a general discussion of eqs.(1.33) and (1.34), let N denote the smallest non-negative number for which $u(\phi)/\phi^{2N}$ is uniformly bounded as

$\phi^2 \to \infty$. According to our classification scheme above, a theory is of Class A if $0 \leqslant N \leqslant 1$, of Class B if $1 < N \leqslant 2$, of Class C if $2 < N < \infty$, and of Class D if $u(\phi)/\phi^{2N}$ is unbounded as $\phi^2 \to \infty$ for all positive values of N. In order for (1.33) to be a welldefined c-number equation, it is evident that

$$\frac{\partial W}{\partial \overline{\phi}}(g(\vec{x},\vec{x}),\overline{\phi}(\vec{x}))$$
$$= 2[\pi g(\vec{x},\vec{x})]^{-1/2} \int_{-\infty}^{\infty} \xi e^{-\xi^2} u(\overline{\phi}(\vec{x}) + g(\vec{x},\vec{x})^{1/2}\xi)d\xi \tag{1.35}$$

must exist as a finite quantity. The existence of (1.35) implies that the function

$$\frac{\partial^2 W}{\partial \overline{\phi}^2}(g(\vec{x},\vec{x}),\overline{\phi}(\vec{x}))$$

is finite in (1.34), and hence that $g(\vec{x},\vec{x})$ is infinite. Thus the existence of (1.35) requires a $u(\phi)$ for which $g_{cr} = \infty$; Class A, B and C theories, but not Class D theories, have the feature that $g_{cr} = \infty$ with (1.31) existing for all non-negative values of g. Moreover, with $g(\vec{x},\vec{x}) \to \infty$ in the limit of an infinite cutoff parameter, the right side of (1.35) can be expressed as

$$\lim_{g\to\infty} 2(\pi g)^{-1/2} \int_{-\infty}^{\infty} \xi e^{-\xi^2} u(\overline{\phi}(\vec{x}) + g^{1/2}\xi)d\xi ,$$

and the existence of the latter quantity can be investigated for Class A, B, and C theories by evoking the asymptotic dependence $u(\phi) \propto \phi^{2N}$ for large values of ϕ^2. Hence, it follows that (1.35) exists as a finite quantity and that (1.33) is a welldefined c-number equation only for Class A theories with $0 \leqslant N \leqslant 1$, and thus only Class A theories admit well-defined quasiclassical stationary states for a Hamiltonian operator (1.3) within the context of the Rayleigh-Ritz approximation procedure. The abbreviated argument presented here has been corroborated by a detailed and rather lengthy study of eqs.(1.33) and (1.34).

For Class A theories we have

$$\lim_{\phi^2\to\infty} [u(\phi)/\phi^2] \equiv \frac{1}{2} m^2 , \tag{1.36}$$

a finite non-negative constant. Since

$$\frac{\partial W}{\partial\overline{\phi}} (\infty,\overline{\phi}) = m^2\overline{\phi}$$

by virtue of (1.35) in the limit $g(\vec{x},\vec{x}) \to \infty$, eq. (1.33) reduces to the linear equation

$$(-\nabla^2 + m^2)\overline{\phi}(\vec{x}) = 0 , \tag{1.37}$$

while the general solution to eq.(1.34) is given by

$$g(\vec{x},\vec{y}) = \frac{1}{(2\pi)^3} \int \frac{e^{i\vec{k}.(\vec{x}-\vec{y})}}{(|\vec{k}|^2 + m^2)^{1/2}} d^3k . \tag{1.38}$$

Also valid for the vacuum state with $\overline{\phi}(\vec{x}) \equiv 0$, the solution (1.38) can be used to compute the energy of the vacuum :

$$E_{vac} = \int [\frac{1}{4} f(\vec{x},\vec{x}) + \frac{1}{4} \vec{\nabla}_x.\vec{\nabla}_y g(\vec{x},\vec{y})|_{y=x}$$
$$+ W(g(\vec{x},\vec{x}),0)]d^3x , \tag{1.39}$$

and from (1.30) and (1.39) it follows that the observable energy of a quasiclassical state is

$$E = E_{q-c} - E_{vac} = \int [\frac{1}{2}|\vec{\nabla}\overline{\phi}|^2 + \frac{1}{2} m^2\overline{\phi}^2]d^3x \tag{1.40}$$

Thus, the Class A theories appear to be effectively linear in the classical limit, according to (1.37) and (1.40), with no macroscopically observable self-interaction of the $\overline{\phi}$ field.

We conclude this section by working out the function (1.31) for some specific theories. Each of the following examples is expressed in natural physical units which leave $u(\phi)$ free of trivial constant parameters :

Example 1 : $u(\phi) = \frac{1}{2} [m^2\phi^2 + a^2(1-e^{-\phi^2})]$,

a Class A theory with :

$$W(g,\overline{\phi}) = \frac{1}{2} m^2(\overline{\phi}^2 + \frac{1}{2} g)$$

$$+ \frac{1}{2} a^2 [1 - (1+g)^{-1/2}\{\exp [-\overline{\phi}^2/(1+g)]\}] .$$

Example 2 : $u(\phi) = \frac{1}{2} [m^2\phi^2 + a^2(\sin\phi)^2]$,

a Class A theory with

$$W(g,\overline{\phi}) = \frac{1}{2} m^2(\overline{\phi}^2 + \frac{1}{2} g) + \frac{1}{4} a^2 (1-e^{-g} \cos^2\overline{\phi}).$$

Example 3 : $u(\phi) = \frac{1}{2} \phi^2 + a^2\phi^4$,

a Class B theory with

$$W(g,\overline{\phi}) = \frac{1}{2} (\overline{\phi}^2 + \frac{1}{2} g) +a^2 (\overline{\phi}^4 + 3\overline{\phi}^2 g + \frac{3}{4} g^2) .$$

Example 4 : $u(\phi) = \phi^6$, a Class C theory with

$$W(g,\overline{\phi}) = \overline{\phi}^6 + (15/2)\overline{\phi}^4 g + (45/4)\overline{\phi}^2 g^2 + (15/8)g^3 .$$

Example 5 : $u(\phi) = \frac{1}{2} a^2 (e^{\phi^2}-1)$, a Class D theory with

$$W(g,\overline{\phi}) = \frac{1}{2} a^2 [(1-g)^{-1/2}\{\exp[\overline{\phi}^2/(1-g)]\}- 1] .$$

2. NONRELATIVISTIC REAL SCALAR-WEYL SPINOR MODEL WITH RENORMALIZATION

In the light of the results of the preceding section, it appears that the "extended objects" or quantized "particlelike" solutions of hadron physics cannot arise via the Rayleigh-Ritz procedure from an essentially nonlinear field theory without renormalization. If one relaxes the requirement of Lorentz-invariance, it is possible to exhibit solvable models which admit singularity-free quantized particlelike solutions with the Rayleigh-Ritz procedure and appropriate renormalization. Consider, for example, the model based on the semi-nonrelativistic Lagrangian

$$\mathcal{L} = - \frac{1}{2} \partial^\mu\phi\partial_\mu\phi + \psi^\dagger(i\sigma^\mu\partial_\mu - m_o)\psi + \lambda\phi^2\psi^\dagger\psi \qquad (2.1)$$

with ϕ a real scalar field, ψ a two-component complex

Weyl spinor field, m_0 a positive bare-mass constant, and λ a positive coupling constant. From (2.1) we obtain the associated Hamiltonian operator

$$H = \int \{-\frac{1}{2}\frac{\delta^2}{\delta\phi(\vec{x})^2} + \frac{1}{2}|\vec{\nabla}\phi(\vec{x})|^2 - i\psi^\dagger(\vec{x})\sigma.\vec{\nabla}\psi(\vec{x})$$

$$+ [m_0 - \lambda\phi(\vec{x})^2]\psi^\dagger(\vec{x})\psi(\vec{x})\}\ d^3x \qquad (2.2)$$

in which the boson field operator $\phi(\vec{x})$ is represented in diagonalized form and the fermion field operator $\psi(\vec{x})$ satisfies the nonrelativistic anticommutation relations $\{\psi(\vec{x}),\ \psi(\vec{x}')\} = 0$, $\{\psi(\vec{x}),\ \psi^\dagger(\vec{x}')\} = \delta(\vec{x}-\vec{x}')$.

The nonrelativistic approximation to the vacuum state, following from the specific ordering in (2.2) is

$$\langle\phi|vac\rangle =$$

$$\{\exp[-\frac{1}{2}\iint \phi(\vec{x})f_{vac}(\vec{x},\vec{y})\ \phi(\vec{y})\ d^3x\ d^3y]\}\Omega_{vac} \qquad (2.3)$$

where Ω_{vac} denotes the nonrelativistic fermion vacuum, $\psi(\vec{x})\ \Omega_{vac} \equiv 0$ with $\Omega_{vac}^\dagger\ \Omega_{vac} = 1$, and $f_{vac}(\vec{x},\vec{y}) = (-\nabla_x^2)^{1/2}\delta(\vec{x}-\vec{y})$ is the real symmetric distribution inverse to

$$g_{vac}(\vec{x},\vec{y}) = \frac{1}{(2\pi)^3}\int e^{i\vec{k}.(\vec{x}-\vec{y})}\frac{d^3k}{|\vec{k}|}$$

$$= \lim_{K\to\infty}(1 - \cos K|\vec{x}-\vec{y}|)/2\pi^2|\vec{x}-\vec{y}|^2 \qquad (2.4)$$

the wave-number integration being restricted to $|\vec{k}| \leq K$, a cutoff constant. Normalized with respect to an appropriate displacement-invariant measure for the inner product functional integration over all fields ϕ, the vacuum state (2.3) is such that the associated energy functionality

$$E_{vac} \equiv \langle vac|\underline{H}|vac\rangle / \langle vac|vac\rangle$$

$$= \int \{\frac{1}{4} f_{vac}(\vec{x},\vec{x}) + \frac{1}{4} \vec{\nabla}_x \cdot \vec{\nabla}_y \, g_{vac}(\vec{x},\vec{y})|_{y=x}\} d^3x \tag{2.5}$$

satisfies the Rayleigh-Ritz equation $\delta E_{vac}/\delta f_{vac} = 0$, an integral equation which yields the expression (2.4). Finally, the vacuum state energy can be obtained by evaluating (2.5),

$$E_{vac} = \lim_{K\to\infty} \int \frac{K^4}{16\pi^2} d^3x \ .$$

Let us seek a physical one-fermion stationary state, a simultaneous eigenstate of (2.2), and the fermion number operator $\int \psi^{\dagger}(\vec{x})\psi(\vec{x})d^3x$ (eigenvalues E_{one} and unity, respectively) with the approximate form

$$\langle\phi|one\rangle = N^{-1} [\exp (- \frac{1}{2} \iint [\phi(\vec{x}) - \phi_1(\vec{x})]$$
$$\times f(\vec{x},\vec{y})[\phi(\vec{y}) - \phi_1(\vec{y})]d^3x \, d^3y]$$
$$\times \int \psi^{\dagger}(\vec{x})\psi_1(\vec{x}) d^3x \ \Omega_{vac} \tag{2.6}$$

where $\phi_1(\vec{x})$, $\psi_1(\vec{x})$ (c-number functional) and $f(\vec{x},\vec{y}) = f(\vec{y},\vec{x})$ (a c-number distribution) are to be determined by the Rayleigh-Ritz procedure and N is a normalization constant. The energy functionality associated with (2.2) and (2.6) is defined and evaluated by functional integration as

$$E_{one} \equiv \langle one|\underline{H}|one\rangle / \langle one|one\rangle$$
$$= \int \{\frac{1}{4} f(\vec{x},\vec{x}) + \frac{1}{4} \vec{\nabla}_x . \vec{\nabla}_y f(\vec{x},\vec{y})|_{y=x} + \frac{1}{2}|\vec{\nabla}\phi_1|^2$$
$$- i\psi_1^\dagger \sigma.\vec{\nabla}\psi_1 + [m_o - \lambda\phi_1^2 - \frac{1}{2}\lambda g(\vec{x},\vec{x})]\psi_1^\dagger\psi_1\} d^3x \tag{2.7}$$

in which $g(\vec{x},\vec{y})$ is the real symmetric distribution inverse to $f(\vec{x},\vec{y})$,

$$\int f(\vec{x},\vec{z}) g(\vec{z},\vec{y}) d^3z \equiv \delta(\vec{x}-\vec{y}) \ . \tag{2.8}$$

From (2.7) we derive the Rayleigh-Ritz equations

$$\frac{\delta E_{one}}{\delta\phi_1} = -\nabla^2\phi_1 - 2\lambda\psi_1^\dagger\psi_1\phi_1 = 0 \tag{2.9}$$

$$\frac{\delta E_{one}}{\delta\psi_1^\dagger} = - i\sigma.\vec{\nabla}\psi_1 + [m_o - \lambda\phi_1^2 - \frac{1}{2}\lambda g(\vec{x},\vec{x})]\psi_1 = 0 \tag{2.10}$$

$$\frac{\delta E_{one}}{\delta f} = \frac{1}{4}\delta(\vec{x}-\vec{y}) \tag{2.11}$$
$$+ \frac{1}{4}\int g(\vec{x},\vec{z})[\nabla_z^2 + 2\lambda\psi_1^\dagger(\vec{z})\psi_1(\vec{z})] g(\vec{z},\vec{y}) d^3z = 0$$

It follows immediately from (2.11) that the singular character of $g(\vec{x},\vec{y})$ as $\vec{y} \to \vec{x}$ is identical to the singular character of $g_{vac}(\vec{x},\vec{y})$ as $\vec{y} \to \vec{x}$, $g(\vec{x},\vec{x}) = g_{vac}(\vec{x},\vec{x}) = \lim_{K\to\infty}(K^2/4\pi^2)$ by (2.4). Hence, in order to have a meaningful c-number equation (2.10), we put

$$m_o \equiv \frac{1}{2}\lambda g(\vec{x},\vec{x}) = \lim_{K\to\infty}(\lambda K^2/8\pi^2) \ . \tag{2.12}$$

Then the singularity-free spherically symmetric "particlelike" solution to the coupled c-number equations (2.9) and (2.10) is obtainable in closed form and given exactly by

$$\phi_1 = \pm (3a/\lambda)^{1/2}(|\vec{x}|^2 + a^2)^{-1/2} \ , \tag{2.13}$$

$$\psi_1 = (3/2\lambda)^{1/2} a(|\vec{x}|^2 + a^2)^{-3/2}(i\sigma.\vec{x} + a)u \tag{2.14}$$

with a denoting a free positive constant of integration a so-called "homology constant" stemming from the scale (dilatation) invariance of eqs.(2.9) and (2.10) with the condition (2.12); in (2.14) u denotes a constant Weyl spinor of unit length, $u^{\dagger}u = 1$. The final Rayleigh-Ritz equation (2.11) with (2.14) is

$$- \int g(\vec{x},\vec{z})[\nabla_z^2 + 3a^2(|\vec{z}| + a^2)^{-2}]g(\vec{z},\vec{y})d^3z = \delta(\vec{x}-\vec{y}) \tag{2.15}$$

or equivalently by (2.8),

$$-[\nabla_x^2 + 3a^2(|\vec{x}|^2 + a^2)^{-2}]\delta(\vec{x}-\vec{y}) = \int f(\vec{x},\vec{z})f(\vec{z},\vec{y})d^3z \tag{2.16}$$

Introducing the Fourier transforms

$$f(\vec{x},\vec{y}) = f_{vac}(\vec{x},\vec{y}) + \frac{1}{(2\pi)^3}\int \alpha(\vec{k},\vec{k}')e^{i\vec{k}.\vec{x}-i\vec{k}'.\vec{y}} d^3k\, d^3k' , \tag{2.17}$$

$$g(\vec{x},\vec{y}) = g_{vac}(\vec{x},\vec{y}) + \frac{1}{(2\pi)^3}\int \frac{\beta(\vec{k},\vec{k}')}{|\vec{k}||\vec{k}|} \times e^{i\vec{k}.\vec{x}-i\vec{k}'.\vec{y}}d^3kd^3k' , \tag{2.18}$$

we see that the equations obtained from (2.16) and (2.8)

$$(|\vec{k}| + |\vec{k}'|)\alpha(\vec{k},\vec{k}') + \frac{3a}{8\pi} e^{-a|\vec{k}-\vec{k}'|} = - \int\alpha(\vec{k},\vec{k}'')\alpha(\vec{k}'',\vec{k}')d^3k'' \tag{2.19}$$

$$\alpha(\vec{k},\vec{k}') + \beta(\vec{k},\vec{k}') = - \int \alpha(\vec{k},\vec{k}'')\beta(\vec{k}'',\vec{k}') \frac{d^3k''}{|\vec{k}''|} \tag{2.20}$$

can be solved by an obvious iteration procedure, the first approximation

$$\alpha(\vec{k},\vec{k}') \cong - \beta(\vec{k},\vec{k}') \cong - \frac{3a}{8\pi} \frac{e^{-a|\vec{k}-\vec{k}'|}}{(|\vec{k}| + |\vec{k}'|)} \tag{2.21}$$

being asymptotic to the exact solution for both

$a|\vec{k}|$, $a|\vec{k}'|$ >> 1. * The energy of the physical one-fermion stationary state can be expressed by evaluating (2.7) with (2.12), (2.13), (2.14), (2.17), and (2.18) :

---

* Formally more direct but difficult to justify with mathematical rigor, an alternative method for solving (2.15) or (2.16) can be based on the fact that the effective Schrödinger potential $-3a^2(|\vec{x}|^2+a^2)^{-2}$ is a smooth singularity-free function which admits no "bound states", the Schrödinger operator $-[\nabla_x^2+3a^2(|\vec{x}|^2+a^2)^{-2}]$ having no eigenfunction with a negative eigenvalue in the space of bounded C^1 piecewise C^2 functions of $\vec{x}$, the nodeless eigenfunction $(|\vec{x}|^2+a^2)^{-1/2}$ possessing the eigenvalue zero. Thus we have the WKB approximations

$$f(\vec{x},\vec{y}) \cong \frac{1}{(2\pi)^3}\int_{|\vec{k}|\geq\mu(\vec{x},\vec{y})}[|\vec{k}|^2-\mu(\vec{x},\vec{y})^2]^{1/2}\,e^{i\vec{k}.(\vec{x}-\vec{y})}d^3k\,,$$

$$\vec{\nabla}_x.\vec{\nabla}_y \cong \frac{1}{(2\pi)^3}\int_{|\vec{k}|\geq\mu(\vec{x},\vec{y})}|\vec{k}|^2[|\vec{k}|^2-\mu(\vec{x},\vec{y})^2]^{-1/2}\times e^{i\vec{k}.(\vec{x}-\vec{y})}d^3k,$$

where $\mu(\vec{x},\vec{y})=\mu(\vec{y},\vec{x})$ is a certain non-negative real function such that $\mu(\vec{x},\vec{x})^2 = 3a^2(|\vec{x}|^2+a^2)^{-2}$. Substituting these WKB approximations into (2.7),we obtain

$$\begin{aligned}E_{one} &= \frac{1}{4}\int\Big[\frac{1}{(2\pi)^3}\int_{|\vec{k}|\geq\mu(\vec{x},\vec{x})}([|\vec{k}|^2-\mu(\vec{x},\vec{x})^2]^{1/2} \\ &\quad + |\vec{k}|^2[|\vec{k}|^2-\mu(\vec{x},\vec{x})^2]^{-1/2})d^3k\Big]d^3x + \frac{9\pi^2}{8\lambda} \\ &= \lim_{K\to\infty}\int\Big[\frac{K^4}{16\pi^2}+\frac{\mu(\vec{x},\vec{x})^4}{32\pi^2}\ln(2K/\mu(\vec{x},\vec{x}))\Big]d^3x + \frac{9\pi^2}{8\lambda} \\ &= E_{vac} + \left(\frac{3}{16}\right)^2\lim_{K\to\infty}a^{-1}\ln(aK) + \frac{9\pi^2}{8\lambda}\end{aligned}$$

which agrees with (2.24).

$$E_{one} = E_{vac} \pm \frac{i}{4} \int [\alpha(\vec{k},\vec{k}) + \beta(\vec{k},\vec{k})] d^3k + \frac{9\pi^2}{8\lambda} ; \tag{2.22}$$

and so the dressed fermion mass value is

$$m = E_{one} - E_{vac} = \frac{9\pi^2}{8\lambda} - \frac{1}{4} \int \alpha(\vec{k},\vec{k}')\beta(\vec{k}',\vec{k}) \frac{d^3k'}{|\vec{k}'|} d^3k \tag{2.23}$$

in view of (2.20). Making use of the approximation (2.21), we finally obtain

$$m = \frac{9\pi^2}{8\lambda} + \left(\frac{3}{16}\right)^2 \lim_{K\to\infty} a^{-1} \ln(aK) \tag{2.24}$$

for $(aK) >> 1$. Since the first approximation (2.21) is asymptotic to the exact solution of eqs.(2.19) and (2.20), eq.(2.24) is an exact consequence of the Rayleigh-Ritz approximation theory.

Now all of the equations in the theory remain perfectly regular in the limit $a \to \infty$, that is, as the homology constant a increases, without bound.† In view of the fact that the homology constant is a free parameter in the theory, the result (2.24) can be made finite by requiring a to manifest a suitable dependence on the cutoff constant K so that $a \to \infty$ as $K \to \infty$, say by putting $a \equiv \lambda^2 K$. With such an $a = a(K)$, the finite, positive, dressed fermion mass value $m = 9\pi^2/8\lambda$ is obtained without ambiguity, independent

---

† Moreover, a straightforward integration calculation with (2.6) and (2.14) shows that the expectation value of the total angular momentum

$$s = \langle one| \int [\tfrac{1}{2}\{\vec{r}\times\vec{\nabla}_\phi, i\tfrac{\delta}{\delta\phi}\} + \psi^\dagger(-i\vec{r}\times\vec{\nabla} + \tfrac{1}{2}\sigma)\psi] d^3r |one\rangle / \langle one|one\rangle$$

$$= \int \psi_1^\dagger(-i\vec{r} \times \vec{\nabla} + \tfrac{1}{2}\sigma)\psi_1 d^3r / \int \psi_1^\dagger \psi_1 d^3r = \tfrac{1}{2} u^\dagger \sigma u,$$

entirely independent of the value assigned to a, is generally consistent for a spin-1/2 particle.

of the infinite, positive, bare-mass constant (2.12).††

3. SELF-INTERACTING COMPLEX SCALAR MODEL WITH RENORMALIZATION

In this final section we employ explicit functional integration and the Rayleigh-Ritz procedure for functionalities to approximate relativistic two-quantum particle-antiparticle bound states and scattering states in a renormalized complex sealar model which features a self-interaction term $\lambda_o(\psi^*\psi)^2$ in the Lagrangian. The mass of a particle-antiparticle bound state can be one or more orders of magnitude less than the mass associated with a one-particle state. This property of the model suggests

---

†† Of course the strong divergence of the normalization constant N in (2.6) is a hard-set feature of the theory. With the appropriate displacement-invariant measure and the relation obtained from (2.14) by ordinary integration

$$\int \psi_1^\dagger(\vec{x})\psi_1(\vec{x})d^3x = \frac{3\pi^2 a}{2\lambda} ,$$

the normalization condition $\langle one|one\rangle = 1$ produces $|N|^2 = (3\pi^2 a/2\lambda)$, where

$$\mathfrak{F} = \int \{\exp [- \int\int \omega(\vec{x})h(\vec{x},\vec{y})\omega(\vec{y})d^3xd^3y]\}$$

$$\times \prod_x \{\exp [-\omega(\vec{x})^2 d^3x]\} (\frac{d^3x}{\pi})^{1/2}\delta\omega(\vec{x})$$

$$h(\vec{x},\vec{y}) \equiv (-\nabla_x^2)^{-1/4}(-\nabla_y^2)^{-1/4}[f(\vec{x},\vec{y}) - f_{vac}(\vec{x},\vec{y})]$$

Now from the approximate solution (2.21) with (2.17) it follows that all iterated kernels constructed from $h(\vec{x},\vec{y})$ [i.e., $\int h(\vec{x},\vec{z})h(\vec{z},\vec{y})d^3z$, etc.] are regular as $\vec{y} \to \vec{x}$, and thus the dominant (divergent) terms in $\mathfrak{F}$ are simply the powers of $\int h(\vec{x},\vec{x})d^3x$ as $K\to\infty$; hence we have

$$\mathfrak{F} = \sum_{n=0}^{\infty} \frac{1}{n!} (- \frac{1}{2} \int h(\vec{x},\vec{x})d^3x)^n = \lim_{K\to\infty} e^{3aK/8}$$

that a theory for a self-interacting quark spinor field may yield a physically realistic quark-antiquark meson with a mass two orders of magnitude less than the mass of a constituent quark.

The Lagrangian is

$$\mathcal{L} = -\partial^{\mu}\psi^{*}\partial_{\mu}\psi - m_o^2\psi^{*}\psi + \lambda_o(\psi^{*}\psi)^2 \qquad (3.1)$$

where the bare mass and coupling constants m_o and λ_o are subject to a specific renormalization in the theory studied here. The abstract Hamiltonian associated with (3.1) is

$$\underline{H} = \int [\pi\pi^{*} + \nabla\psi^{*}.\nabla\psi + m_o^2\psi^{*}\psi - \lambda_o(\psi^{*}\psi)^2]d^3x \qquad (3.2)$$

where the momentum densities π and π^{*} are canonically conjugate to ψ and ψ^{*}, respectively, the canonical commutation relations for a boson field

$$\begin{aligned}[\psi(\vec{x}),\psi(\vec{x}')] &= [\psi(\vec{x}),\psi^{*}(\vec{x}')] = [\psi(\vec{x}),\pi^{*}(\vec{x}')]\\ &= [\psi^{*}(\vec{x}),\pi(\vec{x}')] = [\pi(\vec{x}),\pi(\vec{x}')]\\ &= [\pi(\vec{x}),\pi^{*}(\vec{x}')] = 0 \qquad (3.3)\\ [\psi(\vec{x}),\pi(\vec{x}')] &= [\psi^{*}(\vec{x}),\pi^{*}(\vec{x}')] = i\delta(\vec{x}-\vec{x}'),\end{aligned}$$

being satisfied. Our object is to determine a suitable renormalization of the bare constant m_o and bare coupling constant λ_o in (3.2) which admits vacuum, one-quantum, and particle-antiparticle two-quantum eigenstates for such a model theory. To avoid ambiguities in the definition of functional integrals, real scalar fields denoted by ϕ_1 and ϕ_2, diagonalized for all values of $\vec{x}$ at a fixed instant of time, are evoked to represent the complex scalar field,

$$\psi \equiv (\phi_1 + i\phi_2)/\sqrt{2}, \quad \psi^{*} = (\phi_1 - i\phi_2)/\sqrt{2}, \qquad (3.4)$$

and associated functional differential operators are used to represent their conjugate momentum densities,

$$\begin{aligned}\pi &= -i(\delta/\delta\phi_1 - i(\delta/\delta\phi_2))/\sqrt{2},\\ \pi^{*} &= -i(\delta/\delta\phi_1 + i(\delta/\delta\phi_2))/\sqrt{2}, \qquad (3.5)\end{aligned}$$

so that the Hamiltonian operator takes the form

$$H = \int \{\frac{1}{2} \sum_{k=1}^{2} [- \frac{\delta^2}{\delta\phi_k(\vec{x})^2} + |\nabla\phi_k(\vec{x})|^2 + m_o^2\phi_k(\vec{x})^2]$$

$$- \frac{1}{4} \lambda_o [\phi_1(\vec{x})^2 + \phi_2(\vec{x})^2]^2\} d^3x \qquad (3.6)$$

Since the particles in nature always appear with a definite electric charge, baryon number, etc., we seek stationary states that are simultaneous eigenstates of the <u>number operator</u>

$$N \equiv i \int [\psi^*(\vec{x})\pi^*(\vec{x}) - \psi(\vec{x})\pi(\vec{x})] d^3x$$

$$= i \int [\phi_1(\vec{x}) \frac{\delta}{\delta\phi_2(\vec{x})} - \phi_2(\vec{x}) \frac{\delta}{\delta\phi_1(\vec{x})}] d^3x \qquad (3.7)$$

The vacuum state functional for a theory with a Hamiltonian operator (3.6) is assumed to take the approximate form

$$<\psi|vac> = \exp \{- \frac{1}{2} \iint f(\vec{x},\vec{y})$$

$$\times [\phi_1(\vec{x})\phi_1(\vec{y}) + \phi_2(\vec{x})\phi_2(\vec{y})] d^3x\, d^3y\}, \qquad (3.8)$$

where $f(\vec{x},\vec{y}) = f(\vec{y},\vec{x})$ is a real symmetric distribution to be determined, By applying a Rayleigh-Ritz procedure to the energy functionality

$$E_{vac} = E_{vac}[f] \equiv <vac|\underline{H}|vac>/<vac|vac>, \qquad (3.9)$$

we find

$$f(\vec{x},\vec{y}) = (- \nabla^2 + m^2)^{1/2}\delta(\vec{x}-\vec{y}) \qquad (3.10)$$

where

$$m^2 = \lim_{K\to\infty} V(g_K(\vec{z},\vec{z})) , \qquad (3.11)$$

$$V(g) = m_o^2 - 2 \lambda_o g , \qquad (3.12)$$

$$g_K(\vec{z},\vec{z}) = (1/4\pi^2)[K(K^2+m^2)^{1/2} - m^2\sinh^{-1}(K/m)]$$

$$= (1/4\pi^2)[K^2 - m^2 \ln(2K/m) + \frac{1}{2} m^2 + O(m^2K^{-2})] \tag{3.13}$$

To secure (3.11) as a finite positive constant, we postulate the renormalization formules for m_o and λ_o :

$$m_o^2 = \lambda_o K^2/2\pi^2 - m^2[1 + O(\lambda_o)] , \tag{3.14a}$$

$$\lambda_o = 4\pi^2[\ln(K/\mu) + O(m^2K^{-2})]^{-1} , \tag{3.14b}$$

in which the disposable positive constants m and μ, quantities absolutely independent of the cutoff parameter K, carry the observable physics in the theory. Equations (3.14) complete the specification of our model by prescribing the asymptotic form of the bare mass constant m_o and bare coupling constant λ_o as the cutoff parameter K increases without bound; it follows from (3.14) that the dressed mass m defined by (3.11) is finite (and a free parameter in the theory), notwithstanding the fact that $m_o \to \infty$ and $\lambda_o \to 0$ as $K \to \infty$.

One-quantum state functionals for a theory with a Hamiltonian operator (3.2) are assumed to take the approximate form

$$\langle\psi|\text{one}\rangle = \iint \tilde{\psi}(\vec{x}) f(\vec{x},\vec{y})[\phi_1(\vec{y}) \pm i\phi_2(\vec{y})]d^3x\,d^3y\langle\psi|\text{vac}\rangle, \tag{3.15}$$

where $\langle\psi|\text{vac}\rangle$ is given by (3.8), $f(\vec{x},\vec{y})$ is given by (3.10), and the complex-valued c-number function $\tilde{\psi}(\vec{x})$ is to be determined. By applying the Rayleigh-Ritz procedure to the energy functionality associated with (3.15),

$$E_{one} = E_{one}[\tilde{\psi}^*,\tilde{\psi}] == \langle\text{one}|\underline{H}|\text{one}\rangle/\langle\text{one}|\text{one}\rangle, \tag{3.16}$$

we find that $\tilde{\psi}(\vec{x})$ must satisfy the relativistic wave equation

$$(-\nabla^2 + m^2)^{1/2}\,\tilde{\psi}(\vec{x}) = E\tilde{\psi}(\vec{x}) , \tag{3.17}$$

where $E \equiv E_{one} - E_{vac}$, the observable energy of the quantum, is free to assume all positive real values $\geqslant$ m.

Two-quantum particle-antiparticle state functionals are assumed to take the approximate form

$$<\psi|two> = \iint \tilde{\phi}(\vec{x},\vec{y})[\tau(\vec{x},\vec{y}) - g(\vec{x},\vec{y})]d^3x d^3y <\psi|vac> \tag{3.18}$$

where

$$\tau(\vec{x},\vec{y}) \equiv \phi_1(\vec{x})\phi_1(\vec{y}) + \phi_2(\vec{x})\phi_2(\vec{y}), \tag{3.19}$$

$$g(\vec{x},\vec{y}) = \frac{1}{(2\pi)^3} \int \frac{e^{i\vec{k}.(\vec{x}-\vec{y})}}{(k^2 + m^2)^{1/2}} d^3k \ , \tag{3.20}$$

$<\psi|vac>$ is given by (3.8), and the real-valued symmetric c-number function $\tilde{\phi}(\vec{x},\vec{y}) = \tilde{\phi}(\vec{y},\vec{x})$ in (3.18) is to be determined. Note that the state functional (3.18) is orthogonal to the vacuum state functional (3.8) because

$$<vac|\tau(\vec{x},\vec{y})|vac> = g(\vec{x},\vec{y})<vac|vac>.$$

Moreover, the norm-quared of (3.18) is

$$<two|two> = \iiiint \tilde{\phi}(\vec{w},\vec{x})g(\vec{x},\vec{y})\tilde{\phi}(\vec{y},\vec{z}) \times g(\vec{z},\vec{w})d^3w d^3x d^3y d^3z \tag{3.21}$$

if we prescribe the vacuum state normalization

$$<vac|vac> \equiv 1 \ .$$

By applying the Rayleigh-Ritz procedure to the energy functionality

$$E_{two} = E_{two}[\tilde{\phi}] \equiv <two|\underline{H}|two>/<two|two>, \tag{3.22}$$

we find that $\tilde{\phi}(\vec{x},\vec{y})$ must satisfy the relativistic wave equation

$$\int [\tilde{\phi}(\vec{x},\vec{z})f(\vec{z},\vec{y}) + f(\vec{x},\vec{z})\tilde{\phi}(\vec{z},\vec{y})]d^3z$$
$$+ \frac{1}{2} (\iint g(\vec{x},\vec{r})\tilde{\phi}(\vec{r},\vec{s})g(\vec{s},\vec{x})d^3r d^3s)V'(g(\vec{x},\vec{x}))$$
$$\times \delta(\vec{x}-\vec{y}) = E\tilde{\phi}(\vec{x},\vec{y}) \ , \tag{3.23}$$

where $E \equiv E_{two} - E_{vac}$, the total observable energy of the particle-antiparticle state, is free to assume non-negative real values.

Let us now restrict analysis to the center-of-momentum system, so that the particle-antiparticle wave function $\tilde{\phi}(\vec{x},\vec{y})$ has a Fourier representation

$$\tilde{\phi}(\vec{x},\vec{y}) = \int \omega(\vec{k})\, e^{i\vec{k}\cdot(\vec{x}-\vec{y})}\, d^3k \tag{3.24}$$

that depends only on the relative position vector $\vec{x}-\vec{y}$. A linear equation is obtained for the wave-number amplitude function $\omega(\vec{k})$ in (3.24) by putting the latter equation into (3.23) and recalling (3.10), (3.12) and (3.20) :

$$2(k^2+m^2)^{1/2}\omega(\vec{k}) - \frac{\lambda_0}{(2\pi)^3}\int \frac{\omega(\kappa)}{(\kappa^2+m^2)}\, d^3\kappa = E\omega(\vec{k}) \ . \tag{3.25}$$

Omitting a trivial multiplicative constant, the solution to (3.25) is

$$\omega(\vec{k}) = [(k^2+m^2)^{1/2} - \tfrac{1}{2}E]^{-1} \qquad \text{for } 0 \leqslant E < 2m\ , \tag{3.26a}$$

$$\omega(\vec{k}) = P\,[(k^2+m^2)^{1/2} - \tfrac{1}{2}E]^{-1} + \alpha\delta((k^2+m^2)^{1/2} - \tfrac{1}{2}E) \quad \text{for } E \geqslant 2m, \tag{3.26b}$$

where α is a disposable real constant and the P in (3.26a) signifies that the principal part of the pole-term contribution is to be taken in (3.24). For existence of either the particle-antiparticle bound-state solution (3.26a) or the scattering-state solution (3.26b), it is necessary that the condition

$$\frac{\lambda_0}{2(2\pi)^3}\int \frac{\omega(\kappa)}{(\kappa^2+m^2)}\, d^3k = 1 \tag{3.27}$$

be satisfied. Moreover, for either of the solutions (3.26) to exist in the theory with a large finite cutoff parameter K, it is necessary that the condition

$$I_K \equiv \frac{\lambda_0}{4\pi^2}\int_0^K \frac{\omega(\vec{k})k^2}{(k^2+m^2)}\, dk = 1 \tag{3.28}$$

be satisfied identically in K for large values of the cutoff parameter. Explicit evaluation of the integral in (3.28) with (3.26) produces

$$I_K = (\lambda_0/4\pi^2)[\ln(2K/m) + \mathcal{F}(E/2m) + \alpha G(E/2m) + O(m^2K^{-2}) \quad , \tag{3.29}$$

where

$$\mathcal{F}(\zeta) = \frac{\pi}{2\zeta} - \frac{\pi}{\zeta}(1-\zeta^2)^{1/2} + \frac{(1-\zeta)^{1/2}}{\zeta} \tan^{-1} \frac{(1-\zeta^2)^{1/2}}{\zeta} \quad \text{for } 0 \leqslant \zeta \leqslant 1 \ , \tag{3.30a}$$

$$= \frac{\pi}{2\zeta} - \frac{(\zeta^2-1)^{1/2}}{\zeta} \tanh^{-1} \frac{(\zeta^2-1)^{1/2}}{\zeta} \quad \text{for } \zeta \geqslant 1 \ , \tag{3.30b}$$

and

$$G(\zeta) = 0 \qquad \text{for } 0 \leqslant \zeta \leqslant 1 \ , \tag{3.31a}$$

$$= (\zeta^2-1)^{1/2}/\zeta \qquad \text{for } \zeta \geqslant 1 \ . \tag{3.31b}$$

Consequently, by putting the renormalization formula (3.14b) into (3.29), we obtain

$$I_K = 1 + [\ln(2\mu/m) + \mathcal{F}(E/2m) + \alpha G(E/2m)] / [\ln(K/\mu)] + O(m^2/K^2 \ln(K/\mu)) \ . \tag{3.32}$$

It follows form (3.32) and (3.28) that we have

$$\ln(2\mu/m) + \mathcal{F}(E/2m) + \alpha G(E/2m) = 0 \tag{3.33}$$

as a necessary condition for the solutions (3.26) to exist. Thus admissible values of the center-of-momentum system bound-state energy E (particle-antiparticle bound-state mass) satisfy

$$\mathcal{F}(E/2m) = \ln(m/2\mu) \qquad \text{for } 0 \leqslant E < 2m \tag{3.34}$$

while the scattering constant α in (3.26b) is given by

$$\alpha = [\ln(m/2\mu) - \mathcal{F}(E/2m)]/G(E/2m) \quad \text{for } E > 2m \tag{3.35}$$

The critical case E = 2 m is secured formally by applying the obvious limiting procedure to (3.26b) with (3.35). Finally, we note that the configuration space transform of the wave function

$$\hat{\phi}(m|\vec{x}-\vec{y}|;\ \frac{E}{2m}) \equiv \int g(\vec{x},\vec{z})\tilde{\phi}(\vec{z},\vec{y})d^3z$$

$$= \int \frac{\omega(\vec{k})}{(k^2+m^2)^{1/2}} e^{i\vec{k}.(\vec{x}-\vec{y})} d^3k$$

$$= \frac{4\pi}{|\vec{x}-\vec{y}|}\int_0^\infty \frac{\omega(\vec{k})}{(k^2+m^2)^{1/2}}$$

$$\times (\sin k|\vec{x}-\vec{y}|)k dk \tag{3.36}$$

follows from (3.24), (3.20), and (3.26). In accordance with (3.21), the quantity (3.36) is square-integrable with respect to the relative position vector $(\vec{x}-\vec{y})$ for the bound-state solution (3.26a), but not for the scattering-state solution (3.26b). We discuss the evaluation of (3.36) with (3.26a) and (3.26b) in the Appendix.

Now since (3.30a) produce $F(0) = -1$, $F(1) = \frac{1}{2}\pi$, and an $F(\zeta)$ that increases monotonically over the unit interval, eq.(3.34) admits a non-negative root E/2m less than unity provided that μ satisfies the bounding relation $\frac{1}{2}\, me^{-\pi/2} < \mu \leq \frac{1}{2}\, me$. We obtain a positive root E/2m much less than 1 if μ also satisfies $0 < [(m/2\mu)e-1] \ll 1$, for then (3.30a) and (3.34) yield

$$E/2m \cong (4/\pi)[(m/2\mu)e-1] \ll 1. \tag{3.37}$$

Hence we have a particle-antiparticle bound-state mass E one or more orders of magnitude less than the mass m associated with a one-particle state if the numerical value of μ/m is close to $\frac{1}{2}e \cong 1.359$. The admissibility of a bound-state mass E << 2m in our renormalized complex scalar field theory model suggests that a theory for a self-interacting quark spinor field may yield a quark-antiquark meson with a mass two orders of magnitude less than the mass of a constituent quark.

Some comments are in order regarding the analytical method that we have used to derive approximate stationary states for this model self-interacting complex scalar field theory. First, the Rayleigh-Ritz procedure for functionalities has been employed, with the basic assumption that physically reasonable vacuum, one-quantum, and particle-anti-particle two-quantum stationary states exist for the Hamiltonian operator (3.2) supplemented by the renormalization formulas (3.14). Because the energy is not positive-definite, it is uncertain whether physically reasonable eigenstates actually exist for the Hamiltonian operator (3.2) with finite positive values assigned to m_o and λ_o. More-over, the character of such rigorous eigenstates in the limit prescribed by the renormalization formulas (3.14), $m_o \rightarrow \infty$ and $\lambda_o \rightarrow 0$ as $K \rightarrow \infty$, is totally open to question. However, it is satisfying that the formal Rayleigh-Ritz approximation method leads to reasonable results for the quantities of physical significance, questions of existence notwithstanding.

APPENDIX

By putting the bound-state solution (3.26a) into (3.36), we obtain

$$\hat{\phi}(\rho;\zeta) = \frac{4\pi m}{\rho} \int_0^\infty \frac{(\sin \rho\xi)\xi \, d\xi}{[(\xi^2+1)^{1/2} - \zeta](\xi^2+1)^{1/2}}$$

$$= \frac{2\pi^2 m}{\rho} e^{-(1-\zeta^2)^{1/2}\rho} - \frac{4\pi m\zeta}{\rho} \frac{\partial \mathcal{K}(\rho,\zeta^2)}{\partial \rho}$$

$$\text{for } 0 \leqslant \zeta < 1 , \qquad \text{(A1)}$$

where $\rho \equiv m|\vec{x}-\vec{y}|$ is non-negative, $\zeta \equiv E/2m$, and the transcendental function

$$\mathcal{K}(\rho,\zeta^2) \equiv \int_0^\infty \frac{(\cos \rho\xi)\, d\xi}{(\xi^2 + 1 - \zeta^2)(\xi^2+1)^{1/2}} . \qquad \text{(A2)}$$

We have

$$\mathcal{K}(\rho,\zeta^2) = \sum_{n=0}^{\infty} \zeta^{2n} \int_0^\infty \frac{(\cos \rho\xi)\, d\xi}{(\xi^2+1)^{n+3/2}}$$

$$= \sum_{n=0}^{\infty} \frac{\zeta^{2n}}{1\times 3\times \ldots \times (2n+1)} \rho^{n+1} K_{n+1}(\rho) \qquad \text{(A3)}$$

$$\text{for } 0 \leqslant \zeta < 1 ,$$

with $K_n(\rho)$ denoting the nth-order Bessel function of imaginary argument. In particular, for $\zeta = 0$ we find

$$\mathcal{K}(\rho,0) = \rho K_1(\rho) = (\tfrac{1}{2}\pi\rho)^{1/2} e^{-\rho}[1+\tfrac{3}{8}\rho^{-1}+O(\rho^{-2})] . \qquad \text{(A4)}$$

For $\zeta > 0$ and values of ρ large compared to ζ^{-2}, the function (A2) has the asymptotic expansion

$$\mathcal{H}(\rho,\zeta^2) = \frac{\pi}{2\zeta(1-\zeta^2)^{1/2}} e^{-(1-\zeta^2)^{1/2}\rho} - \zeta^{-2} (\frac{\pi}{2\rho})^{1/2} e^{-\rho}$$

$$\times [1-(\zeta^{-2}+\frac{1}{8})\rho^{-1}+O(\zeta^{-4}\rho^{-2})] , \quad \text{(A5)}$$

as one verifies immediately by noting that (A2) is a particular solution to the equation

$$(1 - \zeta^2 - \frac{\partial^2}{\partial\rho^2})\mathcal{H}(\rho,\zeta^2) = \int_0^\infty \frac{(\cos\rho\xi)\, d\xi}{(\xi^2 + 1)^{1/2}}$$

$$= K_0(\rho) = (\pi/2\rho)^{1/2} e^{-\rho}[1 - \frac{1}{8}\rho^{-1} + O(\rho^{-2})] . \quad \text{(A6)}$$

Hence (A1) is given by

$$\hat{\phi}(\rho;\zeta) = \frac{2\pi^2 m}{\rho} e^{-\rho} \qquad \text{for } \zeta = 0 ,$$

$$= \frac{4\pi^2 m}{\rho} e^{-(1-\zeta^2)^{1/2}\rho} - \frac{m}{\zeta} (\frac{2\pi}{\rho})^{3/2} e^{-\rho} \quad \text{(A7)}$$

$$\times [1 + (\frac{3}{8} - \zeta^{-2})\rho^{-1} + O(\zeta^{-4}\rho^{-2})]$$

$$\text{for } 0 < \zeta < 1 .$$

By putting the scattering-state solution (3.26b) into (3.36), we get

$$\hat{\phi}(\rho;\zeta) = \frac{2\pi^2 m}{\rho} \cos[(\zeta^2-1)^{1/2}\rho] - \frac{4\pi m\zeta}{\rho} \frac{\partial\mathcal{H}_{Re}(\rho,\zeta^2)}{\partial\rho}$$

$$+ (4\pi m\alpha/\rho)\sin[(\zeta^2-1)^{1/2}\rho]$$

$$\text{for } \zeta \geqslant 1 , \quad \text{(A8)}$$

where

$$\mathcal{H}_{Re}(\rho,\zeta^2) \equiv \mathrm{Re}\mathcal{H}(\rho,\zeta^2) \qquad \text{for } \zeta^2 \geqslant 1 . \quad \text{(A9)}$$

It should be noted that the latter quantity (A9), the principal part of the integral (A2) for $\zeta^2 \geqslant 1$, can be obtained from an expression for $\zeta^2 < 1$ by analytic continuation in ζ^2. Thus by virtue of (A5) and the scattering constant formula (3.35), we have

$$\hat{\phi}(\rho;\zeta) = (4\pi^2 m/\rho)[\cos[(\zeta^2-1)^{1/2}\rho]$$
$$+ \pi^{-1}\{[\ln(m/2\mu) - F(\zeta)]/G(\zeta)\}\sin[(\zeta^2-1)^{1/2}\rho]]$$
$$- (m/\zeta)(2\pi/\rho)^{3/2}\, e^{-\rho}[1 + O(\rho^{-1})]$$
$$\text{for} \quad \zeta \geqslant 1 \qquad \text{(A10)}$$

One can read off the (exclusively S wave) scattering cross-section directly from (A10).

REFERENCES

1. See, for example, J.P.Hsu, Phys.Rev.Letters 36, 646 (1976); M.Creutz, "Quantum Mechanics of Extended Objects in Relativistic Field Theory" (preprint of paper to appear in Physical Review D); J.Hietarinta, "On the Existence of Confied Solutions in Nonlinear Chiral Theories" (preprint of paper to appear in Physical Review D), and works cited therein. Alternative functional integration and other approaches to quantized "particlelike" solutions have been proposed by other authors, including : R.F.Dashen et al., Phys. Rev. D 10,4114 (1974); 11, 3424 (1975); J.L.Gervais et al., Phys.Rev. D 12, 1038 (1975); J.Goldstone and R.Jackiw, Phys.Rev.D 11, 1486 (1975).

2. For example, P.M.Morse and H.Feshback, Methods of Theoretical Physics (McGraw-Hill, Inc., New York, 1953) pp.1117-1119, 1696-1698, 1734-1738.

3. G.Rosen, Phys.Rev.Letters 16, 704 (1966); Phys.Rev. 156, 1517 (1967); 160, 1278 (1967); 165, 1934 (1968); 167, 1395 (1968); 172, 1632 (1968); 173, 1680 (1968); J.Math Phys. 9, 804 (1968); Nuovo Cim. 57A, 870 (1968); J.Franklin Inst. 287, 261 (1969).

4. G.Rosen, Formulations of Classical and Quantum Dynamical Theory (Academic Press, Inc., New York, 1969).

Part II

Applications of the Path Integral Approach

Useful Bounds on Interesting Quantities by Path Integrals

J.M. Luttinger

Department of Physics, Columbia University

New York, New York 10027

In these lectures I shall discuss some interesting consequences of the representation of Green's functions by means of path integrals. First we shall discuss a lower bound for the partition function of a system that follows from this representation, and show how it may be used to discuss the density of energy levels of a model disordered system. Second a very natural upper bound will be discussed, and it will be shown that special cases of this comprise all the well-known "isoperimetric inequalities" of physics and geometry.

This research was supported in part by the National Science Foundation under Grant NSF PHY75-05660 A01.

1) The density of electronic levels for a simple model of a disordered system.

In recent years there has been a considerable interest in the electronic structure of disordered systems. In these lectures we shall address ourselves to some relatively simple questions for a very idealized model of a disordered system. The model is the following : an electron is allowed to interact with fixed impurities of which a number N are distributed "at random" in a volume V. (The exact meaning of "at random" will be made clear below.) The interaction of an impurity located at the point $\vec{R}$ and the electron at the point $\vec{r}$ is represented by a short-ranged purely repulsive potential $v(\vec{r}-\vec{R})$. In the thermodynamic limit ($N \to \infty$, $V \to \infty$, but $N/V \to \rho$) it is not difficult to show that the density of electronic energy levels per unit volume in the neighborhood of the energy ε is well defined and vanishes for $\varepsilon < 0$. We denote this quantity by $g(\varepsilon)$. The problem is now to determine $g(\varepsilon)$. The answer is not known in more than one dimension for any reasonable potential v. We shall give in what follows a variational method based on path integrals for the determination of $g(\varepsilon)$ (Friedberg and Luttinger, 1975). Although it does not yield great numerical accuracy for $g(\varepsilon)$, it gives its general behavior as a function of ε rather directly. In particular, the nature of the behavior of $g(\varepsilon)$ as ε approaches zero is obtained easily. This question (which was first considered by Lifshitz (1964) is by no means trivial, and as far as I know has not been answered by non-path integral methods based on the Schroedinger equation. Lifshitz has used an intuitively appealing argument (given below) to obtain the answer : in a sense what follows is a systematic method which puts Lifshitz's argument on a mathematically firmer basis and shows how to obtain the first corrections to it.

We may summarize Lifshitz's idea as follows : Electronic energy levels for arbitrarily small energy can only come from states with wave functions localized in very large regions which are empty of impurities. If the electronic wave function overlaps an impurity appreciably, there will be a finite potential energy of interaction, while the largeness of the region is necessary so that the kinetic energy can be made very small. Now, as is well known, the probability of a large region of volume Ω being free of impurities in a random system of impurities is proportional to

$e^{-\rho\Omega}$. Since the volume Ω is very large, the low-lying levels for states localized in it will be insensitive to the exact conditions on the boundary ($\partial\Omega$) of Ω, and we may take the wave function to be zero on $\partial\Omega$. Now, clearly the main contribution to the probability of finding a low-level ε for the system will be proportional to the probability of finding a region Ω whose lowest level is ε. The probability of finding a region whose second level is ε will be exponentially smaller because of the exponential dependence of the probability on Ω. Further, because of this same exponential dependence, the regions whose shape is such that Ω is smallest for a given lowest level will make the main contribution. By a well-known "isoperimetric" inequality, this will mean spherical regions.* The lowest-level ε in an empty spherical volume of radius R_o with the boundary condition that the wave function vanish on its surface is given by

$$\varepsilon = \pi^2/2R_o^2 \quad \text{or} \quad R_o = (\pi^2/2\varepsilon)^{1/2} \tag{1.1}$$

(units such that $m = \hbar = 1$, which we use throughout this paper). The probability of such a region existing is proportional to $\exp[-\rho(4\pi/3)R_o^3]$, so that the density of low-lying levels will be given by

$$g(\varepsilon) \sim \exp(-c_3/\varepsilon^{3/2}) \quad (\varepsilon \to 0) , \tag{1.2}$$

where

$$c_3 = \frac{1}{3}\, 4\, \pi(\frac{1}{2}\, \pi^2)^{3/2}\, \rho. \tag{1.3}$$

For the system being discussed the hamiltonian is

$$H = \frac{1}{2}\, p^2 + U(\{R\}) \equiv H_o + U(\{R\}) , \tag{1.4}$$

$$U(\{R\}) = \sum_{j=1}^{N} v(\vec{r}-\vec{R}_j) . \tag{1.5}$$

where the $\vec{R}_j$ are the positions of the impurities. To calculate the density of states, we consider first the "partition function" $Z_t(\{R\})$ defined by

---

* We shall hear more about these isoperimetric inequalities later in these lectures.

$$Z_t(\{R\}) = \text{Tr}\, e^{-tH} , \tag{1.6}$$

where t is real and positive. Clearly, for a large system

$$Z_t(\{R\}) = \int_0^\infty d\varepsilon e^{-t\varepsilon} Vg(\varepsilon,(\{R\})) , \tag{1.7}$$

since $V_g(\varepsilon,(\{R\}))$ is the density of energy levels for the system. Now let the probability that $\vec{R}_1$ be in $d\vec{R}_1$, $\vec{R}_2$ be in $d\vec{R}_2$, etc., etc. be $d\vec{R}_1/V$, $d\vec{R}_2/V \ldots d\vec{R}_N/V$. For each arrangement $Z_t(\{R\})$ will have a certain value, so that it is also a random variable with a certain probability distribution function. Now it is not difficult to show that in the thermodynamic limit, as would be expected on physical grounds, $Z_t\{R\}/V$ is certainly equal to its mean value z(t) defined by

$$z(t) \equiv \lim_{\substack{V,N \to \infty \\ N/V \to \rho}} \int_V \frac{dR_1}{V} \frac{dR_2}{V} \ldots \frac{dR_N}{V} \frac{Z_t(\{R\})}{V} , \tag{1.8}$$

so that there is a well-defined density of states per unit volume, $g(\varepsilon)$ given by

$$\int_0^\infty g(\varepsilon) e^{-\varepsilon t} d\varepsilon = z(t) . \tag{1.9}$$

That is, to find $g(\varepsilon)$ we need only find z(t) and invert the Laplace transform.

To find z(t) we proceed as follows. Define

$$I(t) = \frac{1}{Z_t^o} \int_V \frac{d\vec{R}_1}{V} \ldots \frac{d\vec{R}_N}{V} z_t(\{R\})$$

$$\equiv \frac{< z_t(\{R\}) >_{imp}}{z_t^0} \tag{1.10}$$

where z^0 is the partition function in the absence of impurities. Z_t^0 may be calculated easily and yields

$$Z_t^0 = V/(2\pi t)^{3/2} \quad , \tag{1.11}$$

$$I(t) = \frac{1}{Z_t^0} \ll \int_V d\vec{r}_0 < \vec{r}_0 | e^{-tH} | \vec{r}_0 > \gg_{imp} \tag{1.12}$$

in the coordinate representation. Now we interchange the impurity average and the $\vec{r}_0$ integration. Since after averaging over the positions of the impurities there is no preferred position and the impurity average of $< \vec{r}_0 | e^{-tH} | \vec{r}_0 >$ cannot depend on $\vec{r}_0$, we may write ($\vec{r}_0$ now some arbitrary point of V)

$$\begin{aligned} I(t) &= \frac{1}{Z_t^0} V \ll \vec{r}_0 | e^{-tH} | \vec{r}_0 \gg_{imp} \\ &= \frac{\ll \vec{r}_0 | e^{-tH} | \vec{r}_0 \gg_{imp}}{\ll \vec{r}_0 | e^{-tH_0} | \vec{r}_0 \gg_{imp}} \end{aligned} \tag{1.13}$$

The right-hand side of (1.13) may now be written as a path (or Wiener) integral according to the well-known Feynman-Kac formula (Kac, 1959). For the sake of pedagogy and as a convenient way of introducing the notation we shall use, we now give of "physicists" derivation of this expression. Consider the identity

$$\begin{aligned} < \vec{r}_0 | e^{-\tau H} | \vec{r}_0 > &= \int d\vec{r}_1 \, d\vec{r}_2 \ldots d\vec{r}_n < \vec{r}_0 | e^{-\tau H} | \vec{r}_1 > \\ &< \vec{r}_1 | e^{-\tau H} | \vec{r}_2 > \ldots < \vec{r}_n | e^{-\tau H} | \vec{r}_o > \end{aligned} \tag{1.14}$$

where $\tau = t/(n+1)$. Now suppose n is very large (so τ very small). Then

$$\begin{aligned} &< \vec{r}_i | e^{-\tau H} | \vec{r}_j > = \\ &< \vec{r}_i | e^{-\tau(p^2/2 + U)} | \vec{r}_j > \xrightarrow[\tau \to 0]{} < \vec{r}_i | e^{-\tau \, p^2/2} \, e^{-\tau U} | \vec{r}_j > \end{aligned} \tag{1.15}$$

Since the error made due to the non commutativity of ($\tau \, p^2/2$ and (τU) is of the order of τ^2. Therefore

$$< \vec{r}_i | e^{-\tau H} | \vec{r}_j > \underset{\tau\to 0}{\longrightarrow} < \vec{r}_i | e^{-\tau\; p^2/2} | \vec{r}_j > e^{-\tau U(\vec{r}_j)}$$

$$= \frac{e^{-(\vec{r}_i - \vec{r}_j)^2/2\,\tau}}{(2\pi\tau)^{3/2}}\, e^{-\tau U(\vec{r}_j)} \equiv P_{ij} e^{\,-\,\tau U(r_j)} \tag{1.16}$$

The evaluation of the matrix element in (1.16) is trivial, since it is just the free particle propagator or Green's function. Thus, the right hand side of (1.13) becomes, before the impurity average,

$$= \lim_{n\to\infty} \frac{\int d\vec{r}_1 \ldots d\vec{r}_n P_{01} P_{12} \ldots P_{n0}\, e^{-\tau \sum_i U(r_i)}}{\int d\vec{r}_1 \ldots d\vec{r}_n\; P_{01} P_{12} \ldots P_{n0}} \tag{1.17}$$

$$\equiv < e^{-\int_0^t U(\vec{r}_{t'}) dt'} | \vec{r}_0 >_t \tag{1.18}$$

The right hand side of (1.17) can be thought of as the average of the functional of $\vec{r}_t$, $\exp\,(-\int_0^t U(\vec{r}_{t'}) dt')$ over all continuous paths (all $\vec{r}_{t'}$) which begin and end at the point $\vec{r}_0$, with a special measure (or probability) for each path $\vec{r}_t$, by the P_{ij} through (1.17). This is the (conditional) Wiener average over "paths", and (1.18) is the Feynman-Kac formula. Thus

$$I(t) = << \exp\,(-\int_0^t U(\vec{r}_{t'}) dt') | \vec{r}_0 >_t >_{imp} \tag{1.19}$$

Now the great advantage of this representation is that it enables us to carry out the impurity average and thermodynamic limit rather simply :

$$< \exp\,(-\int_0^t U(\vec{r}_{t'}) dt') >_{imp}$$

$$= \prod_{j=1}^{N} \int_V \exp\,(-\int_0^t \nu\; \vec{r}_{t'} - \vec{R}_j)\; dt'\; \frac{d\vec{R}_j}{V}$$

$$= [\frac{1}{V} \int_V d\vec{R} \exp (- \int_0^t \nu(\vec{r}_{t'} - \vec{R})dt')]^N \qquad (1.20)$$

$$= [1 - \frac{1}{V} \int_V d\vec{R} (1 - \exp (\int_0^t \nu(\vec{r}_{t'} - \vec{R})dt'))]^N$$

For a $\nu(\vec{r})$ which goes to zero sufficiently rapidly as $|\vec{r}|$ does to infinity, the integral over $\vec{R}$ converges if extended over all space. Therefore, in the thermodynamic limit we may write

$$\lim_{\substack{N,V\to\infty \\ N/V\to\rho}} < \exp (- \int_0^t U(r_{t'})dt') >_{imp}$$

$$= \lim_{N\to\infty} \{1 - \frac{\rho}{N} \int d\vec{R} [1 - \exp (- \int_0^t \nu(\vec{r}_{t'},-\vec{R})dt'] \}^N \qquad (1.21)$$

$$= e^{-\rho W} \quad ,$$

where W is the functional

$$W = \int d\vec{R} [1 - \exp (- \int_0^t \nu_{\vec{R}}(\vec{r}_{t'})dt')] \quad ,$$

$$\nu_{\vec{R}}(\vec{r}) = \nu(\vec{r}-\vec{R}) \qquad (1.22)$$

and the $\vec{R}$ integration is extended over all space.

Putting all these results together, we have

$$z(t) = \frac{1}{(2\pi t)^{3/2}} < e^{-\rho W}|o>_t \quad , \qquad (1.23)$$

where we have chosen the origin to be the point $\vec{r}_o$. This type of formula seems first to appear in the work of Edwards and Gulyaev (1962). The difficulty in using it lies in the fact that except for some very simple functionals, path integrals are extremely difficult to evaluate.

Now one of the principal advantages of the path-integral formulation as given by (1.23) is that it allows us in a rather natural way to formalize Lifshitz's argument. That is the low-lying states, and therefore the large t behavior, come from electrons trapped in a potential well of large radius. Let us be a little more general and not specify the shape of the potential well for the moment but simply call it $\phi(\vec{r})$. If we only had such a well present, $z(t)$ would be, by the Kac formula, proportional to $< e^{-\Phi}|o>_t$, where

$$\Phi = \int_o^t \phi(\vec{r}_{t'})dt' \quad . \tag{1.24}$$

Therefore, we write

$$< e^{-\rho W}|o>_t = < e^{-\Phi}|o>_t \frac{< e^{-(\rho W-\phi)}e^{-\phi}|_o>_t}{< e^{-\phi}|o>_t}$$

$$\equiv < e^{-\Phi}|o>_t < e^{-A}|o;\phi>_t \tag{1.25}$$

where

$$A = \rho W - \Phi \quad . \tag{1.26}$$

The first factor in (1.25) may be thought of as giving the contribution to $z(t)$ when the electron is trapped in the potential ϕ, while the second factor represents the "probability" of finding such a potential. We shall determine the "best" ϕ by the following argument. Since the second factor is an average of the functional $\exp(-A)$, we have, by Jensen's well-known theorem for convex functions,

$$< e^{-A}|o,\phi>_t \geqslant e^{-<A|o,\phi>_t} \quad . \tag{1.27}$$

Therefore

$$< e^{-\rho W}|o>_t \geqslant < e^{-\phi}|o>_t \, e^{-<A|o,\phi>_t} \quad . \tag{1.28}$$

We choose ϕ such that the right hand side of (1.28) is as large as possible, so that we will have the strongest possible inequality of this form. That is, we choose ϕ such that

$$\frac{\delta}{\delta\phi(\vec{r})} < e^{-\phi}|o>_t \, e^{-<A|o,\phi>_t} = 0 \; , \tag{1.29}$$

where $\delta/\delta\phi(\vec{r})$ is the variational derivative of a functional of ϕ with respect to ϕ at the point $\vec{r}$.

So far all we have from this procedure is an inequality; so one may ask if any progress has been made towards an evaluation. However, (1.27) is actually the first term of the exact (assuming convergence) semi-invariant or cumulant expansion (see for example Kubo (1962)) :

$$< e^{-A} > = \exp\{- <A> + \frac{1}{2!}(<A^2> - <A>^2) \tag{1.30}$$

$$- \frac{1}{3!}[<A^3> - 3<A>(<A^2> - <A^2>) - <A>^3] + \ldots\} .$$

Therefore, if we show that as $t \to \infty$ (for example) and ϕ is given from (1.24) $<A>$ increases more rapidly with t than the other semi-invariants, then (1.18) will an asymptotic evaluation rather than an inequality. This proves to be the case.

In order to carry out the procedure described above, we introduce some notation. Suppose we have a Hamiltonian H given by

$$\tilde{H} = \frac{1}{2}p^2 + \tilde{\phi}(\vec{r}) . \tag{1.31}$$

Then in terms of the Green's function,

$$G_\tau(\vec{r},\vec{r}'|\tilde{\phi}) \equiv < \vec{r}|e^{-\tau\tilde{H}}|\vec{r}' > , \tag{1.32}$$

we have at once, making use of the Kac formula

$$< e^{-\Phi}|o >_\tau = (2\pi t)^{3/2}\, G_t(o,o|\phi) , \tag{1.33}$$

$$< A|o,\phi >_t = \rho <W|o,\phi >_t - < \Phi|o,\phi >_t, \tag{1.34}$$

$$< W|o,\phi >_t =$$

$$= \frac{\int d\vec{R} <[1-\exp(-\int_o^t \nu_{\vec{R}}(\vec{r}_{t'})dt')]\exp(-\int_o^t \phi(\vec{r}_{t'})dt')|o>_t}{<\exp(-\int_o^t \phi(\vec{r}_{t'})dt')|o>_t}$$

$$= \int d\vec{R}\,[G_t(o,o|\phi) - G_t(o,o|\phi + \nu_{\vec{R}})]/G_t(o,o|\phi) , \tag{1.35}$$

$$< \Phi|o,\phi >_t = \frac{< \int_o^t \phi(\vec{r}_{t'})dt' \exp(- \int_o^t \phi(\vec{r}_{t'})dt')|o >_t}{< \exp (- \int_o^t \phi(\vec{r}_{t'})dt')|o >_t}$$

$$= - \frac{\partial}{\partial\mu} \ln <\exp (- \int_o^t \mu\phi(\vec{r}_{t'})dt')|o >_t \Big|_{\mu=1}$$

$$= \frac{\partial \ln (2\pi t)^{3/2} G_t(o,o|\mu\phi)}{\partial\mu}\Big|_{\mu=1} \qquad (1.36)$$

$$= - \frac{\partial G_t(o,o|\mu\phi)}{\partial\mu}\Big|_{\mu=1} / G_t(o,o|\phi) .$$

From these formulas it is not at all difficult to work out the condition (1.29), making use of the elementary perturbation theoretic result :

$$G_t(\vec{r}_1,\vec{r}_2|\tilde{\phi} + \delta\phi) = G_t(\vec{r}_1,\vec{r}_2|\phi)$$
$$- \int_o^t dt' \int d\vec{r}\; G_{t-t'}(\vec{r}_1,\vec{r}|\phi)\delta\tilde{\phi}(\vec{r})G_{t'}(\vec{r},\vec{r}_2|\phi)+O(\delta\tilde{\phi})^2$$
$$(1.37)$$

The resulting condition is quite complicated, leading to a non-linear integro-differential equation for $G_\tau(\vec{r}_1,\vec{r}_2|\phi)$. This equation has not been investigated in detail (though it seems possible using an iterative procedure with a large computer). Instead the limiting cases of large t (which gives the distribution of energy levels at small energy) and small t (which gives the distribution of energy levels at large energy) have been studied. The former is the most interesting case physically (due to its connection with Lifshitz conjecture and the non-applicability of perturbation theory), and we shall devote most of our effort to studying it. To do this, we need the following well-known representation of the Green's function [which follows at once from the definition (1.32)] :

$$G_\tau(\vec{r},\vec{r}'|\tilde{\phi}) = \sum_j e^{-\tau E_j(\tilde{\phi})}\psi_j(\vec{r}|\phi)\psi_j(\vec{r}'|\tilde{\phi}) \ , \qquad (1.38)$$

where the $\psi_j(\vec{r}|\phi)$ and the $E_j(\phi)$ are the (real) normalized eigenfunctions and eigenvalues of $\tilde{H}$, i.e.,

$$\tilde{H}\psi_j(\vec{r}|\tilde{\phi}) = E_j(\tilde{\phi})\psi_j(\vec{r}|\tilde{\phi}) \ . \qquad (1.39)$$

Now we expect from the Lifshitz picture (and will find) that ϕ is a large potential well in which the lower states form a discrete spectrum. Therefore, for large τ we expect the ground state ψ_o, E_o to dominate (1.38) with an error which is exponentially small; so we may write

$$G_t(o,o|\phi) \cong \psi_o^2(o|\phi)e^{-\tau E_o(\phi)} \ . \qquad (1.40)$$

$$G_t(o,o|\phi+\nu_R) = \psi_o^2(o|\phi+\nu_R)e^{-tE_o(\phi+\nu_R)} \ , \qquad (1.41)$$

and

$$\frac{\partial}{\partial\mu} G_t(o,o|\mu\phi)\Big|_{\mu=1}$$

$$= \frac{\partial}{\partial\mu} \psi^2(o|\mu\phi)e^{-tE_o(\mu\phi)}\Big|_{\mu=1}$$

$$= - t\psi^2(o|\phi)e^{-tE_o(\phi)} \frac{\partial E_o(\mu\phi)}{\partial\mu}\Big|_{\mu=1}$$

$$+ 2 e^{-tE_o(\phi)}\psi_o(o|\phi) \frac{\partial\psi_o(o|\mu\phi)}{\partial\mu}\Big|_{\mu} \qquad (1.42)$$

The derivatives in (1.42) are directly calculable using perturbation theory ($\mu = 1 + \varepsilon, \varepsilon \to 0$) ,

$$\frac{\partial E_o(\mu\phi)}{\partial\mu}\Big|_{\mu=1} = \int dr\psi_o^2(\vec{r}|\phi)\phi(\vec{r}) \ , \qquad (1.43)$$

$$\frac{\partial\psi_o(o|\mu\phi)}{\partial\mu}\Big|_{\mu=1} = \sum_{j\neq o} \frac{\int \psi_o(\vec{r}|\phi)\phi(\vec{r})\psi_j(\vec{r}|\phi)d\vec{r})\psi_j(o|\phi)}{E_o(\phi) - E_j(\phi)} \qquad (1.44)$$

The leading term of (1.42) is the first term on the right-hand side (because of the factor of t). We shall drop the second term. (After the calculation is complete, we can go back and evaluate this correction term with the ϕ we have obtained. It is then found that this correction is smaller than the terms considered in this paper, though it is not exponentially small.) Therefore, we write to the accuracy considered,

$$\frac{1}{G_t(o,o|\phi)} \left.\frac{\partial G_t(o,o|\mu\phi)}{\partial\mu}\right|_{\mu=1} = -t \int d\vec{r}\psi_o^2(\vec{r}|\phi)\phi(\vec{r}) \,, \tag{1.45}$$

$$< A|o,\phi >_t = \rho\int d\vec{R}\left(1-\frac{\psi_o^2(o|\phi+\nu_{\vec{R}})}{\psi_o^2(o|\phi)} \exp \{-t[E_o(\phi+\nu_{\vec{R}})-E_o(\phi)]\}\right)$$

$$- t \int d\vec{r}\psi_o^2(\vec{r}|\phi)\phi(\vec{r}) \,. \tag{1.46}$$

Again, the factor $\psi_o^2(o|\phi+\nu_{\vec{R}})/\psi_o^2(o|\phi)$ may be replaced by unity in the limit of large t, to the order which interests us. The reason is essentially that because of the factor $\exp\{-t[E_o(\phi+\nu_R) - E_o(\phi)]\}$, only large R (of the order of the size of the well ϕ) where the wave function is small contributes. Since $\nu_{\vec{R}} = \nu(\vec{r}-\vec{R})$ is a short-range potential, this has little influence on the wave function at the origin. Thus we may write

$$< e^{-\Phi}|o>_t \, e^{-<A|o,\phi>_t}$$

$$= e^{-tQ+\ln\psi_o^2(o|\phi) + \ln(2\pi t)^{3/2}} \,, \tag{1.47}$$

where

$$Q \equiv E_o(\phi) - \int \psi_o^2(\vec{r}|\phi)\phi(\vec{r})d\vec{r}$$

$$+ \frac{\rho}{t} \int d\vec{R}(1 - \exp\{-t[E_o(\phi+\nu_{\vec{R}}) - E_o(\phi)]\})$$

$$= \int \psi_o(\vec{r}|\phi)1/2\, p^2\psi_o(\vec{r}|\phi)d\vec{r}$$

$$+ \frac{\rho}{t} \int d\vec{R}(1 - \exp\{-t[E_o(\phi+\nu_{\vec{R}}) - E_o(\phi)]\}) \tag{1.48}$$

We may neglect the term $\ln \psi_o^2(o|\phi)$ in the exponent since it is found from the same type of self-consistency argument as we have been using to be of order lnt, which again is negligible compared to the terms we are retaining. Finally then, we may write

$$\ln [<(e^{-\Phi}|_o>_t \; e^{-<A|o,\phi)>}t] \cong -tQ \qquad (1.49)$$

The stationarity condition (1.29) becomes

$$\delta Q/\delta\phi(\vec{r}) = 0 \qquad (1.50)$$

To proceed further in a simple manner, we must make some assumption about the potential $\nu_{\vec{R}}$, so as to be able to evaluate the energy shift in (1.48). The simplest assumption (which we shall make) is that the range of $\nu(\vec{r})$ is extremely short, shorter than any other length in the problem. This enables us to use the Fermi method of pseudopotentials. (Blatt and Weisskopf, (1956)). A straightforward application of this method gives the following result : to calculate $E_o(\phi + \nu_{\vec{R}}) - E_o(\phi)$, we may replace $\nu(\vec{r}-\vec{R})$ by $2\pi a\delta(\vec{r}-\vec{R})$ [where a is the scattering length for the potential $\nu(\vec{r})$] and use first-order perturbation theory. That is,

$$E_o(\phi + \nu_{\vec{R}}) - E_o(\phi) = 2\pi a\psi_o^2(\vec{R}|\phi) \; , \qquad (1.51)$$

so that

$$Q = \int \psi\frac{p^2}{2}\psi d\vec{r} + \frac{\rho}{t}\int d\vec{r}\{1 - \exp[-2\pi at\psi^2(\vec{r})]\} \qquad (1.52)$$

[We now use the simplified notation $\psi_o(\vec{r}|\phi) = \psi(\vec{r})$, $E_o(\phi) = E$] .

Since Q is a functional of ψ alone and ψ is normalized, the condition (1.50) is equivalent to

$$\frac{\delta}{\delta\psi(\vec{r})}(Q - \lambda \int\psi^2(\vec{r})d\vec{r}) = 0 \; , \qquad (1.53)$$

where λ is the Lagrange multiplier for the normalization constant. This gives at once

$$\frac{1}{2}p^2\psi(\vec{r}) + 2\pi a\rho \, e^{-2\pi at\psi^2(\vec{r})}\psi(\vec{r}) = \lambda\psi(\vec{r}) \qquad (1.54)$$

$$\int \psi^2(\vec{r})d\vec{r} = 1 \ . \qquad (1.55)$$

Comparing (1.54) with (1.39) when $\tilde{\phi} = \phi$ and $j = 0$, we have

$$\lambda = E \ , \qquad (1.56)$$

$$\phi(\vec{r}) = 2\pi a\rho \ e^{-2\pi a t\psi^2(\vec{r})} \qquad (1.57)$$

Informally, we may regard (1.57), together with the Schrödinger equation,

$$\frac{1}{2} p^2\psi(\vec{r}) + \phi(\vec{r})\psi(\vec{r}) = E\psi(\vec{r}) \ , \qquad (1.58)$$

as a pair of relations by which the electron and the impurities influence each other in a "self-consistent" approximation. If the impurities were distributed uniformly, we might approximate their effect on the electron by a smoothed-out potential $2\pi a\rho$. But if the electronic state is chosen first, the impurity density ρ is multiplied by a Boltzmann factor $\exp[-2\pi a t\psi^2(\vec{r})]$, leading to (1.57). Putting (1.57) and (1.58) together, we get (1.54) which describes the electron moving in the well from which impurities have been excluded by their interaction with that same electron.

It remains to solve (1.54) subject to (1.55) in the limit of large t. The complete analysis of this problem is fairly straightforward but rather long and uninteresting, and we shall not give it here. To get an idea of what is involved, we mention first that one can show rigorously and without difficulty the solution $\psi(\vec{r})$ with the smallest Q is spherically symmetric. Putting $\psi(\vec{r}) = \chi(r)/r$ we obtain :

$$Q = \frac{1}{2}\int_0^\infty dr(\chi'(r))^2 + \frac{4\pi\rho}{t}\int_0^\infty r^2 dr(1-e^{-2\pi a t\chi^2(r)/r^2}) \qquad (1.59)$$

Now let us take as a trial function $\chi = \tilde{\chi}$, where $\tilde{\chi}$ corresponds to the crudest Lifshitz picture. That is, take $\tilde{\chi}$ to correspond to a potential well which is zero until $r = b$ and then infinite. The b is then determined by the condition that the resultant $\tilde{Q}$ be minimized. The normalized wave function for such a well is :

$$\tilde{\chi} = \sin\left(\frac{\pi}{b} r\right)/\sqrt{2\pi b} \ , \tag{1.60}$$

and

$$\tilde{Q} = \frac{\pi^2}{2b^2} + \frac{4\pi\rho}{t} \int_0^b dr\, r^2 \left[1 - \exp\left(- \frac{t \sin^2(\pi r/b)}{br^2}\right)\right] \tag{1.61}$$

$$= \frac{\pi^2}{2b^2} + \frac{4\pi\rho}{3t} b^3 + \tilde{Q}' \ , \tag{1.62}$$

where

$$\tilde{Q}' \equiv - \frac{4\pi\rho}{t} \int_0^b dr\, r^2 \exp\left(- \frac{t \sin^2(\pi r/b)}{br^2}\right) . \tag{1.63}$$

First, assume $\tilde{Q}'$ to be negligible (for large t). Then the best value of b (say b_o) is determined by :

$$\frac{\partial}{\partial b}\left(\frac{\pi^2}{2b^2} + \frac{4\pi\rho}{3t} b^3\right) = 0 \ ,$$

or

$$b_o^5 = \pi t/4\rho \tag{1.64}$$

The leading term of $\tilde{Q}(\tilde{Q}_o)$ is then given by

$$\tilde{Q}_o = \frac{5}{3} \frac{\pi^2}{2b_o^2} = \frac{5}{6} (4\pi^4)^{2/5} \rho^{2/5} t^{-2/5} \ , \tag{1.65}$$

which is already accurate enough to give the Lifshitz conjecture as we shall see below.

To verify that $\tilde{Q}'$ is negligible compared to the term we have calculated, we compute it as follows. Put $r = b_o(1-\xi)$, and replace t by (1.64) in the exponent,

$$\tilde{Q}' = - \frac{4\pi\rho b_o^3}{t} \int_0^1 d\xi (1-\xi)^2 \exp\left(-4\pi\rho a b_o^2 \frac{\sin^2\pi\xi}{(1-\xi)^2\pi^2}\right) \tag{1.66}$$

Since $b_o \to \infty$ as $t \to \infty$ the integral can easily be calculated by saddle-point methods, the main contribution coming from the neighborhood of $\xi = 0$. This yields :

$$\tilde{Q}' = -\tilde{Q}_o \frac{1}{5} 3 \sqrt{\pi}\lambda_o/b_o \, , \tag{1.67}$$

where λ_o is the "skin depth" defined by

$$\lambda_o \equiv (1/4\pi a\rho)^{1/2} \quad . \tag{1.68}$$

Therefore

$$\tilde{Q} = \tilde{Q}_o(1 - \frac{3\sqrt{\pi}}{5} \frac{\lambda_o}{b_o} + \ldots) \tag{1.69}$$

If we solve the equations for $\chi(r)$ rigorously for large t we obtain

$$Q = \tilde{Q}_o \ (1 - \frac{12C}{5} \frac{\lambda_o}{b_o} + \ldots) \tag{1.70}$$

where

$$C \equiv \int_0^\infty dy \, [1 - (1 - e^{-y^2})^{1/2}] = 0.628\ldots \tag{1.71}$$

Comparing (1.70) and (1.69) we see that the leading term agree. All the more correct solution does is replace the numerical coefficient $3\sqrt{\pi}/5 = 1.06$ of the correction term λ_o/b_o by $12C/5 = 1.51$ which lowers Q as it must. Therefore to the order we are working, we find for large t :

$$z(t) \geqslant \exp \, (- d_3 t^{3/5} - d_3' t^{2/5} + \ldots) \tag{1.72}$$

where

$$d_3 = \frac{5}{6} \, [(4\pi)^4]^{2/5} \, \rho^{2/5} \tag{1.73}$$

$$d_3' = - 2 \, (\frac{\pi}{4\rho})^{2/5} \, \frac{c}{a\lambda_o} \tag{1.74}$$

To find the error involved in replacing the inequality in (1.72) by equality, we have to consider the higher semi-invariants. Although it is again a tedious process, it is possible to prove (Friedberg and Luttinger, 1975)) that for large t, all the semi-invariants are smaller than $O(t^{2/5})$. Therefore to the order indicated in (1.72) we may replace the inequality by an equality. To obtain the density of states $g(\varepsilon)$ we have to invert the Laplace transform in (1.9). This can be done in the small ε limit by elementary saddle point methods, and we find (ε approaches zero)

$$\ell n\ g(\varepsilon) = -c_3/\varepsilon^{3/2} - c_3'/\varepsilon - \ldots \tag{1.75}$$

where

$$C_3 = \frac{1}{3}(4\pi)\left(\frac{1}{2}\pi^2\right)^{3/2}\rho \tag{1.76}$$

$$C_3' = -\pi^2 c/a\lambda_o \tag{1.77}$$

The first term of (1.75) is exactly Lifshitz's result, and the second term is the first correction to it.

For simplicity we have discussed the (somewhat unrealistic) case where the impurities are uncorrelated i.e. the probability of any given impurity being in $d\vec{R}$ is just $d\vec{R}/V$. With very slight modification the analysis goes through for correlated impurities (Luttinger, 1976). We now sketch this calculation. The hamiltonian is again given by (1.4) and (1.5). The positions of the scattering centers are governed by a probabilistic law. That is, the probability that the first scattering center is in $d\vec{R}_1$, about $\vec{R}_1$, the second in $d\vec{R}_2$ about $\vec{R}_2$, ..., the Nth in $d\vec{R}_N$ about $\vec{R}_N$ is given by

$$P_V(\vec{R}_1, \vec{R}_2, \ldots, \vec{R}_N)\, d^N\vec{R}\quad , \tag{1.78}$$

where V is the volume of the container and

$$d^N\vec{R} = d\vec{R}_1\ d\vec{R}_2, \ldots, d\vec{R}_N\ . \tag{1.79}$$

(It is trivial to generalize the discussion to several different kinds of scattering centers, as in a random alloy, but we omit this here).

Following the method used earlier, we first construct the partition function per unit volume for fixed positions of the scattering centers, average over these positions, and then go to the thermodynamic limit. Calling the resulting quantity $z(t)$ again, we have

$$z(t) = \lim_{\substack{V,N\to\infty \\ N/V\to\rho}} \int_N d^N\vec{R} P_V(\vec{R}_1,\ldots,\vec{R}_N) \frac{1}{5} \operatorname{Tr}(e^{-tH}). \tag{1.80}$$

It is not difficult to see that for ɛeasonable P_V this limit exists and may be written

$$z(t) = \int_0^\infty e^{-\varepsilon t} g(\varepsilon) d\varepsilon . \tag{1.81}$$

To find $g(\varepsilon)$ for small ε we need only find $z(t)$ for large t.

We may write (1.80) as a Wiener or path integral. That is, just as before.

$$z(t) = (2\pi t)^{-3/2} I(t),$$

$$I(t) = \lim_{\substack{V,N\to\infty \\ N/V\to\rho}} \int_V d^N\vec{R} P_V \cdot \int d\vec{r}_0 \frac{\langle \vec{r}_0 | e^{-tH} | \vec{r}_0 \rangle}{V/(2\pi t)^{3/2}}$$

$$= \lim_{\substack{V,N\to\infty \\ N/V\to\rho}} \int_V d^N\vec{R} P_V \int d\vec{r}_0 \langle \exp(-\int_0^t U(\vec{r}_{t'})dt') | \vec{r}_0 \rangle \frac{1}{V} \tag{1.82}$$

Now if we interchange the $\vec{R}$ and $\vec{r}_0$ integrations, and assume that the system is on the average specially uniform, we may write

$$I(t) = \lim_{\substack{V,N\to\infty \\ N/V\to\rho}} \int_V \frac{d\vec{r}_0}{V} [\int_V d^N\vec{R} \langle \exp(-\int_0^t U(\vec{r}_{t'})dt') | \vec{r}_0 \rangle_t] P_V$$

$$= \lim_{\substack{V,N\to\infty \\ N/V\to\rho}} \int_V \frac{d\vec{r}_0}{V} [\int_V d^N\vec{R} \langle \exp(-\int_0^t U(\vec{r}_{t'})dt') | o \rangle_t] P_V$$

$$= \lim_{\substack{V,N\to\infty \\ N/V\to\rho}} \int_V d^N\vec{R} \langle \exp(-\int_0^t U(\vec{r}_{t'})dt') | o \rangle_t P_V. \tag{1.83}$$

since the term in the large square bracket is independent of $\vec{r}_0$ ("o" is just an arbitrary point, chosen to be the origin).

Using (1.5) we may write

$$\exp\left(-\int_0^t U(\vec{r}_{t'})dt'\right) = \prod_j (1 + W_j) , \tag{1.84}$$

where

$$W_j = W(\vec{R}_j) = \exp\left(-\int_0^t v(\vec{r}_{t'} - \vec{R}_j)dt'\right) - 1 \tag{1.85}$$

Expanding the right-hand side of (1.84) we have :

$$\prod_j (1 + W_j) = 1 + \sum_j W_j + \sum_{(jk)} W_j W_k + \sum_{(jk\ell)} W_j W_k W_\ell + \ldots , \tag{1.86}$$

where $\sum_{(jk)}$ means summation over all distinct pairs, $\sum_{(jk1)}$ means summation over all distinct triplets, etc.
Therefore

$$\int_V \prod(1 + W_j)P_V d^N\vec{R} = 1 + \frac{1}{1!}\int_V d\vec{R}_1 n_1(\vec{R}_1)W(\vec{R}_1)$$

$$+ \frac{1}{2!}\int_V d\vec{R}_1 d\vec{R}_2 n_2(\vec{R}_1,\vec{R}_2)W(\vec{R}_1)W(\vec{R}_2) + \ldots \tag{1.87}$$

where the ℓ-particle distribution function n_1 is defined by

$$n_\ell(\vec{R}_1,\ldots,\vec{R}_1) = \frac{N!}{(N-\ell)!}\int_V P_V d\vec{R}_{\ell+1},\ldots,d\vec{R}_N \tag{1.88}$$

Now we assume n_ℓ has a thermodynamic limit $\bar{n}_\ell$, so that (1.83) becomes [since $W(\vec{R})$ approaches zero rapidly as $\vec{R}$ goes to infinity, we may extend the integrals over all space]

$$I(t) = \langle (1 + \int \bar{n}_1(\vec{R}_1)W(\vec{R}_1)d\vec{R}_1$$

$$+ \frac{1}{2!}\int \bar{n}_{12}(\vec{R}_1,\vec{R}_2)W(\vec{R}_1)W(\vec{R}_2)d\vec{R}_1 d\vec{R}_2 + \ldots)|o\rangle_t \tag{1.89}$$

Introduce the irreducible ℓ-particle distribution functions $\bar{\chi}_\ell$ in the usual way (Uhlenbeck and Ford, 1962) by

$$\bar{n}_1(\vec{R}_1) = \bar{\chi}_1(\vec{R}_1) ,$$

$$\bar{n}_2(\vec{R}_1,\vec{R}_2) = \bar{\chi}_2(\vec{R}_1,\vec{R}_2) + \chi_1(\vec{R}_1)\bar{\chi}_1(\vec{R}_2) ,$$

$$\bar{n}_3(\vec{R}_1,\vec{R}_2,\vec{R}_3) = \chi_3(\vec{R}_1,\vec{R}_2,\vec{R}_3) + \bar{\chi}_1(\vec{R}_1)\bar{\chi}_2(\vec{R}_2,\vec{R}_3)$$

$$+ \bar{\chi}_1(\vec{R}_2)\bar{\chi}_2(\vec{R}_1,\vec{R}_3) + \bar{\chi}(\vec{R}_3)\bar{\chi}_2(\vec{R}_1,\vec{R}_2)$$

$$+ \bar{\chi}_1(\vec{R}_1)\bar{\chi}_1(\vec{R}_2)\bar{\chi}_1(\vec{R}_3) \tag{1.90}$$

and so on. Then (1.89) becomes

$$I(t) = < \exp (\int \bar{\chi}_1(\vec{R}_1)W(\vec{R}_1)d\vec{R}_1$$

$$+ \frac{1}{2!} \int \bar{\chi}_2(\vec{R}_1,\vec{R}_2)W(\vec{R}_1)W(\vec{R}_2)$$

$$+ \frac{1}{3!} \int \bar{\chi}_3(\vec{R}_1,\vec{R}_2,\vec{R}_3)W(\vec{R}_1)W(\vec{R}_2)W(\vec{R}_3)+\ldots)|o>_t$$

$$\equiv < e^F|o >_t \quad . \tag{1.91}$$

The proof of (1.91) is trivial, the combinatorics being identical with those involved in the Mayer-Ursell theory of the virial expansion. For the case of uncorrelated scattering centers $\bar{\chi}_\ell = 0(\ell \geqslant 2)$. Since $\bar{\chi}_1 = \rho$ always, we obtain

$$I(t) = < \exp \{-\rho \int d\vec{R} [1 - \exp(- \int_o^t \nu(\vec{r}_{t'},-\vec{R})dt')]\}|o >_t \tag{1.92}$$

which is exactly the result previously used. Just as before the use of the path integral representation has enabled us to carry out the thermodynamic limit explicitly.

The expression F in (1.91) is a functional of the path $\vec{r}_t'$. Just as before we can approximate I(t) as follows : Introduce a "potential" $\phi(\vec{r})$ again via (1.24) and put

$$< e^F | o >_t = (<e^{F+\Phi} e^{-\Phi} | o>_t / <e^{-\Phi} | o>_t) <e^{-\Phi} | o>_t$$

$$\equiv < e^{F+\Phi} | o,\phi >_t < e^{-\Phi} | o >_t . \qquad (1.93)$$

Using Jensen's inequality again we have

$$< e^{F+\Phi} | o,\phi >_t \geqslant \exp < F + \Phi | o,\phi >_t , \qquad (1.94)$$

$$I(t) \geqslant [\exp < F + \Phi | o,\phi >_t] < e^{-\Phi} | o >_t . \qquad (1.95)$$

The potential ϕ is again chosen so that the right-hand side of (1.95) is maximized, giving the best approximation possible with this method. The approximation given by the right-hand side of (1.95) is the first term of an expansion in higher semi-invariants. For the case of uncorrelated scatterers it was shown that these higher semi-invariants do not contribute to the leading term and first correction to I(t) for large t. This discussion, already quite intricate then, is even more involved in the present case and will be omitted here.

The right-hand side of (1.95) is easily evaluated just as before in terms of Green's functions and then their asymptotic limit for large t. The result may be written as follows, accurate enough for the calculation of the leading term and first correction.

$$I(t) \cong e^{-tQ} \qquad (1.96)$$

$$Q = K + D \qquad (1.97)$$

$$K = \int \psi \frac{1}{2} p^2 \psi \, d\vec{r} , \qquad (1.98)$$

$$D = - \frac{1}{t} (\int d\vec{R}_1 \bar{\chi}_1(\vec{R}_1)(e^{-t(E_1-E)} - 1)$$

$$+ \frac{1}{2!} \int d\vec{R}_1 d\vec{R}_2 \bar{\chi}_2(\vec{R}_1, \vec{R}_2)(e^{e-t(E_{12}-E)}$$

$$- e^{-t(E_1-E)} - e^{-t(E_2-E)} + 1)$$

$$+ \frac{1}{3!} \int d\vec{R}_1 d\vec{R}_2 d\vec{R}_3 \bar{\chi}_3(\vec{R}_1, \vec{R}_2, \vec{R}_3) [e^{-t(E_{123}-E)}$$

$$- (e^{-t(E_{12}-E)} + e^{-t(E_{23}-E)} + e^{-t(E_{31}-E)})$$

$$+ (e^{-t(E_1-E)} + e^{-t(E_2-E)} + e^{-t(E_3-E)}) - 1] + \ldots). \tag{1.99}$$

In (1.99), E is the ground-state energy if the potential is ϕ, E_1 is the ground state energy if the potential is $\phi + \nu(\vec{r}-\vec{R}_1)$, E_{12} is the ground state energy if the potential is $\phi + \nu(\vec{r}-\vec{R}_1) + \nu(\vec{r}-\vec{R}_2)$, etc.

The right-hand side of (1.99) can be simplified further. Put

$$\begin{aligned} E_1 &= E + \Delta_1 , \\ E_{12} &= E + \Delta_1 + \Delta_2 + \Delta_{12} , \\ E_{123} &= E + \Delta_1 + \Delta_2 + \Delta_3 + \Delta_{12} + \Delta_{23} + \Delta_{31} + \Delta_{123} . \end{aligned} \tag{1.100}$$

Since the first result for ϕ will be of the same general form as it was before (to be expected since the Lifshitz result turns out to be valid), we may use the old ϕ to estimate the contribution to (1.99) of $t\Delta_{12}$, $t\Delta_{123}$,..., which turns out to be negligible to the order we are retaining. Neglecting them, we obtain

$$D = + (\int d\vec{R}_1 \bar{\chi}_1(\vec{R}_1)(1-e^{-t\Delta_1})$$

$$- \frac{1}{2!} \int d\vec{R}_1 d\vec{R}_2 \bar{\chi}_2(\vec{R}_1,\vec{R}_2)(1-e^{-t\Delta_1})(1-e^{-t\Delta_2})$$

$$+ \frac{1}{3!} \int d\vec{R}_1 d\vec{R}_2 d\vec{R}_3 \bar{\chi}_3(\vec{R}_1,\vec{R}_2,\vec{R}_3)(1-e^{-t\Delta_1})(1-e^{t\Delta_2})$$

$$(1-e^{-t\Delta_3}) + \ldots) \tag{1.101}$$

Finally, the pseudopotential approximation (not necessary for establishing the Lifshitz term, but necessary to calculate the first correction)

$$\Delta_1 = 2\pi a \psi^2(\vec{R}_1) \tag{1.102}$$

is made, where $\psi(\vec{r})$ is the ground-state wave function in the presence of the potential ϕ.

Q is now a functional of ψ alone, and we must find the minimum value of ϕ subject only to the condition that ψ is normalized. This immediately gives the following equation :

$$\frac{1}{2} p^2\psi + \phi\psi = E\psi \;, \tag{1.103}$$

where

$$\phi(\vec{r}) = 2\pi a e^{-t\Delta(\vec{r})}\Big(\bar{\chi}_1(\vec{r}) - \frac{1}{1!}\int d\vec{R}_1 \bar{\chi}_2(\vec{r},\vec{R}_1)(1-e^{-t\Delta_1})$$
$$+ \frac{1}{2!}\int d\vec{R}_1 d\vec{R}_2 \bar{\chi}_3(\vec{r},\vec{R}_1,\vec{R}_2)$$
$$\times (1-e^{-t\Delta_1})(1-e^{-t\Delta_2}) - \ldots\Big). \tag{1.104}$$

Since we have assumed the system to be homogeneous on the average $\bar{\chi}_1(r)$ is independent of $\vec{r}$ and is clearly the average density of scatterers ρ. If we put $\bar{\chi}_j = 0$ $(j \geqslant 2)$, we recover the previous result for ϕ. In the general case, we must calculate ψ from (1.102), (1.103) and (1.104) in the limit of large t, and then calculate Q with the result. Although we cannot carry this procedure through analytically for arbitrary $\bar{\chi}_j$, we shall show next how the Lifshitz prescription gives the leading term correctly. We shall also obtain the first correction in one particularly simple limiting case.

We recall first that in the uncorrelated case we showed the leading term was obtained correctly if we assume that ϕ is a spherical potential well of radius b. The radius b was determined so that Q is minimum. We may proceed in exactly the same way here. The kinetic energy K is again given by

$$K = \frac{1}{2}\pi^2/b^2 \quad . \tag{1.105}$$

To obtain the leading term of D we assume b is very large (b approaches ∞ as t approaches ∞, as we shall see). The quantity $1 - \exp[-t\Delta(R)]$ is very close to unity till R gets to within a "skin depth" (a number independent of t) of b, and then drops off to zero rapidly within the skin depth. Therefore if we neglect surface terms the leading term of (1.101) is given by

$$D = \frac{1}{t}\Big(\int_{R_1<b} d\vec{R}_1\bar{\chi}_1(\vec{R}_1)$$

$$-\frac{1}{2!}\int_{R_1,R_2<b} d\vec{R}_1 d\vec{R}_2\bar{\chi}_2(\vec{R}_1,\vec{R}_2) + \ldots\Big). \quad (1.106)$$

The fact that the system is specially homogeneous on the average implies that χ_j depends only on the differences of its arguments. Further, since $\bar{\chi}_j$ is the irreducible correlation function it is roughly characterized by a correlation length r_0 which measures how close all its arguments have to be in order for it to differ appreciably from zero. Since b approaches ∞ as t does and r_0 is fixed and finite we have, for the leading term of the integrals in (1.106)

$$\int_{R_1,R_2,\ldots,R_j<b} d\vec{R}_2\ldots d\vec{R}_j\bar{\chi}_j(\vec{R}_1,\vec{R}_2,\ldots,\vec{R}_j)$$

$$\int d\vec{R}_2\ldots d\vec{R}_j\bar{\chi}_j(\vec{R}_1,\vec{R}_2,\ldots,\vec{R}_j) \equiv \alpha_j \ . \quad (1.107)$$

α_j is of course independent of $\vec{R}_1$ because of the assumed spacial homogeneity. Therefore the leading term of D is

$$D = (1/t)\ \frac{4}{3}\ \pi b^2\gamma \quad (1.108)$$

where

$$\gamma = \rho - \alpha_2/2! + \alpha_3/3! - \ldots \ . \quad (1.109)$$

Q is now given by (1.105) plus (1.108). This only differs from the old calculation in that ρ there is replaced here by γ. Therefore we get a leading term of exactly the Lifshitz form with ρ replaced by γ in the formula for the constant. The quantity γ has a known physical meaning. If we ask for the probability P(Ω) that a given macroscopic volume Ω is free of scattering centers (in the thermodynamic limit) we easily find (Kac and Luttinger, 1973)

$$P(\Omega) = e^{-\gamma\Omega} \quad (1.110)$$

This is, however, exactly Lifshitz's prescription for determining the constant γ that goes into his result.

If $P_V(\vec{R}_1,\ldots,\vec{R}_N)$ is the distribution function of a fluid in equilibrium, then we also have (Kac and Luttinger, 1973)

$$\gamma = \beta p \qquad (1.111)$$

where p is the pressure of the fluid, $\beta = 1/k_BT$, and T is the temperature.

It has not proven possible to obtain the first correction to the Lifshitz result in general. Instead we consider a simple and physically interesting case where it may be obtained. The skin depth λ_o which measures the distance over which $1 - \exp[-t\Delta(R)]$ changes from unity to zero was previously given by $(1/4\pi c\rho)^{1/2}$. In the general case we just replace ρ by γ as above. Now consider the correlation length r_o. Except very near a phase transition it will be of the order of the range of the intermolecular forces. If we assume again a very small scattering length a we will have $\lambda_o >> r_o$, and it is this case we shall consider here. Clearly in this limit (1.101) and (1.104) become

$$D = \frac{1}{t}\left(\rho \int d\vec{R}_1(1-e^{-t\Delta_1}) - \frac{\alpha_2}{2!}\int d\vec{R}_1(1-e^{-t\Delta_1})^2 + \frac{\alpha_3}{3!}\int d\vec{R}_1(1-e^{-t\Delta_1})^3 + \ldots\right) \qquad (1.112)$$

and

$$\phi(\vec{r}) = 2\pi a e^{-t\Delta(\vec{r})}[\rho - \alpha\,(1-e^{-t\Delta(\vec{r})}) + (\alpha_3/2!)(1-e^{-t\Delta(\vec{r})})^2 - \ldots] \,. \qquad (1.113)$$

The discussion of the solution of (1.103) with ϕ given by (1.113) is practically identical to the case of uncorrelated scatterers. This yields exactly the same expression for the correction to the Lifshitz term that we had before except that ρ must be replaced by γ and C by $\tilde{C}$ where

$$\tilde{C} \equiv \frac{1}{\sqrt{\gamma}} \int_0^\infty dh \, [\gamma^{1/2} - (\rho(1-e^{-h^2}) - \frac{\alpha_2}{2!}(1-e^{-h^2})^2$$

$$+ \frac{\alpha_3}{3!}(1-e^{-h^2})^3 - \ldots)^{1/2}] \quad . \tag{1.114}$$

The integral (1.114) must be done numerically in general. In the case of an equilibrium distribution of scattering centers, the α_j are completely calculable from the equation of state (Hemmer, 1968). We easily find

$$\tilde{C} = \int_0^\infty dh \, [1 - (\frac{p(z,\beta) - p(ze^{-h^2},\beta)}{p(z,\beta)})^{1/2}] \quad . \tag{1.115}$$

where $p(z,\beta)$ is the pressure of the scatterers expressed as a function of the fugacity (z) and β.

Recently related methods have been applied by Gross (unpublished) to study the density of states problem when the potential $U(\vec{r})$ is not given by the sum of interactions with fixed scattering centers, but is a gaussian random variable of mean zero and variance

$$< U(\vec{r})U(\vec{r}\,') > = w(\vec{r}-\vec{r}\,') \tag{1.116}$$

In this case it is easy to see that (1.23) is replaced (in one dimension) by

$$z(t) = \frac{1}{(2\pi t)^{1/2}} < \exp(-\frac{1}{2}\int_0^t\int_0^t w(x_{t'}-x_{t''})dt'dt'')|o>_t \tag{1.117}$$

The case of a "white noise" potential $(w(\vec{r}-\vec{r}\,') = w_0\delta(\vec{r}-\vec{r}\,'))$ is particularly simple, because it only contains one parameter and in one dimension the problem can be solved exactly and compared to the variational treatment. Gross finds that excellent results can be obtained by approximating the functional in (1.117) not by a potential as we have, but by a "two-time quadratic functional"

$$g = \frac{w^2}{4t} \int_0^t \int_0^t dt' dt'' \, (x_{t'} - x_{t''})^2 \tag{1.118}$$

where w is the adjustable parameter which is used to obtain the best variational bound. (This is similar to the Feynman method in the theory of the ground state energy of the polaron.) We shall only quote his results in the low and high temperature limits in one dimension :

Low temperature

$$z_{gross}(t) = \frac{w_o^2}{\sqrt{2\pi t}} \frac{t^3}{2\pi} \frac{4}{\sqrt{e}} \exp\left(\frac{w_o^2 t^3}{8\pi}\right) \tag{1.119}$$

This yields for the density of states per unit length as $\varepsilon \to -\infty$

$$g_{gross}(\varepsilon) = \frac{16}{(3)^{3/2}\, \pi^{1/2}} \frac{|\varepsilon|}{w_o} \exp\left(-\frac{2}{3}\sqrt{\frac{8\pi}{3}} \frac{|\varepsilon|^{3/2}}{w_o}\right) \tag{1.120}$$

The exact value is (as $\varepsilon \to -\infty$)

$$g_{exact}(\varepsilon) = \frac{4}{\pi} \frac{|\varepsilon|}{w_o} \exp\left(-\frac{4\sqrt{2}}{3} \frac{|\varepsilon|^{3/2}}{w_o}\right) \tag{1.121}$$

The coefficient of $|\varepsilon|^{3/2}$ in the exponential is almost correct (the ratio of the Gross value to the exact coefficient is $\sqrt{\pi/3}$) and the energy dependence of the prefactor is correct. Further, if we calculate the correction coming from the first neglected cumulant we find that $\sqrt{\frac{64\pi}{201}} = 1.00015$, very close to the exact value of unity. Finally, the effect of the first neglected cumulant on the numerical value of the prefactor is to bring it within 1 percent of the exact value.

High Temperature

In the high temperature region high electronic energy levels are involved, and these can be calculated by standard statistical mechanical perturbation theory on the potential. Gross' variational method turns out to give the first <u>two</u> orders of perturbation theory correctly. Therefore it is excellent in both the high and low temperature region, and probably pretty good inbetween since z(t) is a smooth function of t. This has not yet been investigated in detail, however. In passing, I should remark that the variational method of myself and Friedberg gives only the <u>first</u> order of perturbation theory correctly and therefore is not as good at high temperatures. At extremely low temperatures though it is a little better, since it gives the Lifshitz result exactly whereas Gross' method gives only a close approximation to it.

Addendum

There is some typical physicists' loose mathematics and inelegance in our derivation of the Lifshitz result. What we have actually shown is that each term in the cummulant expansion (beyond the $\langle A \rangle$ term) is negligible in the low temperature limit compared to the two terms kept, but we have not shown that this is true for the whole infinite series or even that the series converges. In this addendum I would like to point out that the Lifshitz result (but not the first correction) can be obtained simply and with full mathematical rigor from the path integral expression (1.23). The method is due to the mathematicians Donsker and Varadhan, and enables one to evaluate a whole class of path integrals in the limit of large "time". We shall not give the proof of their results here as they are seemingly based on rather delicate estimates from the modern theory of Brownian motion, but simply state their prescription and show how it may be applied to our case. Their result may be stated (in a slightly modified form) as follows : consider a continuous "path" $\vec{r}_t$, which starts ($\tau' = 0$) at the origin and ends ($\tau' = \tau$) at the origin. Define a functional $\ell(\vec{r})$ of the path by

$$\ell(r) = \frac{1}{\tau} \int_0^\tau \delta(\vec{r}_{\tau'} - \vec{r})\, d\tau' \qquad (A.1)$$

and let $F(\ell,\tau)$ be any functional of $\ell(\vec{r})$ and function of τ. Then

$$\lim_{\tau\to\infty} \frac{1}{\tau} \ln \langle e^{-\tau F(\ell,\tau)} | 0 \rangle_\tau = -\mathrm{Min}_\psi \left[\frac{1}{2} \int (\vec{\nabla}\psi)^2 d\vec{r} + F(\psi^2,\infty)\right] \qquad (A.2)$$

where it is assumed that $F(\ell,\infty)$ exists and Min_ψ means the minimum with respect to all normalized $\psi(\vec{r})$, i.e. all ψ such that

$$\int \psi^2(\vec{r})d\vec{r} = 1 \ . \qquad (A.3)$$

To apply this to (1.23) we proceed as follows :

$$I(t) = \langle \exp\{\rho \int d\vec{R}\,(1-e^{-\int_0^t \nu(\vec{r}_{t'},-\vec{R})dt'})\}|0\rangle_t$$

$$= \langle \exp\{\rho\lambda^3 \int d\vec{R}'(1-e^{-\lambda^2 \int_0^{t/\lambda^2} \nu(\vec{r}_{\lambda^2\tau'},-\lambda\vec{R}')d\tau'})\}|0\rangle_t \tag{A.4}$$

where we have put $\vec{R} = \lambda\vec{R}'$ and $t' = \lambda^2\tau'$. Now from the definition of the path average (1.17) and (1.18) we have the following trivial and well-known identity

$$\langle f\{\vec{r}_{\sigma t'}\}|0\rangle_{\sigma t} = \langle f\{\sqrt{\sigma}\,\vec{r}\,'_{t''}\}|0\rangle_t \tag{A.5}$$

where $f\{\vec{r}_{\sigma t'}\}$ is some functional of the path $\vec{r}_{\sigma t'}$; on the left hand side the path $\vec{r}_{\sigma t'}$ vanishes at $t' = 0,t$ and on the right hand side $r'_{t''}$ vanishes at $t'' = 0,t$.

This may also be written

$$\langle f\{\vec{r}_{\lambda^2\tau}\}|0\rangle_t = \langle f(\vec{r}\,'_{t''})|0\rangle_{t/\lambda^2} \tag{A.6}$$

Using (A.6), (A.4) becomes

$$I(t) = \langle \exp\{-\rho\lambda^3 \int d\vec{R}'(1-e^{-\lambda^2 \int_0^{t/\lambda^2} \nu(\lambda(\vec{r}_{\tau''}-\vec{R}'))d\tau''})\}|0\rangle_{t/\lambda^2} \tag{A.7}$$

Now choose λ such that $\rho\lambda^3 = t/\lambda^2 \equiv \tau$,

$$\lambda = (\frac{t}{\rho})^{1/5} \tag{A.8}$$

$$\tau = \rho(\frac{t}{\rho})^{3/5} \tag{A.9}$$

and(with a slight change of notation)

$$I(t) = \langle \exp\{-\tau \int d\vec{R}(1-e^{-(\rho\tau^2)^{1/3}\frac{1}{\tau}\int_0^\tau \nu_\lambda(\vec{r}_\tau,-\vec{R})d\tau'})\}|0\rangle_\tau \tag{A.10}$$

where

$$\nu_\lambda(\vec{r}) \equiv \lambda^3\nu(\lambda\vec{r}) \tag{A.11}$$

writing

$$\frac{1}{\tau}\int_0^\tau \nu_\lambda(\vec{r}_\tau,-\vec{R})d\tau' = \int d\vec{r}\,\frac{1}{\tau}\int_0^\tau d\tau'\;\delta(\vec{r}_\tau,-\vec{r})\nu_\lambda(\vec{r}-\vec{R}) \quad (A.12)$$

$$= \int d\vec{r}\;\ell(\vec{r})\nu_\lambda(\vec{r}-\vec{R})$$

We see that I(t) may be written in the form

$$< \exp\;(-\tau F(\ell,\tau)\,|\,0\;>_\tau \quad ,$$

where

$$F(\ell,\tau) = \int d\vec{r}\;1-e^{-(\rho\tau^2)^{1/3}\int d\vec{r}\ell(\vec{r})\nu_\lambda(\vec{r}-\vec{R})}) \quad (A.13)$$

To apply (A.2) we must first take the limit for large τ. The following somewhat crude argument can be justified completely when ν is non-negative, which we shall assume. For large λ, changing variables,

$$\lim_{\lambda\to\infty}\int d\vec{r}\ell(\vec{r})\nu_\lambda(\vec{r}-\vec{R}) = \lim_{\eta\to\infty}\int d\vec{r}\,'\ell(\vec{R}+\frac{\vec{r}\,'}{\lambda})\nu(\vec{r}\,')$$

$$= \ell(\vec{R})\int \nu(\vec{r}\,')d\vec{r}\,' = A\ell(\vec{R}) \quad (A.14)$$

where $A > 0$. Since $(\rho\tau^2)^{1/3} \to \infty$ as $\tau \to \infty$ (A.13) becomes in the limit

$$F(\ell,\infty) = V\;[\,\ell(\vec{R})] \quad (A.15)$$

where $V[\,\ell(\vec{R})]$ is the support of $\ell(\vec{R})$, i.e. the volume of the domain in which $\ell(\vec{R}) \neq 0$. Therefore (A.2) becomes

$$\lim_{\tau\to\infty}\frac{1}{\tau}\ln I(t) = -\mathrm{Min}_\psi\;[\frac{1}{2}\int (\vec{\nabla}\psi)^2 d\vec{r} + V(\psi^2(\vec{r}))] \quad (A.16)$$

Clearly finite support is necessary to make the right hand side of (A.16) finite. Now in fact the answer is known to the minimization problem (A.16); it is a famous <u>isoperimetric</u> problem. Imagine first the magnitude of V to be given, then the problem is to find the shape such that $\frac{1}{2}\int_V(\vec{\nabla}\psi)^2 d\vec{r}$ is minimum, subject to (A.3). This is proportional to the square of the lowest natural frequency sound made in a rigid cavity of a given volume. The answer, first conjectured by Lord Rayleigh and proved later (see last lecture), is that the lowest frequency comes when V is a sphere. Call it's radius b', Then one sees, as earlier, that

$$\psi(r) = \frac{\sin \frac{\pi}{b'} r}{\sqrt{2\pi b' r}}$$

if $r < b'$ and zero otherwise, and for given b'

$$\text{Min}_{\psi} \left(\frac{1}{2} \int (\vec{\nabla}\psi)^2 dr + V(\psi^2(\vec{r}))\right) = \frac{\pi^2}{2b'^2} + \frac{4\pi}{3} b'^3 \tag{A.17}$$

Minimizing the right-hand side of (A.17) with respect to b' gives $5\pi^{8/5}/3.2^{1/5}$ which together with (A.16) and the definition of τ (A.9) is the precise mathematical statement of the Lifshitz result (1.65).

2. Generalized Isoperimetric Inequalites

In this lecture we shall discuss a type of bound (Luttinger (1973)) which, when applied to the partition function, gives a useful upper bound. These bounds, which have a very simple interpretation in path integral language, turn out to be a far ranging generalization of the well-known isoperimetric inequalities (an invaluable reference in this field is Polya and Szegö (1951)). The classical isoperimetric inequality (known already to the Greeks) states that, of all curves with given perimeter, the circle has the largest area. Similarly, of all solids with a given surface area, the sphere has the largest volume. To these (and other) purely geometrical "isoperimetric" inequalities, some of a more physical nature have been added. For example, Lord Rayleigh conjectured (in 1877) that, of all membranes with a given area and fixed boundary, the circular one has the minimum lowest natural frequency. This was not fully proved for about 50 years. Again, Poincaré (1903) stated and gave a partial proof of the conjecture that of all solids with a given volume the sphere has the minimum electrostatic capacity. (This was not fully proved until 1930, by C.Szegö).

Now the circle is the most symmetrical of all domains in the plane. J.Steiner (in 1836) invented a geometric operation (which we shall call "Steiner symmetrization") which increases the symmetry of any domain. In the plane, it preserves area and does not increase the length of the boundary of a domain. We shall describe this process in detail below, but mention at this point that Steiner symmetrization never increases the lowest natural frequency of a membrane or the electrostatic capacity of a solid. (These and other similar results were first proved by Polya and Szegö). Such results represent a very considerable generalization of the classic isoperimetric inequalities.

Here we shall be concerned with a still greater generalization of these ideas. There are essentially two new elements in this generalization. The first is that instead of just a quantity like the lowest natural frequency of a membrane, the entire Green's function is involved. The second is that the inequalities involve a function rather than a domain, the domain type of results arising when the function involved is specialized to a characteristic function of the domain. As an example of what our results are like, consider a particle (in two dimensions) interacting with a potential $\phi(x,y)$. If the potential approaches infinity at infinity, the allowed energy states will be discrete, with energies $\varepsilon_o, \varepsilon_1, \varepsilon_2 \ldots$ (We use quantum mechanical language only for convenience; we are really considering the spectra of certain differential operators.) Let us define the "partition function" Z(t) by (t is a real positive parameter)

$$Z(t) = \sum_{j=o}^{\infty} e^{-t\varepsilon_j} . \tag{2.1}$$

(It is a certain integral over the Green's function for the system.) Let the potential $\phi(x,y)$ be replaced by a "Steiner symmetrized" potential $\phi^*(x,y)$, the exact definition of which will be given later. Call the corresponding energy levels ε_j^*, and the corresponding partition function $Z^*(t)$. Then, for this case, our inequality takes the form

$$Z(t) \leqslant Z^*(t) . \tag{2.2}$$

Now suppose $\phi(x,y)$ is taken to be zero inside a certain domain D and infinite outside D. The "Steiner symmetrized" potential $\phi^*(x,y)$ will be zero in a domain D^* and infinite outside D^*, and in fact D^* will be the Steiner symmetrization of the domain D. Clearly apart from constants all of which may be absorbed into t, the ε_j are the squares of the natural frequencies of a uniform membrance having the shape of D with fixed boundary. By letting t approach infinity, only the ε_o contributes and (2.2) may be written

$$e^{-\varepsilon_o t} \leqslant e^{-\varepsilon_o^* t} ,$$

or (2.3)

$$\varepsilon_o^* \leqslant \varepsilon_o .$$

This is just Polya and Szegö's generalization of Rayleigh's conjecture. On the other hand, when t is small, the leading terms of Z(t) are (Kac, 1966, for example) :

$$Z(t) = \frac{\Omega(D)}{2\pi t} - \frac{L(D)}{4} \frac{1}{(2\pi t)^{1/2}} \tag{2.4}$$

where $\Omega(D)$ is the area of the domain D and $L(D)$ is the length of its boundary. Similarly

$$Z^*(t) = \frac{\Omega(D^*)}{2\pi t} - \frac{L(D^*)}{4} \frac{1}{(2\pi t)^{1/2}} \tag{2.5}$$

Since $\Omega(D) = \Omega(D^*)$, (2.2) becomes

$$L(D) \geqslant L(D^*) \ , \tag{2.6}$$

which is just Steiner's generalization of the classic isoperimetric inequality.

That is, the usual Steiner type of isoperimetric inequalities are just extreme specializations of (2.2). Consider first a one-dimensional particle in a potential $\phi(x)$, where $\phi(x)$ approaches infinity as $|x|$ does. The hamiltonian is

$$H = \frac{p^2}{2} + \phi(x) \tag{2.7}$$

and the Green's function is given by (1.31) and (1.32). Expressed as a path integral (as in Section 1), the Green's function becomes

$$G_t(x,x'|\phi) = \lim_{n\to\infty} \int_{-\infty}^{\infty} dx_2 \ldots dx_{n-1} P_\tau(x-x_1) e^{-\tau\phi(x_2)}$$
$$P_\tau(x_2-x_3) e^{-\tau\phi(x_3)} \ldots P_\tau(x_{n-1}-x') e^{-\tau\phi(x')} \tag{2.8}$$

where $\tau \equiv t/n-1$ and

$$P_\tau(x) = (2\pi\tau)^{-1/2} \exp\left(-\frac{x^2}{2\tau}\right) \tag{2.9}$$

Consider next the quantity

$$J \equiv \int_{-\infty}^{\infty} dx_1 dx_n G_t(x_1,x_n|\phi) \Gamma(x_n-x_1) \gamma(x_1)$$
$$= \lim_{n\to\infty} \int_{-\infty}^{\infty} dx_1 \ldots dx_n P_\tau(x_1-x_2) e^{-\tau\phi(x_2)} \ldots P_\tau(x_{n-1}-x_n) e^{-\tau\phi(x_n)}$$

$$\Gamma(x_n - x_1)\gamma(x_1) \tag{2.10}$$

where Γ,γ are real non-negative quantities. Now the integral in (2.10) satisfies a "rearrangement" inequality (Hardy, Littlewood and Polya (1952)). Intuitively, a rearrangement $F_R(x)$ of a function $F(x)$ means that $F_R(x)$ takes on the same values as $F(x)$, but at different locations. That is, for arbitrary W (such that the integral converges)

$$\int_{-\infty}^{\infty} W(F(x))dx = \int_{-\infty}^{\infty} W(F_R(x))dx \ . \tag{2.11}$$

The symmetrically decreasing rearrangement when it exists is the $F_R(x)$ ($\equiv [F(x)]^*$) which is a non-increasing function of $|x|$, where the origin is chosen arbitrarily. Similarly the symmetrically increasing rearrangement of $F(x)$ is the $F_R(x)(\equiv {}^*[F(x)])$ which is a non-decreasing-function of $|x|$, when it exists. Using these definitions, and letting $H^{(j)}(x)$, $F^{(j)}(x)$ be real valued non-negative functions which go to zero as $|x|$ goes to infinity sufficiently rapidly for all integrals to exist, we have (Luttinger and Friedberg (1976))

$$\begin{aligned}
&\int_{-\infty}^{\infty} dx_1 \ldots dx_n H^{(1)}(x_1-x_2)H^{(2)}(x_2-x_3) \\
&\qquad \ldots H^{(n)}(x_n-x_1)F^{(1)}(x_1)F^{(2)}(x_2)\ldots F^{(n)}(x_n) \\
&\leqslant \int_{-\infty}^{\infty} dx_1 \ldots dx_n \ [H^{(1)}(x_1-x_2)]^* \ [H^{(2)}(x_2-x_3)]^* \\
&\qquad \ldots \ [F^{(1)}(x_1)]^* \ [F^{(2)}(x_2)]^* \ \ldots
\end{aligned} \tag{2.12}$$

This theorem has an almost trivial intuitive meaning in terms of diffusion. Because of the linearity of both sides of (2.12) in $F^{(j)}$, $H^{(j)}$ we can without loss of generality take the $H^{(j)}$ and the $F^{(j)}$ to be normalized to unity, i.e. :

$$\int_{-\infty}^{\infty} H^{(j)}(x)dx = \int_{-\infty}^{\infty} F^{(j)}(x)dx = 1 \tag{2.13}$$

From the definition (2.11), choosing $W(x) = x$, we see that this is also true of rearranged functions. Let the real axis be uniformly covered with particles at concentration C. Further, let $H^{(1)}(x_1-x_2)dx_2$ be the probability that a particle initially at x_1 finds itself after one interval of time in (x_2, x_2+dx_2). Similarly, let $H^{(2)}$ represent the same probability, the "jump" taking place during the second interval of time, etc., etc. Let $F^{(2)}(x_2)$ be the probability that a particle at x_2 survives absorption before it jumps to x_3 (in the second interval), $F^{(3)}(x_3)$ be the probability that a particle at x_3 survives absorption before it jumps to x_4 (in the third interval) etc., etc. Then

$$\ell(Cdx_1) \int_{-\infty}^{\infty} H^{(1)}(x_1-x_2)F^{(2)}(x_2)H^{(2)}(x_2-x_3)F^{(3)}(x_3)\ldots$$

$$\times\ H^{(n)}(x_n-x_1)F^{(n+1)}(x_1)dx_2\ldots dx_n$$

represents the number of particles which start in the interval dx_1 around x_1 and end after n jumps within the (very small) interval ℓ around x_1. The total number of particles returning to within ℓ of this starting point (ℓ very small) is therefore [defining $F^{(n+1)}(x)$ as $F^{(1)}(x)$]

$$C\ell \int_{-\infty}^{\infty} dx_1\ldots dx_n H^{(1)}(x_1-x_2)\ldots$$

$$\times\ H^{(n)}(x_n-x_1)F^{(1)}(x_1)\ldots F^{(n)}(x_n)\ ,$$

which is proportional to the left-hand side of (2.12). Now it is intuitively obvious why (2.12) is valid : The right-hand side of (2.12) is proportional to the same probability with the absorbing material rearranged to increase as we go away from the origin, and the jumping probabilities rearranged to favor short jumps. That is, for the particles which survive anyway (those near the origin), the short jumps are favored, tending to keep them in the region of high survival, so that in the end more survive.

Accepting (2.13), we apply it to (2.10). Since $P_\tau(x)$ is already a symmetrically decreasing function,

$$[P_\tau(x)]^* = P_\tau(x) \quad . \tag{2.14}$$

Further, if for j = 2,...,n

$$F^{(j)}(x) \equiv e^{-\tau\phi(x)} \tag{2.15}$$

then

$$[F^{(j)}(x)]^* = e^{-\tau^*[\phi(x)]} \tag{2.16}$$

where $^*[\phi(x)]$ is the symmetrically increasing rearrangement of $\phi(x)$. Therefore, (2.13) tells us

$$\int_{-\infty}^{\infty} G_t(x,x'|\phi)\Gamma(x'-x)\gamma(x)dx\ dx'$$

$$\leqslant \int_{-\infty}^{\infty} G_t(x,x'|^*[\phi])\ [\Gamma(x'-x)]^*\ [\gamma(x)]^* \tag{2.17}$$

The inequality (2.17) is the basic result. Some special cases however are of particular interest :

(a) Letting $\Gamma(x'-x)$ approach the Dirac δ function $\delta(x'-x)$ and $\gamma(x)$ approach unity, we obtain (since both of these are symmetric and not increasing)

$$\int_{-\infty}^{\infty} G_t(x,x|\phi)dx \leqslant \int_{-\infty}^{\infty} G_t(x,x|^*[\phi])\ dx \tag{2.18}$$

Using the representation (1.38), (2.18) becomes an inequality for the partition function

$$\sum_j e^{-tE_j(\phi)} \leqslant \sum_j e^{-tE_j(^*[\phi])} \tag{2.19}$$

(b) Letting both Γ and γ approach unity we obtain

$$\int_{-\infty}^{\infty} G_t(x,x'|\phi)dx\ dx' \leqslant \int_{-\infty}^{\infty} G_t(x,x'|^*[\phi]) \tag{2.20}$$

which becomes, using (1.38) again

$$\sum_j e^{-tE_j(\phi)} [\int_{-\infty}^{\infty} dx\ \psi_j(x|\phi)]^2 \leq \sum_j e^{-tE_j(*[\phi])} [\int_{-\infty}^{\infty} \psi_j(x|^*[\phi])]^2 \tag{2.21}$$

(c) Letting $\Gamma(x) = \delta(x-x' + \varepsilon-b)$, $\gamma(x) = \delta(x-b)$ we find $[\Gamma(x)]^* = \delta(x)$, $[\gamma(x)]^* = \delta(x)$ and (2.17) becomes

$$G_t(b,a|\phi) \leq G_t(0,0|^*[\phi]) \tag{2.22}$$

with a,b arbitrary.

We conclude this section with a discussion of two limiting cases :

(1) t very large : In this case the essential result comes from (2.19) which tells us

$$E_0(\phi) \leq E_0(^*[\phi]) \tag{2.23}$$

That is, our inequality provides a lower bound on the lowest eigenvalue of H. This may prove of use since the usual variational principle (Rayleigh-Ritz principle) provides a convenient upper bound, and then (2.23) enables us to bracket $E_0(\phi)$.

(2) t very small : For this case the leading term of the Green's function is trivially obtained by the method of Kac (1959). The result is

$$G_t(x,x'|\phi) = \frac{e^{-(x-x')^2/2t}}{(2\pi t)^{1/2}}\ e^{-t\phi(x)} \ . \tag{2.24}$$

[$\phi(x)$ may be replaced by $\phi(x')$ in this expression, since first factor is essentially $\delta(x-x')$ for small t.] Similarly,

$$G_t(x,x'|[\phi]) = \frac{e^{-(x-x')^2/2t}}{(2\pi t)^{1/2}}\ e^{-t^*[\phi(x)]} \ . \tag{2.25}$$

By using (2.24) and (2.25) (2.17) becomes

$$\int_{-\infty}^{\infty} dx dx' \frac{e^{-(x-x')^2/2t}}{(2\pi t)^{1/2}} \Gamma(x'-x)\gamma(x)$$

$$\leq \int_{-\infty}^{\infty} dx dx' \frac{e^{-(x-x')^2/2t}}{(2\pi t)^{1/2}} e^{-t^*[\phi(x)]} \Gamma^*(x'-x)\gamma^*(x) \ . \tag{2.26}$$

If $\Gamma, \Gamma^*, \gamma, \gamma^*$ are smooth enough (2.26) becomes

$$\Gamma(0) \int_{-\infty}^{\infty} dx \ e^{-t\phi(x)} \gamma(x)$$

$$\leq \Gamma(0) \int_{-\infty}^{\infty} dx \ e^{-t^*[\phi(x)]} \gamma^*(x) \ . \tag{2.27}$$

The result is an immediate consequence of a well-known rearrangement inequality (Hardy et al (1952))

$$\int_{-\infty}^{\infty} \sigma(x)\gamma(x)dx \leq \int_{-\infty}^{\infty} \sigma^*(x)\gamma^*(x)dx \tag{2.28}$$

since $\Gamma^*(0) \geq \Gamma(0)$. [$\Gamma^*(0)$ being the maximum value of the function $\Gamma(x)$.] If we wish to calculate the "partition function" from (2.18), (2.24) and (2.25) are inadequate. Because $^*[\phi]$ is a rearrangement of ϕ (2.18) reduces to a trivial equality. Again, the next term is easily obtained by the method of Kac (it is essentially the first quantum correction to the "classical" partition function obtained by Wigner and Kirkwood), and yields

$$\int_{-\infty}^{\infty} G_t(x,x|\phi) = \frac{1}{(2\pi t)^{1/2}}$$
$$\left[\int_{-\infty}^{\infty} dx \ e^{-t\phi} - \frac{t^3}{24} \int_{-\infty}^{\infty} dx \ e^{-t\phi} \left(\frac{d\phi}{dx}\right)^2\right] \tag{2.29}$$

Making use of (2.18), we have

$$\int_{-\infty}^{\infty} dx \left(\frac{d\phi}{dx}\right)^2 e^{-t\phi} \geq \int_{-\infty}^{\infty} dx \left(\frac{d^*[\phi]}{dx}\right)^2 e^{-t^*[\phi]} \tag{2.30}$$

It is not difficult to prove this inequality directly for arbitrary non-negative t.

Higher Dimensions

Exactly the same reasoning which led to (2.17) now leads to

$$\int G_t(\vec{r},\vec{r}|\phi)\Gamma(\vec{r}'-\vec{r})\gamma(\vec{r})d\vec{r}\,d\vec{r}'$$

$$\leq \int G_t(\vec{r},\vec{r}'|{}^*_z[\phi])\,[\Gamma(\vec{r}'-\vec{r}]^*_z\,[\gamma(\vec{r})]^*_z \tag{2.31}$$

where the "z-direction" is some arbitrary direction in space : $[\Gamma(\vec{r})]^*_z$, $[\gamma(\vec{r})]^*_z$ are the symmetrically decreasing rearrangements of Γ and γ viewed as functions of z with x,y (coordinates in a plane perpendicular to the z-axis) held fixed, and ${}^*_z[\phi]$ is the corresponding symmetrically increasing rearrangement of ϕ. Specializing Γ and γ as before we obtain

$$\text{(a)}\quad \sum_j e^{-tE_j(\phi)} \leq \sum_j e^{-tE_j({}^*_z[\phi])} \tag{2.32}$$

$$\text{(b)}\quad \sum_j e^{-tE_j(\phi)}\,[\int \psi_j(\vec{r}|\phi)d\vec{r}]^2$$

$$< \sum e^{-tE_j({}^*_z[\phi])}.[\int \psi_j(\vec{r}|{}^*_z[\phi])d\vec{r}]^2 \tag{2.33}$$

$$\text{(c)}\quad G_t(\vec{r},\vec{r}'|\phi) \leq G_t(0,0|{}^*_z[\phi]) \tag{2.34}$$

The relationship of (2.31) to the process of "Steiner symmetrization" may be seen as follows. Choose the potential $\phi(\vec{r})$ to be zero if $\vec{r}$ is a point of some bounded domain D and infinite if it is not. Then the characteristic values and functions of H are given by

$$-\frac{1}{2}\nabla^2\psi_i = \varepsilon_i\psi_i \quad (\vec{r}\text{ in D}), \tag{2.35}$$

$$\psi_i = 0\ (\vec{r}\text{ not in D}) \tag{2.36}$$

and

$$\psi_i = 0\ (\vec{r}\text{ on the boundary of D}). \tag{2.37}$$

What is the potential ${}^*_z[\phi(\vec{r})]$? It must satisfy the condition (for "arbitrary" W such that the integral converges)

$$\int_{-\infty}^{\infty} W(\phi(\vec{r}))dz = \int_{-\infty}^{\infty} W(^{*}_{z}[\phi(\vec{r})]\, dz \qquad (2.38)$$

and be a nondecreasing function of $|z|$. For convergence we must assume $W(\infty)$ is zero and $W(o)$ finite. Then the left-hand side of (2.38) is just $W(o)\ell(x,y)$, where $\ell(x,y)$ is the length of the intersection of a line parallel to the z axis and passing through the point (x,y,o), with the domain D. Let us define a domain D^*_z by the following conditions :

(a) D^*_z is a symmetric with respect to the plane $z = 0$.

(b) Any straight line perpendicular to the plane $z = 0$ which intersects one of the domains D and D^*_z also intersects the other, and these intersections have the same length.

(c) The intersection with D^*_z consists of just one line segment (the intersection with D could consist of several segments), which, because of (a), is bisected by the plane $z = 0$).

Now choosing $^{*}_{z}[\phi(\vec{r})]$ to be zero if $\vec{r}$ is in the interior of D^*_z and infinite otherwise, we see at once that (2.38) is satisfied [because of (b)] and that it is a symmetric increasing function of z because of (a) and (c). Thus the characteristic value problem for this potential is

$$-\frac{1}{2}\Delta^2[\psi_j]^*_z = [\varepsilon_j]^*_z\,[\psi_j]^*_z \quad (r \text{ in } D^*_z) \qquad (2.39)$$

$$[\psi_j]^*_z = 0 \quad (\vec{r} \text{ not in } D^*_z)\ , \qquad (2.40)$$

and

$$[\psi_j]^*_z = 0 \ (\vec{r} \text{ on the boundary of } D^*_z). \qquad (2.41)$$

The domain D^*_z satisfies exactly the definition of Steiner symmetrization of the domain D with respect to the plane $z = 0$ (Polya and Szegö (1951)). Choosing to apply this to (2.32) for very large t, we obtain the result of Polya and Szegö (mentioned in the introduction for the two-dimensional case) that Steiner symmetrization decreases the lowest natural frequency of a vibrating fluid confined to a domain D, with Dirichlet boundary conditions on the boundary of D. The inequality (2.31) may be thought of as a generalization of their results.

We also mention that if we integrate (2.33) over t from zero to infinity and consider the two-dimensional case, the left-hand side of (2.33) becomes (apart from a factor of 4) the torsional rigidity of the domain D (Polya and Szegö (1951)) and the right-hand side becomes the torsional rigidity of the Steiner symmetrized domain. Therefore, we have show that Steiner symmetrization increases the torsional rigidity, a result first proved by Polya in 1948.

The symmetrization process leading to (2.31) may be applied again to some other direction (rather than the z direction). By continuing this procedure with respect to "all possible directions" we will finally obtain

$$\int G_t(\vec{r},\vec{r}\,'|\phi)\Gamma(\vec{r}\,'-\vec{r})\gamma(\vec{r})d\vec{r}d\vec{r}\,'$$

$$\leqslant \int G_t(\vec{r},\vec{r}\,'|{}^*_s[\phi])[\Gamma(r'-r)]^*_s[\gamma(\vec{r})]^*_s \qquad (2.42)$$

where $[\Gamma(\vec{r})]^*_s$ is the rearrangement of $\Gamma(\vec{r})$ which is a non-decreasing function of $(\vec{r})$ alone, and ${}^*_s[\phi]$ is the corresponding non-increasing rearrangement of ϕ. Therefore upper bounds of the form (3.42) can be obtained by solving a problem with spherical symmetry.

Finally we mention (without proof) another very similar class of generalized isoperimetric inequalities which hold when the potential ϕ has no bound states and goes to zero rather than infinity at infinity (scattering problems rather than bound state problems). The simplest of these relationships is the following :

$$\int d\vec{r}d\vec{r}\,'\,[G_t(\vec{r},\vec{r}\,'|0) - G_t(\vec{r},\vec{r}\,'|\phi)]$$

$$\geqslant \int d\vec{r}d\vec{r}\,'\,[G_t(\vec{r},\vec{r}\,'|0) - G_t(\vec{r},\vec{r}\,'|[\phi]^*_z] \qquad (2.43)$$

In (2.43), $G_t(\vec{r},\vec{r}\,'|0)$ is the unperturbed Green's function

$$G_t(\vec{r},\vec{r}\,'|0) = \frac{1}{(2\pi t)^{3/2}} \exp\left(-\frac{(\vec{r}-\vec{r}\,')^2}{2t}\right) \qquad (2.44)$$

and $[\phi]^*_z$ is the symmetrically decreasing rearrangement (with respect to the plane $z = 0$) of ϕ. (Equation (2.43) is also valid if $[\phi]^*_z$ is replaced by $[\phi]^*_s$.). If we apply this to the potential

$$\begin{aligned} \phi(\vec{r}) &= \infty, \ \vec{r} \text{ in } D \\ &= 0, \ \vec{r} \text{ not in } D \end{aligned} \tag{2.45}$$

(D a bounded domain), we find

$$\begin{aligned} [\phi(\vec{r})]^*_z &= \infty, \ \vec{r} \text{ in } D^*_z \\ &= 0, \ \vec{r} \text{ not in } D^*_z \end{aligned} \tag{2.46}$$

where D^*_z is the Steiner symmetrization of the domain D with respect to the plane $z = 0$. It is not difficult to evaluate (2.43) in the limit of small and large times. For t small, the left-hand side of (2.43) is trivial to evaluate, and becomes $V(D) + \sqrt{\frac{2}{\pi}}\, A(D)\sqrt{t}$, where V(D) is the volume of the domain D and A(D) is its surface area. Since $V(D) = V(D^*)$, (2.43) yields $A(D) > A(D^*_z)$ the usual theorem about the effect of Steiner symmetrization on the surface area. For t large, the calculation is more complicated, but it has been shown (Spitzer (1964)) that as $t \to \infty$ the left-hand side of (2.43) becomes $2\pi t C(D)$ where C(D) is the electrical capacity with respect to infinity of a perfect conductor in the shape of the domain D. Therefore (2.43) becomes

$$C(D) \geqslant C(D^*_z) \tag{2.47}$$

another result due to Polya and Szegö (1951). If instead of a domain potential we have a more general one, it is easy to see (Kac and Luttinger (1975)) that the left-hand side of (2.43) is just $2\pi t a(\phi)$, where $a(\phi)$ is the scattering length for the potential ϕ. Thus (2.43) becomes

$$a(\phi) \geqslant a([\phi]^*_z) \tag{2.48}$$

which is a generalization of the result (2.47).

References

Blatt, J. and Weisskopf, V., Nuclear Physics (Wiley, N.Y., 1956).

Donsker, M.D. and Varadhan, S.R.S., Comm. Pure and Applied Math. 28, 1 (1975); 28, 279 (1975); 28, 525 (1975). The last paper also contains the evaluation of expressions like (1.23). See also Proc. Int. Conf. on Function Space Integration (Oxford, 1974) for a preliminary report.

Edwards, S.F. and Gulyaev, V.B., Proc. Phys. Soc. London 83, 495 (1965). Their result was for time rather than temperature Green's functions, i.e. $t \to it$.

Friedberg, R. and Luttinger, J.M., Phys. Rev. B12 4460 (1975).

Hardy, G.H., Littlewood, J.E. and Polya, G., Inequalities, Cambridge University Press, London and New York, 2nd Edition, p.276 ff (1952).

Hemmer, P.C., Phys. Norv. 3, 9 (1968).

Kac,, M., Probability and Related Topics in Physical Science (Interscience, N.Y., 1959), Vol.1, p.161 ff.

Kac, M., Amer. Math. Monthly, 73. 1 (1966).

Kac, M. and Luttinger, J.M., Journ. of Math. Phys. 14, 583 (1973).

Kac, M. and Luttinger, J.M., Annales de l'Institute Fourrier, 25, 3 et 4, 317 (1975).

Kubo, R., J. Phys. Soc. Japan, 17, 1100 (1962).

Lifshitz, I.M., Usp. Fiz. Nauk 83, 617 (1964) [Soviet Physics Usp. 7, 549 (1965)].

Luttinger, J.M., Proc. Nat. Acad. of Sciences USA, 70, 1005 (1973)
J. Math. Phys. 14, 586 (1973).
J. Math. Phys. 14, 1444 (1973).
J. Math. Phys. 14, 1448 (1973).

Luttinger, J.M., Phys. Rev. B13, 2596 (1976).

Luttinger, J.M. and Friedberg, R., Archive for Rat. Mech. and Anal., 61, 45 (1976); Archive for Rat. Mech. and Anal., 61, 35 (1976). A more general theorem and much simpler proof may be found in Brascamp, H.J., Lieb, E.H. and Luttinger, J.M., Jour. of Funct. Anal. 17, 227 (1974).

Spitzer, F., Zeit.für Wahrscheinlichkeitrechnung 3, 110 (1964).

Uhlenbeck, G. and Ford, G.W., Studies in Stat. Mech. (North-Holland, Amsterdam, 1962) Vol.I, Part B.

PATH INTEGRALS AND POLYMER PROBLEMS

S.F. Edwards

Science Research Council and Cavendish Laboratory
Madingley Road, Cambridge CB3 OHE

1. INTRODUCTION

Many physical problems can be written as path integrals. Polymer problems have a unique status since the representation is identical with the physical problem : the polymer is the path. For example the development of the perturbation theory of the interaction between polymers, and internally with the self interaction of a single polymer when represented by Feynman diagrams is both a representation of the mathematical problem and of what actually occurs in physics. This has a wonderful consequence in that one can use an intuitive approach quite directly to the mathematics.

The polymerised state is central to the world around us, both in biological material and in the ubiquitous glassy amorphous state. It is important to realise that the perfect crystalline state is not that common in nature, perhaps the only common example being in metals. Our clothes, food, furniture, windows and indeed ourselves are all polymerised, and whereas physics (in distinction to technology) works only with a world it can understand and so creates a world it can understand, the time is rapidly approaching in which the ordered world of crystals will be fully understood, so that physicists must study the well defined polymerised states to progress further.

These notes will only be able to set the scene of the problem, for many of the essential difficulties are as yet unresolved, but I hope to show that whereas much of physics research is hunting more and more obscure effects, central problems in polymerised materials remain unsolved.

2. Statistics of Free Chains

Polymers consist of repeated units, called monomers, which can be simple as in polythene, or extremely complex as in DNA. For a discussion of the relation of chemistry to the following mathematics see Flory [1]. We shall take the simplest case based on a Markov chain.

2.1. The Kuhn Effective Step Length

This is of course classic ground but we go over it briefly to establish notations and a point of view. Start with the Markov chain : let the probability of finding $\underline{r}_n$ given $\underline{r}_{n-1}$ be $P(\underline{r}_n-\underline{r}_{n-1})$. Then:

$$P_N(\underline{r}_N-\underline{r}_o) = \int P(\underline{r}_N-\underline{r}_{N-1})\, P(\underline{r}_{N-1}-\underline{r}_{N-2})\ldots$$

$$\ldots P(\underline{r}_1-\underline{r}_o) \prod_{1}^{N-1} d^3r_\alpha \tag{2.1.1}$$

One can write P in terms of its Fourier transform

$$P_k = (2\pi)^{-3} \int e^{-i\underline{k}\cdot\underline{r}}\, P(\underline{r})\, d^3r \tag{2.1.2}$$

$$P(\underline{r}) = \int e^{i\underline{k}\cdot\underline{r}}\, P_k d^3k \tag{2.1.3}$$

Then :

$$P_N(\underline{r}_N-\underline{r}_o) = \int \prod_{o}^{N-1} d^3k_\alpha \exp\{i\underline{k}_{N-1}\cdot(\underline{r}_N-\underline{r}_{N-1})+\ldots+i\underline{k}_o\cdot(\underline{r}_1-\underline{r}_o)\}$$

$$\times P_{k_{N-1}} P_{k_{N-2}} \ldots P_{k_o} \prod_{1}^{N-1} d^3r_\beta \tag{2.1.4}$$

$$= (2\pi)^{3N} \int \delta(\underline{k}_{N-1}-\underline{k}_{N-2}) \ldots \delta(\underline{k}_1-\underline{k}_o) \prod_{k_{N-1}}^{k_o} P_\alpha$$

$$\times \exp(i\,\underline{k}_N\cdot\underline{r}_N - i\,\underline{k}_o\cdot\underline{r}_o) \tag{2.1.5}$$

$$= (2\pi)^{3N} \int P_k^N \exp [i\, \underline{k}.(\underline{r}_N - \underline{r}_0)]\, d^3k \qquad (2.1.6)$$

Note that since P_k is a probability

$$\int P(\underline{r}) d^3r = 1 \qquad P_o = (2\pi)^{-3} \qquad (2.1.7)$$

Hence, since

$$P_N(\underline{R}) = \int d^3k \exp [i\, \underline{k}.\underline{R} - N \log ((2\pi)^3 P_k)] \quad (2.1.8)$$

as R or N becomes large P_k can be approximated by expansion

$$(2\pi)^3\, P_k = 1 - (\alpha/2)\, \underline{k}^2$$

$$P_N(\underline{R}) = \int d^3k \exp [i\underline{k}.\underline{R} - N_\alpha \underline{k}^2 /2]$$

$$= \frac{e^{-R^2/(2N\alpha)}}{(2\pi N\alpha)^{3/2}}$$

For freely hinged rods

$$P(\underline{r}) = \frac{\delta(|r| - \lambda)}{4\pi\lambda^2} \qquad \alpha = \frac{\lambda^2}{3}$$

The Gaussian distribution has log $2\pi\, P_k$ exactly equal to $- (\alpha k^2/2)$ and so exactly produces the Rayleigh distribution. The freely jointed chain is rather exceptional, and in general one can expect the probability of finding the point $\underline{r}_n$ related to $\underline{r}_{n-1}$, $\underline{r}_{n-2}$, ... A simple solution on the lines above is not then possible, although it is not difficult to get solutions in special cases, for example a random walk on a lattice, and in the case of infinitesimal steps, considered further below.

A characteristic of all solutions is that for large N i.e. small k the form of $P_n(r)$ is the same, so that one can interpret $\alpha = \lambda^2/3$ as defining an effective step length. The result is physically obvious. Mathematically it follows from the fact that however many r's appear in P, the final form for P_N $(\underline{r}_N, \underline{r}_0)$ must be P_N $(\underline{r}_N - \underline{r}_0)$ by translational invariance, and since

$$\int P_N(\underline{R})\, d^3R = 1$$

$$P_N(k) = (2\pi)^{-3}\, e^{-\psi(k)} \qquad \psi(o) = 0$$

The function ψ will be analytic in k for small k, so can be expanded, and the result follows.

2.2. Relationship to Brownian Motion

One can use the mathematics of § 2.1 to describe the random walk in time of a particle if the probability of finding the particle at r_n at time t given that it was at r_{n-1} at t - τ is $p(r_n - r_{n-1})$, then the mathematics above and in the subsequent sections apply with trifling changes.

2.3. Fourier Transform of Configurations

A particular set of $(r_n \ldots r_0)$ we call a configuration. It will be useful to introduce Fourier coefficients of the configuration,

$$r(m) = \sum_{o}^{N} e^{2\pi i n m/N} r_n \qquad (2.3.1)$$

There are an infinite set of $\underline{r}(m)$, so they are not independent. However, for a very long chain, $N \to \infty$ they become the natural coordinates to use. If one considers the value of nl as an arc length s so that the configuration is represented by the intrinsic equation $\underline{r} = \underline{r}(s)$, we can consider the Fourier integral transform.

$$\underline{r}(q) = \frac{1}{\sqrt{2\pi}} \int e^{iqs}\, \underline{r}(s)\, \frac{ds}{\ell} \qquad (2.3.2)$$

$$\underline{r}(s) = \frac{1}{\sqrt{2\pi}} \int e^{-iqs}\, \underline{r}(q)\, \ell\, dq \qquad (2.3.3)$$

There will never be any ambiguity in using the same symbol for $\underline{r}(s)$ and $\underline{r}(q)$. In the forms chosen both $\underline{r}(q)$ and $\underline{r}(s)$ have the dimensions of length. The coordinate $\underline{r}(q)$ has some analogy with collective coordinates in statistical physics. If one has a series of particles labelled R_a, the density is

$$\sum_a \delta(\underline{r} - \underline{R}_a) \qquad (2.3.4)$$

an extremely singular function. For many purposes one only needs the properties of long wave modulations of the density, so one introduces :

$$\rho_{\underline{k}} = \sum_a e^{i\underline{k}.R_a} \qquad (2.3.5)$$

Again there are an infinite number of ρ_k, but this seldom causes difficulty. For example as $k \to \infty$ one can expect (and prove) that the probability distribution of the ρ_k will be gaussian. Indeed, averaged over a random distribution in a box volume V

$$< \rho_k \rho_k^* > = \sum_{ab} < \exp [ik(R_a - R_b)] > \qquad (2.3.6)$$

$$= N \text{ since only } a = b \text{ contributes.} \qquad (2.3.7)$$

Hence if

$$P([\rho]) = \exp [- \int d^3k \, [\frac{\rho_k \rho_k^*}{C(k)}]] \qquad (2.3.8)$$

$$< \rho_k \rho_j^* > = C(k) \, \delta(k-j) \qquad (2.3.9)$$

$$< \rho_k \rho_k^* > = \frac{V}{(2\pi)^3} C(k) \qquad (2.3.10)$$

since

$$\delta(\underline{k}) = \frac{1}{(2\pi)^3} \int e^{-\underline{k}.\underline{r}} d^3r \qquad (2.3.11)$$

$$\delta(o) = \frac{V}{(2\pi)^3} \qquad (2.3.12)$$

Hence

$$C(\underline{k}) = (2\pi)^3 \frac{N}{V} \qquad (2.3.13)$$

One may now write the general probability distribution for the ρ_k when (soft) forces are present, for this is generally

$$e^{-H/uT} \to [- \int (v_k \frac{\rho_k \rho_k^*}{kT}) d^3k - \int \rho_k o_k^* C(k) d^3k] \qquad (2.3.14)$$

where v_k is the Fourier transformation of the potential :

$$v(r_\alpha - r_\beta) = \frac{1}{(2\pi)^3} \int \exp [-i\underline{k}.(r_\alpha - r_\beta)] v_k d^3k \qquad (2.3.15)$$

2.4. The Rayleigh Distribution

The distribution has many remarkable properties : It is isotropic, i.e. the probability in y, z is independent of end points in x. Thus if we draw a density distribution, projecting the probability into the y axis when the chain goes through A, B and trough A', B' the projection is the same, it does not shrink in y when "pulled" in x.

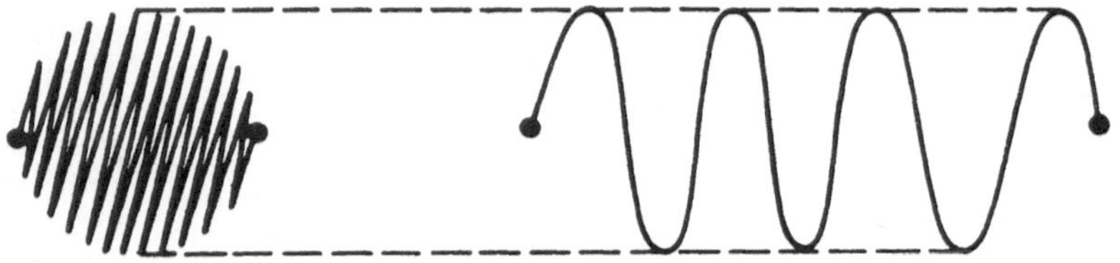

The probability of finding some point of the polymer at r when it starts at the origin is coulombic for a long chain

$$\int_0^\infty \frac{\exp\left(-\frac{3}{2\ell}\frac{r^2}{s}\right)}{\left(\frac{2\pi\ell}{3}\right)^{3/2} s^{3/2}}\, ds = \frac{1}{v\ell}\,\frac{3}{2\pi}\ . \tag{2.4.1}$$

Fluctuations of any quantity do not fit into any closed pattern e.g. the density

$$\rho(\underline{r}) = \int \delta(\underline{r} - \underline{R}(s))\, \frac{ds}{\ell} \tag{2.4.2}$$

from ρ has the mean

$$< \rho(r) > = \left(\frac{3}{2\pi}\right) \frac{1}{r\ell^2} \tag{2.4.3}$$

$$< \rho_k > = \frac{6}{k^2\ell^2} \tag{2.4.5}$$

for a long chain, but

$$< \rho(r)\ \rho(r') > - < \rho(r) >< \rho(r') > \tag{2.4.6}$$

or more easily

$$< \rho_k\rho_j > - < \rho_k >< \rho_j > \alpha \ \frac{2\ kj}{(k\ \sin)^2}\,\frac{1}{k^2}\,\frac{1}{j^2}\,\frac{1}{\ell^4} \tag{2.4.7}$$

is not small, being of order $<\rho><\rho>$. Likewise any higher moments.

Many means can be carried out with this distribution, i.e.

$$f\,(\underline{r}(s_1),\underline{r}(s_2)\ldots) = \int \tilde{f}\,(k_1 k_2\ldots)\,\exp\,[\,i\,\Sigma\,k_\alpha r(s_\alpha)]\;\Pi\,d^3k \qquad (2.4.8)$$

where $\tilde{f}$ is the Fourier transform of f.

Suppose s_1, s_2 ... are arranged in order, as s_α, s_β... then the Rayleigh distribution, being Markovian, can be used between each segment

$$P(r_\alpha, r_\beta) = \left(\frac{3}{2\pi|s_\alpha - s_\beta|}\right)^{3/2} \exp\left[-\frac{3}{2\ell}\,\frac{(\underline{r}_\alpha - \underline{r}_\beta)^3}{|s_\alpha - s_\beta|}\right] \times \ldots \qquad (2.4.6)$$

and the value of $<f>$ thus readily obtained. For example :

$$\left\langle \frac{1}{|\underline{r}(s) - \underline{r}(s')|}\right\rangle = \frac{1}{(2\pi)^3}\,\frac{1}{4\pi}\int \frac{d^3k}{\underline{k}^2} < \exp\,[\,i\underline{k}(\underline{r}(s) - \underline{r}(s'))] >$$

$$= \frac{1}{(2\pi)^3}\,\frac{1}{4\pi}\int \frac{d^3k}{\underline{k}^2}\,\exp\left[-\frac{k^2\ell}{6}\,|s-s'|\right]$$

$$= \frac{\sqrt{\pi/2}}{(2\pi)^3}\left(\frac{6}{\ell|s-s'|}\right)^{1/2} \qquad (2.4.7)$$

Likewise

$$< \delta(\underline{r}(s) - \underline{r}(s')) > = \left(\frac{3}{2\pi\,\ell|s-s'|}\right)^{3/2} \qquad (2.4.8)$$

The first result in this field due to Einstein was of course

$$< (\underline{r}(s) - \underline{r}(s'))^2 > = \ell\,|s-s'| \qquad (2.4.9)$$

which happens to be true for rods as well as gaussian distributions for the steps of the walk. In general however whereas one can expect all distributions to settle down to the Rayleigh for large s - s', it is unreliable at small distances.

The Rayleigh distribution is a solution of Fick's equation

$$\left(\frac{\partial}{\partial s} - \frac{\ell}{6} \nabla^2\right) \rho(\underline{r},\underline{r}';s,s') = \delta(\underline{r}-\underline{r}')\delta(s-s') \quad (2.4.10)$$

which can be derived directly from the original Markov specification, and will only apply for large (s - s')

2.5. The Wiener Distribution [2]

Since over arc lengths it is irrelevant which short scale distribution is used, let us take it to be the gaussian. Then one has

$$P(r_1 \ldots r_N) = \left(\frac{2\pi}{3} \ell^2\right)^{-3N/2} \exp\left[-\frac{3}{2\ell^2} \sum_n (\underline{r}_n - \underline{r}_{n-1})^2\right] \quad (2.5.1)$$

Use the Fourier coefficients to get

$$= \exp\left[-\frac{3}{\ell} \sum_q \sin\frac{q\ell}{2} \left|\underline{r}(q)\right|^2\right] \quad (2.5.2)$$

The transformation $r_n \rightarrow r(q)$ will have a Jacobian which will tend to unity as N and the total number of q's used both tend to infinity i.e. just as with the analysis of the ρ_k above

$$f(r_n) \rightarrow r(q) \rightarrow \exp\left(-\frac{\Sigma \left|\underline{r}_q\right|^2}{N}\right) \rightarrow 1, \ N \rightarrow \infty \quad (2.5.3)$$

Since the distribution will be employed only for the large N i.e. small q, the $\sin^2$ can be approximated and we are left with

$$P(\ldots r_q \ldots) = N \exp\left[-\frac{3}{2} \int \ell \, dq \, q^2 \left|\underline{r}(q)\right|^2\right] \quad (2.5.4)$$

which is the Wiener distribution. It is normally written in the back Fourier transform i.e.

$$P([r]) = N \exp\left[-\frac{3}{2\ell} \int \underline{r}'^2 \, ds\right] , \quad (2.5.5)$$

but it must be understood that this form implies a random walk structure at arbitrarily small step length and cannot be used at small distances. For example for rods

$$< (\underline{r}_1 - \underline{r}_2)^2 > = \ell^2 \tag{2.5.6}$$

$$< (\underline{r}_1 - 2\underline{r}_2 + \underline{r}_3)^2 > = 2\ \ell^2 \tag{2.5.7}$$

To use the Wiener integral we see that

$$< (\underline{r}(s) = \underline{r}(s'))^2 > =$$

$$= \frac{1}{2\pi} \int dq\ d\eta < \underline{r}_q \underline{r}_\eta (e^{iqs} - e^{iqs'})(e^{i\eta s} - e^{i\eta s'}) > \tag{2.5.8}$$

But

$$< \underline{r}_q^\alpha\ \underline{r}_\eta^\beta > = \delta(q + \eta)\ \delta^{\alpha\beta} \frac{2}{3\ell q^2} \tag{2.5.9}$$

so that

$$< (\underline{r}(s) - \underline{r}(s'))^2 > = 2 \int \frac{dq}{q^2} \sin^2 [\frac{q(s-s')}{2}] \ \frac{1}{\pi \ell} \tag{2.5.10}$$

$$= \ell |s - s'| \ . \tag{2.5.11}$$

Similarly other averages can be computed.

The diffusion equation can be directly related to the Wiener distribution by assuming differentiability i.e.

$$\frac{3\ (\underline{r}(s) - \underline{r}(s+ds))^2}{2\ ds} \to \frac{3\ (ds)^2\ \underline{r}'^2(s)}{2\ ds} \tag{2.5.12}$$

so that given a set of intervals

$$\sum_i \frac{3\ (\underline{r}(s_i) - \underline{r}(s_{i+1}))^2}{2\ |s_i - s_{i+1}|} \to \frac{3}{2} \int ds\ \underline{r}'^2(s) \ . \tag{2.5.13}$$

The process of proving the mathematical validity of this process is outside our scope here, and in practice it is the Fourier form which will always be employed. Much more general differential equations can be put in the functional integral form. For example it is easy to see that

$$\left(\frac{\partial}{\partial s} - \frac{\ell}{6}\nabla^2 + U(c)\right) G(\underline{r},\underline{r}';s,s') = \delta(\underline{r}-\underline{r}')\ \delta(s-s') \tag{2.5.14}$$

can be represented by

$$N \int \exp\left[-\frac{3}{2\ell}\int_{s'}^{s} \underline{r}'^2(s_1)ds_1 - \int_{s'}^{s} U(\underline{r}(s_1))ds_1\right](\delta\underline{r}) \tag{2.5.15}$$

and

$$N \int \exp\left[-\frac{3}{2\ell}\int_{s'}^{s} (\underline{r}' + \varepsilon\nabla\omega)^2\right](\delta r) \tag{2.5.16}$$

represents

$$\left(\frac{\partial}{\partial s} - \frac{\ell}{6}\nabla(\nabla + \frac{6\varepsilon}{\ell}\nabla\omega)\right)G = \delta(\underline{r}-\underline{r}')\delta(s-s') \tag{2.5.17}$$

where (δr) is used for $\Pi dr(s) = \lim \Pi\, dr_n$.

The symbol N is used for the awkward normalisation. Note this awkwardness does not arise if the following parametrisation is used

$$G = \int \prod_s dk(s) \prod_s dr(s) \exp\left[-i\int \underline{k}(s).\underline{r}'(s)ds - \frac{\ell}{6}\int \underline{k}(s)^2 ds - \int U\, ds\right] \tag{2.5.18}$$

which is directly derivable from writing

$$G(\underline{r}_1,\underline{r}_2;s_1,s_2) =$$

$$= \tilde{G}(\underline{r}_1 + \underline{r}_2,\ \underline{r}_1 - \underline{r}_2;\ \delta_1 - \delta_2)$$

$$= \int dk \exp\left[i\underline{k}.(\underline{r}_1 - \underline{r}_2)\right]\tilde{G}(\underline{r}_1 + \underline{r}_2, \underline{k};\ s_1 - s_2) \tag{2.5.19}$$

Then as $\underline{r}_1 \to \underline{r}_2$, $s_1 \to s_2$ one can think of $\underline{r}_1 + \underline{r}_2 \to r$ and for example :

$$\left(\frac{\partial}{\partial s} - \frac{\ell}{6}\nabla^2 + U(r)\right) G(\underline{r}_1\underline{r}_2\ s_1 s_2) = \delta\delta \tag{2.5.20}$$

becomes

$$\left(\frac{\partial}{\partial s} + \frac{\ell}{6}k^2 + U(r)\right)\tilde{G}(\underline{r},\underline{k},\ s_1 - s_2) = \delta \tag{2.5.21}$$

and the form (2.5.18) above results.

Clearly we can use this formulation for any differential operator

$$(\frac{\partial}{\partial s} + \mathcal{H}(\underline{r},\frac{\partial}{\partial \underline{r}}))\ G(\underline{r}\underline{r}';ss') = \delta(\underline{r}-\underline{r}')\delta(s-s') \tag{2.5.22}$$

Put

$$G = \int \exp\,[i\underline{k}.(\underline{r}-\underline{r}')]\,\tilde{G}(\underline{r}+\underline{r}',\underline{k};s,s')\ d^3k \tag{2.5.23}$$

$$(\frac{\partial}{\partial s} + \mathcal{H}(\underline{r},\underline{k}))\ \tilde{G}(\underline{r},\underline{k};\ s,s') = \delta(s-s') \tag{2.5.24}$$

(Care must be taken to have $\mathcal{H}$ well ordered with all $\frac{\partial}{\partial r}$ on right, on left).

$$\tilde{G} = \exp\,[-\ i \int \underline{k}.\underline{r}'ds - \int \mathcal{H}(\underline{r},\underline{k})ds] \tag{2.5.25}$$

and as before

$$G = \int \Pi\ dk\ \Pi dr\ \exp\,[-\ i \int \underline{k}.\underline{r}'ds - \int \mathcal{H}(\underline{r},\underline{k})ds] \tag{2.5.26}$$

In quantum mechanics $3/2\ell$ is i/h, $\mathcal{H}$ is the Hamiltonian, and if k can be integrated out, what remains is the Lagrangian.

It will be noted that we have used symbol G in this last section rather than P. This is because in general the solutions to these differential equations are not normalised, and the fact that P satisfied a differential equation is a consequence of the simplicity of Fick's equation. In general G represents the total number of configurations relative to the free chain, and if the number of configurations per link is $e^{+\sigma_o}$, an equation for the entropy of the system is :

$$(\frac{\partial}{\partial s} + \sigma_o - \frac{\ell}{6}\nabla^2)G = \delta(r-r')\ \delta(s-s') \tag{2.5.27}$$

$$S = K \log G \tag{2.5.28}$$

Equally if we have some complicated law of diffusion

$$S = K \log[\int \Pi\ dk\ \Pi\ dr\ \exp\,[-i \int \underline{k}.\underline{r}'ds - \int \mathcal{H}(\underline{r},\underline{k})ds]] \tag{2.5.29}$$

or, should part of $\mathcal{H}$ belong to a potential energy and the expression contain kT, e.g. $W(r_\alpha - r_\beta)/kT$

$$e^{-A/kT} = [\int \Pi dk\ \Pi dr \exp[-i \int \underline{k}.\underline{r}'ds - \int \mathcal{H}(\underline{r}_1\underline{k})ds]] \tag{2.5.30}$$

We are still not at the full level of generality, for

$$ikr' - \mathcal{H}(\underline{r},k) = \mathcal{L}(k,\underline{r},\underline{r}')$$

does not contain r", r''' etc.

In general $\mathcal{H}$ may contain many different variables. Let us write this in terms of time rather than arc length when it can be considered as a problem in quantum mechanics. Suppose one has a Hamiltonian

$$\mathcal{H}(q_1,q_2,\ldots,\ i\hbar\frac{\partial}{\partial q_1},\ i\hbar\frac{\partial}{\partial q_2},\ \ldots)$$

in the Schrödinger equation

$$(i\hbar\frac{\partial}{\partial t} + \mathcal{H}(q_1,\ldots,\ i\hbar\frac{\partial}{\partial q_1}\ \ldots))\ G(q_1,q_1'\ldots)$$

$$= \Pi\delta(q-q')\ \delta(t-t') \tag{2.5.32}$$

The same methods show that

$$G = \int \Pi dp\ \Pi dq \exp[\frac{i}{\hbar}\int \mathcal{H}(q_1\ldots q_n,p_1\ldots p_n)dt - \frac{i}{\hbar}\sum_\alpha p_\alpha \dot{q}_\alpha] \tag{2.5.33}$$

If one writes

$$-L(q_1\ldots q_n,\ p_1\ldots p_n) = \mathcal{H} - \Sigma p\dot{q} \tag{2.5.34}$$

one can identify L with the Lagrangian written in terms of q,p. The classical form of Lagrangian dynamics writes

$$L \text{ as } L([q]) \quad \text{or} \quad L(q,\dot{q},\ddot{q}).$$

The relation between the $p_1 \ldots p_n$ and the $q_1 \ldots q_n$ derivatives $\dot{q}$, $\ddot{q}$ ... is given in Whittaker's Analytic Dynamics [3] and we merely note the theory is fully developed. As an example we return to polymer problems and study a chain with stiffness.

2.6. The Worm-like Chain and the Fokker-Planck Equation [4]

Suppose there is an energy of bending in our chain and the random walk is not due to abrupt changes in direction, but due to the thermal fluctuation of the potential energy of bending. Such chains were studied in the polymer context by Kratky and Porod. We could argue that

$$\exp\left[-\frac{3}{2\ell}\int \underline{r}'^2(s)\,ds\right]$$

representing the inextensibility of the chain might still appear, but also now a factor

$$\exp\left[-\frac{\eta}{2kT}\int \underline{r}''^2(s)\,ds\right] \qquad (2.6.1)$$

Thus we have

$$\exp\left[-\frac{\eta}{2kT}\int \underline{r}''^2(s)\,ds - \frac{3}{2\ell}\int \underline{r}'^2(s)\,ds\right] \qquad (2.6.2)$$

This comes back to the previous form if we introduce $v = r'$:

$$\exp\left[-\left(\frac{\eta}{2kT}\right)\int \underline{v}'^2(s)ds - \frac{3}{2\ell}\int \underline{v}^2(s)ds\right] \prod_s \delta(v-r') \qquad (2.6.3)$$

which represents the differential equation

$$\left(\frac{\partial}{\partial s} + \underline{v}\frac{\partial}{\partial \underline{r}} + \frac{kT}{2\eta}\frac{\partial^2}{\partial v^2} + \frac{3}{2\ell}\underline{v}^2\right) G(v,v';r,r';s,s') = \delta\delta\delta \qquad (2.6.4)$$

for proceeding as before

$$G = \int \Pi dk\, dq\, dr\, dv \exp\left[-i\int \underline{k}\cdot\underline{r}' + i\int \underline{k}\cdot\underline{v} - \frac{kT}{\eta}\int q^2 - \int \underline{q}\cdot\underline{v}' - \frac{3}{2\ell}\int \underline{v}^2\right] . \qquad (2.6.5)$$

We have an

$$\mathcal{L}(\underline{k},\underline{r},\underline{v},\underline{r}',\underline{q},\underline{v}') \qquad (2.6.6)$$

Clearly the case of r''' etc. can be included if desired.

Equation (2.6.4) would be the Fokker-Planck equation, if derived with time instead of s, $\underline{v}$ being the velocity of a Brownian particle suffering a weak rapidly fluctuating random force. This leads to a persistence of velocity (direction of polymer) and to a gradual change in direction of the particle (gradual bending of the polymer). In the quadratic case one may evaluate

$$< (\underline{r}(s)-\underline{r}(s'))^2> = \int \frac{(1 - e^{iq(s-s')})dq}{(\frac{\eta}{2kT} q^4 + \frac{3}{2\ell} q^2)} \tag{2.6.7}$$

$$= \frac{kT}{\eta} (s - \sqrt{\frac{\ell\eta}{3kT}} (1 - \exp (- \sqrt{\frac{3kT}{\ell\eta}}(s-s')))) \tag{2.6.8}$$

At small s - s' this gives $(s-s')^2$, at large (s - s') this gives $\ell|(s - s')|$.

One can indeed give the complete Green function for this case because

$$(\frac{\partial}{\partial s} - \frac{1}{2} \frac{\partial^2}{\partial \underline{v}^2} - \frac{v^2}{2}) G(\underline{v},\underline{v}'; s,s') = \delta(\underline{v}-\underline{v}') \delta(s-s') \tag{2.6.9}$$

has the solution

$$G = [\operatorname{cosech} \frac{(s-s')}{2}]^{3/2} \exp [- \frac{(\underline{v}^2+\underline{v}'^2)\cosh(s-s') - 2 \underline{v}.\underline{v}'}{2 \sinh (s-s')}] \tag{2.6.10}$$

from which it is easy to deduce that

$$(\frac{\partial}{\partial s} + \underline{v} \frac{\partial}{\partial \underline{r}} - \frac{1}{2} \frac{\partial^2}{\partial \underline{v}^2} - \frac{v^2}{2}) G(\underline{r},\underline{r}'; \underline{v},\underline{v}'; s,s')$$

$$= \delta(\underline{r}-\underline{r}') \delta(\underline{v}-\underline{v}') \delta(s-s') \tag{2.6.11}$$

has the solution

$$G = [s - s' - \tanh \frac{(s-s')}{2}]^{-3/2} \times$$

$$\exp [- \frac{(r - r' - (v+v') \tanh (s-s')/2}{2 (s - s' - \tanh (s-s')/2}] \tag{2.6.12}$$

3.1. Dynamics of Chains

Another way of studying Brownian motion is to start with equations of motion rather than probabilities. Suppose a particle satisfies the equation of motion

$$\nu \dot{\underline{r}} = \underline{f} \qquad (3.1.1)$$

where $\nu\dot{\underline{r}}$ is the Stokes drag, and $\underline{f}$ a random force. Then if

$$< f(t)\ f(t') > = \frac{1}{2}\ h\ \delta(t-t') \qquad (3.1.2)$$

it can be taken to have a functional distribution

$$\exp\ (-\ \frac{1}{2h}\ \int f^2(t)\ dt) \qquad (3.1.3)$$

so that

$$P([r]) = N \exp\ (-\ \int \dot{r}^2/2h) \qquad (3.1.4)$$

and immediately implies

$$(\frac{\partial}{\partial t} - \frac{h}{2}\frac{\partial^2}{\partial r^2})\ G(\underline{r},\underline{r}';\ t,t') = \delta(\underline{r}-\underline{r}')\ \delta(t-t') \qquad (3.1.5)$$

Similarly if inertia be included

$$m\ddot{\underline{r}} + \nu\dot{r} = f \qquad (3.1.6)$$

leads to mathematics as in section 2.6.

Now consider a particle diffusing whilst subject to a potential well U(r)

$$\nu\dot{r} + \nabla U = f \qquad (3.1.7)$$

or

$$P([r]) = N \exp\ (-\ \frac{1}{2h}\ \int\ (\nu\dot{r} + \nabla U)^2) \qquad (3.1.8)$$

This is equivalent to the differential equation

$$(\frac{\partial}{\partial t} - \frac{h}{2}\frac{\partial}{\partial r}\ (\frac{\partial}{\partial r} + \frac{\nu}{h}\ \nabla U))G = \delta\delta \qquad (3.1.9)$$

The equilibrium distribution of such a particle must be

$$\exp\ (-\ U/kT) \qquad (3.1.10)$$

so we can identify $kT = h/\nu$, the usual Einstein relationship.

In particular for harmonic wells we have

$$\left(\frac{\partial}{\partial t} - \frac{h}{2}\frac{\partial}{\partial \underline{r}}\left(\frac{\partial}{\partial \underline{r}} + \frac{1}{2}\frac{\omega_o^2}{kT}\underline{r}\right)\right)G = \delta\delta \qquad (3.1.11)$$

This we can apply directly to the individual Fourier components of the polymer.

Consider an ensemble of particles labelled q in wells. Then

$$\left(\frac{\partial}{\partial t} - \sum_q \frac{h}{2}\frac{\partial}{\partial r_q}\left(\frac{\partial}{\partial \underline{r}_q} + \frac{1}{2}\omega_q^2\, r_q\right)\right)G = \delta\Pi\delta \qquad (3.1.12)$$

We can identify r_q with the former component

$$r(q) = \sum_n e^{2\pi iqn/N}\, r_n \qquad (3.1.13)$$

and the equilibrium distribution (2.5.4) gives :

$$\omega_q^2 = \frac{\ell}{2}\, q^2 \;.$$

To put it another more formal way, we can invoke the general format of dissipative systems.

3.2. Lagrangian and Rayleighian Formalism

It is worth going briefly over the mathematics required to handle the dynamic problem. The dynamics of a conservative system can be derived from Hamilton's Principle Function, the time interval of the Lagrangian

$$\mathbf{A} = \int dt\, \mathcal{L}(r,\dot{r}) = \int dt\, \mathcal{L}([r]) \qquad (3.2.1)$$

When a constraint is present, this is handled by adding a Lagrange multiplier times the constraint to $\mathbf{A}$ i.e. if

$$Z^i(r,\dot{r}) = 0 \qquad (3.2.2)$$

$$\mathbf{A} \to \int dt\, (\mathcal{L} + \Sigma\, \lambda^i\, Z^i) \qquad (3.2.3)$$

The equations of motion are :

$$\delta\mathbf{A}/\delta\underline{r} = 0 \qquad (3.2.4)$$

and the Lagrange multipliers are determined by fitting the constraint. When friction is present one uses a Rayleighian

$$\mathcal{R} = \mathrm{A} + \mathrm{B} \tag{3.2.5}$$

where

$$\mathrm{B} = \int dt \quad \mathrm{M}(v) \qquad (v \equiv \dot{r}) \tag{3.2.6}$$

and the equations of motion become

$$\delta \mathrm{A}/\delta \underline{r} + \delta \mathrm{B}/\delta v = 0 \tag{3.2.7}$$

These same equations are obtained when a Lagrange multiplier is present, but a decision must be made to include the Lagrange multiplier term either in A or in B. Thus the no slip boundary condition between a polymer and fluid can be expressed as $\underline{u} = \underline{v}$ or $\underline{\dot{r}} = \underline{\dot{R}}$ (coordinates of fluid and polymer velocity, or fluid and polymer position, respectively). One must not put it as

$$\underline{u} = \underline{\dot{r}} \quad \text{or} \quad \underline{\dot{R}} = \underline{v} \tag{3.2.8}$$

The variational approach is very convenient and unambiguous.

The equilibrium distribution function suggests that for a fluctuation r(q) the free energy due to the configuration is

$$F = F_0 - \frac{3\ell}{2} kT \int |\underline{r}(q)|^2 q^2 dq \quad \begin{pmatrix} = - TS \\ = k \log \Omega \end{pmatrix} \tag{3.2.9}$$

so that provided we study $q \sim 0$ and very slow processes the free energy F can be regarded as a potential energy. Suppose the chain is embedded in a medium which exerts a (Stokes) viscosity at each point of $\dot{r}$ and a random force from a potential Φ $\nabla\Phi = F$ which is random at each point s, and therefore also each q. The Rayleighian is diagonal in q

$$\mathcal{R} = \int dq \left(\frac{1}{2} m|\dot{r}(q)|^2 + \frac{1}{2} \nu|v(g)|^2 - \frac{3\ell}{2} kT\,|r(q)|^2 + \Phi\right) \tag{3.2.10}$$

so that

$$m r_q + \nu \dot{r}_q - 3\ \ell kT\ q^2 r_q = F_q \tag{3.2.11}$$

For slow motion the inertia can be ignored and

$$\nu \dot{r}_q - 3\ \ell kT\ r_q q^2 = F_q \tag{3.2.12}$$

Now F is supposed to vary instantly in time so that

$$< F_q(t)\ F_{-q'}(t') > = h\delta(t-t')\ \delta(q-q') \tag{3.2.13}$$

This corresponds to a functional probability distribution

$$P(F) = N \exp [-\frac{1}{2h} \int |F_q|^2\ dt] \tag{3.2.14}$$

and hence

$$P([r]) = N \exp [-\frac{1}{2h} \sum_{\nu} \int |\nu \dot{r}_q + 3\ \ell kT\ q^2 r_q| dt \tag{3.2.15}$$

This is a Wiener integral but now with t rather than s as the basic variable. It follows that

$$(\frac{\partial}{\partial t} - \frac{\partial h}{2\nu^2} \int dq \frac{\partial}{\partial r_q} (\frac{\partial}{\partial \underline{r}_q} + \frac{6\ kT\nu}{h}\ \ell q^2 r_q))f = 0 \tag{3.2.16}$$

which is the Rouse equation.

3.3. Solution of the Rouse Equation

The equation can be solved from the properties of Hermite's equation, but we also see from the path integral that

$$< \underline{r}(q,\omega)\ \underline{r}(q',\omega') > = \frac{\nu\ kT}{\nu^2\omega^2 + (kT)^2 q^4}\ \delta(q+q')\delta(\omega+\omega') \tag{3.3.1}$$

Hence for the Rouse model, making everything general functions of s and t for the moment

$$< (\underline{r}(s,t) - \underline{r}(s',t'))^2 >$$

$$= \int \frac{d\omega\ dq}{(2\pi)^2} \frac{(1 - e^{i\omega(t-t')+iq(s-s')})}{\nu^2\omega^2 + (3\ kT\ \ell q^2)^2} \frac{h(q)}{2} \tag{3.3.2}$$

For t = t', we must get

$$< (r(s,t) - r(s',t))^2 > = \ell \, |s - s'| \qquad (3.3.3)$$

so that, since this value of the integral is correctly given with $h(q) = h$

$$\frac{h\ell}{6kT\nu} \, |s - s'|$$

we thus regain the Einstein relation

$$h = 6 \, kT\nu$$

However, for $s = s'$ we have

$$< (r(st) - r(s,t'))^2 > = \int \frac{d\omega \, dq}{(2\pi)^2} \frac{6 \, kT\nu.(1-\cos\omega(t-t'))}{(\nu^2\omega^2 + (3kT \, \ell q^2)^2)}$$

$$= (t-t')^{1/2} \, (\frac{kT\ell}{\nu})^{1/2} \qquad (3.3.4)$$

This is the diffusion law implicit in the simplest Rouse equation. For very long times of course the whole chain moves and diffuses like $(t - t')$, but the local effects go like $(t - t')^{1/2}$, but with a much larger coefficient. When the hydrodynamic interaction is included $\nu \propto q^{1/2}$ and the integral will give $(t - t')^{2/3}$. Again the whole chain will diffuse very slowly with $t - t'$.

When the chain is in highly restrictive surroundings and these surroundings are taken as fixed, the progress of the chain can only be by wriggling throught the available space, a process called reptation by de Gennes. The diffusion law $t^{1/2}$ derived above acts independently in the three cartesian dimensions, and so will act equally along the confining pipe

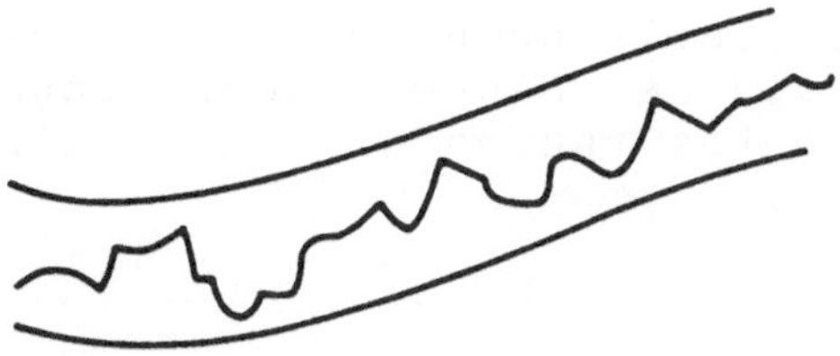

Thus if a coordinate system is chosen in which the axis of the pipe is chosen as one coordinate, S say,

$$< (S(s,t) - S(s,t'))^2 > \quad \propto (t - t')^{1/2} \qquad (3.3.5)$$

But S itself is the arc length of the pipe, itself a random walk, so in an absolute coordinate system

$$\begin{aligned} < (\underline{r}(S) - \underline{r}(S'))^2 > \quad &\propto |S - S'| \\ &\propto (t - t')^{1/4} \end{aligned} \qquad (3.3.6)$$

The reptation diffusion is therefore slower again. Note that the hydrodynamic interaction will not be appropriate in this case.

3.4. Melt Diffusion

If the surroundings of a chain are themselves in a state of flux as in a melted plastic, one can show that

$$h = h_o + q^2 h_1 \qquad (3.4.1)$$

where

$$h_o = h_{oo} - h_{o1}(q) \qquad (3.4.2)$$

h_{o1} arising from the entanglement density confining chains and damping fluctuations. At some critical density will reach zero and from then on

$$h \sim q^2 h_1 \qquad (3.4.3)$$

This leads to a diffusion

$$< (\underline{r}(st) - \underline{r}(st'))^2 > \; \propto \; |t - t'|^{1/4} \qquad (3.4.4)$$

It follows that in addition to a chain reptating, the general motion of the melt will cause a 1/4 law diffusion. It is difficult however to see how these curious new laws could be checked experimentally, since they fall into the region of time scales which is most inaccessible.

3.5. Constraints

This is a very difficult problem, but it can be resolved in two dimensions by various methods and is an elegant example of path integral methods. A powerful constraint can be applied by a point in two dimensions, or a line in three, when the condition that the polymer cannot cross the point or line is applied. This is indeed the case physically when a polymer is so long that its ends can be considered infinitely far off, or in the physically possible, though unusual case, of a closed loop. Clearly if we have two closed loops the configurations shown are not accessible to one another by motion of the molecules.

If one is infinite, the diagram becomes

In two dimensions one can ask for the probability of a large group from A to B in configurations with an angle swept around O, the "winding number" :

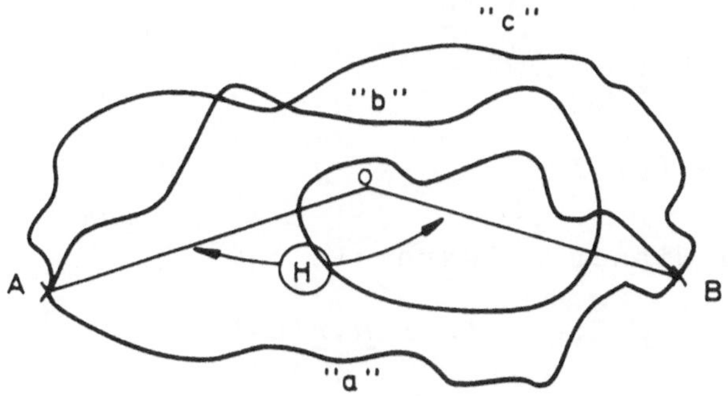

If $\hat{BOA} = \Theta$ then if $\underline{r}(s)$ is written $r \cos \theta$, $r \sin \theta$

$$\int_a d\theta = \Theta \qquad \int_b d\theta = \Theta - 2\pi \qquad \int_o d\theta = 2\pi + \Theta \tag{3.5.1}$$

We are here using $\int d\theta$ as a single valued variable

$$\text{i.e. } -\infty < \theta < \infty \quad , \theta + 2\pi n$$

is equivalent to θ as in ordinary multivalued polar coordinates. Thus the problem is to rework section 1 subject to the constraint

$$\int d\theta = \Theta + 2\pi m \qquad m \text{ integer} \tag{3.5.2}$$

One method would be to employ the power of complex analysis in two dimensions and consider Riemann surfaces on which O is the commencement of a branch cut, a method extendable in principle to simultaneous constraints around amny points of the plane (See example Ito and McKeans book [2]). But we prefer to use methods which can be extended to 3 dimensions and so will directly study the Wiener integral and show that this particular problem can be solved in a way which suggests methods in 3 dimensions.

The constraint $\int d\theta = \theta$ can be exponentiated

$$\delta(\theta - \int d\theta) = \frac{1}{2\pi} \int_{-\infty}^{\infty} \exp\,(-\,i\lambda \int d\theta + i\lambda\theta)\,d\lambda \tag{3.5.3}$$

and so can be written in terms of cartesians taking O as origin, as

$$d\theta = \frac{x\,dy - y\,dx}{x^2 + y^2} = \underline{A}.d\underline{r} \tag{3.5.4}$$

where

$$\underline{A} = \frac{1}{x^2 + y^2}\,(y, -x) \tag{3.5.5}$$

The constrained Wiener integral is now

$$N \exp\,[-\,\frac{1}{\ell} \int \underline{r}'^2 ds - i\lambda \int \underline{A}.\underline{r}' ds] \tag{3.5.6}$$

i.e.

$$N \exp \left[- \frac{1}{\ell} \int (r' + i\lambda \frac{\ell A}{2})^2 ds - \int \frac{\lambda^2 \ell^2}{4} \underline{A}^2 ds \right] \tag{3.5.7}$$

which is equivalent to the differential equation

$$\left(\frac{\partial}{\partial s} - \frac{\ell}{4} \underline{\nabla} (\underline{\nabla} - 2i\lambda \underline{A}) - \frac{\lambda^2 \ell^2}{4} A^2\right) G_\lambda = \delta\delta \tag{3.5.8}$$

or :

$$\left(\frac{\partial}{\partial s} - \frac{1}{4} \left(\frac{\partial}{\partial \underline{r}} - i\lambda \underline{A}\right)^2\right) G_\lambda = \delta(s-s')\ \delta(\underline{r}-\underline{r}') \tag{3.5.9}$$

and we are looking for

$$G_v = \frac{1}{2\pi} \int_{-\infty}^{\infty} e^{i\lambda v}\ G_\lambda\ d\lambda \tag{3.5.10}$$

As was remarked in § 2.5 for the free chain G is p within the entropy per link factor, so that G_v is in fact p_v, where

$$\int_{-\infty}^{\infty} p_v\ dv = \exp\left(- \frac{(R_1 - R_2)^2}{L\ell}\right)\left(\frac{1}{\pi L \ell}\right) \tag{3.5.11}$$

where A,B have coordinates R_1, R_2 and L is the length i.e. the probability that the angle swept out is θ.

Equation is just Bessel's equation. Using polar coordinates

$$\left(\frac{\partial}{\partial s} - \frac{\ell}{4} \left(\frac{\partial^2}{\partial r^2} + \frac{1}{\nu}\frac{\partial}{\partial r} + \frac{1}{\nu^2}\left(\frac{\partial}{\partial \varphi} + i\lambda\right)^2\right)\right) G_\lambda$$

$$= \frac{\delta(r - r')}{\nu r'}\ \delta(\varphi - \varphi')\quad \delta(s-s') \tag{3.5.12}$$

and writing

$$p = \sum_m P_m \exp\left[im\ (\varphi - \varphi')\right] \tag{3.5.13}$$

$$\left(\frac{\partial}{\partial s} - \frac{\ell}{4} \left(\frac{\partial^2}{\partial r^2} + \frac{1}{r}\frac{\partial}{\partial r} - \frac{(\lambda+m)^2}{r^2}\right)\right) P_m = \frac{\delta(r-r')\delta(s-s')}{rr'}$$

$$P_m = \left(\frac{1}{\pi L \ell}\right) I_{|\lambda+m|} \left(\frac{2 R_1 R_2}{L\ell}\right) \exp\left[- \left(\frac{R_1^2 + R_2^2}{L\ell}\right)\right] \tag{3.5.14}$$

where I satisfies

$$\left(\frac{\partial^2}{\partial a^2} + \frac{1}{a}\frac{\partial}{\partial a} - 1 - \frac{(\lambda+m)^2}{a^2}\right) I_{|\lambda+m|}(a) = 0 \qquad (3.5.15)$$

Then

$$P_v = \frac{1}{2\pi}\int_{-\infty}^{\infty} d\lambda \sum_m \exp(im\varphi + i\lambda v) I\left(\frac{2R_1R_2}{L\ell}\right)\exp\left(-\frac{(R_1^2 + R_2^2)}{L\ell}\right) \qquad (3.5.16)$$

Hence, writing λ for $\lambda + m$ and employing

$$\sum_{-\infty}^{\infty} \exp[im(\varphi - \theta)] = 2\pi \sum_{-\infty}^{\infty} \delta(\varphi - \theta + 2\pi n) \qquad (3.5.17)$$

$$P_v = \int d\lambda \exp\left[i\lambda\left(+2\pi(n+\tfrac{1}{2}) - \pi\right)\right] I_{|\lambda|} \exp\left(-\frac{(R_1^2 + R_2^2)}{L\ell}\right) \times \delta(\theta - -2\pi n) \qquad (3.5.18)$$

i.e. $v = \varphi + 2\pi n$ is obviously the case

$$P_\theta = \sum_n \delta(\theta - \varphi - 2\pi n)\, p_n \qquad (3.5.19)$$

The integral can be further reduced, but we stop considering this case at this point, since it is only a model of the three dimensional case.

It is also worth remarking that this analysis does not exhaust the complexities of two dimensions if more than one obstacle is present, e.g. paths a, b cannot be deformed into one another even though angles swept around O_1 O_2 are the same, so that more elaborate functions than the two angles are needed to decompose the system into topological invariants.

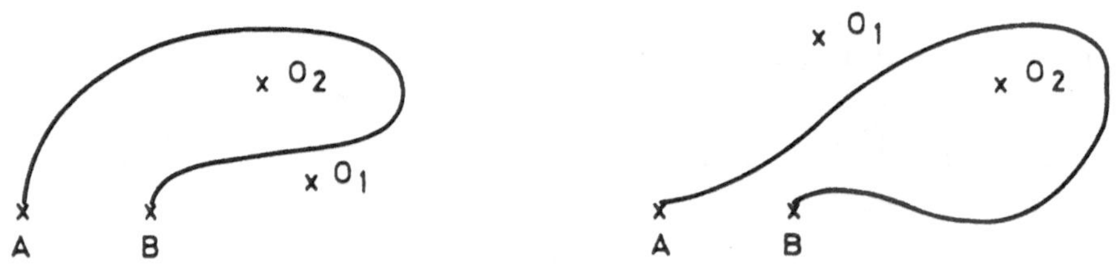

3.6. Topological Constraints in Three Dimensions [8]

The three dimensional invariant corresponding to the angle is

$$I_{12} = \iint (dr_1 \times dr_2).\nabla \frac{1}{r_{12}} \tag{3.6.1}$$

which is readily shown to be an invariant, by Stokes' Theorem, and reduces to the angle when say r_2 is a straight line and r_1 is a locus in a plane perpendicular to r_2. Such expressions are familiar in Ampere's treatment of electric currents in wires, and go back to Gauss. Now if r_2 is a straight line, or any given simple locus the problem resembles the preceding discussion, for example Alexander-Katz has treated the entropy of a random-walk inside a helix. But in general it will be the case of two random walks which is of the greatest interest. For this one needs to study

$$G_v = N \iint \exp\left[-\frac{3}{2\ell}\int r_1^{2} ds_1 - \frac{3}{2\ell}\int r_2'^{2} ds_2\right] \delta(I_{12}-\theta) \, \delta r_1 \, \delta r_2 \tag{3.6.2}$$

where again since

$$I = \theta_0 + 4\pi n \tag{3.6.3}$$

n an integer, θ_0 the angle subtended by the points at infinity of the two chains; note that the invariant in 3 dimensions is only a true invariant for infinite or closed chains, whereas in 2 dimensions the invariant always has a strict meaning

$$G_v = \sum_n \delta(\theta - \theta_0 - 4\pi n) G_n \tag{3.6.4}$$

As before one can write

$$G_v = \frac{1}{2\pi} \int_{-\infty}^{\infty} d\lambda \left[-\frac{3}{2\ell}\int r_1'^{2} ds_1 - \frac{3}{2\ell}\int r_2'^{2} ds_2 + i\lambda \iint (d\underline{r}_1 \times d\underline{r}_2).\nabla^{\perp}_{\gamma_{12}} + i\lambda\theta\right] \delta r_1 \, \delta r_2 \tag{3.6.5}$$

The evaluation of such expressions is no longer possible in closed terms, and is in fact a mathematical problem equivalent to quantum field theory, to be precise rather like the interaction of two highly relativistic electrons via a magnetic field. Some modest headway can be made with the problem, but it turns out that the complete solution is not required in polymer applications, since chains will only really box one another in if they are reasonably dense.

Consider the diagram

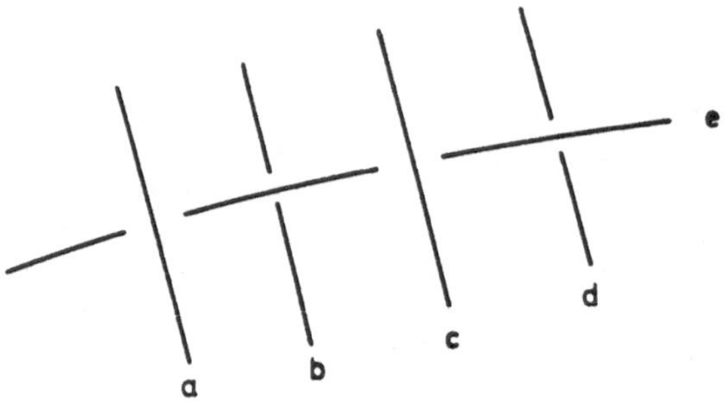

Chain (e) will suffer a pressure from chains (a) and (c) in an opposite sense to (b) and (d). The mean effects of all the chains boxing in chain (e) will be zero, and only the fluctuations will give rise to its localisation and to phenomena like rubber elasticity when the system is disturbed from equilibrium. Thus in spite of the rather ferocious look to the invariant, its effect will be rather gentle and one can start by treating it as a perturbation. The first step is to note that the invariant takes on discrete values, so that if we wish to say we must not write $\delta(I - 4\pi)$ but $\delta_{I,4\pi}$ where $\delta_{a,b}$ is the Kronecker delta $\delta_{a,b} = 1,\ a = b = 0,\ a = b$. The use of the Kronecker δ avoids the emergence of subsidiary (Dirac) deltas in the course of calculation. In an exact calculation it does not matter which is used, but in an approximate calculation one is much easier than the other.

Now the Kronecker delta can be parametrised as

$$\delta_{a,b} = \frac{1}{2\pi} \int_{-\pi}^{\pi} \exp\,[\,i\mu(a-b)]\,d\mu\ , \qquad (3.6.7)$$

and if a,b consist of sums of many parts one can argue that there is no loss of accuracy in using a more amenable mathematical form e.g.

$$\delta_{a,b} = \frac{1}{2\pi} \int_{-\pi}^{\pi} (1 + i\mu(a-b) - \frac{\mu^2}{2} (a-b)^2 \ldots) \, dp \tag{3.6.8}$$

$$= \frac{1}{2\pi} (2 - \frac{\pi^3}{3} (a-b)^2 \ldots)$$

$$\cong \exp [- \frac{\pi^2}{6} (a-b)^2] \ . \tag{3.6.9}$$

If we now consider a single chain $\underline{r}_o$, which had an initial form $\underline{R}_o$, and which has an invariant topology to a set of chains $\underline{R}_x$, the distribution of $\underline{r}_o$ will be

$$\exp [- \frac{3}{2\ell} \int r_o'^2 ds] \ \delta_{I(R_oR_\alpha) I_o(R_oR_\alpha)}$$

$$\cong \exp [- \frac{3}{2\ell} \int r_o'^2 ds - \frac{\pi^2}{6} \sum_\alpha (I(r_o,R_\alpha) - I(R_oR_\alpha))^2] \tag{3.6.10}$$

If there are many $\underline{R}_\alpha$ so that r_o and R_o are boxed in, as in 3.6.8, one can expand to get :

$$\int (r_o - R_o)^2 \sum_\alpha (\nabla I(R_o,R_\alpha))^2 \tag{3.6.11}$$

in the exponent, and replace by its mean i.e.

$$\exp [\frac{q_o^2}{6\ell} \int (\underline{r}_o(s) - \underline{R}_o(s))^2 \, ds] \tag{3.6.12}$$

where

$$\frac{q_o^2}{6\ell} = \frac{\pi^2}{6} \sum_\alpha < (\nabla I)^2 > = C\rho_o/\ell^2 \tag{3.6.13}$$

where C is a numerical constraint, and ρ_o the (polymer) density of the system.

The actual evaluation of C can be very difficult since to get a finite answer the fine structure of the chain is needed; the Wiener integral will give an infinite value for C as it is equivalent to permitted infinitesimal changes in direction. But clearly the dimensions will be determined with the use of 1.

CONCLUSION

In this brief report we have set out the straightforward part of the theory. The problems arise further when

(i) There are many chains cross linked together (rubber). What is the dynamics and equation of state?

(ii) Chains interact with forces which change the path integral from

$$\frac{3}{2\ell} \int \underline{r}'^2 ds$$

to

$$\frac{3}{2\ell} \int \underline{r}'^2 ds + \iint W(\underline{r}(s_1) - \underline{r}(s_2))\, ds_1\, ds_2$$

The potential has a profound effect on configuration and dynamics.

(iii) Chains in solution affect one another via a hydro-dynamic interaction. How can this be handled dynamically?

(iv) All dynamics need to take the topological constraints into account.

Although all these problems have extensive literatures, it is true to say that they are as yet unresolved to the level of other branches of physics.

REFERENCES

1. Flory P.J., Statistical Mechanics of Chain Molecules Interscience, New York (1969).

2. Feynman R.P. and Hibbs A.R., Path Integrals in Quantum Mechanics, Mc Graw Hill, New York 1965.
   Freed K.F., Adv.Phys.Chem. 22 1 (1972).
   Ito K. and Mc Kean H.P., Diffusion Processes and Their Sample Paths, Springer Verlag, Berlin (1965).

3. Whittaker E.B., Analytic Dynamics C.U.P.

4. A useful review is given by Freed, ref.2.

5. Edwards S.F. and Grant J.V., J.Phys. A6 1169-1185 (1973).

PATH INTEGRALS AND CONTINUUM FRÖHLICH POLARONS†

J.T. Devreese*

Institute of Applied Mathematics, R.U.C.A.
Groenenborgerlaan 171
B-2020 ANTWERPEN (Belgium)

SYNOPSIS

In this set of lectures, the physics of Fröhlich polarons is discussed in terms of the Feynman approach to the polaron problem. Attention is given to the implicit introduction of the polaron center of gravity in the action by elimination of the phonon variables. The variational principle introduced by Feynman is discussed and a connection with approximants of Stieltjes integrals is indicated. The accuracy of the groundstate energy of the Feynman model is tested, using an exactly soluble one-dimensional model. The linear response of polarons is considered. The dynamical aspects are investigated on physical as well as mathematical consistency requirements. The path integral approach to linear response is compared with self-consistent solutions of the Heisenberg equations of motion for the polaron. The static aspects of the linear response are considered and the $\frac{3}{2}$ kT controversy of the mobility is discussed critically. The generalization to a nonlinear response theory is given and compared with a Boltzmann equation approach to the nonlinear behavior of polarons in high electric fields.

---

† Work performed in the framework of the project E.S.I.S. (Electronic Structure in Solids) of the University of Antwerpen and the University of Liège.

* Also at: Dept. Natuurkunde, Universitaire Instelling Antwerpen, B-2610 Wilrijk (Belgium).

Table 1 CALCULATION METHODS OF THE POLARON SELF ENERGY AND EFFECTIVE MASS

Perturbation theory		
Second order [2]	Weak coupling $\frac{\Delta E}{\hbar\omega} = -\alpha$	$\frac{m^\star}{m} = (1 - \frac{\alpha}{6})^{-1}$
Canonical transformations		
Elimination of the electron coordinates [6]	Weak coupling $\frac{\Delta E}{\hbar\omega} \leqslant -\alpha$	$\frac{m^\star}{m} = 1 + \frac{\alpha}{6}$
Ibidem and diagonalization of recoil [7]	Weak coupling $\frac{\Delta E}{\hbar\omega} \leqslant - -1.26(\frac{\alpha}{10})^2 -1.875(\frac{\alpha}{10})^3$	$\frac{m^\star}{m} = 1+\frac{\alpha^2}{6}+2.24(\frac{\alpha}{10})^2$
	Strong coupling $\frac{\Delta E}{\hbar\omega} \leqslant -1.05\,\alpha^2$	
Elimination of the center of gravity [8]	Strong coupling $\frac{\Delta E}{\hbar\omega} \leqslant -\frac{\alpha^2}{3\pi} -3\log 2 - \frac{3}{4}$	
Adiabatic approximation		
Localized trial wave functions[9]	Strong coupling $\frac{\Delta E}{\hbar\omega} = -0.106\,\alpha^2$	$\frac{m^\star}{m} = 0.0208\,\alpha^4$
Translational invariant trial wave function[10]	Weak coupling	same results as [6]
	Strong coupling $\frac{\Delta E}{\hbar\omega} = -0.106\,\alpha^2 -2.73- \frac{8.6}{\alpha^2}$	$\frac{m^\star}{m} = 155(\frac{\alpha}{10})^4(1+\frac{67}{\alpha^2})^{-1}$

Table 1 (continued)

Green's functions		
Summation of diagrams [12]	Weak coupling	Same results as [6]
Variational solution of Dyson equation [11]	Weak coupling	Same results as [6]
	Strong coupling $\frac{\Delta E}{\hbar\omega} = -0.108\alpha^2 - 3\ln 2 - O(\alpha)^2$	
Path integrals		
Elimination of phonon variables and quadratic approximation [13]	All coupling $\frac{\Delta E}{\hbar\omega} \leqslant \frac{3}{4v}(v-w)^2 - \frac{\alpha v}{\sqrt{\pi}} \int_0^\infty \frac{du\ e^{-u}}{[w^2u + v^{-1}(v^2-w^2)(1-e^{-uv})]^{1/2}}$ $\frac{m^*}{m} = 1 + \frac{1}{3}\pi^{1/2}\alpha \times$ $\times \int_0^\infty \frac{du\ u^2 e^u}{[w^2u + v^{-1}(v^2-w^2)(1-e^{-uv})]^{3/2}}$ v and w are variational parameters	
Heisenberg equations of motion		
Perturbative solution [14]	Weak coupling	same results as [2]
Self-consistent solution [15]	Weak and strong coupling $\frac{\Delta E}{\hbar\omega}$: same result as [6] and [9] respectively $\frac{m^*}{m} = [1 - \frac{\alpha}{3\sqrt{\pi}}(\frac{m^*}{m})^{1/2} \int_0^\infty \frac{du\ u^2\ e^{-u}}{[C(1-e^{\xi u})+u]^{3/2}}]^{-1}$ C and ξ are determined self-consistently	

I. INTRODUCTION

The polaron concept dates back to L. Landau [1] who introduced it in a two-page paper in 1933. The polaron is an electron, together with the self-induced polarization it carries around in a polarizable medium. As one of his pioneering contributions to the use of quantum field theory in solid state physics, Fröhlich (1950) [2] introduced the following Hamiltonian to describe the polaron.

$$H = \frac{p^2}{2m} + \sum_{\vec{k}} \hbar\omega\, a^+_{\vec{k}} a_{\vec{k}} + \sum_{\vec{k}} (V_k a_{\vec{k}}\, e^{i\vec{k}.\vec{r}} + h.c.) \qquad (1a)$$

$$V_k = \frac{i\hbar\omega}{k} \left(\frac{\hbar}{2m\omega}\right)^{1/4} \left(\frac{4\pi\alpha}{V}\right)^{1/2} \qquad (1b)$$

$$\alpha = \frac{e^2}{\hbar} \left(\frac{m}{2\hbar\omega}\right)^{1/2} \left(\frac{1}{\varepsilon_\infty} - \frac{1}{\varepsilon_0}\right) \qquad (1c)$$

where $\vec{r}$ and $\vec{p}$ are the canonical operators describing the electron coordinate and momentum, $a_{\vec{k}}$ and $a^+_{\vec{k}}$ are the field operators for the longitudinal optical phonon field. The Fröhlich Hamiltonian (1a) contains 4 parameters taken from experiment: ε_∞, ε_0 (the electronic, respectively the static dielectric constant of the crystal), ω (the frequency of the longitudinal optical phonons at k=o) and m (the band mass of the conduction electron). V is the volume of the crystal. α is by now called the Fröhlich or polaron coupling constant.

The Fröhlich Hamiltonian describes the polaron under the assumption that the polarizable medium around the electron can be treated as a continuum. Several review papers [3] and the proceedings of two meetings devoted to polarons [4][5] provide detailed information on our present knowledge of polarons. As the Fröhlich Hamiltonian (1) describes a divergency-free interaction between a fermion and a scalar field, it attracted the attention of physicists, working in field theory. In this context especially the renormalization of the electron mass (resulting in the polaron mass m^*) and the self-energy ΔE of the polaron received attention. In table 1 a review is displayed of a number of methods which have been applied to the Fröhlich polaron Hamiltonian, together with the main approximation, the results and their region of validity.

II. COMMENTS ON THE GROUNDSTATE OF POLARONS AND THE FEYNMAN MODEL

The investigations of Fröhlich on polarons were related to his attempts towards further progress in the theory of superconductivity. At one stage of his investigations, Fröhlich considered the possibility that the intermediate ($\alpha \approx 6$) solutions for polarons would be connected to superconductivity. This was one of the motivations for Feynman to develop an "all coupling" polaron method [13], given in his famous 1955 paper. I will comment on the following topics in the Feynman paper:

- The polaron is described using the path integral formulation of quantum mechanics. From the path integral referring to the groundstate, the *phonon coordinates* are formally *eliminated*. This is similar to the elimination of the photon variables in Quantum Electrodynamics ("there are no fields, only particles").
- A *variational principle* for path integrals is introduced which allows one to choose a "trial-action" much like in the Schrödinger formulation of quantum mechanics, a "trial wave function" can be chosen.
- A trial action is introduced which is *quadratic* in the electron coordinates and velocities and which contains variational parameters (v,w).
- The self-energy ΔE and the polaron mass m^{*} are calculated via the *minimization of the energy* with respect to the parameters v and w. (From here on the units $\omega = m = \hbar = 1$ are used, where ω denotes the frequency of the longitudinal optical phonons and m the electron bandmass).

For the details of the calculation, I refer to the original paper and to a didactical lecture of T.D. Schultz [16].

a) The Elimination of the Phonon Variables and the Center of Gravity of the Polaron

After the phonons are eliminated the path integral describing the polaron takes the form:

$$e^{S} \sim e^{-\frac{1}{2}\int\left(\frac{d\vec{r}}{d\tau}\right)^{2}d\tau \;+\; \frac{\alpha}{2^{3/2}\sqrt{\pi}}\int\int\frac{e^{-|\tau-\sigma|}}{|\vec{r}(\tau)-\vec{r}(\sigma)|}d\tau d\sigma} \qquad (2)$$

The action S corresponds to imaginary times and is only valid for the groundstate. The form of the action S in (2), describing the interaction between the electron and itself at "previous times", suggests that for relatively slow electrons the polaron physics is contained in the *fluctuations* of the electron around the "center of gravity" of the polaron. The center of gravity can be defined as the point described by the position operator $\vec{R}$ canonically conjugate to the total momentum operator $\vec{P} = \vec{p} + \sum_{\vec{k}} \vec{k}\, a^+_{\vec{k}} a_{\vec{k}}$.

If the following coordinates are introduced:

$$\begin{aligned} \vec{r}(\tau) &= \vec{R}(\tau) + \vec{\xi}(\tau) \\ \vec{r}(\sigma) &= \vec{R}(\sigma) + \vec{\xi}(\sigma) \end{aligned} \tag{3}$$

then one obtains for small time differences $|\tau-\sigma|$ that the change of the position of the polaron center $|\vec{R}(\tau)-\vec{R}(\sigma)|$ will be small compared to $|\vec{\xi}(\tau)-\vec{\xi}(\sigma)|$ if the polaron moves slowly. For relatively large time differences $|\tau-\sigma|$, the exponential $e^{-|\tau-\sigma|}$ leads to a negligible contribution of the integrand. Therefore the fluctuations of the electron around the polaron center will be the crucial ingredient in describing the polaron dynamics [17]. Obviously, if the *polaron mass* has to be calculated or if the polaron is subjected to an external electric field, the *translation* of the *center of gravity* will, nevertheless, have to be taken into account.

b) The Center of Gravity and Approximations for the Groundstate

Other well known calculations of the polaron groundstate energy are also related to the introduction (explicitly or implicitly) of the center of gravity of the polaron. E.g. the "first" canonical transformation used by Lee, Low and Pines [6]:

$$U = \exp(-i \sum_{\vec{k}} \vec{k}.\vec{r}\; a^+_{\vec{k}} a_{\vec{k}}) \tag{4}$$

amounts to choosing the electron position as the center of gravity; the subsequent approximations make this choice only meaningful if negligible oscillations of the electron occur i.e. if the coupling with the phonons is weak. The canonical transformation of

Bogolubov and Tyablikov [8] corresponds to describing the fluctuations of the electron around a point fixed in space. This is only valid in the case of strong coupling.

The Feynman trial action to simulate S (eq.2) is:

$$S_o = -\frac{1}{2}\int\left(\frac{d\vec{r}}{d\tau}\right)^2 d\tau + C\int\int e^{-w|\tau-\sigma|}\,|\vec{r}(\tau)-\vec{r}(\sigma)|^2 d\tau d\sigma \tag{5}$$

where C and w are variational parameters. One of the characteristics of this approximate action is that it involves $|\vec{r}(\tau)-\vec{r}(\sigma)|$ in the interaction term and consequently the polaron center is implicitly eliminated from the problem.

Attempts to introduce the polaron center of gravity explicitly in the description of the polaron via the Schrödinger equation or the Heisenberg equations of motion, have to deal with the problem of subsidiary variables [18] [19], which is unsolved up to now. The arguments presented above show that the difficulties arising from the explicit introduction of the polaron center are avoided if approximations involving only $|\vec{r}(\tau)-\vec{r}(\sigma)|$ are made, because then the electron approximately oscillates around the polaron center.

c) The Feynman Variational Principle and Stieltjes Integrals

Feynman's variational principle is based on the following inequality:

$$\langle e^{-sx}\rangle \geqslant e^{-s\langle x\rangle} \tag{6}$$

where the brackets denote the expectation value forthe random variable x. Explicitly:

$$\langle e^{-sx}\rangle = \frac{\int_a^\infty L(x)e^{-sx}dx}{\int_a^\infty L(x)dx}$$

where L(x) is the non-normalized probability density of x. In ref. [15] it has been shown that (6) can be related to expansions of Stieltjes integrals in continued fractions. Indeed, taking the Laplace transform of (6), one obtains:

$$\int_0^\infty e^{-sz} \langle e^{-sx} \rangle ds \geqslant \int_0^\infty e^{-sz}\, e^{-s\langle x\rangle} ds \qquad (8)$$

Using eq.(7) the inequality becomes:

$$\int_a^\infty \frac{L(x)}{x+z}\, dx \geqslant \int_a^\infty \frac{L(x)dx}{\langle x\rangle + z} = \frac{a_0}{a_1 + z} \qquad (9)$$

where

$$a_0 = \int_a^\infty L(x)dx \qquad (10a)$$

and

$$a_1 = \frac{\int_a^\infty xL(x)dx}{\int_a^\infty L(x)dx} \qquad (10b)$$

The inequality remains valid under the condition that $L(x)$ is positive-definite with x real and $z > -a$. The theory of Stieltjes shows that the inclusion of successive moments and approximants in the continued fraction (of which $\frac{a_0}{a_1+z}$ contains the first approximant [20]) leads to alternating upper and lower bounds. This indeed, then suggests the possibility of generalizing the Feynman variational principle to higher order moments and upper and lower bounds.

A more complete discussion, as well as the application to the linear response of polarons, is presented in ref. [15].

d) The Accuracy of the Groundstate Energy of the Feynman Polaron Model

Although the variational model of Feynman results in lower energies for the groundstate of the polaron for the whole range of the coupling constant α, this, in itself, does not prove that the method provides

accurate energies for all α. Arguments like the one presented above (in section a) concerning the fluctuations of the electron in the polaron, tend to justify the model further. Nevertheless, in view of the bold step between the exact polaron action (2) and the approximate one (5), it seemed necessary to further investigate the precision of the Feynman model. Such an investigation had been carried out:
The exact (numerical) and approximate solution of a one-dimensional polaron model are compared. A reprint of this calculation is given in the Appendix. The fact that the Feynman polaron model is more accurate than 1.5 % at all coupling for this one-dimensional model, suggests that the Feynman polaron model for the groundstate is quite accurate indeed.

Conclusion

In this lecture, I emphasized that the elimination of the phonon variables casts the action for the groundstate of the polaron in a form for which the center of gravity is implicitly and approximately introduced. Moreover the quadratic approximation for the trial action conserves this property of the exact groundstate action. The variational approach ensures that the resulting groundstate energy is larger than the exact groundstate energy. The relation between the approximants of continued fractions and the inequality (6) offers the possibility to generalize the variational principle used in the 1955 paper.

It should be remarked that the Feynman approach, discussed in this lecture, has been applied to the so-called Lifschitz problem [21].

III. ON THE PATH INTEGRAL TREATMENT OF LINEAR RESPONSE FOR POLARONS AND ITS CONSISTENCY

The correlation function, which is most frequently studied for polarons [22][23][24][25], is the velocity-auto-correlation function. This function is related to physical quantities such as the optical absorption and the mobility. Because the mobility can also be studied using the Boltzmann equation, the dynamical and static response of polarons will be discussed separately. For the dynamical response I will give a short review of the methods used to calculate the polaron response functions and discuss the physical as well as the mathematical consistency of the resulting expression for the optical absorption of polarons.

A. Dynamical Response and Optical Absorption

The velocity auto-correlation function is defined as the linear coefficient that relates the change in velocity $\vec{v}$ (current) of the polaron with the vector potential $\vec{A}$ applied to the system:

$$\Delta\vec{v}_{\Omega} = \chi_{\vec{v}\vec{v}}(\Omega)\ \vec{A}_{\Omega} \tag{11}$$

where Ω is the frequency of the applied vector potential. The relation between the velocity-auto-correlation function and the impedance used by Feynman, Hellwarth, Iddings and Platzman [22] (henceforth denoted by FHIP) follows from the relation between the vector potential and the electric field

$$\Delta\vec{v}_{\Omega} = \frac{1}{Z(\Omega)}\ \vec{E} \tag{12}$$

or

$$\chi_{\vec{v}\vec{v}}(\Omega) = \frac{-i\Omega}{Z(\Omega)} \tag{13}$$

The expression for the impedance $Z(\Omega)$ can be rewritten as

$$i\Omega Z(\Omega) = \Omega^2 - \chi(\Omega) \tag{14}$$

where the first term describes the response of a free electron and the second term describes the deviation of this behavior due to the interaction. The auxiliary correlation function $\chi(\Omega)$ plays the same role as a self energy for a one-particle Green's function. In terms of $\chi(\Omega)$, the absorption coefficient of electro-magnetic radiation by the polaron is given by

$$\Gamma(\Omega) = \kappa \frac{\Omega \operatorname{Im}\chi(\Omega)}{(\Omega^2 - \operatorname{Re}\chi(\Omega))^2 + (\operatorname{Im}\chi(\Omega))^2} \tag{15}$$

where κ is a constant related to the specific material.

The work of FHIP represented the first attempt to study the linear response functions of polarons at all couplings and all temperatures. For details of the derivations which by now are standard, I refer to the original paper [22] and to a didactical expose by P.M. Platzman [26]. However, some simplifying tricks are given in notes by Thornber [27]. Our comments are concerned with the following topics in the FHIP paper:

- The impedance function $Z(\Omega)$ is calculated using the *adiabatic approximation*. It is assumed that at $t=-\infty$ the density matrix is given by $\exp(-\beta \sum_{\vec{k}} \hbar\omega\, a^{+}_{\vec{k}} a_{\vec{k}})$ where $\beta = 1/kT$.
- After elimination of the phonon coordinates, the response function is expressed as a double path integral. In that path integral, the *Coulomb like terms* in the action are replaced by *quadratic terms* as is done in the 1955 paper for the groundstate and the parameters v and w, that minimized the groundstate energy are used throughout the calculation.
- The following expansion is used:

$$\tilde{G}(\Omega) = \int\int e^{i\Phi} DxDx' = \int\int e^{i\Phi_0} DxDx' + \int\int e^{i\Phi_0} i(\Phi-\Phi_0) DxDx' + \ldots \tag{16}$$

where Φ is the action of the polaron, supplemented by source functions and Φ_0 is the quadratic approximation for Φ.
From $\tilde{G}(\Omega)$ the inverse of the impedance can be obtained by taking functional derivatives with respect to the source functions. In FHIP the zeroth and first order term of the expansion (16) are calculated. Writing $\Omega Z(\Omega)$ as $1/G(\Omega)$ one obtains from the first two terms of (16):

$$G(\Omega) = G_0(\Omega) + G_1(\Omega) + \ldots \tag{17}$$

- The application of this expansion to an exactly soluble model (an electron bound harmonically to a point in space) shows that the expansion for $G(\Omega)$ has no physical resonances, but that the *geometrical progression* of a series like (17) gives the correct result for the model. Therefore FHIP uses the following conjecture for $Z(\Omega)$:

$$Z(\Omega) = G_0^{-1}(\Omega) - G_0^{-2}(\Omega)G_1(\Omega) \qquad (18)$$

Comparing these results (18) with equation (14), the following expression for $\chi(\Omega)$ is obtained:

$$\chi(\Omega) = \int_0^\infty (1-e^{i\Omega\mu}) \, \mathrm{Im}S(\mu)d\mu \qquad (19a)$$

where

$$S(\mu) = \frac{2\alpha}{3\sqrt{\pi}} [D(\mu)]^{3/2}(e^{i\mu} + 2P(\beta)\cos\mu) \qquad (19b)$$

and

$$P(\beta) = 1/(e^{\beta}-1) \qquad (19c)$$

$$D(\mu) = \frac{w^2}{v^2} [\frac{v^2-w^2}{w^2 v} (1-e^{iv\mu} + 4P(\beta v)\sin^2(\frac{v\mu}{2})) - i + \frac{\mu^2}{\beta}] \qquad (19d)$$

v and w are the variational parameters introduced to minimize the groundstate energy and β is $1/kT$.

As nobody has been able to diagonalize the Fröhlich Hamiltonian exactly, studies on the validity of the FHIP approximation have been indirect:
a) The physical consistency of the FHIP model from the study of the optical absorption.

Some physical response properties of polarons have been studied in the weak coupling limit $\alpha\to 0$ and the strong coupling limit $\alpha\to\infty$ (e.g. optical absorption). The optical absorption can also be calculated from the function $\chi(\Omega)$ (as is shown in ref. [24]). If the FHIP response function is adequate to obtain these limiting cases (which are understood physically) this gives further justification to the validity of the FHIP response function in globo (i.e. without references to each of the approximations made separately by FHIP).

Considering first the absorption spectrum calculated in the weak and strong coupling limit by other non-path integral methods:

- For *weak coupling* (and to order α) it can be shown [31] that only the process shown in fig. (1) contributes. I.e. absorption of a photon followed by emission of a real optical phonon. Using the Kubo formula for the velocity auto-correlation function one obtains for $\Gamma(\Omega)$ (see ref. [23] for details)

$$\Gamma(\Omega) = \kappa \frac{2}{3} \alpha\pi\Omega^{-3}(\Omega - 1)^{1/2} \quad \text{for } \Omega > 1 \qquad (20)$$

$$= 0 \quad \text{for } 1 \geqslant \Omega > 0$$

- In ref. [32], it was shown, using the golden rule and Pekar type [33] *strong coupling* wave functions for the polaron that the main excitations dominating the optical absorption at strong coupling are:

- transitions from the groundstate to the so-called "relaxed excited state" (R.E.S.)
- transitions from the groundstate to the same "relaxed excited state" with simultaneous emission of a finite number of real phonons

This second type of transitions results in a side band of the R.E.S. transition. The transitions towards scattering states (fig. (1)) are of negligible oscillator strength as the coupling constant becomes large.

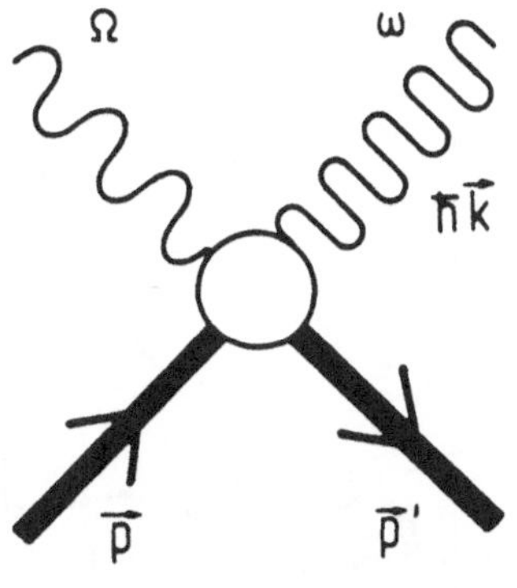

Fig. 1

Transition towards scattering states: Ω is the photon frequency, ω is the frequency of the L.O. phonon. $\vec{p}$ and $\vec{p}'$ are the initial and respectively final polaron momentum.

Compare now the absorption spectrum calculated from $\chi(\Omega)$ [24] using the FHIP expression and absorption spectrum calculated in the weak and strong coupling limit [23], [32]. The shape of the optical absorption coefficient $\Gamma(\Omega)$ (15) will depend on the behavior of

$\mathrm{Im}\chi(\Omega)$ as well as on the behavior of $\Omega^2 - \mathrm{Re}\chi(\Omega)$. Using the expression (19) for $\chi(\Omega)$, it is seen that the equation

$$\Omega^2 = \mathrm{Re}\chi(\Omega) \qquad (21)$$

seems to have only solutions for Ω real if the coupling constant α is great enough.

A plot of $\Omega^2 - \mathrm{Re}\chi(\Omega)$ is given for different α in figure (2). If $\alpha > 6$ pronounced structure is present in the polaron optical absorption. This structure is typical for relatively localized polaron states. For $\alpha < 6$ the optical spectrum is characterized by a broad structure.

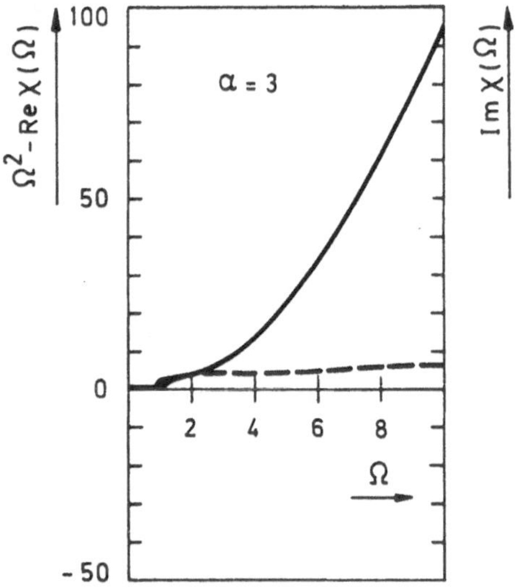

Fig. 2(a)

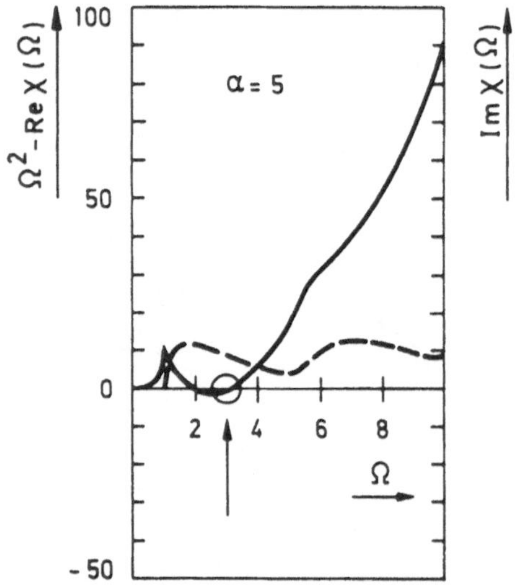

Fig. 2(b)

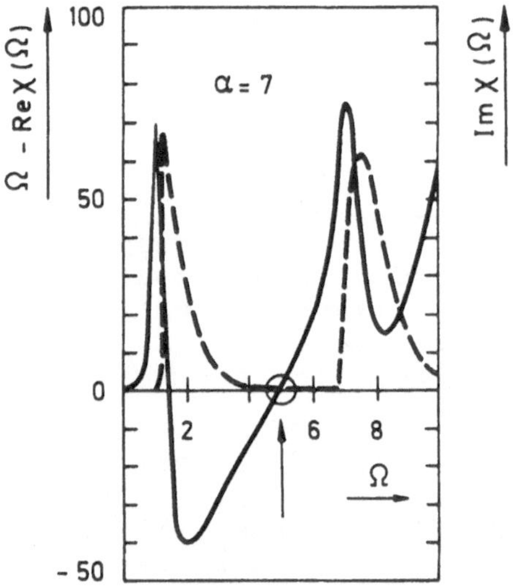

Fig. 2(c)

A plot of $\Omega^2 - \mathrm{Re}\chi(\Omega)$ (——) and $\mathrm{Im}\chi(\Omega)$ (---) is given for a) $\alpha=3$. b) $\alpha=5$. c) $\alpha=7$. Note that the occurence of a zero in $\Omega^2 - \mathrm{Re}\chi(\Omega)$, together with $\mathrm{Im}\chi(\Omega)$ small, determines the oscillator strength of the R. E. S.

As illustrated in fig. 2(a), the denominator of $\Gamma(\Omega)$ is of little importance for the weak coupling , taking the limit $\alpha \to o$ of $\text{Im}\chi(\Omega)$, and retaining the dominant term in the denominator, the weak coupling result (20) is recovered. In fig. (3) the optical absorption spectrum for strong coupling [24] is illustrated. The FHIP model reveals for large α all the physical transitions predicted in ref. [32].

From the preceding remarks, it is concluded that the FHIP model reproduces the characteristics for the optical absorption of polarons which had been derived with other methods for weak and strong coupling. Especially the occurence of the R.E.S. in the FHIP model is remarkable as this delicate property of the polaron system (related to the relaxation of the lattice when the electron is excited in its own potential well) is not constructed ad hoc in the Feynman model. Moreover, this property of the FHIP model gives support for the applicability of the variational parameters of the Feynman model for excited states.

b) Mathematical consistency of the linear response function.

It is well known that response functions have to satisfy a number of sum rules. These sum rules express the short time behavior of the interacting system and can therefore be written as averages over the groundstate wave function (or equilibrium ensemble for $T \neq o$). For the response of polarons two sum rules [34][35][36] together with the fluctuation dissipation theorem[28] have been analyzed. They show the consistency between the response function calculated by FHIP and the groundstate of the Feynman model.

- *The effective mass sum rule* [36]

Connecting the large time and small time behavior of

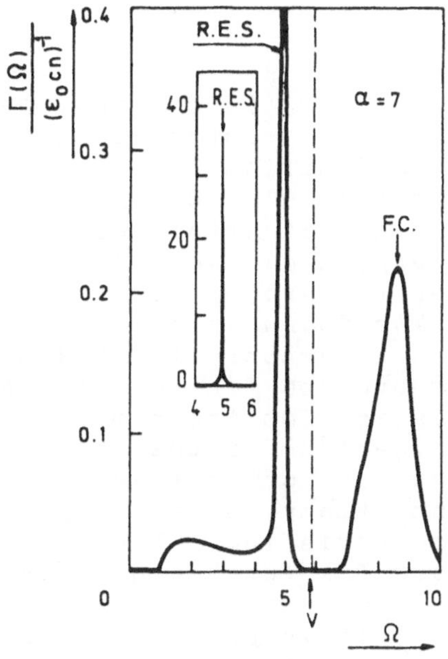

Fig. 3

Optical absorption for strong coupling.

the polaron system, the following relation has been proposed in ref. [36]

$$\frac{\pi e^2}{2m^\star} + \int_1^\infty d\Omega \ \mathrm{Re}\ \chi(\Omega) = \frac{\pi e^2}{2m} \tag{22}$$

where Re $\sigma(\Omega)$ is given by

$$\mathrm{Re}\ \sigma(\Omega) = \frac{e^2}{m} \frac{\Omega\ \mathrm{Im}\ \chi(\Omega)}{(\Omega^2 - \mathrm{Re}\ \chi(\Omega))^2 + (\mathrm{Im}\ \chi(\Omega))^2} \tag{23}$$

This relation enables one to calculate for a given optical absorption the corresponding polaron effective mass $m^\star$. For $\chi(\Omega)$, resulting from the FHIP model, it is shown that the effective mass obtained from eq. (22) is numerically equal to the effective mass obtained by Feynman [13].

- *The groundstate theorem* [34]

Using the first moment of Re $\chi(\Omega)$, a relation between the groundstate energy and $\chi(\Omega)$ was obtained:

$$E(\alpha) = E(o) - \frac{3m}{e^2} \int_0^\infty \frac{d\alpha'}{\alpha'} \int_0^\infty \frac{d\Omega}{\pi}\ \Omega\ \mathrm{Re}\ \sigma(\Omega,\alpha') \tag{24}$$

where $E(\alpha)$ is the groundstate energy, $E(o)$ is the energy of the non-interacting electron and $\sigma(\Omega,\alpha')$ is considered as a function of the frequency Ω and the coupling constant α'. Using expression (23) for $\sigma(\Omega,\alpha)$, it was shown that within numerical accuracy, the groundstate energy $E(\alpha)$ (eq. 24) coincides with the Feynman result for the groundstate energy [13].

- *The Thornber criterion*[28]

K.K. Thornber [28][29] has generalized the groundstate theorem to finite temperature. Using the fluctuation dissipation theorem, he was able to derive an exact criterion for the optical absorption at *each frequency*. The work of this author gives even more stringent conditions for the Feynman model which again, he shows, are satisfied. The reader is referred to Thornber's work for further details.

It is concluded that the application of sum rules and the fluctuation dissipation theorem shows the consistency between the Feynman groundstate model and the FHIP response calculation. In the next section the inversion of $G(\Omega)$ will be discussed.

c) Formulation of the linear response of polarons with Heisenberg's equation of motion in connection with the FHIP approach [15].

In ref. [15] (henceforth denoted by DEK) an approximation is formulated to describe the linear response of polarons in the framework of the Heisenberg equations of motion. An outline of the DEK treatment and a discussion of the approximations made by DEK is presented here. For the details I refer to the original paper.

- DEK start from the Heisenberg equations of motion for the operators $\vec{p}$, $\vec{r}$, $a_{\vec{k}} a^{+}_{\vec{k}}$ describing the electron, respectively the phonon variables. The equations for the phonon-field operators can be formally solved leading to:

$$\frac{d\vec{p}}{dt} = -i\, e^{-\varepsilon|t|} \sum_{\vec{k}} \vec{k}\, V_{\vec{k}}\, e^{i\vec{k}.\vec{r}} a_{+}(\vec{k})$$
$$- e^{-\varepsilon|t|} \sum_{k} \vec{k} |V_{\vec{k}}|^{2} e^{i\vec{k}.\vec{r} - i\omega t} \int_{-\infty}^{t} d\tau\, e^{-\varepsilon|\tau|} e^{-i\vec{k}.\vec{r}(\tau) + i\omega\tau}$$
$$+ hc \qquad (25)$$

- A trial expression for the *electron position operator* is introduced:

$$\vec{r}(t) = \vec{R} + \frac{\vec{P}^{\star}}{m^{\star}} t + \frac{i}{m} \sum_{\vec{k}} \vec{k}\ f_{\vec{k}} \frac{e^{-i\gamma_{\vec{k}} t}}{\vec{k}} e^{i\vec{k}.R} a_{+}(\vec{k}) -$$
$$f^{\star}_{\vec{k}}\, a^{+}_{+}(\vec{k})\, e^{-i\vec{k}.\vec{R}} \frac{e^{i\gamma_{\vec{k}} t}}{\vec{k}} \qquad (26)$$

$\vec{R}$ is the position operator of the polaron center, $\vec{P}^{\star}$ is the conjugate momentum, $m^{\star}$ is the polaron mass. The operators $a_{+}(\vec{k})$ and $a^{+}_{+}(\vec{k})$ are annihilation, respectively creation, operators for "incoming solutions" as introduced in scattering theory. $f_{\vec{k}}, \gamma_{\vec{k}}$, $m^{\star}$ are c numbers which have to be determined. The choice of the form of the trial operator (26) is analogous to the choice of the quadratic polaron action in Feynman's model. This particular form of the trial operator was chosen, taking into consideration:

1) the solution for small α: the functions f_{k}, λ_{k} and the polaron mass $m^{\star}$ are given by:

$$\gamma_{\vec{k}} = \omega - \frac{\vec{p}(t_o).\vec{k}}{m} + \frac{\hbar k^2}{2m} \qquad (27a)$$

$$m^\star = m\ (1 - \frac{\alpha}{6})^{-1} \qquad (27b)$$

$$f_{\vec{k}} = V_k\ (\gamma_{\vec{k}} + i\varepsilon)^{-1} \qquad (27c)$$

2) the solution for strong coupling: the form (26) for $\vec{r}(t)$ yields the strong coupling solution asymptotically

- The functions $f_{\vec{k}}, \gamma_{\vec{k}}$ and the effective mass $m^\star$ are determined *self consistently*: For that purpose the expression (26) for $\vec{r}(t)$ and the resulting $\frac{d\vec{p}}{dt}$ are introduced in the equation of motion (25). Matrix elements between states with zero or one real phonon (any number of virtual phonons) are then calculated. This leads to a set of equations determining $\gamma_{\vec{k}}$, $m^\star$, $f_{\vec{k}}$ under the further assumption that translation and oscillation components of the motion of the electron act independently. The equation for $\gamma_{\vec{k}}$ only involves mass renormalization.

- In solving for $f_{\vec{k}}$ and $m^\star$ an expansion in continued fractions is used for some integrals over the wave vector (after transforming them to integrals over $\gamma_{\vec{k}}$). This expansion is shown to be related to the inequality $\langle e^{-x}\rangle \geqslant e^{-\langle x\rangle}$ used in the Feynman model.

If the results are put together to calculate the polaron optical absorption, it follows:

$$\Gamma(\Omega) = \kappa \frac{\frac{2}{3}\ \alpha\pi\Omega\sqrt{\Omega-1}\quad \exp\ [C(\Omega-1)]}{(\Omega^2 - \mathrm{Re}\chi(\Omega))^2 + (\mathrm{Im}\chi(\Omega))^2} \qquad (28)$$

The equation for the effective mass $m^\star$ being closely related to that obtained by Feynman

$$\frac{m^\star}{m} = [1 - \frac{\alpha}{3\sqrt{\pi}}\ (\frac{m^\star}{m})^{1/2} \int_0^\infty dx\ \frac{x^2 e^{-x}}{[C(1-e^{-\xi x})+x]^{3/2}}]^{-1} \qquad (29)$$

The similarity between the optical absorption calculated by DEK and the optical absorption calculated from FHIP is striking indeed. The analytic structure (except the specific value of m^*) of the denominator of (28) is exactly the same as that following from the FHIP model. However the numerator of (28) is only the first term of Im $\chi(\Omega)$. Therefore the self-consistent equations of motion scheme:

- provides results (effective mass, optical absorption) for $\alpha < 4$ and $\alpha \to \infty$ that are very close to the results obtained by Feynman and FHIP.
- provides resonances (cfr. denominator of (28)) of precisely the same structure as in the FHIP model and gives therefore an additional justification of the geometric series assumption.
- does not constitute an equivalent of the FHIP model at present.

This example shows that it might be difficult in some cases to translate an approximation stemming from path integral formulations. In the case of the polaron, the reason is, presumably, that at each stage of the approximations the physical conservation principles are preserved in the path integral calculation. In the self-consistent equation of motion scheme some approximations had to be made to keep the evaluation of the physical quantities tractable. One of these approximations can be the reason why the causal structure of $\Gamma(\Omega)$ present in the FHIP model is not completely recovered in DEK. In the case of the equation of motion treatment, a canonical transformation would be helpful to provide a complete transliteration and a one-to-one correspondence of the FHIP model.

B. The Static Response of the FHIP Model and the $\frac{3}{2}$ kT Factor

In the FHIP model the mobility is defined as follows:

$$\mu^{-1}_{\text{FHIP}} = \lim_{\Omega\to 0} \frac{\chi(\Omega)}{\Omega} \tag{30a}$$

$$= \frac{2}{3}\frac{\alpha}{\sqrt{\pi}}\left(\frac{v}{w}\right)^3 \frac{\beta}{2}^{5/2} \frac{1}{\sinh(\beta/2)}\int_0^\infty du \frac{\cos u}{(u^2+a^2-b\cos vu)^{3/2}}$$

where

$$a^2 = \frac{1}{4}\beta^2 + R\beta \coth\left(\frac{\beta v}{2}\right) \tag{30b}$$

$$b = \frac{R\beta}{\sinh\left(\frac{\beta v}{2}\right)} \tag{30c}$$

$$R = \frac{v^2 - w^2}{w^2 v} \tag{30d}$$

and v, w are the variational parameters of the Feynman model. This formula has given rise to quite some dispute. If the limit $\alpha \to o$ of (30) is taken, it differs with *a factor* $\frac{3}{2}$ *kT* from the result obtained by several other authors [38][39][40][41]

$$\mu_{FHIP} = \frac{3}{2\beta}\mu^\circ \tag{31a}$$

where

$$\mu^\circ = \frac{e^\beta}{2\alpha} \qquad (\omega = m = \hbar = 1) \tag{31b}$$

The question does not seem to be completely settled to this date. Indeed for most of the theories resulting in μ° a *relaxation time approximation* has been made [41]. It is by no means obvious that this approximation can be justified in the case of polarons and the FHIP model does not rely on it. It is tempting to assume that the FHIP result is the most reliable one as it takes *interference between successive collisions* of the electrons with the phonons into account. If on the other hand one calculates the mobility, using the Heisenberg equation of motion [41], then one finds with a model, which is analogous to FHIP for the optical response of the polaron, the result μ°. This is what other authors also obtained with the relaxation time approach.

A possible difficulty with the result (30) is realized from an analysis of the *Kubo formula* [42]. One has for the polaron conductivity tensor:

$$\sigma(\Omega) = \frac{i}{\Omega+i\varepsilon} - \frac{1}{(\Omega+i\varepsilon)^2} \int_0^\infty dt\ e^{-i(\Omega+i\varepsilon)t} \left(\int_0^\beta d\lambda <[V,P_z](-i\lambda),[V,p_z](t)>\right) \quad (32)$$

If the Ω=o limit is taken the Kubo formula becomes (formally):

$$\sigma(o) = \frac{1}{\varepsilon} + \frac{1}{\varepsilon^2} \int_0^\infty dt\ e^{+\varepsilon t} \int_0^\infty d\lambda <[V,p_z](-i\lambda)[V,p_z](t)> \quad (33)$$

where V is the interaction term in the Hamiltonian (1). Calculating σ(o) to order α, eq. (33) has an apparent divergency in the limit $\varepsilon \to$o. If, however, one inverts the series:

$$\sigma(o) = \frac{1}{\varepsilon} + \frac{1}{\varepsilon^2} \chi_{FF} \simeq \frac{1}{\varepsilon - \chi_{FF}} \quad (34)$$

one finds after taking the limit $\varepsilon \to$o:

$$\mu = \frac{3}{2\beta} \mu^\circ \quad (35)$$

which is the FHIP result as $\alpha \to$o. The formal summation of $\frac{1}{\varepsilon} + \frac{1}{\varepsilon^2} \chi_{FF}$ as a geometric (divergent) series is not justified.* Further investigation is necessary here, especially the order of the limits that have been taken, should be investigated for eq. (30).

The present discussion illustrates that, although path integrals have been very powerful in providing an all coupling polaron model, certain difficulties relating to D.C. transport do not seem to have been resolved as yet. These difficulties presumably are related to the absence of a Kubo formula for the resist-

*Of course, the fact that a non-justified derivation leads to (35) does not prove that this result is not correct. Nevertheless it seems useful to realize this difficulty with the Kubo formula.

ivity rather than for the conductivity. Indeed, as the conductivity tends to infinity for $\alpha \to o$ an expansion for small α of σ is meaningless. Further study of (19) (which does not present any difficulties if $\Omega \neq o$ and is valid for all α) and the path integral formulation leading to it, might contribute to establish a *Kubo formula for the resistivity*. To examine further the significance of (30) it would be useful to *dispose* of a *rigorous* treatment of the Boltzmann equation for low applied electric field. As far as we know such a treatment for the Boltzmann equation is, at present, not available. In [43] it was shown that only under quite strong approximations (at most valid if T=o) the result $\mu = \mu^{\circ}$ is rederived. It also became apparent that α/E is the natural parameter in treating the Boltzmann equation.

IV. NONLINEAR RESPONSE OF POLARONS: PATH INTEGRALS AND BOLTZMANN EQUATION

a) Note on the Thornber-Feynman treatment

A powerful contribution to the nonlinear response of polarons is the Thornber-Feynman theory (which generalizes the FHIP model to high applied electric fields [30]. See also K.K. Thornber's present school lecture notes). For details I refer to the original paper [30]. Only the main steps and approximations of the Thornber-Feynman approach are analyzed here:

- It is assumed that a steady state is reached:

$$\langle \dot{\vec{p}} \rangle = 0 \tag{36}$$

where $\vec{p}$ is the electron momentum. The so-called Fröhlich run-away solution is therefore not considered (see the contribution of K.K. Thornber to the present volume for further discussion of the run-away solution).

- The phonons are (as always for a Feynman model) formally eliminated from the problem resulting in an equation of the type:

$$\vec{E} = \sum_{\vec{k}} \vec{k} \langle R_{\vec{k}} \rangle \tag{37}$$

The $\langle R_{\vec{k}} \rangle$ can be interpreted as the net rate of emission of longitudinal optical phonons with wave vector $\vec{k}$. ($\vec{k}$ is the change of momentum of the electrons). $\vec{E}$ is the applied electric field and $\langle R_{\vec{k}} \rangle$ is expressed as a double path integral.

- The Feynman model is introduced i.e. the retarded Coulomb interactions are replaced by retarded quadratic interactions.

- A reference frame is chosen, which moves with the electrons. The technical difficulties are in evaluating the resulting ("double") path integrals.

Several difficulties pointed out by the authors of [30] remain to be solved:

- presumably due to the order of the limits which are taken, the weak coupling limit of the mobility, calculated with in this approach, differs also with a factor $\frac{3}{2}$ kT from the other weak coupling results.

- The equation for the energy balance

$$\vec{E}.\vec{v} = \sum_{\vec{k}} \omega \langle R_{\vec{k}} \rangle \tag{38}$$

does not lead to the same result as the equation for the momentum balance (37).

Nevertheless, the approach of ref. [30] constitutes a major step forward to nonlinear transport theory in general.

b) Note on the Boltzmann equation

In order to understand the Thornber-Feynman approximation and to make further progress concerning the high field problem, J. Devreese and R. Evrard [43] have discussed the linearized Boltzmann equation for polarons at weak coupling. Although the Boltzmann equation does not allow for interference effects between successive collisions (contrary to the Thornber-Feynman model) it allowed us to obtain the shape of the steady state distribution for small α.

Before analyzing this solution of the Boltzmann equation, it is useful to point out in what respect this approach to the Boltzmann equation is different from the approach considered in ref. [22]. Usually, one calculates the mobility by remarking that the maxwellian distribution $f_o(\vec{p}) = (\frac{2\pi}{\beta})^{-3/2} e^{-\frac{\beta p^2}{2}}$ is an exact solution for the driving electric field equal to zero (the principle of detailed balance) and then considering deviations from $f_o(\vec{p})$, linear in the electric field. The resulting form of the distribution f(p) is given by:

$$f(\vec{p}) = f_o(\vec{p}) \; [1 + E \, e^{i\Omega t} p_z \; h(\vec{p})] \tag{39}$$

The electric field E is chosen in the z direction. The equation for $h(\vec{p})$ becomes after a Fourier transform with respect to the time t:

$$(i\Omega h(p) + \beta)p_z = - \int \frac{d^3p'}{(2\pi)^3} \; \Pi(\vec{p}\rightarrow\vec{p}\,') \; [p_z h(\vec{p}) - p_z' h(\vec{p})] \tag{40}$$

where $\Pi(\vec{p}\to\vec{p}\,')$ is the probability density for electron transition from momentum $\vec{p}$ to momentum $\vec{p}\,'$.
The induced current is given by

$$J(t) = - E\, e^{i\Omega t} \int d^3p\; p_z^2\; h(\vec{p}) f_0(\vec{p}) \qquad (41)$$

Multiplying eq. (40) by $p_z h(\vec{p}) f_0(\vec{p})$ and integrating over all p, it is found that

$$Z(\Omega) = \frac{T(\Omega)}{N} \qquad (42a)$$

where

$$T(\Omega) = \frac{1}{2}\int\int d^3p d^3p' f_0(p)\Pi(\vec{p}\to\vec{p}\,')[p_z h(\vec{p}) - p_z' h(\vec{p})]$$

$$+ i\Omega \int \frac{d^3p}{(2\pi)^3}\; p_z^2\; h^2(\vec{p}) f^0(\vec{p}) \qquad (42b)$$

and

$$N = (2\pi)^3\; \beta[\int d^3p\; p_z^2\; h(\vec{p}) f_0(\vec{p})]^2 \qquad (42c)$$

Taking $h(\vec{p})$ equal to a constant it is found that

$$Z(\Omega) - i\Omega = \Gamma \qquad (43)$$

where

$$\Gamma = \frac{\beta}{2} \int\int \frac{d^3pd^3p'}{(2\pi)^3}\; f_0(\vec{p})\; [p_z - p_z']\; \Pi(\vec{p}\to\vec{p}\,') \qquad (44)$$

This approximation is a relaxation time approximation for high frequency Ω: If the collision term would be small compared to Ω then the assumption that h is a constant becomes correct in the high frequency limit.

In the approach considered in ref. [43], it is assumed that a steady state is reached under the influence of the static external field. Therefore the time independent Boltzmann equation is considered. In doing so, a method was devised to solve the Boltzmann equation analytically if no electrons with kinetic energy $\frac{p^2}{2m} > 2\,\hbar\omega$ are allowed (which is realistic under many physical circumstances). As far as we know this method of solution is new and is therefore briefly outlined here:

- Notice that absorption of a phonon increases the electron energy by $\hbar\omega$, emission lowers the energy by the same amount. This leads to a division of the p-space in spheres of radii $p = (2m\, n\, \hbar\omega)^{1/2}$, where n is an integer. The decomposition of the vector p along the field p_z and orthogonal to the field $p_\perp$ gives rise to a set of critical circles in the $(p_z, p_\perp)$ plane.

- Assuming that the distribution function is zero for $p^2 > 4\,\hbar m\omega$ the emission of phonons always brings electrons from the outside to the inside of the first critical circle (at T=o). This allows for a separation of the distribution function in $f_1(p_\perp, p_z)$ and $f_2(p_\perp, p_z)$ where the index indicates whether the variables lie in the first circle or between the first and second circle.

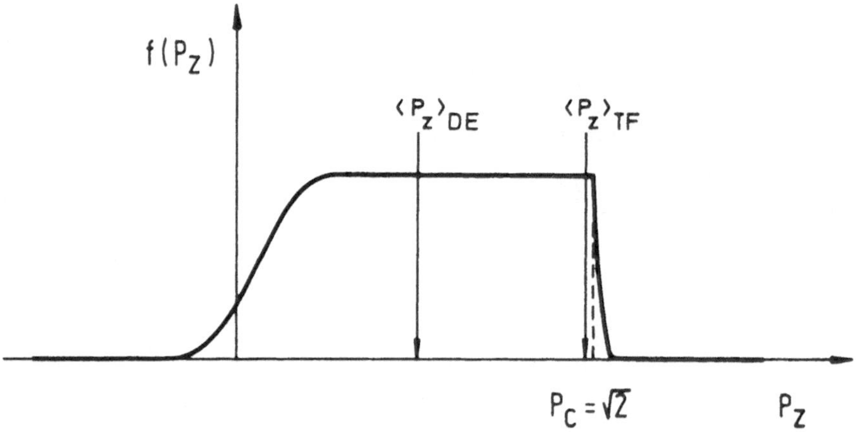

Fig. 4

The section of the distribution function along the $(p_z, p_\perp = o)$ plane is shown; the average momentum according to the Boltzmann equation approach $(\langle p_z \rangle_{DE})$ and the Thornber-Feynman approach $(\langle p_z \rangle_{TF})$ is given for T=o and α = 0,002 (InSb) and E = 12 V/cm.

- It is remarked that $f_1(p_\perp, p_z)$ and $f_2(p_\perp, p_z)$ obey distinct solvable equations:
At T=o the Boltzmann equation reduces to

$$\frac{\partial f(\vec{p})}{\partial p_z} eE = \int d^3p' \, f(\vec{p}')\Pi(\vec{p}'\rightarrow\vec{p}) - f(\vec{p}) \int \Pi(\vec{p}\rightarrow\vec{p}')d^3p' \tag{45}$$

where

$$\Pi(\vec{p}\rightarrow\vec{p}') = \frac{\alpha}{\pi\sqrt{2}} \frac{\delta(\frac{p^2}{2} - 1 - \frac{p'^2}{2})}{|\vec{p}-\vec{p}'|^2} \tag{46}$$

Because emission brings electrons always outside the critical circle, the first term of the right-hand side of equation (45) is identically zero, so that

$$\frac{\partial f_2(p_\perp, p_z)}{\partial p_z} eE = - f_2(p_\perp, p_z) \int d^3p' \, \Pi(\vec{p}\rightarrow\vec{p}') \qquad p > \sqrt{2} \tag{47}$$

On the contrary, the second term is missing when $p < \sqrt{2}$ and

$$\frac{\partial f_1(p_\perp, p_z)}{\partial p_z} eE = \int d^3p' \, f_2(p_\perp', p_z') \, \Pi(\vec{p}'\rightarrow\vec{p}) \tag{48}$$

Moreover in the latter case, the electrons come from outside the circle: $f_2(p_\perp', p_z)$ is the solution of (47).

- After the formal solutions for f_1 and f_2 are obtained, these functions are matched numerically at the first critical circle. When the distribution function is known, the resulting mobility is calculated. In fig. (4) the distribution resulting from the Boltzmann equation is given. The average momentum, obtained from the Boltzmann equation as well as the average momentum resulting from the Thornber-Feynman approach are indicated.

It is hoped that further comparison between the Boltzmann equation results and the Thornber-Feynman approximation will lead us to make further progress with the high field problem.

REFERENCES

[1] L. Landau, Phys. Z. Sowjetun. 3, 664 (1933).
[2] H. Fröhlich, H. Pelzer and S. Zienau Phil. Mag. 41, 221 (1950).
[3] H. Fröhlich, Advances in Physics 3, 325 (1954).
G.R. Allcock, Advances in Physics 5, 412 (1956).
E.P. Gross, in "Mathematical Methods in Solid State and Superfluid Theory", edited by R.C. Clark and G.D. Derrick, Plenum Press, New York (1968).
J. Appel, in "Solid State Physics, Vol. 21", edited by F. Seitz, D. Turnbull and H. Ehrenreich, Academic Press (1958).
[4] C. Kuper and G. Whitfield (editors), "Polarons and Excitons", Oliver & Boyd (1963).
[5] J.T. Devreese (editor), "Polarons in Ionic Crystals and Polar Semiconductors", North Holland (1972).
[6] T.D. Lee, F.E. Low and D. Pines, Phys. Rev. 90, 297 (1953).
[7] A.V. Tulub, Soviet Phys. JETP 14, 1301 (1962).
[8] N.N. Bogolubov and S.V. Tjablikov, Zh. Eksp. Teor. Fiz. 19, 256 (1949).
[9] V.M. Buimistrov and S.I. Pekar, Soviet Phys. JETP 5, 970 (1957).
[10] V.M. Buimistrov and S.I. Pekar, Soviet Phys. JETP 6, 977 (1958).
[11] D. Matz and B.C. Burkey, Phys. Rev. B3, 3487 (1971).
D. Matz, in "Polarons in Ionic Crystals and Polar Semiconductors", edited by J.T. Devreese, North Holland (1972), p. 463.
[12] W. Van Haeringen, Phys. Rev. 137, 1902 (1965).
[13] R.P. Feynman, Phys. Rev. 127, 1004 (1955).
[14] E. Kartheuser in "Polarons in Ionic Crystals and Polar Semiconductors", edited by J.T. Devreese, North Holland (1972).
[15] J.T. Devreese, R. Evrard, E. Kartheuser, Phys. Rev. B12, 3353 (1975).
[16] T.D. Schultz,in "Polarons and Excitons", edited by C. Kuper and G. Whitfield, Oliver & Boyd (1963).
[17] R. Evrard and J. Devreese, in "Lectures on Solid State Physics, Vol. I". Hercegnovi (1963).
[18] R. Evrard, Thesis. Liège (1964).
[19] E.P. Gross, Phys. Rev. 100, 1571 (1955).
[20] H.F. Wall, "Analytic Theory of Continued Fractions", Van Nostrand (1967), p. 15.
[21] J. Luttinger, see his contribution to these proceedings.
V. Samathiyakanit, J. Phys. C7, 2849 (1974).
E.P. Gross, Journal of Stat. Phys. 17, 265 (1977).

[22] R.P. Feynman, R. Hellwarth, C. Iddings and P.M. Platzman, Phys. Rev. 127, 1004 (1962).
[23] J.T. Devreese, W. Huybrechts and L. Lemmens, Phys. Stat. Sol. (b) 48, 77 (1971).
[24] J.T. Devreese, J. De Sitter and M. Goovaerts, Phys. Rev. B5, 2367 (1972).
[25] K.K. Thornber, Phys. Rev. B3, 1929 (1971).
[26] P.M.Platzman, in "Polarons and Excitons", edited by C. Kuper and G. Whitfield, Oliver and Boyd (1963).
[27] K.K. Thornber, in "Polarons in Ionic Crystals and Polar Semiconductors", edited by J.T. Devreese, North Holland (1972).
[28] K.K. Thornber, Phys. Rev. B9, 3489 (1974).
[29] K.K. Thornber, in "Linear and Nonlinear Electron Transport in Solids", edited by J.T. Devreese and V.E. Van Doren, Plenum Press (1976).
[30] K.K. Thornber and R.P. Feynman, Phys. Rev. B1, 4099 (1970).
[31] V.L. Gurevich, I.E. Lang and Yu. A. Firsov, Soviet Phys. Solid State 4, 918 (1962).
[32] E. Kartheuser, R. Evrard and J.T. Devreese, Phys. Rev. Lett. 22, 94 (1969).
[33] S.I. Pekar,"Research in Electron Theory of Solids", AEC-TR-5575 U.S. GPO, Washington, D.C. (1963).
[34] L.F. Lemmens, J. De Sitter and J.T. Devreese, Phys. Rev. B8, 2717 (1973).
[35] L.F. Lemmens and J.T. Devreese, Solid State Commun. 12, 1067 (1973).
[36] J.T. Devreese, L.F. Lemmens and J. Van Royen, Phys. Rev. B15, 1212 (1977).
[37] F.E. Low and D. Pines, Phys. Rev. 98, 914 (1958).
[38] D.C. Langreth and L.P. Kadanoff, Phys. Rev. 133, A1070 (1969).
[39] Y. Osaka, Prog. Theoret. Phys. (Kyoto) 25, 517 (1971).
[40] L.P. Kadanoff, Phys. Rev. 130, 1364 (1963).
[41] J.T. Devreese and R. Evrard, in "Linear and Non-linear Electron Transport in Solids", edited by J.T. Devreese and V.E. Van Doren, Plenum Press (1976).
[42] J.T. Devreese and J. Van Royen, to be published.
[43] J.T. Devreese and R. Evrard, Phys. Stat. Sol. (b) 78, 85 (1976).

APPENDIX

Reprinted from
"PROCEEDINGS OF THE BRITISH CERAMIC SOCIETY"
No. 10, March, 1968

13.—Investigation of the Quadratic Approximation in the Theory of Slow Electrons in Ionic Crystals

By J. T. DEVREESE
Solid State Physics Department, SCK, Mol, and Rijksuniversitair Centrum, Antwerpen (Belgium)

and

R. P. EVRARD*
Institut de Physique, Université de Liège (Belgium)

ABSTRACT

The ground-state properties of polarons have been approximated by Feynman by means of his path integral method. His approximation is of the "harmonic oscillation"-type. Although the Feynman approximation yields a better estimate for the ground-state energy than any other method, there existed no criterion to evaluate the accuracy. In this paper an exactly soluble model for the polaron is used to investigate Feynman's method. Feynman's approximation turns out to be very accurate. His largest error is about 1·42%. It occurs for intermediate coupling and is related to correlation effects. Even at intermediate coupling the precision of the Feynman model is four times better than that of Lee, Low and Pines. There are several arguments to justify an extrapolation of our results to the Fröhlich polaron. It can be concluded that the Feynman model can be safely used to analyse the experimental results concerning the ground-state properties of conduction electrons in ionic crystals (alkali halides, Cu_2O, MgO etc.).

1. INTRODUCTION

The aspects of the continuum polaron which, until now, have been investigated most profoundly from the theoretical point of view are the self energy, the effective mass in the ground state and its mobility.[1–5] Experimental work on polarons has been performed by F. C. BROWN and co-workers.[6–8] They found rather convincing

*Chargé de Recherches du Fonds National Belge de la Recherche Scientifique.

(although indirect) evidence for the existence of continuum polarons, e.g. in AgBr.

More recently attention has been given to a number of more complex properties of the polaron: its different types of excitations,[9,10] its response to light,[11] its properties in static electric[12] and magnetic fields,[13] the relation between its self energy and its momentum in the ground state, etc.,[14] These investigations have been inspired by the desire to arrive at a measurement revealing directly a polaron property. However, these studies are still being developed and are certainly not completed. For example the study of the different excitations (which were first treated in detail in references nine and ten) needs to be completed by a calculation of absorption curves and more generally by the study of optical transitions in polarons. Only after these investigations have been carried out will it be possible to compare the results of FEYNMAN, HELLWARTH, IDDINGS, PLATZMAN[11] (which, in our opinion, do not treat the relaxed internal excited states) with other calculations. The study of the polaron in static electric and magnetic fields has essentially been limited to perturbation treatments. This is also true for the study of the energy *versus* momentum relation which certainly requires further investigation.

As to the earlier work, there is at least one question which merits investigation: what is the accuracy of the methods of LEE, LOW, PINES (L.L.P.), LANDAU-PEKAR (L.P.) and FEYNMAN? This point will be investigated in this paper.

In a forthcoming paper E. Kartheuser and the present authors present results on absorption curves.

2. STUDY OF THE ACCURACY OF THE "CLASSICAL" METHODS

This study will be performed by applying these methods to an exactly soluble polaron model.

In earlier work[15,16] we tested the theories of L.L.P. and L.P. by applying them to the exactly soluble model of GROSS.[17] However, this model presents a number of disadvantages discussed in reference 10, e.g. the minima of its energy *versus* momentum curves do not correspond to a momentum equal to zero. Furthermore, this model does not allow for any correlation between phonons with different **k** vector. Therefore we introduced a symmetrical model which does not present these disadvantages. It is especially adapted to study the Feynman approximation, since this approximation has only been worked out for total momentum equal to zero.

2.1 The Symmetrical Model[18]

2.11 General

Although the results of the calculations of this model have been given in reference 10, the calculations have not been published, and they are therefore presented here.

The Hamiltonian of the symmetrical model, consisting of an electron interacting with two oscillators, with wave vector **k** resp −**k**, is as follows:

$$H=\frac{p^2}{2m}+(a_k^+a_k+a_{-k}^+a_{-k})\hbar\omega+V_k\,(a_k\,e^{ikq}+a_{-k}\,e^{-ikq})$$
$$+V_k^*\,(a_k^+\,e^{-ikq}+a_{-k}^+\,e^{ikq}) \quad . \quad . \quad . \quad (1)$$

This Hamiltonian is directly obtained from the Fröhlich Hamiltonian, which is well known. q and p are the position and momentum operators of the electron with "effective mass" m. $\hbar\omega_k$ is the energy of a phonon with wave vector **k** (one takes $\omega_k=\omega=$constant). $a^+{}_k$ and a_k are the creation and annihilation operators for the phonons. V_k is a Fourier component which measures the strength of the electron phonon interaction.

Let us write the Schrödinger equation as

$$H\Psi=E\Psi \quad . \quad . \quad . \quad . \quad (2)$$

With the canonical transformation operator of L.L.P.

$$S=e^{\frac{i}{\hbar}[\lambda-\hbar k\,(a_k^+a_k-a^+\,a_{-k})]\,q} \quad . \quad . \quad . \quad (3)$$

where λ is an eigen value of the total momentum operator, one can eliminate the electron co-ordinates from the Hamiltonian.

If only the case of zero total momentum is treated, the following transformed Hamiltonian is obtained

$$S^{-1}\,HS=\frac{\hbar^2k^2}{2m}\,(a_{-k}^+a_{-k}-a_k^+a_k)^2+V_k\,(a_k+a_{-k})+V_k^*\,(a_k^++a_{-k}^+)$$
$$+\hbar\omega\,(a_k^+a_k+a_{-k}^+a_k) \quad . \quad . \quad . \quad (4)$$

It is convenient to introduce here the dimensionless parameters $\gamma=2\mathrm{i}V_k/\hbar\omega$ and $\kappa^2=\hbar k^2/m\omega$ and to change the phase of both V_k and V_k^* so as to make them real. With $\hbar=m=\omega=1$, (4) becomes:

$$\mathscr{H}=\frac{\kappa^2}{2}\,(a_{-k}^+a_{-k}-a_k^+a_k)^2+\frac{\gamma}{2}\,(a_k+a_{-k}+a_k^++a_{-k}^+)+$$
$$(a_k^+a_k+a_{-k}^+a_{-k}) \quad . \quad . \quad . \quad (5)$$

$\mathscr{H}$, of course has the same eigen value spectrum as the original Hamiltonian, $\mathscr{H}\varphi=E\varphi$ being the new Schrödinger equation. $\mathscr{H}$ is invariant with respect to a permutation of a_k^+ with a_{-k}^+ and of a_k with a_{-k}. Therefore the parity operator for this permutation commutes with the Hamiltonian. As a consequence one has two types of wave functions:

(1)
$$\varphi_s = \sum_{\substack{m,n \\ m>n}} a_{m,n} \mid m,n>_+ + \sum_m a_{m,m} \mid m.m> . \qquad (6)$$

These wave functions correspond to the eigen value +1 of the parity operator. We call them symmetric wave functions. $\mid i,j>$ are eigen vectors of the unperturbed Hamiltonian. i and j are the numbers of phonons with wave number k resp. $-k$. The notation $\mid m,n>_+$ indicates:

$$\mid m,n>_+ = \frac{\mid m,n> + \mid n,m>}{\sqrt{2}} \qquad (7)$$

(2)
$$\varphi_a = \sum_{\substack{m,n \\ m>n}} p_{m,n} \mid m,n>_- \qquad (8)$$

with

$$\mid m,n>_- = \frac{\mid m,n> - \mid n,m>}{\sqrt{2}} \qquad (9)$$

corresponding to the eigenvalue −1 of the parity operator. These are called antisymmetric wave functions.

One finally has the following orthonormality relations:

$$\begin{aligned} <l,j \mid k,m> &= \delta_l^k \delta_j^m \\ {}_+<i,j \mid m,n>_+ &= \delta_{im}\delta_{jn} \\ {}_+<i,j \mid m,m> &= 0 \qquad \text{etc.} \end{aligned} \qquad (10)$$

2.2 The Ground State

The wave function of the ground state is always symmetric and in this part we are only interested in the ground state.

Let us treat the equation $\mathscr{H}\varphi_s = E\varphi_s$ which describes the symmetric states of the system.

If both sides of this equation are multiplied with ${}_+<p,q \mid$ at the left ($p>q$, a condition which from now on will be included in the notation ${}_+<pq \mid$,) one obtains

$$\sum_{\substack{m,n \\ m>n}} {}_+<p,q \mid \mathscr{H} \mid m,n>_+ a_{m,n} + \sum_m {}_+<p,q \mid \mathscr{H} \mid m,m> a_{m,m} =$$
$$E_s \left[\sum_{\substack{m,n \\ m>n}} a_{m,n} \, {}_+<p,q \mid m,n>_+ + \sum_m a_{m,m} \, {}_+<p,q \mid m,m> \right] \qquad (11)$$

If both sides are multiplied at the left with $<t,t\mid$ one obtains

$$\sum_{\substack{m,n\\m>n}} a_{m,n} <t,t \mid \mathscr{H} \mid m,n>_+ + \sum_{m} <t,t \mid \mathscr{H} \mid m,m> a_{m,m} =$$

$$E_s\Bigg[\sum_{\substack{m,n\\m>n}} a_{m,n} <t,t \mid m,n>_+ + \sum_{m} a_{m,m} <t,t \mid m,m>\Bigg] \quad . \quad (12)$$

Equations (11) and (12) together with the condition

$$\sum_{k\geqslant l} \mid a_{kl} \mid ^2 = 1$$

determine the eigen values and eigen vectors of $\mathscr{H}\varphi_s = E_s\varphi_s$. If the matrix elements are calculated explicitly and substituted in equation (11), one obtains a first type of equation for the coefficients in the development (6):

$$a_{p,q}\left[\frac{x^2}{2}\ (q-p)^2+(p+q)\right] \pm \frac{\gamma}{2}\sqrt{p+1}\, a_{p+1,q} + \frac{\gamma}{2}\sqrt{q+1}\, a_{p,q+1}$$

$$\times(1-\delta_p^{q+1}) + \frac{\gamma}{2}\sqrt{p}\, a_{p-1,q}(1-\delta_p^{q+1}) + \frac{\gamma}{2}\sqrt{q}\, a_{p,q-1}$$

$$+\gamma\frac{\sqrt{2p}}{2}\delta_p^{q+1}\, a_{p,p} + \frac{\sqrt{2}}{2}\gamma\sqrt{p}\,\delta_p^{q+1}\, a_{p-1,\,p-1} = E_s a_{p,q} \qquad (13)$$

If the convenient matrix elements are now substituted in Equation (12) one finds

$$a_{t,t}\,(2t-E_s) + \frac{\sqrt{2(t+1)}}{2}\gamma a_{t+1,t} + \gamma\frac{\sqrt{2t}}{2} a_{t,t-1} = 0 \quad . \quad (14)$$

Equations (13) and (14) together with a normalization condition determine the eigen values and eigen functions of $\mathscr{H}\varphi_s = E_s\varphi_s$.

In the numerical calculations, only a finite number of equations is considered. They have to be classified following a rule based upon the physics of the problem. The number of equations one needs to find a solution with a certain accuracy will depend on the coupling strength, i.e. on the mean number N_s of virtual phonons in the ground state

$$N_s = \sum_{k=0}^{\infty} \sum_{\substack{p+q=k\\p\geqslant q}} (p+q) \mid a_{p,q} \mid ^2 = \sum_{k=0}^{\infty} k \sum_{\substack{p+q=k\\p\geqslant q}} \mid a_{p,q} \mid ^2 \quad . \quad (15)$$

where the components of the wave function with the same total number of phonons are taken together. If N_s^0 is the mean number of phonons in the ground state, the $\mid a^0_{p,q} \mid$ corresponding to

$p+q=k \gg N_s^0$ will be negligible with respect to those for which $p+q \approx N_s^0$. It is then logical to classify the equations corresponding to increasing values of $p+q$ and $2t$ in $_+\langle p,q|$ and $\langle t,t|$. As the coupling strength increases, it will be necessary to consider an increasing number of equations in order to obtain accurate solutions.

The solutions for E_0 (or the different E_s) are found by studying a secular determinant (resulting from Equations (13) and (14)), whose order is determined by the coupling strength.

This numerical problem has been solved by a computer. The programme permitted the treatment of secular determinants of order 210.

In Figure 1 the solution for $E_0/\hbar\omega$ is given as a function of the exact mean number of phonons in the ground state which is a valuable coupling parameter.

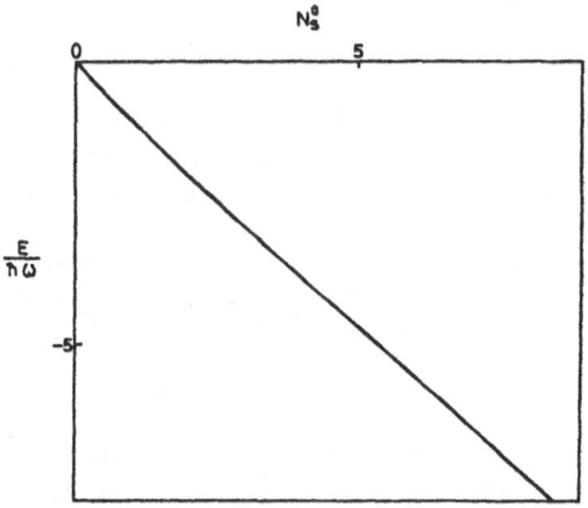

FIGURE 1.—Self-energy of the symmetrical polaron model as a function of the exact number of virtual phonons in the ground state ($\kappa=\sqrt{2}$).

3. APPLICATION OF THE MOST IMPORTANT POLARON THEORIES TO THE CALCULATION OF THE SELF ENERGY OF THE SYMMETRICAL MODEL

The most important contributions to the study of the self-energy of the polaron are those of LANDAU-PEKAR,[1] LEE, LOW, PINES[2] and FEYNMAN.[3] The treatment of Landau-Pekar was the first historically and it was intended to treat the polaron as it appears in ionic crystals.

However, after FRÖHLICH[19] introduced a convenient Hamiltonian and the coupling parameter α describing the strength of the coupling between the electron and the phonons, it was soon realized that the

Landau-Pekar method is only adapted for large α (say $\alpha > 10$). Now it turns out that for typical ionic crystals $\alpha \approx 3$. This was the stimulus for Lee, Low and Pines to develop their canonical transformation method which they hoped to be reasonably valid from $\alpha = 0$ up to $\alpha \approx 6$.

In 1955, Feynman developed an approximation in the framework of his path integral formulation of quantum mechanics. All these methods have a variational character and so it is easy to decide which of them approaches the self-energy nearest. It turns out that the Feynman method provides the lowest approximation for all α. So far as we know there is no other method giving a lower or even the same E_0.

On the other hand, one does not have any *a priori* knowledge about the accuracy of any of these methods, especially at intermediate coupling. Therefore, it seemed to us that the different methods should be tested by a special exactly soluble model. The application of the three methods mentioned above will now be commented.

3.1 Application of the Lee, Low, Pines Method

In reference 15 we studied the model of Gross and the application of the L.L.P. method to it. The application of the L.L.P. method to the symmetrical model proceeds along the same lines and is even simplified, as only total momentum equal to zero has to be considered. One obtains the following expression for E_0^L the energy of the ground state in the L.L.P. approximation

$$E_0^L = -\frac{\gamma^2}{2+\kappa^2}\hbar\omega \quad . \quad . \quad . \quad (16)$$

In Figure 2, the curve L.L.P. gives the relative error on the self energy corresponding to this expression. The relative error is now plotted as a function of $A = \gamma^2/4$, the mean number of virtual phonons in the ground state, as approximated by the L.L.P. method. This parameter is equivalent here to the coupling parameter α of Fröhlich. If $\kappa = \sqrt{2}$ is considered as the characteristic wave number for the symmetrical model, (16) reduces to $E_0^L = -\alpha_{eq}\,\hbar\omega$. This choice of $\kappa = \sqrt{2}$ as the mean wave number is discussed in more detail in reference 15. It should be remembered that only a small region of k values ($k < \sqrt{2m\omega/\hbar}$) is important in the polaron problem. And $\kappa = \sqrt{2}$ characterizes this region.

3.2 Application of the Landau-Pekar Method

The results obtained in reference 16 can easily be extended to the symmetrical model. The ground state energy for the symmetrical

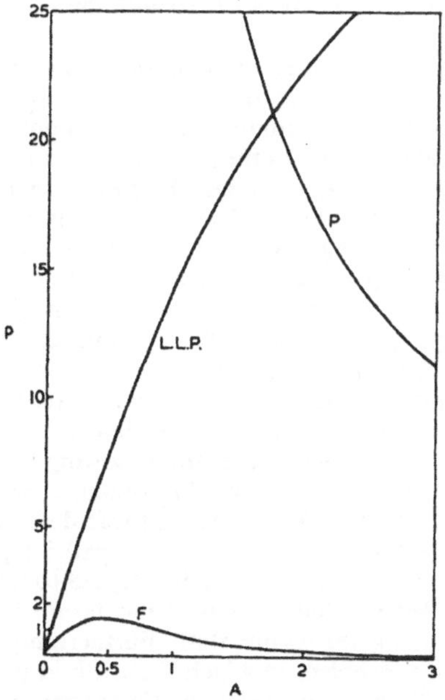

FIGURE 2.—Relative error of different polaron theories on the self-energy of the symmetrical model ($\kappa = \sqrt{2}$) (P=Pekar; L.L.P.=Lee, Low, Pines; F=Feynman).

model as approximated with a localized electron wave function turns out to be

$$\frac{Ep}{\hbar\omega} = \frac{\kappa^2 A}{2}\left(\tfrac{1}{2} - A\right) \quad . \quad . \quad . \quad (17)$$

where A is obtained from:

$$A^2\, e^{\frac{1}{2A}} = \frac{\gamma^2}{\kappa^2} \quad . \quad . \quad . \quad . \quad . \quad (18)$$

In Figure 2 we have also plotted the accuracy of the Pekar method. In accordance with our earlier results it also follows here that the Landau-Pekar method furnishes the exact asymptotic behaviour for strong coupling. This might be surprising at first sight, as in the symmetrical polaron model the electron moves in a periodic potential while here the electron is localized. However, at strong coupling the electron moves in a narrow band and has a very high effective mass which permits one to make a "localized" approximation.

3.3 Treatment of the Symmetrical Model with the Feynman Approximation [20]

Feynman's approximation of the polaron ground state energy is based on a quadratic approximation. A trial potential is introduced which corresponds to the assumption of an harmonic oscillation of the electron around the centre of the polaron. No further approximation is made. The self energy obtained by Feynman gives the exact strong and weak coupling limits. The methods discussed above are only adapted for special regions. The L.L.P. method fails at stronger coupling because it neglects a term in the Hamiltonian which deals with wave functions including at least two phonons. Pekar's method fails at weak coupling as it assumes that the ions cannot respond to the instantaneous motion of the electron.

Although Feynman's method gives accurate results in the two limiting cases, this does not mean *a priori* that it is accurate in the intermediate region. The purpose of the following calculation is to obtain an idea about the accuracy of Feynman's method at intermediate coupling. The theory of Feynman was developed using his special formulation of quantum mechanics: the path integral method.[21]

Although this method is not especially adapted to treat relativistic problems or problems including spin, it has been of historical importance in explaining the Lamb shift. Furthermore, the polaron problem is a typical problem in which the method is adapted as it neither is relativistic nor includes spin. Furthermore, as far as we know, nobody has been able to obtain the results of Feynman by other techniques. Details about the method can be found in reference 21 and in a recent book by Feynman and Hibbs.[22]

To apply Feynman's method we first derive the Lagrangian for the symmetrical polaron model, as the path integral method is based upon a calculation of the propagator in terms of the Lagrangian. Starting from (1) one obtains

$$L=\frac{m\dot{q}^2}{2}-\frac{\omega^2}{2}(Q_1^2+Q_2^2)+\tfrac{1}{2}(\dot{Q}_1^2+\dot{Q}_2^2)+\hbar\omega$$
$$-2iV_k\sqrt{\frac{\omega}{\hbar}}\,Q_1\cos kq-2iV_k\sqrt{\frac{\omega}{\hbar}}\,Q_2\sin kq \quad . \quad . \quad . \quad (19)$$

The Q_i are normal modes associated with the two harmonic oscillators which interact with the electron.

We will now apply the method of Feynman to the symmetrical model, indicating only the essential intermediate steps. After introducing imaginary times $t=-i\tau$ (τ is real) and eliminating the field variables Q_l, the transformation function becomes

$$K(q''\tau'', q'\tau') = \int \mathscr{D}\, q(\tau) \exp\left\{ -\tfrac{1}{2} \int_{\tau'}^{\tau''} \left(\frac{dq}{d\tau}\right)^2 d\tau \right.$$

$$\left. - V_k^2 \int\int_{\tau'}^{\tau''} \cos k[q(\tau) - q(\sigma)] e^{-|\tau-\sigma|} d\tau\, d\sigma \right\} \quad . \quad . \quad (20)$$

As is well known, the asymptotic behaviour of $K(q''\ \tau'', q'\ \tau')$ (which is expressed as an integral over all path $q(\tau)$, indicated by the symbol $\int \mathscr{D}q(\tau) \ldots$) for large $\tau'' - \tau'$, gives the self-energy of the system. In the exponent the "action" now appears (of course for imaginary times).

The quadratic approximation in the case of the symmetrical model corresponds to taking the following trial action:

$$S_o = -\tfrac{1}{2} \int_{\tau'}^{\tau''} \left(\frac{dq}{d\tau}\right)^2 d\tau + C \int\int_{\tau'}^{\tau''} \left[1 - \frac{(q(\tau)+q(\sigma))^2}{2}\right] e^{-w|\tau-\sigma|}\, d\tau d\sigma \quad (21)$$

Feynman's variational principle for path integrals reveals that:

$$E_g \leq E_0 - \mathscr{S} \quad . \quad . \quad . \quad . \quad (22)$$

where E_g is the true ground state energy of the system and

$$E_0 = -\lim_{T\to\infty} \frac{\ln \int e^{S_0} \mathscr{D}\, q(\tau)}{T} \quad . \quad . \quad . \quad (23)$$

and

$$\mathscr{S} = \lim_{T\to\infty} \frac{\int e^{S_0}(S - S_0)\mathscr{D}q(\tau)}{\int e^{S_0}\, \mathscr{D}_q(\tau)} \quad . \quad . \quad (23b)$$

Once E_0 and $\mathscr{S}$ are obtained, they can still be minimized with respect to the variational parameters C and w.

The numerator of (23 b) is given by

$$N = \int\int_{\tau'}^{\tau''} d\tau_1 d\tau_2 \int \mathscr{D}\, q(\tau) \left\{ - V_k^2 \cos k(q(\tau_1) - q(\tau_2)) e^{-|\tau_1-\tau_2|} \right.$$

$$\left. + \frac{C}{2}\left[1 - \frac{(q(\tau_1)-q(\tau_2))^2}{2}\right] e^{-w|\tau_1-\tau_2|} \right\} e^{S_0} \quad . \quad . \quad (24)$$

As in Feynman's calculation the function

$$f_k(\tau) = ik[\delta(\tau - \tau_1) - \delta(\tau - \tau_2)] \quad . \quad . \quad (25)$$

is useful in order to write (24) explicitly in the form of a do-able path integral

$$N=\int_{\tau'}^{\tau''}\!\!\int \mathrm{d}\tau_1\mathrm{d}\tau_2\int \mathscr{D}q(\tau)\Bigg\{-\frac{V_k^2}{2}\left(e^{\int_{\tau'}^{\tau''} f_k(\tau)q(\tau)\mathrm{d}\tau}+e^{-\int_{\tau'}^{\tau''} f_k(\tau)q(\tau)\mathrm{d}\tau}\right)$$

$$+\frac{C}{2}\left[1+\frac{1}{2}\left(\frac{\mathrm{d}^2}{\mathrm{d}k^2}\,e^{\int_{\tau'}^{\tau''} f_k(\tau)q(\tau)\mathrm{d}\tau}\right)_{k=0}\right]e^{-w|\tau_1-\tau_2|}\Bigg\}\,e^{S_0} \quad . \quad (26)$$

A form like, e.g.

$$\frac{\int \mathscr{D}q(\tau)\,e^{S_0+\int_{\tau'}^{\tau''} f_k(\tau)q(\tau)\mathrm{d}\tau}}{\int \mathscr{D}q(\tau)\,e^{S_0}} \quad . \quad . \quad . \quad (27)$$

now simply reduces to

$$e^{\left(S_0+\int_{\tau'}^{\tau''} f_k(\tau)q(\tau)\mathrm{d}\tau\right)_{cl}} \quad . \quad . \quad . \quad . \quad (28)$$

where cl stands to indicate that for $q(\tau)$ the solution of the classical equations of motion corresponding to the action in the brackets has to be taken. Indeed, if in both the numerator and the denominator of (27) $q(\tau)$ are separated in a classical and a quantum mechanical part, one has the same quadratic terms in the numerator and the denominator. Furthermore the linear terms disappear in both of them, as they both correspond to a certain equation of motion. The denominator $e^{(S_0)_{cl}}$ equals 1 as will follow from Equation 29.

From an inspection of the equations of motion one sees that (27) reduces to

$$e^{\frac{1}{2}\int_{\tau'}^{\tau''} f_k(\tau)q_{cl}(\tau)\mathrm{d}\tau} \quad . \quad . \quad . \quad . \quad (29)$$

It is then possible to solve the equations of motion along the lines given by Feynman. Some transients have to be disregarded. In fact they are implicitly related to boundary conditions on the field. This is seen more clearly if one studies the problem in the Heisenberg representation.

It is now easy to obtain the first two terms of Equation (24) and the third one can be obtained by a series expansion of one of the first two. One obtains

$$S=-2V_k^2\,e^{\frac{-2Ck^2}{v^3w}}\int_0^\infty \mathrm{d}\lambda\,e^{-\left(w^2+\frac{w^2k^2}{2v^2}\right)\lambda}\;e^{\frac{2Ck^2}{v^3w}e^{-v\lambda}}+\frac{C}{vw} \quad . \quad (30)$$

The path integral for E_0 is a quadratic one and is easily obtained:

$$E_0 = \tfrac{1}{2}(v-w) \quad . \quad . \quad . \quad . \quad . \qquad (31)$$

Therefore, the exact self energy of the symmetrical model fulfills the following equation

$$E_g \leqslant \left(\frac{v-w}{4v}\right)^2 + 2V_k^2\, e^{-2(v^2-w^2)\frac{k^2}{2v^3}} \int_0^\infty d\lambda\, e^{-\left(1+\frac{w^2k^2}{2v^2}\right)\lambda}\, e^{\left(\frac{v^2-w^2}{2}\right)\frac{k^2}{v^3}e^{-v\lambda}} \qquad (32)$$

with

$$C = (v^2 - w^2)\, w/4$$

Let us examine this result at strong resp. weak coupling after the introduction of dimensionless quantities and for $\kappa = \sqrt{2}$:

$$E_g \leqslant \left(\frac{v-w}{4v}\right)^2 - \frac{\gamma^2}{2}\, e^{-(v^2-w^2)\frac{1}{v^3}} \int_0^\infty d\lambda\, e^{-\left(1+\frac{w^2}{v^2}\right)\lambda}\, e^{\left(\frac{v^2-w^2}{v^3}\right)e^{-v\lambda}} \qquad (33)$$

2.31 Strong Coupling

This corresponds to $v \gg 1$. Equation (33) reduces immediately to:

$$Eg \leqslant \frac{\gamma}{4} - \frac{\gamma^2}{2} e^{-\frac{1}{v}} \quad . \quad . \quad . \quad . \quad . \qquad (34)$$

if this expression is minimized with respect to v one finds Equations (17) and (18).

2.32 Weak Coupling

This corresponds to $C \approx 0$ or $v \approx \omega$. One easily finds for the dominant term:

$$-\frac{2V_k\, V_k^*}{1+k^2/2}$$

which is the same as (16).

At intermediate coupling the calculations were performed with a computer. (Only v was varied and w put equal to 1.)

4. DISCUSSION

From Figure 2 where the relative error of the three methods has been plotted it is seen that the Feynman method is very precise, although only one of the two parameters was varied. The largest error is about 1·42% and occurs for a mean number of phonons equal to about 1/2. This value of $\gamma^2/4$ somehow corresponds to values of $\alpha \approx 3$ to 5 in the real case. (If the intersection of the L.L.P. and Pekar curves is considered to correspond to $\alpha \approx 10$). Even at these coupling strengths, the Feynman method is about a factor 4 more precise than the L.L.P. method.

Two questions arise:

(1) Why is the path integral approach so accurate?

(2) Can these conclusions be extrapolated to the case of a complete **k**-spectrum.

As to the first question two points seem important to us:

The fact that Feynman implicitly introduces a co-ordinate which is the "centre" of the polaron.[23] The advantage of introducing such co-ordinate has also been discussed in reference twenty-four.

Feynman eliminates the phonons. This means that he does not perform any explicit approximation on the field. The implicit approximation for the field follows from the approximation on the electron motion which is an intuitive one.

Concerning the second question, it has to be emphasized that the extrapolation is not direct. This study can be compared with the study of exactly soluble models in other fields of physics, such as the Lee model of the linear chain. They provide a test for the theory.

The conclusion of this study is that the Feynman method is extremely accurate at all coupling strengths of the symmetrical model. This strongly suggests that the method is also accurate in the case of a complete **k** spectrum.

Although no study has been made on the effective mass, the precision on this parameter will almost certainly be related to that on the self energy. Therefore experimentalists performing cyclotron resonance measurements in order to determine the effective polaron mass might prefer the expression given by Feynman.

ACKNOWLEDGMENTS

We wish to thank Dr. L. Schotsmans for programming the computer calculations of the rigorous solutions and Mr. De Corte who programmed the work to minimize expression (33).

REFERENCES

1. PEKAR, S., 1954 "Untersuchungen über die Electronentheorie der Kristalle" (Akademie-Verlag, Berlin).
2. LEE, T. D., LOW, F. E., and PINES, D., *Phys. Rev.*, **90**, 297, 1953.
3. FEYNMAN, R. P., *Phys. Rev.*, **97**, 660 (1955). See also: SCHULTZ, T. D. in "Polarons and Excitons" (Ed. Kuper G. Whitfield)(Oliver Boyd: London, 1963).
4. LANGRETH, D. C., and KADANOF, L. P., *Phys. Rev.*, **133**, A270, 1964.

5. OSAKA, Y., *Progr. Theoret. Phys.*, **25**, 517, 1961.
6. ASCARELLI, G., and BROWN, F. C., *Phys. Rev. Letters*, **9**, 209, 1962.
7. AHRENKIEL, R. K., and BROWN, F. C., *Phys. Rev.*, **136**, A223, 1964.
8. MASUMI, T., AHRENKIEL, R. K., and BROWN, F. C., *Phys. stat. sol.*, **11**, 163, 1965.
9. EVRARD, R., *Phys. Lett.*, **14**, 295, 1965.
10. DEVREESE, J., and EVRARD, R., *Phys. Lett.*, **11**, 278, 1964.
11. FEYNMAN, R. P., HELLWARTH, R. W., IDDINGS, C. K., and PLATZMAN, P. M., *Phys. Rev.*, **127**, 1004, 1962.
12. LARSEN, D. M., *Phys. Rev.*, **133**, A860, 1964.
13. LARSEN, D. M., *Phys. Rev.*, **135**, A419, 1964.
14. WHITFIELD, G., and PUFF, R., *Phys. Rev.*, **139**, A338, 1965.
15. DEVREESE, J., and EVRARD, R., *Phys. stat. sol.*, **3**, 2133, 1963.
16. DEVREESE, J., and EVRARD, R., *Phys. stat. sol.*, **9**, 403, 1965.
17. GROSS, E. P., *Phys. Rev.*, **84**, 818, 1951.
18. DEVREESE, J. Thesis, University of Louvain, 1964.
19. FRÖHLICH, H., PELZER, H., and ZIENAU, S., *Phil. Mag.*, **41**, 221, 1950.
20. DEVREESE, J., and EVRARD, R., *Phys. Lett.*, **23**, 196, 1966.
21. FEYNMAN, R. P., *Rev. Mod. Phys.*, **20**, 367, 1948.
22. FEYNMAN, R. P., and HIBBS, A. R., "Quantum Mechanics and Path Integrals" (McGraw, Hill New York, 1965).
23. EVRARD, R., and DEVREESE, J., "Discussion of the principal methods of polaron theory", 8th Jugoslav Summer School of Physics, Hercegnovi 1963.
24. EVRARD, R. Thesis, Liège 1964.

APPLICATIONS OF PATH INTEGRALS TO PROBLEMS IN DISSIPATION

K. K. Thornber
Bell Laboratories, Incorporated
Murray Hill, New Jersey 07974, U.S.A.

ABSTRACT

Feynman's path-integral method offers a unique approach to problems involving the transport of electrons in dissipative media in electric and magnetic fields. This is because actual, physical dissipative systems can be approximated by similar dissipative systems which can be solved exactly using path integrals. The difference between the exact and approximate systems can then be treated as a perturbation. Several examples are considered including both a.c.-linear and d.c.-nonlinear response as well as the problem of electron acceleration in sub-threshold fields. The latter problem appears to require a more detailed understanding of the scattering in the presence of the field than do transport problems.

1. INTRODUCTION

While it is true that the Feynman path integral contains the same physics as the Schroedinger equation, it contains this information in a different way.[1-6] If we could solve our problems exactly, it would make little difference which method we used. However, as nearly all problems of interest must be solved approximately, this difference can be exploited to advantage. In solving a problem approximately, one often tries to represent the actual situation as closely as possible by a model problem which can be solved exactly, and then treating the difference between the two as a perturbation. Our success or failure then hinges quite critically on our ability to model the situation accurately.

The problem of the motion of an electron interacting with dissipative media under the influence of electric and magnetic fields is a problem to which the path-integral approach is well-suited.[7-17] There are several reasons for this: (1) The electric and magnetic fields enter in a natural way and do not significantly enhance the difficulty of the problem. (2) Exactly solvable models of dissipative processes exist. (3) One can often pass smoothly from free-particle-like solutions for weakly-coupled problems to harmonic-oscillator-like solutions for strongly-coupled problems. (4) One can treat the fluctuation-dissipation terms in a perturbation expansion as readily as propagator terms, which are normally treated in the context of very weakly dissipative systems.

In what follows we shall take advantage of various exactly solvable models of electrons interacting with dissipative media in the presence of finite electric and magnetic fields in order to discuss several problems of interest in the understanding of the transport of electrons in insulators used in semiconductor devices. First, however, we shall consider the ingredients of a general expansion in the difference between the exact and an approximate influence functional. Then we shall focus on impedance, drift and acceleration of electrons under the influence of electric fields. Magnetic-field problems will not be discussed apart from the general expansions.

2. THE FUNDAMENTAL PROBLEM

The path integral to which we shall devote most of our attention is the following:[14]

$$\int\!\!\int \exp(S^e_{t,t'}(\mathbf{x},\mathbf{x}') + i\int_{-\infty}^{\infty} dt\,(\mathbf{f}_t\cdot\mathbf{x}_t - \mathbf{f}'_t\cdot\mathbf{x}'_t))\,D(\mathbf{x}_t)\,D(\mathbf{x}'_t) \tag{2.1a}$$

where

$$\begin{aligned} S^e \equiv\ & i\int_{-\infty}^{\infty} dt\,(\frac{1}{2}\dot{\mathbf{x}}_t\cdot\mathbf{m}\cdot\dot{\mathbf{x}}_t + \frac{1}{2}\dot{\mathbf{x}}_t\cdot\mathbf{H}\times\mathbf{x}_t) \\ & - i\int_{-\infty}^{\infty} dt\,(\frac{1}{2}\dot{\mathbf{x}}'_t\cdot\mathbf{m}\cdot\dot{\mathbf{x}}'_t + \frac{1}{2}\dot{\mathbf{x}}'_t\cdot\mathbf{H}\times\mathbf{x}'_t) \\ & - \int_{-\infty}^{\infty} dt_i \int_{-\infty}^{t_i} dt_i'\,\Sigma_{\mathbf{k}_i}(V_{i,x',x'} - V_{i,x,x'} + V^*_{i,x,x} - V^*_{i,x',x}) \end{aligned} \tag{2.1b}$$

$$V_{i,x,x'} \equiv |C_{\mathbf{k}_i}|^2 T_{\omega_{\mathbf{k}_i}}(t_i - t'_i)\exp(-i\mathbf{k}_i\cdot(\mathbf{x}_{t_i} - \mathbf{x}'_{t'_i})),\ldots \tag{2.1c}$$

$$T_\omega(\tau) = e^{i\omega\tau}(1-e^{-\beta\omega})^{-1} + e^{-i\omega\tau}(e^{\beta\omega}-1)^{-1} \tag{2.1d}$$

Here $\mathbf{m}$ is an effective mass tensor, $\mathbf{H}$ a constant applied magnetic field, and $\mathbf{f}_t$ and $\mathbf{f}'_t$ are arbitrary functions of time which can include an applied, time-varying electric field $\mathbf{E}_t$, and hence need not be included in S^e. The V arise from the interaction of the electron with the lattice phonons, which have been exactly eliminated from the problem using either ordered-operator[3] or path-integral techniques.[7–15] $\beta \equiv (kT)^{-1}$, T being the lattice temperature. We assume that there are so few electrons available that electron-electron interactions and over-all lattice heating can be neglected.

Expression (2.1) is very useful because from it we can obtain, by appropriately choosing $\mathbf{f}$ and $\mathbf{f}'$, most expectation values involving electron coordinates $\mathbf{x}$ and $\mathbf{x}'$. If in addition phonon variables arise, they can be included by referring back to the derivation of (2.1). For example, single operators $a_{\mathbf{k},t}$ and $a^\dagger_{\mathbf{k},t}$ entering an expectation value give rise to

$$-iC^*_{\mathbf{k}} \int_{-\infty}^{t} d\tau \left[\frac{\exp(-i\mathbf{k}\cdot\mathbf{x}_\tau)}{1-\exp(-\beta\omega_{\mathbf{k}})} - \frac{\exp(-i\mathbf{k}\cdot\mathbf{x}'_\tau)}{\exp(\beta\omega_{\mathbf{k}})-1}\right]\exp(-i\omega_{\mathbf{k}}(t-\tau)) \tag{2.2a}$$

$$iC_{\mathbf{k}} \int_{-\infty}^{t} d\tau \left[\frac{\exp(i\mathbf{k}\cdot\mathbf{x}'_\tau)}{1-\exp(-\beta\omega_{\mathbf{k}})} - \frac{\exp(i\mathbf{k}\cdot\mathbf{x}_\tau)}{\exp(\beta\omega_{\mathbf{k}})-1}\right]\exp(i\omega_{\mathbf{k}}(t-\tau)) \tag{2.2b}$$

respectively, when the oscillators are eliminated from the problem. Throughout we shall consider only one phonon branch. More branches can be included simply by replacing $\Sigma_{\mathbf{k}}$ by $\Sigma_{n,\mathbf{k}}$, $C_{\mathbf{k}}$ by $C_{\mathbf{k},n}$, $\omega_{\mathbf{k}}$ by $\omega_{\mathbf{k},n}$, etc.[14]

To calculate a typical expectation value, we proceed as follows. We define

$$<\exp(i\int_{-\infty}^{\infty} dt(\mathbf{f}_t\cdot\mathbf{x}_t - \mathbf{f}'_t\cdot\mathbf{x}'_t))> \equiv \int\int \exp(i\int_{-\infty}^{\infty} dt(\mathbf{f}_t\cdot\mathbf{x}_t - \mathbf{f}'_t\cdot\mathbf{x}'_t))e^{S^e}/\int\int e^{S^e} \qquad (2.3a)$$

$$= A\cdot B/C\,, \qquad (2.3b)$$

where

$$A \equiv \frac{\int\int \exp(i\int_{-\infty}^{\infty} dt(\mathbf{f}_t\cdot\mathbf{x}_t - \mathbf{f}'_t\cdot\mathbf{x}'_t))e^{S^o}e^{S^e-S^o}}{\int\int \exp(i\int_{-\infty}^{\infty} dt(\mathbf{f}_t\cdot\mathbf{x}_t - \mathbf{f}'_t\cdot\mathbf{x}'_t))e^{S^o}} \equiv <e^{S^e-S^o}>_{S^o} \qquad (2.3c)$$

$$B \equiv \int\int \exp(i\int_{-\infty}^{\infty} dt(\,f_t\cdot\mathbf{x}_t - \mathbf{f}'_t\cdot\mathbf{x}'_t))e^{S^o}/\int\int e^{S^o} \qquad (2.3d)$$

$$C \equiv \int\int e^{S^e}/\int\int e^{S^o} \qquad (2.3e)$$

It is important to note that $C = A\,(f{=}f'{=}0)$, and that by expanding the $\exp(S^e{-}S^o)$ in (2.3c), A can be obtained from a knowledge of B. In (2.3) S^o is an approximate influence functional over which the path integrals in B can be performed. A useful choice is to replace the third term in (2.1b) by a distribution of harmonic oscillators:

$$+\int_{-\infty}^{\infty} dt_i \int_{-\infty}^{t_i} dt'_i(V^o_{i,x',x'} - V^o_{i,x,x'} + V^{o*}_{i,x,x} - V^{o*}_{i,x',x}) \qquad (2.4a)$$

$$V^o_{i,x,x'} \equiv (\mathbf{x}_{t_i} - \mathbf{x}'_{t'_i})\cdot\mathbf{G}^o(t_i - t'_i)\cdot(\mathbf{x}_{t_i} - \mathbf{x}'_{t'_i}) \qquad (2.4b)$$

$$\mathbf{G}^o(t-t') = \int_{-\infty}^{\infty} d\Omega\; \mathbf{G}^o(\Omega)\exp(-i\Omega(t-t')) \qquad (2.4c)$$

Setting $\mathbf{G}$ to zero results in the S^o for a free particle. Clearly $\mathbf{G}$ is symmetric.

For simplicity we first state the result for B:

$$B = \exp(\int_{-\infty}^{\infty} dt \int_{-\infty}^{t} dt'(\mathbf{f}_t - \mathbf{f}'_t)(\mathbf{L}^*_{t-t'}\mathbf{f}_{t'} - \mathbf{L}_{t-t'}\mathbf{f}'_{t'})) \qquad (2.5a)$$

$$\mathbf{L}_t \equiv \int_{-\infty}^{\infty} \frac{d\nu}{2\pi i}\, \frac{1}{\mathbf{Z}^o_\nu}\, (-i4\pi\mathbf{G}^o_{-\nu})\, \frac{1}{\mathbf{Z}^{o\dagger}_\nu}\, e^{-i\nu t} \qquad (2.5b)$$

$$\mathbf{Z}^o_\nu \equiv -\mathbf{m}(\nu+i\epsilon)^2 - i(\nu+i\epsilon)\underline{\epsilon}\cdot\mathbf{H} - 4(\nu+i\epsilon)\int_{-\infty}^{\infty} d\Omega\, \frac{P}{\Omega}\, \frac{\mathbf{G}^o(\Omega)}{\Omega^2-(\nu+i\epsilon)^2} \qquad (2.5c)$$

$$= -\mathbf{m}(\nu+i\epsilon)^2 - i(\nu+i\epsilon)\underline{\epsilon}\cdot\mathbf{H} + 4\int_0^{\infty} d\zeta(1-e^{i\nu\zeta})\mathrm{Im}(\mathbf{G}^{o*}(\zeta)) \qquad (2.5d)$$

$\underline{\epsilon}$ is defined by $\mathbf{A}\cdot\underline{\epsilon}\cdot\mathbf{B}\cdot\mathbf{C} \equiv A_i\epsilon_{ijk}B_jC_k = \mathbf{A}\cdot\mathbf{B}\times\mathbf{C}$. Details are presented elsewhere.[14] To obtain A from B we note that, if we expand $\exp(S^e - S^o)$ in a power series in $(S^e - S^o)$, then

$$A = \Sigma_{n=0}^{\infty} A_n/n! \,, \quad A_n \equiv \langle (S^e - S^o)^n \rangle_{S^o} \tag{2.6}$$

To obtain A_n, which we can calculate, we proceed as follows. Note that $A_0 = 1$.

First in the numerator of A we let $\mathbf{f} \to \mathbf{f} + \mathbf{g}$ and $\mathbf{f}' \to \mathbf{f}' + \mathbf{g}'$. By choosing $\mathbf{g}$, $\mathbf{g}'$ appropriately we can calculate any term arising in A_n. Thus to determine A we shall need to make use of the relation

$$\langle \exp(i\int_{-\infty}^{\infty} dt\,(\mathbf{g}_t \cdot \mathbf{x}_t - \mathbf{g}'_t \cdot \mathbf{x}'_t)) \rangle_{S^o} = \exp\{\int_{-\infty}^{\infty} dt \int_{-\infty}^{t} dt' ((\mathbf{g}_t - \mathbf{g}'_t)(\mathbf{L}^*_{t-t'}\mathbf{g}_{t'} - \mathbf{L}_{t-t'}\mathbf{g}'_{t'})$$
$$+ (\mathbf{f}_t - \mathbf{f}'_t)(\mathbf{L}^*_{t-t'}\mathbf{g}_{t'} - \mathbf{L}_{t-t'}\mathbf{g}'_{t'}) + (\mathbf{g}_t - \mathbf{g}'_t)(\mathbf{L}^*_{t-t'}f_{t'} - \mathbf{L}_{t-t'}f'_{t'}))\} \tag{2.7}$$

which follows from (2.5a) From (2.7) we can in principle determine each A_n. Thus for example $A_1 = \langle (S^e - S^o) \rangle_{S^o}$ is given by

$$A_1 = -\int_{-\infty}^{\infty} dt \int_{-\infty}^{t} dt' \, \Sigma_{\mathbf{k}}(|C_{\mathbf{k}}|^2 - |C^o_{\mathbf{k}}|^2)$$
$$\cdot\{T_{\omega_{\mathbf{k}}}(t-t')(\langle \mathbf{g}_\tau = 0,\ \mathbf{g}'_\tau = \mathbf{k}(\delta(\tau - t) - \delta(\tau - t')) \rangle$$
$$- \langle \mathbf{g}_\tau = -\mathbf{k}\delta(\tau - t),\ \mathbf{g}'_\tau = -\mathbf{k}\delta(\tau - t') \rangle)$$
$$+ T^*_{\omega_{\mathbf{k}}}(t-t')(\langle \mathbf{g}_\tau = \mathbf{k}(\delta(\tau - t) - \delta(\tau - t')), \mathbf{g}'_\tau = 0 \rangle$$
$$- \langle \mathbf{g}_\tau = -\mathbf{k}\delta(\tau - t'),\ \mathbf{g}'_\tau = -\mathbf{k}\delta(\tau - t) \rangle)\} \tag{2.8}$$

where $\langle \mathbf{g}, \mathbf{g}' \rangle$ is (2.7) evaluated for the specified $\mathbf{g}$, $\mathbf{g}'$. The only awkwardness in (2.7) as it stands is the restriction that $t > t'$. This can be over come by noting the following properties of $\mathbf{L}_t$:

$$\mathbf{L}^*_{t'-t} = \tilde{\mathbf{L}}_{t-t'} \tag{2.9a}$$
$$\mathbf{L}_{\langle t'-t \rangle} = \tilde{\mathbf{L}}_{\langle t-t' \rangle} \quad \textit{where we define} \tag{2.9b}$$
$$\mathbf{L}_{\langle \tau \rangle} \equiv \mathbf{L}_\tau (\tau > 0), \ \equiv \mathbf{L}^*_\tau (\tau < 0) \tag{2.9c}$$

$\langle \tau \rangle$ is a necessary generalization of $|\tau|$ for tensor $\mathbf{L}$. Equation (2.9a) is essentially Onsager's relation. It follows at once from (2.7) and (2.9) that

$$\langle \exp i \int_{-\infty}^{\infty} dt\,(\mathbf{g}_t \cdot \mathbf{x}_t - \mathbf{g}'_t \cdot \mathbf{x}'_t) \rangle_{S^o} =$$
$$\exp(\int_{-\infty}^{\infty} dt \int_{-\infty}^{\infty} dt' ((\mathbf{g}_t \mathbf{L}^*_{\langle t-t' \rangle} - \mathbf{g}'_t \mathbf{L}^*_{t-t'})\mathbf{f}_{t'} - (\mathbf{g}_t \mathbf{L}_{t-t'} - \mathbf{g}'_t \mathbf{L}_{\langle t-t' \rangle})\mathbf{f}'_{t'}))$$
$$\exp(\int_{-\infty}^{\infty} dt \int_{-\infty}^{\infty} dt' (\frac{1}{2}\, \mathbf{g}_t \mathbf{L}^*_{\langle t-t' \rangle} \mathbf{g}_{t'} + \frac{1}{2}\, \mathbf{g}'_t \mathbf{L}_{\langle t-t' \rangle} \mathbf{g}'_{t'} - \mathbf{g}_t \mathbf{L}_{t-t'} \mathbf{g}'_{t'})) \tag{2.10}$$

where the last term in (2.10), $\mathbf{g}_t\mathbf{L}_{t-t'}\mathbf{g}'_{t'}$, may be replaced by $\mathbf{g}'_t\mathbf{L}^*_{t-t'}\mathbf{g}_{t'}$ if convenient.

Using (2.10) in (2.8), A_1 follows, which we choose to write after some manipulation in the form

$$A_1 = -\int_{-\infty}^{\infty} dt_j \int_{-\infty}^{t_j} dt'_j \Sigma_{\mathbf{k}_j}(P_{j+} - Q_{j+} + P^*_{j-} - Q^*_{j-}) \tag{2.11a}$$

$$P_{j+} \equiv (|C_{\mathbf{k}_j}|^2 - |C^o_{\mathbf{k}_j}|^2) T_{\omega_{\mathbf{k}_j}}(t_j - t'_j)\exp(-\mathbf{k}_j\bar{\mathbf{L}}_{t_j-t'_j}\mathbf{k}_j)\exp(-i\mathbf{k}_j \int_{-\infty}^{\infty} d\eta (\mathbf{Y}^o_{t-\eta} - \mathbf{Y}^o_{t'-\eta})\mathbf{f}^+_\eta)$$
$$\exp(-\int_{-\infty}^{\infty} d\eta \frac{1}{2}\mathbf{f}^-_\eta((\mathbf{L}_{\eta-t_j} - \mathbf{L}_{\eta-t'_j}) + (\mathbf{L}_{<\eta-t_j>} - \mathbf{L}_{<\eta-t'_j>}))\mathbf{k}_j) \tag{2.11b}$$

$$Q_{j+} \equiv (|C_{\mathbf{k}_j}|^2 - |C^o_{\mathbf{k}_j}|^2) T_{\omega_{\mathbf{k}_j}}(t_j - t'_j)\exp(-\mathbf{k}_j\bar{\mathbf{L}}_{t_j-t'_j}\mathbf{k}_j)\exp(-i\mathbf{k}_j \int_{-\infty}^{\infty} d\eta (\mathbf{Y}^o_{t_j-\eta} - \mathbf{Y}^o_{t'_j-\eta})\mathbf{f}^+_\eta)$$
$$\exp(-\int_{-\infty}^{\infty} d\eta \frac{1}{2}\mathbf{f}^-_\eta((\mathbf{L}^*_{<\eta-t_j>} - \mathbf{L}_{\eta-t'_j}) + (\mathbf{L}^*_{\eta-t_j} - \mathbf{L}_{<\eta-t'_j>}))\mathbf{k}_j) \tag{2.11c}$$

$$i\mathbf{Y}^o_\tau = \mathbf{L}^*_\tau - \mathbf{L}_{<\tau>} = \mathbf{L}^*_{<\tau>} - \mathbf{L}_\tau = (\mathbf{L}^*_\tau - \mathbf{L}_\tau)u_\tau \tag{2.11d}$$

$$\bar{\mathbf{L}}_\tau = \mathbf{L}_\tau - \mathbf{L}_{0+} \tag{2.11e}$$

$$\mathbf{f}^+_\tau \equiv (\mathbf{f}_\tau + \mathbf{f}'_\tau)/2, \quad \mathbf{f}^-_\tau \equiv \mathbf{f}_\tau - \mathbf{f}'_\tau \tag{2.11f}$$

$$P_{j-} \equiv P_{j+}(\mathbf{f}^- \rightarrow -\mathbf{f}^-) \tag{2.11g}$$

$$Q_{j-} \equiv Q_{j+}(\mathbf{f}^- \rightarrow -\mathbf{f}^-) \tag{2.11h}$$

We write A_1 in the form (2.11) for several reasons. First, we keep the potential strength $|C^o_{\mathbf{k}_j}|^2$ explicit because if we wish to obtain in A_1, V^o in place of V, we need only effect the transformation

$$\Sigma_{\mathbf{k}} |C^o_{\mathbf{k}}|^2 T_{\omega_{\mathbf{k}}}(t) \rightarrow 2\mathbf{G}^o(t)\partial^2/\partial\mathbf{k}\,\partial\mathbf{k} \quad (\mathbf{k}=0) \tag{2.12}$$

Second, $\mathbf{f}$ and $\mathbf{f}'$ have been replaced by $\mathbf{f}^+$ and $\mathbf{f}^-$. This simplifies many calculations because electric fields $\mathbf{E}_t$ or $\mathbf{e}_t$ and initial momenta $\mathbf{p}_i$ which enter both $\mathbf{f}$ and $\mathbf{f}'$ equally, enter only $\mathbf{f}^+$. On the other hand, quantities evaluated at the time of interest are included only in $\mathbf{f}^-$ since they enter anti-symmetrically in $\mathbf{f}$ and $\mathbf{f}'$. The P and Q will enter into all A_n, $n > 1$.

We have nearly all we need for writing down the A_n. Consider $A_2 = \langle (S^e - S^o)(S^e - S^o) \rangle_{S^o}$ and (2.10). As in A_1 (2.8), the $\mathbf{g}$ and $\mathbf{g}'$ from which each term in A_2 will arise, will contain the corresponding $\mathbf{k}_1$ and $\mathbf{k}_2$. Now, however, not only will factors P_1P_2, P_1Q_2, etc., arise from (2.10), but, in addition, cross terms in $\mathbf{k}_1$, $\mathbf{k}_2$ will enter. To determine A_3, similar cross terms in $\mathbf{k}_1$, $\mathbf{k}_2$; $\mathbf{k}_1$, $\mathbf{k}_3$; $\mathbf{k}_2$, $\mathbf{k}_3$ will arise, but no $(\mathbf{k}_1, \mathbf{k}_2, \mathbf{k}_3)$ terms are present. In general, no higher order terms than $\mathbf{k}_i, \mathbf{k}_j$ arise and there are only 16 possible. For our purposes we reduce the number to 8 needed as follows. In A_2, for example, we transform the time integrals:

$$\frac{1}{2}\int_{-\infty}^{\infty} dt_1 \int_{-\infty}^{t_1} dt'_1 \int_{-\infty}^{\infty} dt_2 \int_{-\infty}^{t_2} dt'_2 = \int_{-\infty}^{\infty} dt_1 \int_{-\infty}^{t_1} dt'_1 \int_{-\infty}^{t_1} dt_2 \int_{-\infty}^{t_2} dt'_2 \tag{2.13}$$

As the arguments of the integrals are symmetric in $k_1t_1t'_1, k_2t_2t'_2$ there is no problem in doing this. It also greatly facilitates a subsequent factorization. If R is either a P or a Q, then the cross terms $ct_{R_iR_j}$ between R_i and R_j, $t_i > t_j$, are given by

$$ct_{P_iP_j} = ct_{Q_iP_j} = \exp(\mathbf{k}_i((\mathbf{L}_{t_i-t_j}-\mathbf{L}_{t_i-t'_j}) - (\mathbf{L}_{<t'_i-t_j>}-\mathbf{L}_{<t'_i-t'_j>}))\mathbf{k}_j) \tag{2.14a}$$

$$ct_{P_iQ_j} = ct_{Q_iQ_j} = \exp(\mathbf{k}_i((\mathbf{L}^*_{t_i-t_j}-\mathbf{L}_{t_i-t'_j}) - (\mathbf{L}^*_{t'_i-t_j}-\mathbf{L}_{<t'_i-t'_j>}))\mathbf{k}_j) \tag{2.14b}$$

$$ct_{P_iP_j^*} = ct_{Q_iP_j^*} = \exp(-\mathbf{k}_i((\mathbf{L}^*_{t_i-t_j}-\mathbf{L}^*_{t_i-t'_j}) - (\mathbf{L}^*_{t'_i-t_j}-\mathbf{L}^*_{t'_i-t'_j}))\mathbf{k}_j) \tag{2.14c}$$

$$ct_{P_iQ_j^*} = ct_{Q_iQ_j^*} = \exp(-\mathbf{k}_i((\mathbf{L}_{t_i-t_j}-\mathbf{L}^*_{t_i-t'_j}) - (\mathbf{L}_{<t'_i-t_j>}-\mathbf{L}^*_{t'_i-t'_j}))\mathbf{k}_j) \tag{2.14d}$$

$$ct_{P_i^*P_j} = ct_{Q_i^*P_j} = \exp(-\mathbf{k}_i((\mathbf{L}_{t_i-t_j}-\mathbf{L}_{t_i-t'_j}) - (\mathbf{L}_{t'_i-t_j}-\mathbf{L}_{t'_i-t'_j}))\mathbf{k}_j) \tag{2.14e}$$

$$ct_{P_i^*Q_j} = ct_{Q_i^*Q_j} = \exp(-\mathbf{k}_i((\mathbf{L}^*_{t_i-t_j}-\mathbf{L}_{t_i-t'_j}) - (\mathbf{L}^*_{<t'_i-t_j>}-\mathbf{L}_{t'_i-t'_j}))\mathbf{k}_j) \tag{2.14f}$$

$$ct_{P_i^*P_j^*} = ct_{Q_i^*P_j^*} = \exp(\mathbf{k}_i((\mathbf{L}^*_{t_i-t_j}-\mathbf{L}^*_{t_i-t'_j}) - (\mathbf{L}^*_{<t'_i-t_j>}-\mathbf{L}^*_{<t'_i-t'_j>}))\mathbf{k}_j) \tag{2.14g}$$

$$ct_{P_i^*Q_j^*} = ct_{Q_i^*Q_j^*} = \exp(\mathbf{k}_i((\mathbf{L}_{t_i-t_j}-\mathbf{L}^*_{t_i-t_j}) - (\mathbf{L}_{t'_i-t_j}-\mathbf{L}^*_{<t'-t'_j>}))\mathbf{k}_j) \tag{2.14h}$$

Note that the $\pm$ subscript on P and Q is not necessary as no Γ terms appear in the cross terms. These eight cross terms (actually only four as the second four are complex conjugates of the first four, *e.g.*, $ct^*_{PQ^*} = ct_{P^*Q}$) along with P and Q are all that we need in order to write down any A_n. The interchangability of P_i and Q_i, $t_i > t_j$, will facilitate this further. Thus,

$$A_2 = \int_{-\infty}^{\infty} dt_1 \int_{-\infty}^{t_1} dt'_1 \int_{-\infty}^{t_1} dt_2 \int_{-\infty}^{t_2} dt'_2\, \Sigma_{\mathbf{k}_1,\mathbf{k}_2}$$
$$((P_{1+}-Q_{1+})(ct_{P_1P_2}P_{2+} - ct_{P_1Q_2}Q_{2+} + ct_{P_1P_2^*}P^*_{2-} - ct_{P_1Q_1^*}Q^*_{2-})$$
$$+ (P^*_{1-}-Q^*_{1-})(ct_{P_1^*P_2}P_{2+} - ct_{P_1^*Q_2}Q_{2+} + ct_{P_1^*P_2^*}P^*_{2-} - ct_{P_1^*Q_1^*}Q^*_{2-})) \tag{2.15}$$

and

$$A_3 = -\int_{-\infty}^{\infty} dt_1 \int_{-\infty}^{t_1} dt'_1 \int_{-\infty}^{t_1} dt_2 \int_{-\infty}^{t_2} dt'_2 \int_{-\infty}^{t_2} dt_3 \int_{-\infty}^{t_3} dt'_3\, \Sigma_{\mathbf{k}_1,\mathbf{k}_2,\mathbf{k}_3}$$
$$\{(P_{1+}-Q_{1+})((ct_{P_1P_2}P_{2+}-ct_{P_1Q_2}Q_{2+})(ct_{P_1P_3}ct_{P_2P_3}P_{3+}-ct_{P_1Q_3}ct_{P_2Q_3}Q_{3+}$$
$$+ ct_{P_1P_3^*}ct_{P_2P_3^*}P^*_{3-} - ct_{P_1Q_3^*}ct_{P_2Q_3^*}Q^*_{3-}) + (ct_{P_1P_2^*}P^*_{2-} - ct_{P_1Q_2^*}Q^*_{2-})$$
$$(ct_{P_1P_3}ct_{P_2^*P_3}P_{3+} - ct_{P_1Q_3}ct_{P_2^*Q_3}Q_{3+} + ct_{P_1P_3^*}ct_{P_2^*P_3^*}P^*_{3-} - ct_{P_1Q_3^*}ct_{P_2^*Q_3^*}Q^*_{3-}))$$
$$+ (c.c., + \rightleftarrows -)\} \tag{2.16}$$

In the last of (2.16), $(c.c., + \rightleftarrows -)$ means to take the complex conjugate of the foregoing and innerchange $+$ and $-$ in the P and Q. This saves writing out the $(P^*_{1-} - Q^*_{1-})$ term in A_3.

These expansions, though complicated will permit rapid evaluation of a number of quantities of interest, as we shall see.

3. AC Transport, Impedance

One of the most important properties of a dissipative system is its optical absorption. This can be obtained from the velocity response,[8,11] usual admittance,[18–20] or from the position response to a small a.c. electric field, which we shall refer to as admittance. Thus

$$\mathbf{Y}_{t_f t_i} \equiv i<\mathbf{x}_{t_f}(\mathbf{x}_{t_i}-\mathbf{x}'_{t_i})> \tag{3.1a}$$

$$= -i\,\frac{\partial}{\partial\lambda}\,\frac{\partial}{\partial \mathbf{e}}\,<\mathbf{f}_\tau^- = \lambda\delta(\tau-t_f),\ \mathbf{f}_\tau^+ = \mathbf{e}\delta(\tau-t_i)>|_{\lambda=\mathbf{e}=0} \tag{3.1b}$$

where the $<>$ is to be evaluated according to (2.3). If we consider an expansion of $\mathbf{Y}$ of the form (I.10), then we can use (2.3) and (2.6) to evaluate each term of the $\mathbf{Y}$ series. As $A_o = 1$, $\mathbf{Y}^o$ comes directly from (2.5a):

$$\mathbf{Y}^o_{t_f t_i} = -i(\mathbf{L}^*_{t_f-t_i} - \mathbf{L}_{t_f-t_i})u(t_f-t_i) \tag{3.2a}$$

(Realizing this was the case, we took the liberty back in (2.11d) to label this combination as we did.) We should also note that

$$\mathbf{Y}^o_\nu \equiv \int_{-\infty}^{\infty} dt\ e^{+i\nu t}\,\mathbf{Y}^o_t = \mathbf{Z}^{o^{-1}}_\nu \tag{3.2b}$$

where again in (2.5c) we anticipated this result in defining $\mathbf{Z}^o_\nu$.

The first order $\mathbf{Y}^1$ is more interesting, now we must use A_1 as well as B. (Actually the B term does not enter, for $\lambda = 0$ implies $A_n = 0$, $n \geqslant 1$. This is most easily seen from (2.15) and (2.16) as $\lambda = 0$ implies $P_1 = Q_1$.) Using (2.11a) we obtain

$$\mathbf{Y}^1_{t_f t_i} = -i\,\frac{\partial}{\partial\lambda}\,\frac{\partial}{\partial\mathbf{e}}\,(-)\int_{-\infty}^{\infty} dt_1 \int_{-\infty}^{t_1} dt'_1\ \Sigma_{\mathbf{k}_1}\ (|C_{\mathbf{k}_1}|^2 - |C^o_{\mathbf{k}_1}|^2)$$

$$(P_{1+} - Q_{1+} + P^*_{1-} - Q^*_{1-})|_{\lambda=\mathbf{e}=0} \tag{3.3a}$$

$$= -\int_{-\infty}^{\infty} dt \int_{-\infty}^{t} dt'\ \Sigma_{\mathbf{k}}(|C_{\mathbf{k}}|^2 - |C^o_{\mathbf{k}}|^2)[\mathbf{Y}^o_{t_f-t}\mathbf{k}\mathbf{k}(-i)(\mathbf{Y}^o_{t-t_i} - \mathbf{Y}^o_{t'-t_i})$$

$$\cdot T_{\omega_\mathbf{k}}(t-t')\exp(-\mathbf{k}\bar{\mathbf{L}}_{t-t'}\mathbf{k}) + c.c.] \tag{3.3b}$$

$$= -\int_{-\infty}^{\infty} \frac{d\nu}{2\pi}\ \exp(-i\nu(t_f-t_i))\mathbf{Y}^o_\nu\mathbf{Z}^1_\nu\mathbf{Y}^o_\nu \tag{3.3c}$$

where

$$\mathbf{Z}^1_\nu = 4\int_0^{\infty} d\zeta\ (1-e^{i\nu\zeta})\mathrm{Im}(\frac{1}{2}\ \Sigma_\mathbf{k}(|C_\mathbf{k}|^2 - |C^o_\mathbf{k}|^2)\mathbf{k}\mathbf{k}T_{\omega_\mathbf{k}}(\zeta)e^{-\mathbf{k}\bar{\mathbf{L}}_\zeta\mathbf{k}}) \tag{3.3d}$$

This choice of $\mathbf{Z}^1_\nu$ was dictated both by (I.12b) and (2.5d). Combining $\mathbf{Z}^o_\nu$ and $\mathbf{Z}^1_\nu$ we obtain from (2.5d) and (2.12),

$$\mathbf{Z}^o_\nu + \mathbf{Z}^1_\nu = -\mathbf{m}(\nu+i\epsilon)^2 - i(\nu+i\epsilon)\underline{\epsilon}\cdot\mathbf{H}$$

$$+4\int_0^\infty d\zeta(1-e^{i\nu\zeta})\cdot\mathrm{Im}(\frac{1}{2}\ \Sigma_{\mathbf{k}}|C_{\mathbf{k}}|^2\mathbf{kk}T_{\omega_{\mathbf{k}}}(\zeta)e^{-\mathbf{k}\mathbf{L}\zeta\mathbf{k}}) \tag{3.3e}$$

with *no* explicit dependence on S^o.

We now make the following observation. Suppose that in (3.3) we were to carry out (2.12) on $C_{\mathbf{k}}$ as well as on $C^o_{\mathbf{k}}$. This is equivalent to saying that we are expanding the admittance of one distribution in terms of that of another. Then,

$$\mathbf{Z}^1_\nu = 4\int_0^\infty d\zeta(1-e^{i\nu\zeta})\mathrm{Im}(G^*(\zeta) - G^{o*}(\zeta)) \tag{3.4a}$$

$$= \mathbf{Z}_\nu - \mathbf{Z}^o_\nu \tag{3.4b}$$

In other words an expansion in $(S^e{-}S^o)$ for $\mathbf{Z}_\nu$ to only first order gives the correct impedance to all orders when S^e is harmonic. In fact even more is true, $\mathbf{Z}^n_\nu$ is zero for all $n > 1$. This is clear from (I.12a), for if $\Sigma_2^\infty\ \lambda^n\mathbf{Z}_n = 0$ for all λ, then $Z_n = \mathbf{Z}^n_\nu$ is zero for all $n > 1$. This in turn means we do not have to evaluate Y_n for harmonic potentials from (3.1) but rather can use (I.12b) obtaining

$$\mathbf{Y}^n_\nu = \mathbf{Y}^o_\nu(-\mathbf{Z}^1_\nu)\mathbf{Y}^o_\nu(-\mathbf{Z}^1_\nu)\mathbf{Y}^o_\nu\cdots\mathbf{Y}^o_\nu \tag{3.5}$$

with n, Z^1_ν factors. Of course, in general $C_{\mathbf{k}}$ cannot be made harmonic; however, the insight gained in (3.4) and (3.5) by assuming it to be so is most helpful.

Returning now to (3.1) and our desire to calculate $\mathbf{Y}_\nu$, we note that in case of high dissipation, strong coupling, the character of the effective potential seen by the electron will resemble a harmonic distribution. (See (2.1c) in the limit of small spacial differences.) The $\mathbf{Z}_\nu$ expansion contains all such terms by first order, whereas the $\mathbf{Y}_\nu$ expansion contains contributions from the harmonic portion of the potential in all orders. It is, therefore, not surprising that we choose to calculate $\mathbf{Z}_\nu$ and then calculate $\mathbf{Y}_\nu = 1/\mathbf{Z}_\nu$. Further advantages will be apparent when we expand $\mathbf{Z}_\nu$ schematically below.

Consider the following representation. (Drop the tensor and frequency-dependence for simplicity.) From (3.3d) we note that $-Z^1$ can be characterized by $(S^e{-}S^o)$. From (3.3c) we then characterize Y^1 as $Y^o(S^e{-}S^o)\,Y^o$. Similarly, Y^n can be represented as $Y^o(S^e{-}S^o)^n Y^o$. The two Y_o's come from the $\partial/\partial\lambda$ to give the $\mathbf{x}_{t_l}$ and the $\partial/\partial\mathbf{e}$ to give the $(\mathbf{x}_{t_l}{-}\mathbf{x}'_{t_l})$ in (3.1a). Thus

$$(-Z^1) \rightarrow (S^e{-}S^o)$$

$$(-Z^2) \rightarrow (S^e{-}S^o)(S^e{-}S^o) - (S^e{-}S^o)\,Y^o(S^e{-}S^o)$$

The difference between $S_1S_2S_3S_4$ and $S_1S_2Y^oS_3S_4$, for example, is that in the latter the cross terms (2.14) between S_1 and S_4, S_2 and S_3, and S_2 and S_4 have been expanded to lowest nonvanishing order in $\mathbf{k}_1{\cdot}\mathbf{k}_3$, etc. While this is a smaller modification than what is done in passing from a general $C_{\mathbf{k}}$ to a harmonic distribution via (2.12), one sees how Z^2 vanishes in this and, therefore, in the harmonic limit as SS goes to SY^oS.

It is very important to notice that S^o being harmonic already implies a quadratic expansion. Thus $-Z^2$ expanded is

$$S^eS^e - S^oS^e - S^eS^o + S^oS^o - S^eY^oS^e + S^oY^oS^e + S^eY^oS^o - S^oY^oS^o$$

which reduces to

$$-Z^2 \rightarrow S^eS^e - S^eY^oS^e$$

In words this says that we are to subtract from S^eS^e that portion which arises from the harmonic part of the interaction, $S^eY^oS^e$. This is what we shall find in general for higher Z^n. We also note that not only to first order (3.3e) but also to second order, the approximate influence functional S^o does *not* enter our computation of Z_ν explicitly. It is still present implicitly, of course, in **L**. It does enter explicitly for higher order Z^n as we shall see.

The third order expansion is represented by

$$\begin{aligned}-Z^3 \rightarrow\ & S^eS^eS^e - S^eS^eY^oS^e - S^eY^oS^eS^e \\ & + S^eY^oS^eY^oS^e - S^eS^oS^e + S^eS^oY^oS^e \end{aligned} \tag{3.6}$$

The $S_1^eS_2^oS_3^e$ term still contains the full cross term ct_{13} even though the ct_{12} and ct_{23} are to be expanded. By contrast the $S_1^eS_2^oY^oS_3^e$ term has split 3 from 1, and thus is the same as $S_1^eY^oS_2^oS_3^e$. Now suppose we start with the general $-Z^3$ term above with full ct_{ij} between 12, 23, and 13. Then if we expand ct_{12} and ct_{13} we would have

$$\begin{aligned} & S^eY^oS^eS^e - S^eY^oS^eY^oS^e - S^eY^oS^eS^e \\ & + S^eY^oS^eY^oS^e - S^eY^oS^oS^e + S^eY^oS^oY^oS^e = 0 \end{aligned}$$

The last two terms cancel because S_i^o implies all ct_{ij} and ct_{ki} are to be expanded. If we expand ct_{13} and ct_{23} instead, then $-Z^3$ would also vanish. If we expand ct_{12} and ct_{23}, however, then cancellation is achieved if and only if S^o is S^{sc}, the self-consistent influence functional.[14,16,17] The S^o obtained by expanding S^e to quadratic order will *not* achieve this as this requires in $S^eS^eS^e$ expanding all terms, not just the ct_{ij}, to lowest order in the **k**'s. The worst problem that arises is that in expanding S^e one obtains scattering rates for the electron as if recoil were negligible. This is avoided using S^{sc}. In addition, for $S^o = S^{sc}$, the fourth and sixth terms in (3.6) are equal.

The general expansion for $-Z^n$ is obtained as follows. First write down 2^{n-1} products $S^e...S^e$, n factors each. Then for $l = 0,...,n-1$ insert l, Y^o's between pairs of S^e's in all possible ways, and multiply each term by $(-1)^l$. Call this sum $J(S^e \cdots S^e)$. Now in J replace each S^e by an S^o in all (2^n) ways and multiply by $(-1)^m$ where m is the number of S^o's. This latter operation can be greatly simplified in a number of ways. First note that

$$J(S^oS^e...S^e) = J(S^oY^oS^e...S^e) = -J(S^oS^e...S^e) \tag{3.7a}$$

or

$$J(S^oS^e...S^e) = 0 = J(S^e...S^eS^o) \tag{3.7b}$$

Thus no terms arise with S^o on either end. Second note that of the three terms

$$\pm[+(...S^eS^oY^oS^e...) + (...S^eY^oS^oS^e...) - (...S^eY^oS^oY^oS^e...)]$$

the first and third or second and third will cancel leaving only one term. Thus one need insert a Y^o only on one side of an S^o.

We now return to a point raised in discussing Z^3. If at any one position k between S's we split each term in Z^n by a Y^o, then Z^n vanishes. This means that we expand the ct_{ij} for all $i \leqslant k$ and $j > k$. If on the other hand, we only expand ct_{ij} and ct_{ji} for i fixed and all j, then unless i is 1 or n, Z^n will not vanish in general. It will vanish if S^o is chosen to be the self-consistent influence functional S^{sc} developed elsewhere[14,16,17] and discussed below. Thus S^{sc} provides the capability of yielding the correct $\mathbf{Z}$ to lowest order in any expansion of any set of $\mathbf{k}_i$.

The self-consistent influence functional S^{sc} can be defined in many equivalent ways. Perhaps the simplest in the context of our discussion here is to require that $\mathbf{Z}_\nu^1$ (3.3d) vanish. While this choice does eliminate many terms from an expansion such as that of (I.12b), it does not appreciably simplify expansions of the form (3.6). For example, in $-Z^4$, setting $S_o = S^{sc}$ will reduce the total number of distinct terms by 4 to 14 from 18 all told. In general, self-consistency leads to combinations of similar terms containing S's with no unexpanded cross terms.

The interest in the self-consistent influence functional arose originally[14] in calculating the appropriate approximate influence functional S^o to use in calculating nonlinear dc transport properties. It was the self-consistent oscillator distribution $\mathbf{G}^{sc}$ (2.5c), which when inserted into (2.5c) for $\mathbf{Z}^o$ and (2.5b) for $\mathbf{L}$, reproduced itself when the impedance was calculated from the expectation value of the ac rate-of-change-of-momentum operator. The same result is obtained by setting Z^1 to zero. Later it was found that this self-consistent influence functional produced the same admittance when the latter was calculated in six different (and inequivalent) ways.[16,17] For zero applied electric and magnetic fields, these admittances satisfied not only the ground-state energy[19,20] and the free-energy sum rule (as in fact does any S^o provided the self-energy admittance is chosen), but also satisfies each step of the derivation of this sum rule[16] and a more demanding alternate, frequency-by-frequency comparison derived from the fluctuation-dissipation theorem.[12] As these topics have been discussed fully elsewhere,[16,17] we need not elaborate them further here. We note only that we believe that this represents the only example known of where a nontrivial, approximate solution to a nontrivial many-body problem exactly satisfies nontrivial sum rules.

4. D.C. TRANSPORT, STEADY-STATE AND ACCELERATION

As we mentioned in the introduction, the over-riding reason for using path integrals to study dissipative systems was that we can approximate the dissipative system of interest by a dissipative system which we can solve exactly even in the presence of arbitrary constant magnetic field and time-varying, electric field. We shall now make use of this feature of the path-integral method to discuss several aspects of electronic transport in insulators.

It would be out of place to elaborate the importance to the design of modern, small-scale devices of understanding the behavior of electrons in insulators under the influence of large electric fields. Such electrons have a tendency to either transfer appreciable amounts of energy locally to the lattice through the coupling with optical phonons or accelerate and gain sufficient energy from the field to ionize impurity atoms in the material or to create electron-hole pairs. In the latter case, if sufficiently many electrons can be induced or created, the chain-reaction can lead to an unstable current and breakdown. While at present many processes contribute to these effects, and they are best interpreted empirically, nonetheless some progress has been made in interpreting certain basic features underlying them, to which we now turn our attention.

The simplest problem concerns the relation between the expectation value of the steady-state velocity $\mathbf{v}$ and the constant, applied electric field $\mathbf{E}$. The surprising experimental finding which motivated this problem was that in attempting to accelerate electrons injected into the conduction band of an insulator (Al_2O_3) by applying large electric fields (2 to 4 MV/cm), it was found that the polar (longitudinal optical) phonons were capable of dissipating the electron's energy on the order of .02 to .04 eV/Å as required to maintain a steady-state average velocity. The theoretical understanding of this effect is outlined below.

The next problem of interest relates to current instabilities and breakdown. The current empirical lore, unchanged in forty years, remains primitive and will not be discussed. Rather, we focus on the processes leading to ionization. To ionize, an electron must have an energy of about the bandwidth, approximately 7 to 12 eV. In steady-state just below the threshold field above which steady-state is not possible, the electron has an energy of about .1 to .2 eV. To acquire the ionization energy, the electron must accelerate gaining the needed kinetic energy from the applied electric field. The problem, therefore, divides into two parts, acceleration and ionization, and depending on conditions either can be the rate limiting process. While ionization has been treated in some detail using various scattering theories, not necessarily path integrals, the problem of acceleration has been largely ignored. Thus we consider only the latter.[22]

The most interesting feature of acceleration is that it can occur for applied electric fields *less* than the threshold field. To be sure above threshold, all electrons will accelerate. Below threshold, while they will tend to a steady-state velocity, there remains a probability that they will make a transition to an accelerating state and accelerate to ionization energies. Naturally this probability drops rapidly with decreasing electric field, but for large, sub-threshold fields is not negligible. Following a brief discussion of the steady-state problem, we shall discuss this acceleration problem in some detail.

We now desire the relation between the steady-state velocity $\mathbf{v}$ of an electron under the influence of a large electric field $\mathbf{E}$ in a dissipative material characterizable by (2.1). As this problem has been discussed in some detail before,[12–15] we shall focus only on the important features of the calculation. One can, of course, write

$$\mathbf{v} = \underset{t\to\infty}{L} <\dot{\mathbf{x}}>_t = \underset{t\to\infty}{L}\; Tr(\dot{\mathbf{x}}\rho_t)/Tr(\rho_t) \tag{4.1}$$

where ρ_t is the density matrix of the system propagated from some appropriate initial value to the time t of interest. Such a calculation was originally carried out[12] in the spirit of the impedance expansion of Section 3, but, of course, without linearizing in the field. However, as also stressed in that section, one should choose an expansion which focuses attention on the form of what is being calculated and what terms are expected to be most important. The approach chosen, which resulted in the same answer as the (4.1) calculation but which utilized the more basic expansion developed in Section 2, follows.

It is a general feature of loss that it is maximum for a particular energy or velocity, and decreases as the velocity either increases or decreases. This loss, which in steady-state is balanced by the electric field, is, therefore, a single-valued function of the velocity, whereas the velocity is a doubled-valued function of the loss. Rather than calculating the velocity, we focus on the loss, specifically the rate of loss of momentum. Thus we focused on the expectation value of

$$d(\mathbf{m}\dot{\mathbf{x}})/dt = i[H, \mathbf{m}\dot{\mathbf{x}}] \tag{4.2a}$$

$$= q\mathbf{E} + q\mathbf{e}_t - \Sigma_{\mathbf{k}}\mathbf{k}\hat{R}_{\mathbf{k}} \tag{4.2b}$$

$$\hat{R}_{\mathbf{k}} = -i(C_{\mathbf{k}}^{*}a_{\mathbf{k}}^{\dagger}\exp(\underset{t}{-}i\mathbf{k}\cdot\mathbf{x}) - C_{\mathbf{k}}a_{\mathbf{k}}\exp(i\mathbf{k}\cdot\mathbf{x})) \tag{4.2c}$$

the operator relation governing conservation of momentum. (For simplicity we have set $\hbar = 1$, $\mathbf{H} = 0$ and consider only one phonon branch.) We have included the small, time-varying, probe field $\mathbf{e}_t$, although originally no use was made of it. Note that from $<\hat{R}_\mathbf{k}>$ we obtain an expression for $\mathbf{E}$ from (4.2b).

In calculating the expectation value of (4.2) first the phonon coordinates were eliminated exactly. Using (2.2) this transformed $<\hat{R}_\mathbf{k}>$ into a path integral over $R_{\mathbf{k},t}$ given by

$$R_{\mathbf{k},t} = |C_\mathbf{k}|^2 \int_{-\infty}^{t} d\tau \Big(\frac{\exp(-i\omega_\mathbf{k}(t-\tau))\exp(i\mathbf{k}\cdot(\mathbf{x}_t-\mathbf{x}_\tau))}{1-\exp(-\beta\omega_\mathbf{k})} - \frac{\exp(i\omega_\mathbf{k}(t-\tau))\exp(-i\mathbf{k}\cdot(\mathbf{x}_t-\mathbf{x}_\tau))}{\exp(\beta\omega_\mathbf{k})-1}$$
$$+ \frac{\exp(i\omega_\mathbf{k}(t-\tau))\exp(-i\mathbf{k}\cdot(\mathbf{x}_t-\mathbf{x}'_\tau))}{1-\exp(-\beta\omega_\mathbf{k})} - \frac{\exp(-i\omega_\mathbf{k}(t-\tau))\exp(i\mathbf{k}(\mathbf{x}_t-\mathbf{x}'_\tau))}{\exp(\beta\omega_\mathbf{k})-1}\Big) \tag{4.2d}$$

Rather than expanding in $(S^e - S^o)$ at this point, it was noted that in as much as the electron is on the average translating with a velocity $\mathbf{v}$, we should first shift variables in the path integral (2.1a) from $\mathbf{x}_t$, $\mathbf{x}'_t$ to $\mathbf{y}_t$, $\mathbf{y}'_t$ given by $\mathbf{x}_t = \mathbf{y}_t + \mathbf{v}t$ and $\mathbf{x}'_t = \mathbf{y}'_t + \mathbf{v}t$. We then approximate S^e by an S^o appropriate to the drifting frame of reference. Originally at this point it was felt that the result resembled the true result sufficiently accurately that the Feynman one-oscillator model would suffice for S^o, the two parameters of this model being evaluated to minimize the zero-temperature, ground state energy. More recently[14] one would use the a.c. behavior of (4.2b) to determine an S^{sc} to use for S^o, which characterizes the fluctuations or equivalently the a.c. response of the electron about its steady-state motion. While the field $\mathbf{E}$ appears in the $\int\int R_\mathbf{k}$, this dependence could be dropped along with a compensating term, since in the drifting frame, the electron would experience only relative motion without acceleration. The final result for $\mathbf{E}(\mathbf{v})$ was

$$\mathbf{E} = \Sigma_\mathbf{k} |C_\mathbf{k}|^2 \mathbf{k} \int_{-\infty}^{\infty} d\zeta \; T_{\omega_\mathbf{k}}(\zeta) e^{-i\mathbf{k}\cdot\mathbf{v}\zeta} e^{-\mathbf{k}\bar{L}_\zeta \mathbf{k}} \tag{4.3}$$

where all these quantities have been defined in Section 2. Finite $\mathbf{H}$ and other phonon branches are easily included.[14]

We now turn to the question of acceleration. Classically, sustained acceleration is impossible at electric fields below threshold. Thus if the equation of motion is

$$\dot{p} + p/\tau_p = f_t \, , \tag{4.4a}$$

then for a constant field $f_t = qE = F$ and starting from p_o at t_o, p_t at time t is given implicitly by

$$\int_{p_o}^{p_t} dp\,(1 - p/F\tau_p)^{-1} = (t - t_o)F \tag{4.4b}$$

Now typically p/τ_p has some nonzero slope for $p = 0$, increases to a maximum p_m/τ_m at some p_m, then decreases to zero as $p \to \infty$. If $0 \leqslant p_o < F\tau_{p_o}$ then p will increase until $p = F\tau_p$. This is possible so long as $p_m/\tau_m > F$. For sufficiently large F, $p_m/\tau_m < F$, $p/F\tau_p < 1$ for all p. Then p_t will continue to grow, ultimately at rate F. Classically, therefore, the threshold $F_T \equiv \max_p(p/\tau_p)$ represents a sharp demarcation between the steady state and the acceleration state.

Quantum mechanically, of course, no such sharp transition exists, acceleration being possible for any $F > 0$. Physically any electron which can avoid an energy or momentum loss transition for sufficiently long, can acquire sufficient velocity that further losses are very unlikely. Figuratively speaking the electron may be imagined as tunneling through a barrier in momentum space (rather than through one in real space). The only problem with this analogy is that the electron is not making a discrete transition from one well-defined state on one side of the barrier to another such state on the other side, acceleration states in dissipative media being rather hard to describe. Other approaches are, therefore, necessary.

One simple approach appropriate to weak coupling and small electric field is to solve the Boltzmann equation for the distribution in momentum in the presence of an electric field and note its time dependence for momenta much larger than p_m. This provides considerable physical insight into the problem and motivates the more general path-integral solution. Thus starting from

$$\frac{\partial f(\mathbf{p},t)}{\partial t} + \mathbf{e}_t \cdot \frac{\partial f(\mathbf{p},t)}{\partial \mathbf{p}} = -\Sigma_{\mathbf{p}'} R_{\mathbf{p}',\mathbf{p}} f(\mathbf{p},t) + \Sigma_{\mathbf{p}'} R_{\mathbf{p},\mathbf{p}'} f(\mathbf{p}',t) \tag{4.5a}$$

where $R_{\mathbf{p}',\mathbf{p}}$ is the rate of scattering from $\mathbf{p}$ to $\mathbf{p}'$, we find, solving by the method of characteristics,[23] that $f(\mathbf{p},t)$ satisfies

$$f(\mathbf{p},t) = \int_{t_o}^{t} dt' \exp\Big(-\int_{t'}^{t} dt''/\tau\big(\mathbf{p} - \int_{t''}^{t} dt''' \mathbf{e}_{t'''}\big)\Big) \Sigma_{\mathbf{p}'} R_{\mathbf{p}-\int_{t'}^{t} \mathbf{e}_{t''} dt'',\, \mathbf{p}'} \mathbf{f}(\mathbf{p}',t')$$

$$+ f\Big(\mathbf{p} - \int_{t_o}^{t} \mathbf{e}_{t'}\, dt', t_o\Big) \exp\Big(-\int_{t_o}^{t} dt''/\tau\big(\mathbf{p} - \int_{t''}^{t} dt''' \mathbf{e}_{t'''}\big)\Big) \tag{4.5b}$$

$$1/\tau_{\mathbf{p}} \equiv \Sigma_{\mathbf{p}'} R_{\mathbf{p}',\mathbf{p}} \tag{4.5c}$$

Expression (4.5b) admits a straightforward physical interpretation. Consider first the second term therein. The first factor is just the initial distribution translated by $\mathbf{e}_t$ applied between t_o and t. This corresponds to pure acceleration without scattering. The second factor represents the probability of transition that the electron at $\mathbf{p}$ at t went from t_o to t without scattering. (It corresponds to the "tunneling probability" through the $1/\tau_{\mathbf{p}}$ barrier in momentum space.) While this second term represents the initial transient, the first term in (4.5b) is the heart of the matter: a particle at $(\mathbf{p},t)$ has arrived there from $(\mathbf{p}',t')$ by scattering to $\mathbf{p} - \int_{t'}^{t} \mathbf{e}_{t''} dt''$ at t' and then propagating to $(\mathbf{p},t)$ without scattering. Ordinarily such an integral equation would have to be solved for $f(\mathbf{p},t)$. However, with some physical insight we can simplify matters greatly.

First we assume that $\mathbf{e}_t \,||\, \hat{\mathbf{x}}$ and that $e_t = F = qE$, a constant force F. Let $(p_y,p_z) \equiv \mathbf{p}_\perp$ and change the variable of integration in the transition probability from t'' to $p''_x = p_x - F(t-t'')$. Then

$$f(\mathbf{p},t) = \int_{t_o}^{t} dt' \exp\Big(-\int_{p_x - F(t-t')}^{p_x} dp''_x / F\tau_{(p''_x,\mathbf{p}_\perp)}\Big) \Sigma_{\mathbf{p}'} R_{\mathbf{p}-F(t-t')\hat{\mathbf{x}},\,\mathbf{p}'} f(\mathbf{p}',t')$$

$$+ f(\mathbf{p}-F(t-t_o)\hat{\mathbf{x}},t_o) \exp\Big(-\int_{p_x - F(t-t_o)}^{p_x} dp''_x / F\tau_{(p''_x,\mathbf{p}_\perp)}\Big) \tag{4.6}$$

This form of $f(\mathbf{p},t)$ indicates one striking feature. As $p_x \to \infty$, $1/\tau_{\mathbf{p}} \to 0$ as we noted above. However, does $L_{(p_x \to \infty)} \int_0^{p_x} dp'_x/\tau_{(p'_x,\mathbf{p}_\perp)}$ tend to infinity or to a finite number. If the former holds, $\int_0^\infty dp_x/\tau_{\mathbf{p}} = \infty$, then acceleration is impossible. If on the other hand $\int_0^\infty dp_x/\tau_{\mathbf{p}}$ is some finite number, as it is for optical phonon scattering, for example, then acceleration is possible for any $F > 0$. Note the form of the transition probability from some p_x to very large momentum.

$$\exp(-\int_{p_x}^{\infty} dp'_x/F\tau(p'_x,\mathbf{p}_\perp))$$

The inverse exponential dependence on the electric force is very reminiscent of Zener tunneling. Note also that had we included only the relaxation term in (4.5a) that only the second term in (4.5b) and (4.6) would be obtained, missing the whole effect. Note also that one cannot set $\partial f/\partial t = 0$ in (4.5a) in the presence of $\mathbf{e}_t$.

We proceed to calculate the tunneling rate as follows. Let $F^>(p_x^l,t)$ be the probability at t that $p_x > p_x^l$. Take $t_o = 0$. Then using (4.6) we obtain

$$F^>(p_x^l,t) \equiv \int_{p_x^l}^{\infty} dp_x \int d\mathbf{p}_\perp f(\mathbf{p},t) \tag{4.7a}$$

$$= \int_0^t d\tau \int_{p_x^l - F\tau}^{p_x^l} dp_x \int d\mathbf{p}_\perp \, \Sigma_{\mathbf{p}'} \, R_{\mathbf{p},\mathbf{p}'} f(\mathbf{p}', \tau - (p_x^l - p_x)/F)$$

$$\cdot \exp(-\int_{p_x}^{p_x^l + F(t-\tau)} dp''_x/F\tau(p''_x,\mathbf{p}_\perp)) \tag{4.7b}$$

$$+ \int_{p_x^l - Ft}^{\infty} dp_x \int d\mathbf{p}_\perp f(\mathbf{p},0) \exp(-\int_{p_x}^{p_x+Ft} dp'_x/F\tau(p'_x,\mathbf{p}_\perp))$$

where p_x^l is chosen sufficiently large that $\Sigma_{\mathbf{p}'} f_{(\mathbf{p}')} R_{\mathbf{p},\mathbf{p}'}$ is negligible compared to its value for $p_x < p_x^l$. The rate per electron at which electrons transition from the quasi-steady state to the accelerating state is simply

$$r \equiv dF^>(p_x^l,t)/dt/\Sigma_{\mathbf{p}} f^{qss}(\mathbf{p},t) = 1/\tau_d \tag{4.8}$$

After initial transients, we assume that $f(\mathbf{p},t)$ for the electrons in the quasi-steady state $f^{qss}(\mathbf{p},t)$ has the form $f^{qss}(\mathbf{p})\exp(-t/\tau_d)$. Then $f^{qss}(\mathbf{p})$ satisfies

$$f^{qss}(\mathbf{p}) = \int_{-\infty}^{p_x} \frac{dp_x'}{F} \exp(-\int_{p'_x}^{p_x} \frac{dp''_x}{F} (\frac{1}{\tau_{(p''_x,\mathbf{p}_\perp)}} - \frac{1}{\tau_d})) \Sigma_{\mathbf{p}''} R_{p'_x,\mathbf{p}_\perp,\mathbf{p}''} f^{qss}(\mathbf{p}'') \tag{4.9}$$

and we obtain

$$r = \Sigma_{\mathbf{p}} f^{qss}(\mathbf{p})/\tau_d(p)/\Sigma_{\mathbf{p}} f^{qss}(\mathbf{p}) = 1/\tau_d \tag{4.10a}$$

$$1/\tau_d(\mathbf{p}) = \int_{-\infty}^{\infty} d\mathbf{p}' R_{\mathbf{p}',\mathbf{p}} \exp(\frac{p_x^l - p'_x}{F\tau_d}) \exp(-\int_{p'_x}^{\infty} dp''_x/F\tau(p''_x,\mathbf{p}_\perp)) \tag{4.10b}$$

Here $1/\tau_d(\mathbf{p})$ has the meaning of the transition rate of an electron in state $\mathbf{p}$. For small F, τ_d can be expected to be rather large so that $\tau_d(\mathbf{p})$ is relatively independent of τ_d, and hence $\tau_d(\mathbf{p})$ can be calculated without knowledge of the distribution f. For larger fields τ_d will become appreciable hence one must solve for τ_d self-consistently between (4.9) and (4.10). The simplest procedure is to assume $\tau_d = \infty$ in (4.9) and (4.10b), get a τ_d from (4.10a), etc. To be sure, this expression is most meaningful when $F\tau_d >> p_x^l$, which is usually the case somewhat below threshold. Excellent solutions of (4.9) exist for τ_d infinite,[24] and this modification for $\tau_d < \infty$ should be straightforward, e.g., replace $\partial f_t/\partial t = 0$ by $\partial f_t/\partial t = -f_t/\tau_d$.

The foregoing, while providing a remarkably clear physical picture of the problem of electron acceleration below threshold, is nonetheless based on the Boltzmann equation and hence is valid only for small electron-lattice coupling and small electric fields. In Appendix II, (4.5b) and, hence, the subsequent relations through (4.10b), are generalized somewhat to include the modification of propagation and scattering due to large electric fields and electron interaction with the solid. However, these modifications are only expected to be useful for $\alpha < 1$. For $\alpha > 1$ a direct evaluation of $1/\tau_d(\mathbf{p})$ is more meaningful.[22]

In choosing an influence functional S^o to model the acceleration problem, we must use different criteria than in choosing S^o, e.g., S^{sc}, for steady-state transport, especially for weak coupling. For S^{sc} the electron drifts with a certain velocity determined by the magnitude of the electric field. For the one-oscillator S^o of Feynman, S^F, the electron accelerates, albeit with an altered mass and with relative, oscillatory motion. The ability to accelerate is, of course, essential to the acceleration problem, whereas it can be glossed over for steady-state calculations, except for weak coupling. Physically, S^{sc} over-constrains the electron's motion by including oscillator coupling at frequencies down to d.c. and over-averages the *finite* energy gain in the applied field and loss in scattering. The criterion for choosing S^o for the acceleration problem demands a more detailed understanding of this problem.[22]

5. CONCLUSION

In the foregoing we have outlined several applications of path integrals to problems involving substantial dissipation. We have noted how the existence of exactly solvable, dissipative models within the path-integral formalism in addition to the facility with which electric and magnetic fields can be treated using path integrals renders a number of interesting problems approachable.

In approaching the problem of instability in electron transport in high but subthreshold fields, we focused on acceleration. Ionization *per se* can be treated using an influence functional similar to (2.16). The number of electrons can be increased in accordance with the ionization rate. However, at present the characteristic features of breakdown phenomena seem to be most readily understood in terms of instabilities, of which the transition from a steady-state to an accelerating state, as discussed in Section 4 is one example.

Appendix I, General Expansions

Perturbation expansions can be performed in many ways. If, for example, an expectation value A = <.> is desired and it is known that A has the functional form f(.), then the identity

$$A = f(f^{-1}(A)) \tag{I.1}$$

and the expansion

$$f^{-1}(A) = f^{-1}(A)|_o + \frac{\delta f^{-1}(A)}{\delta S_1}|_o S_1 + \frac{1}{2}\frac{\delta^2 f^{-1}(A)}{\delta S_1 \delta S_2}|_o S_1 S_2 + \cdots \tag{I.2}$$

$$\frac{\delta f^{-1}(A)}{\delta S} = \frac{df^{-1}(A)}{dA}\frac{\delta A}{\delta S} \tag{I.3a}$$

$$\frac{\delta^2 f^{-1}(A)}{\delta S_1 \delta S_2} = \frac{df^{-1}(A)}{dA}\frac{\delta^2 A}{\delta S_1 \delta S_2} + \frac{d^2 f^{-1}(A)}{dA^2}\frac{\delta A}{\delta S_1}\frac{\delta A}{\delta S_2} \tag{I.3b}$$

permits a result of the anticipated form to be obtained as a series in S, the perturbation. (The $|_o$ refers to the calculation to the 0-order Hamiltonian or influence functional.) By way of contrast, a direct expansion of A,

$$A = A|_o + \frac{\delta A}{\delta S_1}|_o S_1 + \frac{1}{2}\frac{\delta^2 A}{\delta S_1 \delta S_2}|_o S_1 S_2 + \cdots \tag{I.4}$$

would have to be carried out to all orders and the significant terms giving rise to f(.) selected out in each order.

One of the most useful expansions of this form is the cumulate expansion where f(.) =exp(.). Typical is

$$\langle e^{\lambda \hat{A}}\rangle = \langle \Sigma_{n=0}^{\infty} \lambda^n \hat{A}^n/n!\rangle = 1 + \Sigma_{n=1}^{\infty} \lambda^n A_n/n! \tag{I.5}$$

$$= exp\,(\Sigma_{m=1}^{\infty} \lambda^m B_m/m!) \tag{I.6a}$$

$$= 1 + \Sigma_{r=1}^{\infty} \left[\Sigma_{m=1}^{\infty} \lambda^m B_m/m!\right]^r/r! \tag{I.6b}$$

where $A_n \equiv \langle \hat{A}^n \rangle$ and <.> refers to a calculable (0-order) problem. Solving for A_n in terms of $B_n, B_{n-1},...B_1$, then writing $B_n = A_n - (...)$ yields.

$$B_1 = A_1 \tag{I.7a}$$

$$B_2 = A_2 - B_1^2 \tag{I.7b}$$

$$B_3 = A_3 - 3B_2B_1 - B_1^3 \tag{I.7c}$$

$$B_4 = A_4 - 4B_3B_1 - 3B_2B_2 - 6B_2B_1^2 - B_1^4 \tag{I.7d}$$

The simple way to remember this expansion is to note that B_n is just A_n less contributions from all possible ways in which independent clusters can be made. As B_m's, m< n appear in these sums rather than A_m's, one is assured of not double counting similar terms in lower-order clusters. For example, solving

$$\Sigma_{m=1}^{\infty} \lambda^m B_m/m! = \ln (I + \Sigma_{n=1}^{\infty} \lambda^n A_n/n!)$$
$$= \Sigma_{s=1}^{\infty} (-1)^{s+1} (\Sigma_{n=1}^{\infty} \lambda^n A_n/n!)^s/s$$

would give B_n in terms of $A_n, A_{n-1}, \ldots A_1$, with a much less direct physical interpretation.

Another very useful expansion is the inverse expansion. For example,

$$Z = Z|_o + \frac{\delta Z}{\delta S_1}|_o S_1 + \frac{1}{2} \frac{\delta^2 Z}{\delta S_1 \delta S_2}|_o S_1 S_2 + \cdots \tag{I.8}$$

and

$$ZY = 1, \quad \frac{\delta Z}{\delta S_1} = - Z \frac{\delta Y}{\delta S_1} Z \tag{I.9a}$$

$$\frac{\delta^2 Z}{\delta S_1 \delta S_2} = -Z \frac{\delta^2 Y}{\delta S_1 \delta S_2} Z + 2Z \frac{\delta Y}{\delta S_1} Z \frac{\delta y}{d S_2} Z \tag{I.9b}$$

is very useful for calculating the impedance Z in terms of terms arising in the expansion of the admittance Y. Again an expansion of the form

$$Y = Y|_o + \frac{\delta Y}{\delta S_1}|_o S_1 + \frac{1}{2} \frac{\delta^2 Y}{\delta S_1 \delta S_2}|_o S_1 S_2 + \cdots \tag{I.10}$$

would have to be carried out to all orders to recover resonant terms which enter at once in first order in (I.8). Here $Z_o = Y_o^{-1}$. Thus if we write

$$Y = Y_o + \Sigma_{n=1}^{\infty} \lambda^n Y_n = Z^{-1} = (Z_o + \Sigma_{m=1}^{\infty} \lambda^m Z_m)^{-1} \qquad \text{I.11a)}$$

$$= Z_o^{-1}(1 - \Sigma_{m=1}^{\infty} \lambda^m (-Z_m) Z_o^{-1})^{-1} \tag{I.11b}$$

$$= Z_o^{-1}(1 + \Sigma_{r=1}^{\infty} (\Sigma_{m=1}^{\infty} \lambda^m (-Z_m) Z_o^{-1})^r) \tag{I.11c}$$

we find solving for Y_n in terms of Z_n, $Z_{n-1}, \ldots, Z_1$, and writing $-Z_n = Z_o Y_n Z^o - (\ldots)$,

$$Z = Z_o - \Sigma_{m=1}^{\infty} \lambda^n (-Z_n) \tag{I.12a}$$

$$- Z_n = Z_o Y_n Z_o - \Sigma_{i_1 \ldots i_m = 1}^{m \geq 2, \Sigma_{j=1} i_j = n} (-Z_{i_1}) Y_o (-Z_{i_2}) Y_o \cdots Y_o (-Z_{i_m})$$

Writing out the first few terms we have

$$-Z_1 = Z_o Y_1 Z_o$$

$$-Z_2 = Z_o Y_2 Z_o - (-Z_1) Y_o (-Z_1)$$

$$-Z_3 = Z_o Y_3 Z_o - (-Z_1) Y_o (-Z_2) - (-Z_1) Y_o (-Z_1) Y_o (-Z_1) - (-Z_2) Y_o (-Z_1)$$

As with the cumulate expansion, by solving for what is known, A or Y, in terms of B or Z, which we in fact desire, we obtain the most useful expression. Were we to solve

$$Z = Z_o + \Sigma_{m=1}^{\infty} \lambda^m Z_m = Y^{-1} = (Y_o - \Sigma_{n=1}^{\infty} \lambda^n (-Y_n))^{-1}$$

we would obtain a very similar expansion but we would be stuck with Z_n in terms of $Y_n, Y_{n-1}, \ldots Y_1$. (The "$-Z_n$" could be changed to Z_n at the expense of having $Z = Z_o - \Sigma_{m=1}^{\infty} \lambda^m Z_m$, but the expressions would be neater.) Unlike the cumulate in

(f_l, f'_l), the inverse expansion for $Z_{ll'}$ of one distribution of oscillators for another yields the correct Z by first order and all higher orders vanish: $Z_n = (-Z_1) Y_o \cdots Y_o(-Z_1)$, nZ_1's. By contrast an expansion of Y to all orders would be necessary to achieve a similar result.

Appendix II, Boltzmann Equation

It is a straightforward matter to derive the Boltzmann equation for the relaxation of a weakly coupled system toward equilibrium in the absence of electric and magnetic fields.[25] The task is somewhat more of a challenge in the presence of a driving electric field. For example, even if the fields are small, the Boltzmann equation has the defect of predicting only decaying solutions when the system is known to have absorption structure. For large fields, both below and above threshold, the scatterings rates are modified from their simple Fermi-Golden-Rule values. The path-integral technique can handle these fields as a matter of course.

In this appendix we shall generalize the Boltzmann equation to arbitrary electric fields. While it would be simplest to work with the free particle S^o, we can use a more general S^o to include the change in the interscattering propagation (essentially the effective mass) and in the (intra)scattering rates due to the electron-medium interaction. This does, however, provide only a gross account of the interference terms present at finite coupling and does not pretend to offer nearly as adequate a treatment for zero field as already exists.[26] On the other hand, at finite coupling the utility of the concept of the electron distribution function is greatly diminished, and alternate methods focusing on the quantities of interest are usually sought.[7–17] Our result, while possessing a Boltzmann-like character, will include the absorption character of a dissipative system, the electric-field modification of the scattering rate, and the medium modification of the propagation and scattering.

We shall make use of the basic assumption usually made in deriving a Boltzmann result: for any pair of self-energy terms, $V_{ixx} V_{jxx}$, $t_i > t_i' > t_j > t_j'$ or $t_i > t_j > t_j' > t_i'$, and for any pair of fluctuation-dissipation terms, $V_{ixx'} V_{jxx'}$, $t_i > t_j$, $t_i' > t_j'$. Other approximations about the form of the propagation will not be made. We shall not consider the case of a magnetic field nor tensorial Y,L, etc.

To proceed, we desire $f(\mathbf{p}_f, t_f)$ given $f(\mathbf{p}_i, t_i)$.

$$f(\mathbf{p}_f, t_f) = \int d\mathbf{p}_i f(\mathbf{p}_i, t_i) < \delta(\hat{\mathbf{p}} - \mathbf{p}_f) >_{t_f; \mathbf{p}_i, t_i} \tag{II.1a}$$

$$= \int d\mathbf{p}_i f(\mathbf{p}_i, t_i) \int d\bar{\lambda} e^{-i\lambda \cdot \mathbf{p}_f} < e^{i\lambda \cdot \hat{\mathbf{p}}} >_{t_f; \mathbf{p}_i, t_i} \tag{II.1b}$$

To evaluate the expectation value in (II.1b) we shall make use the results of section 2. In particular we have

$$\mathbf{f}_i^+ = \mathbf{e}_i u(t-t_i) + \mathbf{p}_i \delta(t-t_i) \tag{II.1c}$$

$$\mathbf{f}_i^- = -m\lambda\dot{\delta}(t-t_f) \tag{II.1d}$$

These values for $\mathbf{f}^+$ and $\mathbf{f}^-$ can be inserted into (2.11) and (2.14) to obtain P, Q and ct. It is at this point that we must consider our approximations. For the Feynman one-oscillator model at zero temperature

$$\bar{L}_\zeta = \frac{1}{2m}\frac{w^2}{v^2}\left[-i\zeta + \frac{v^2-w^2}{w^2 v}(1-e^{iv\zeta})\right] \tag{II.2a}$$

This form is characteristic of $G(\Omega)$'s lacking oscillator strength near $\Omega = 0$. Here the first term corresponds to the drift of a quasi-particle with effective mass mv^2/w^2, while the second term represents the memory, so to speak, of the electron-oscillator interaction. In the $\mathbf{f}^+$ and $\mathbf{f}^-$ terms in (2.11), and in (2.14) where we expect to see the particle propagate from t_i to t_j or t_j to t_f, we expect the effective mass portion to dominate. Thus the form

$$\bar{L}_\zeta = -i\zeta/2m^* \tag{II.2b}$$

can be used. In $L_{t_i-t_i'}$, however, which appears in the scattering, we cannot use this asymptotic form, but rather must use the full L. Also in $\mathbf{f}^-$ in fact is a time derivative (II.1d) which also precludes such a naive picture once $\alpha > 1$. [To be sure, these difficulties are not encountered if one uses the S^o of a free particle for which $L_\zeta = -i\zeta/2m$. However, then the independent-scattering approximations break down at even lower α, that is for $\alpha > .3$. The approximate S^o adjusts the mass, a trivial correction, and the scattering, a very nontrivial correction.] A more accurate L_t must also be maintained in a portion of the $\mathbf{f}^+$ term in $\mathbf{e}_t$.

Understanding these approximations, we write down the P, Q and the ct's we need

$$P_{r+}(\tau_r, \tau'_r) = (|C_{\mathbf{k}_r}|^2 - |C^o_{\mathbf{k}_r}|^2)\, T_{\omega_{\mathbf{k}_r}}(\tau_r - \tau'_r)\exp(-k_r^2 \bar{L}_{\tau_r-\tau'_r})$$

$$\cdot\, exp(-i\mathbf{k}_r\cdot(\int_0^\infty d\zeta (Y^o_\zeta - \zeta/m^*)(\mathbf{e}_{\tau-\zeta} - \mathbf{e}_{\tau'-\zeta}) - \int_{\tau'_r}^{\tau_r} \mathbf{e}_\eta (\eta - \tau'_r)/m^* d\eta))$$

$$\cdot\, exp(-i\mathbf{k}_r\cdot(\mathbf{p}_i + \int_{t_i}^{\tau_r} \mathbf{e}_\eta d\eta)(\tau_r - \tau'_r)/m^*) \tag{II.3a}$$

$$Q_{r+}(\tau_r, \tau'_r) = P_{r+}(\tau_r, \tau'_r)\exp(-i\lambda\cdot\mathbf{k}_r\gamma) \tag{II.3b}$$

$$\gamma \equiv m/m^* \tag{II.3c}$$

$$P^*_{r-}(\tau_r, \tau'_r) = P^*_{r+}(\tau_r, \tau'_r) \tag{II3.d}$$

$$Q^*_{r-}(\tau_r, \tau'_r) = P^*_{r-}(\tau_r, \tau'_r)\exp(-i\lambda\cdot\mathbf{k}_r\gamma) \tag{II.3e}$$

$$ct_{P_1P_2} = ct_{P_1P_2^*} = ct_{Q_1P_2} = ct_{Q_1P_2^*} = 1 \tag{II.3f}$$

$$ct_{P_1Q_2} = ct_{P_1Q_2^*} = ct_{Q_1Q_2} = ct_{Q_1Q_2^*} = \exp(-i\mathbf{k}_r\cdot\mathbf{k}_j(\tau_r - \tau'_r)/m^*) \tag{II.3g}$$

where in (II.3g) $(\tau_r > \tau_j,\ \tau'_r > \tau'_j)$. Finally we need the zero-order (overall-factor) term:

$$exp(i\lambda\cdot(\mathbf{p}_i + \int_{t_i}^{t_f}\mathbf{e}_\eta d\eta)\gamma) \tag{II.3h}$$

(Here we dropped the factor $\exp(-\ddot{L}_{0+}\lambda^2 m^2/2)$ which otherwise would result in an energy broadening. Again this is unsatisfactory, but part of the Boltzmann picture.)

Rather than the term by term expansion developed in Section 2, we shall make use of a modification more appropriate to the distinction between self-energy and fluctuation-dissipation terms. Writing

$$S^e - S^o = -\Sigma_{\mathbf{k}_r}\int_{t_i}^{t_f} d\tau_r \int_{t_i}^{\tau_r} d\tau'_r (V_{rx'x} - V^o_{rx'x'} + V^*_{rxx} - V^{o*}_{rxx})$$

$$+ \Sigma_{\mathbf{q}_j}\int_{t_i}^{t_f} dt_j \int_{t_i}^{t_f} dt'_j (V_{jxx'} - V^o_{jxx'}) \tag{II.4a}$$

$$\equiv -S_{se} - S^*_{se} + S_{fd} \tag{II.4b}$$

we can expand $\exp(S^e - S^o)$ in the form

$$e^{S^e - S^o} = e^{-S^*_{se}(t_f,t_i)} e^{-S_{se}(t_f,t_i)}$$

$$+ \Sigma_{\mathbf{q}_1}\int_{t_i}^{t_f} dt_1 \int_{t_i}^{t_f} dt'_1 e^{-S^*_{se}(t_f,t'_1)} e^{-S_{se}(t_f,t_1)} (V_{1xx'} - V^o_{1xx'}) e^{-S^*_{se}(t'_1,t_i)} e^{-S_{se}(t_1,t_i)}$$

$$+ \Sigma_{\mathbf{q}_1\mathbf{q}_2}\int_{t_i}^{t_f} dt_1 \int_{t_i}^{t_1} dt'_1 \int_{t_i}^{t_1} dt_2 \int_{t_i}^{t'_1} dt'_2 e^{-S^*_{se}(t_f,t'_1)} e^{-S_{se}(t_f,t_1)}$$

$$(V_{1xx'} - V^o_{1xx'}) e^{-S^*_{se}(t'_1,t'_2)} e^{-S_{se}(t_1,t_2)} (V_{2xx'} - V^o_{2xx'})$$

$$e^{-S^*_{se}(t'_2,t_i)} e^{-S_{se}(t_2,t_i)} + \cdots \tag{II.4c}$$

To evaluate $<\exp(S^e - S^o)>$ for $\mathbf{f}^+, \mathbf{f}^-$ of (II.1cd), we consider the nth order term of (II.4c). As $V_{rxx'}$ and V_{rxx} in $S^*_{se}(t_{j-1},t_j)$ and $S_{se}(t_{j-1},t_j)$ couple only with the preceding $V_{lxx'}$, $l = j,\ldots,n$, and not to each other (II.3fgh) we can expand the *se* propagators in (II.4c) and calculate them. The result is

$$e^{-S^*_{se}(t'_{j-1},t'_j)} \rightarrow$$

$$U^*_{n,\mathbf{p}}(t'_{j-1},t'_j) = 1 - \Sigma_{\mathbf{k}_1}\int_{t'_j}^{t'_{j-1}} d\tau_1 \int_{t'_j}^{\tau_1} d\tau'_1 U^*_{n,\mathbf{p}-k_1}(\tau_1,\tau'_1)\bullet \tag{II.5a}$$

$$\bullet P_1^{j,n}(\tau_1,\tau'_1)\, U^*_{n,\mathbf{p}}(\tau'_1,t'_j)$$

$$P_r^{j,n}(\tau_r,\tau'_r) \equiv P_{r+}(\tau_r,\tau'_r)\exp(i\mathbf{k}_r(\Sigma^n_{l=j}\mathbf{q}_l)(\tau_r - \tau'_r)/m^*) \tag{II.5b}$$

$$dU^*_{n,\mathbf{p}}(t'_{j-1},t_j)/dt'_{j-1} = -\Sigma_{\mathbf{k}_1} \int_{t'_j}^{t'_{j-1}} U^*_{n,\mathbf{p}-\mathbf{k}_1}(t'_{j-1},\tau) P_1^{j,n}(t'_{j-1},\tau)\, U^*_{n,\mathbf{p}}(\tau,t'_j)\, d\tau \quad \text{(II.5c)}$$

$$Q_{j,n}(t_j,t'_j) \equiv Q_{j+}(t_j,t'_j)\exp(i\mathbf{q}_j(\Sigma_{l=j+1}^{n}\mathbf{q}_l)(t_j-t'_j)/m^* \quad \text{(II.5d)}$$

Putting this all together we can write

$$<\exp(S^e-S^o)> = U_o^*(t_f,t_i)\, U_o(t_f,t_i)$$

$$+ \Sigma_{\mathbf{q}_1} \int_{t_i}^{t_f} dt_1 \int_{t_i}^{t_f} dt'_1\, U_1^*(t_f,t'_1)\, U_1(t_f,t_1)\, Q_{1,1}(t_1,t'_1)\, U_1^*(t'_1,t_i)\, U_1(t_1,t_i)$$

$$+ \Sigma_{\mathbf{q}_1\mathbf{q}_2} \int_{t_i}^{t_f} dt_1 \int_{t_i}^{t_f} dt'_1 \int_{t_i}^{t_1} dt_2 \int_{t_i}^{t'_1} dt'_2\, U_2^*(t_f,t'_1)\, U_2(t_f,t_1)\, Q_{1,2}(t_1,t'_1)$$

$$U_2^*(t'_1,t'_2)\, U_2(t_1,t_2)\, Q_{2,2}(t_2,t_2)\, U_2^*(t'_2,t_i)\, U_2(t_2,t_i) + \cdots \quad \text{(II.6)}$$

In order to convert (II.6) into an integral equation it is expedient to write

$$W_{\mathbf{k}_r}(\gamma(\mathbf{p}_i + \int_{t_i}^{\tau_r} \mathbf{e}_\eta d\eta - \Sigma_{l=j}^{n}\mathbf{q}_l);\ \tau_r,\tau'_r) \equiv P_r^{j,n}(\tau_r,\tau'_r) \quad \text{(II.7a)}$$

$$Q_{j,n}(t_j,t'_j) = W_{\mathbf{q}_j}(\gamma(\mathbf{p}_i + \int_{t_i}^{t_j} \mathbf{e}_\eta d\eta - \Sigma_{l=j+1}^{n}\mathbf{q}_l);\ t_j,t'_j)\, e^{-i\lambda\cdot\mathbf{q}_j\gamma} \quad \text{(II.7b)}$$

$$U(\gamma(\mathbf{p}_i - \int_{t_i} \mathbf{e}_\eta d\eta - \Sigma_{l=j}^{n}\mathbf{q}_l);\ t_{j-1},t_j) = U_n(t_{j-1},t_j) \quad \text{(II.7c)}$$

in order to exhibit the momentum dependence explicitly. The nth order term contains the factor, (II.7b), (II.3h) and (II.1b)

$$\exp(i\lambda\cdot((\mathbf{p}_i + \int_{t_i}^{t_f} \mathbf{e}_\eta d\eta - \Sigma_{l=1}^{n}\mathbf{q}_l)\gamma - \mathbf{p}_f)) \quad \text{(II.8)}$$

Upon integration over λ this will yield a momentum conserving delta function. Thus we can replace the first argument in (II.7abc) by

$$\mathbf{p}_f - \gamma\int_{\tau_r}^{t_f} \mathbf{e}_\eta d\eta + \gamma\Sigma_{l=1}^{j-1}\mathbf{q}_l \quad \text{(II.9a)}$$

$$\mathbf{p}_f - \gamma\int_{t_j}^{t_f} \mathbf{e}_\eta d\eta + \gamma\Sigma_{l=1}^{j}\mathbf{q}_l \quad \text{(II.9b)}$$

$$\mathbf{p}_f - \gamma\int^{t_f} \mathbf{e}_\eta d\eta + \gamma\Sigma_{l=1}^{j-1}\mathbf{q}_l \quad \text{(II.9c)}$$

Inserting (II.9) into (II.7), the latter into (II.6), inserting this result into (II.1) and noting the obvious recursion, we obtain

$$f(\mathbf{p}_f,t_f) = f(\mathbf{p}_f,t_f,t_f) \quad \text{(II.10a)}$$

$$f(\mathbf{p}_f,t_f,t'_f) = U^*(\mathbf{p}_f - \gamma\int^{t_f} \mathbf{e}_\eta d\eta;t'_f,t_i)\, U(\mathbf{p}_f - \gamma\int^{t_f} \mathbf{e}_\eta d\eta;t_f,t_i)$$

$$\cdot f(\mathbf{p}_f - \gamma\int_{t_i}^{t_f} \mathbf{e}_\eta d\eta,t_i) + \Sigma_{\mathbf{q}_1} \int_{t_i}^{t_f} dt_1 \int_{t_i}^{t'_f} dt'_1\, U^*(\mathbf{p}_f - \gamma\int^{t_f} \mathbf{e}_\eta d\eta;t'_f,t'_1)$$

$$U(\mathbf{p}_f - \gamma\int^{t_f} \mathbf{e}_\eta d\eta;t_f,t_1)\, W_{\mathbf{q}_1}(\mathbf{p}_f - \gamma\int_{t_1}^{t_f} \mathbf{e}_\eta d\eta + \gamma\mathbf{q}_1;t_1,t'_1)$$

$$f(\mathbf{p}_f - \gamma\int_{t_1}^{t_f} \mathbf{e}_\eta d\eta + \gamma\mathbf{q}_1;t_1,t'_1) \qquad \text{(II.10b)}$$

for our generalized Boltzmann result.

I believe that (II.10) is about as general as one can be and still retain the flavor of the Boltzmann equation. If in (II.10b) we let $t_f = t'_f$ and $t_1 = t'_1$, the latter *except* in W, and U^*U be a quasi-particle propagator, relaxation mode, then we obtain (4.5b), with the important electric field and influence-functional modification of the scattering rates and propagators. Setting $\gamma = 1$ and S^o to the free particle case yields (4.5b) itself.

If on the other hand, we evaluate $\partial f(\mathbf{p}_f,t_f)/\partial t_f$ from (II.10) using

$$dU^*(\mathbf{p}_f - \gamma\int^{t_f} \mathbf{e}_\eta d\eta;t'_f,t_1)/dt'_f =$$

$$-\Sigma_{\mathbf{k}} \int_{t_1}^{t'_f} d\tau\, U^*(\mathbf{p}_f - \gamma\int^{t_f} \mathbf{e}_\eta d\eta - \gamma\mathbf{k};t'_f,\tau) \qquad \text{(II.11)}$$

$$W^*_{\mathbf{k}}(\mathbf{p}_f - \gamma\int_{t'_f}^{t_f} \mathbf{e}_\eta d\eta;t'_f,\tau)\, U^*(\mathbf{p}_f - \gamma\int^{t_f} \mathbf{e}_\eta d\eta;\ \tau,t_1)$$

then we obtain

$$\partial f(\mathbf{p}_f,t_f)/\partial t_f + \gamma \mathbf{e}_{t_f}\cdot \partial f(\mathbf{p}_f,t_f)/\partial \mathbf{p}_f =$$

$$- \Sigma_{\mathbf{k}} \int_{t_i}^{t_f} d\tau\, U^*(\mathbf{p}_f - \gamma \int_{\tau}^{t_f} \mathbf{e}_\eta d\eta - \gamma \mathbf{k}; t_f, \tau)\, W_{\mathbf{k}}(\mathbf{p}_f; t_f, \tau)\, U(\mathbf{p}_f - \gamma \int^{t_f} \mathbf{e}_\eta d\eta; t_f, \tau)$$

$$f(\mathbf{p}_f - \gamma \int_{\tau}^{t_f} \mathbf{e}_\eta d\eta, \tau)$$

$$- \Sigma_{\mathbf{k}} \int_{t_i}^{t_f} d\tau\, U(\mathbf{p}_f - \gamma \int_{\tau}^{t_f} \mathbf{e}_\eta d_\eta - \gamma \mathbf{k}; t_f, \tau)\, W^*_{\mathbf{k}}(\mathbf{p}_f; t_f, \tau)\, U^*(\mathbf{p}_f - \gamma \int^{t_f} \mathbf{e}_\eta d\eta; t_f, \tau)$$

$$f(\mathbf{p}_f - \gamma \int_{\tau}^{t_f} \mathbf{e}_\eta d\eta, \tau)$$

$$+ \Sigma_{\mathbf{k}} \int_{t_i}^{t_f} d\tau\, W^*_{\mathbf{k}}(\mathbf{p}_f + \gamma \mathbf{k}; t_f, \tau)\, U(\mathbf{p}_f - \gamma \int^{t_f} \mathbf{e}_\eta d\eta; t_f, \tau)$$

$$U^*(\mathbf{p}_f - \gamma \int^{t_f} \mathbf{e}_\eta d\eta + \gamma \mathbf{k}; t_f, \tau) f(\mathbf{p}_f - \gamma \int_{\tau}^{t_f} \mathbf{e}_\eta d\eta + \gamma \mathbf{k}, \tau)$$

$$+ \Sigma_{\mathbf{k}} \int_{t_i}^{t_f} d\tau\, W_{\mathbf{k}}(\mathbf{p}_f + \gamma \mathbf{k}; t_f, \tau)\, U^*(\mathbf{p}_f - \gamma \int^{t_f} \mathbf{e}_\eta d\eta; t_f, \tau)$$

$$U(\mathbf{p}_f - \gamma \int^{t_f} \mathbf{e}_\eta d\eta + \gamma \mathbf{k}; t_f, \tau) f(\mathbf{p}_f - \gamma \int_{\tau}^{t_f} \mathbf{e}_\eta d\eta + \gamma \mathbf{k}, \tau) \quad \text{(II.12)}$$

This is the generalization of the Boltzmann equation for finite electric field.

Expression (II.12) reduces to the nearly-free-electron result in the presence of an arbitrary electric field simply by setting $U = U^* = 1$, $\gamma = 1$; it reduces to a result known for some time[27,28] in the presence of a weak electric field of arbitrary frequency; it reduces to the standard Boltzmann equation in the limit of weak and very low frequency electric field. In these two latter cases, the electric field $\mathbf{e}_t$ must be sufficiently small that its alteration of the scattering W is negligible. In general the field-dependent term in W gives rise to an energy broadening in the energy conserving "delta function" of the order of $\Delta E \approx \hbar(\mathbf{k}\cdot\mathbf{F}/m)^{1/2}$, where $\mathbf{k}$ is the momentum transfer during the collision. While normally small, it reaches .1 eV in Al_2O_3 near threshold.[12,13]

We conclude that the validity of the Boltzmann equation (4.5a) is restricted to weak-scattering in small, slowly-varying fields. Should any of these conditions be violated, the equation, if used at all, should be generalized in a nontrivial manner; altering the effective mass is insufficient.

While in the foregoing we have focused on the Boltzmann equation, it should be clear that by letting $f(\mathbf{p}_i,t_i) = \delta(\mathbf{p}_i - \mathbf{p}_o)$ we could derive the corresponding master equation for $P(\mathbf{p}_f,t_f;\mathbf{p}_o,t_i)$ the probability that $\mathbf{p} = \mathbf{p}_f$ at t_f given $\mathbf{p} = \mathbf{p}_o$ at t_i.

References

1. R. P. Feynman, Ph.D. Thesis, Princeton University (1942), unpublished.
2. R. P. Feynman, Rev. Mod. Phys. *20*, 367 (1948).
3. R. P. Feynman, Phys. Rev. *84*, 108 (1951).
4. R. P. Feynman, F. L. Vernon, Jr., Ann. Phys. (N.Y.) *24*, 118 (1963).
5. R. P. Feynman, A. R. Hibbs, *Quantum Mechanics and Path Integrals*, New York: McGraw-Hill (1965).
6. R. P. Feynman, *Statistical Mechanics*, Reading, Mass.: W. A. Benjamin (1972).
7. R. P. Feynman, Phys. Rev. *97*, 660 (1955).
8. R. P. Feynman, R. W. Hellwarth, C. K. Iddings, P. M. Platzman, Phys. Rev. *127*, 1004 (1962). FHIP.
9. P. M. Platzman, Phys. Rev. *125*, 1961 (1962).
10. R. W. Hellwarth, P. M. Platzman, Phys. Rev. *128*, 1599 (1962).
11. P. M. Platzman, in *Polarons and Excitons*, C. G. Kuper, G. D. Whitfield, eds., New York: Plenum, 1963.
12. K. K. Thornber, Ph.D. Thesis, Part II, California Institute of Technology (1966), unpublished.
13. K. K. Thornber, R. P. Feynman, Phys. Rev. *B1*, 4099 (1970), *B4*, 674E (1971).
14. K. K. Thornber, Phys. Rev. *B3*, 1929 (1971), *B4*, 675E (1971).
15. K. K. Thornber, in *Polarons in Ionic Crystals and Polar Semiconductors*, J. T. Devreese, ed., Amsterdam: North-Holland (1972).
16. K. K. Thornber, Phys. Rev. *B9*, 3489 (1974).
17. K. K. Thornber, in *Linear and Nonlinear Electron Transport in Solids*, J. T. Devreese, V. E. van Doren, eds., New York: Plenum (1976).
18. J. T. Devreese, J. deSitter, M. Goovarts, Phys. Rev. *B5*, 2367 (1972).
19. L. F. Lemmens, J. T. Devreese, Solid-State Commun. *12*, 1067 (1973).
20. L. F. Lemmens, J. de Sitter, J. T. Devreese, Phys. Rev. *B8*, 2717 (1973).
21. H. B. Callen, T. A. Welton, Phys. Rev. *83*, 34 (1951).
22. K. K. Thornber, in preparation.
23. R. Courant, D. Hilbert, *Methods of Mathematical Physics II*, New York: Interscience (1962) Ch. 2.
24. J. T. Devreese, R. Evrard, in Ref. 17.
25. L. van Hove, Physica *21*, 517 (1955).
26. L. van Hove, Physica *23*, 441 (1957).
27. W. Kohn, J. M. Luttinger, Phys. Rev. *108*, 590 (1957).
28. Price, IBM J. Research and Devel. *10*, 395 (1966).

FUNCTIONAL INTEGRAL APPROACH TO SOME MODELS OF SOLID STATE PHYSICS

Bernhard Mühlschlegel

Institut für Theoretische Physik

Universität zu Köln, 5000 Köln 41, Germany

Summary

The method of Gaussian functional averages is applied to the local Anderson model of a magnetic impurity in a metal and to the extended Ising model of ferromagnetism in greater detail. Approximations for these systems are used also in the functional treatment of normal and superconducting many-body problems, and of multi-side models of Coulomb-correlated electrons in bands.

1. Introduction

Consider a system which consists of several or many subsystems (individuals). The behavior of the system shall completely be characterized by direct interactions between individuals. We may eliminate this direct interaction and, instead, couple all individuals to an external influence. The result will, in general, be a poor realization of the original rich behavioral structure. Only when we introduce many different external influences which all are permitted to couple with the individuals may we expect to find in summa again the true original behavior. Though from a social point of view the decoupling of individuals appears to be quite unhuman, for physical many-body systems the behavior of the whole system and not the fate of the individuals is of interest. Therefore, the replacement of direct interactions between bodies by an external influence (auxiliary fields, or fluctuating fields) should bring some advantages which shall be described on the following pages for a number of selected systems.

We are led to functional integrals of a special type which should be distinguished from Feynman path integrals [1]. Feynman integrals are over classical paths of the actual system and contain its dynamics ab initio. Functional averages considered below are over continuous auxiliary variables which have nothing to do with the classical behavior of the system. They are solely introduced to replace the direct interaction between particles. Sometimes both functional integrals are used simultaneously [2], but we will not consider such cases here.

The method of functional averages over auxiliary variables stems from quantum field theory [3]. The application to problems of statistical physics was influenced by short notes of Stratonovich and of Hubbard [4], but was especially promoted by many contributions, beginning with the late 1950's, of Edwards in England and Siegert in the USA, and their students. Their influence may not be seen so much in the following presentation but it is certainly contained in some of the works of our reference list [5].

The procedure of interaction elimination is discussed in the next section for the most simple system of an isolated one-orbital "atom". It becomes non-trivial when brought into contact with its surroundings in section 3. This is the functional integral for the partition function of the Anderson model of a local magnetic moment. At the beginning of section 3 the general properties of Gaussian functional averages are explained. These are needed further in a somewhat different context in section 4, which deals with the Ising model of ferromagnetism. At the end of this section contact is made with the Landau functional integral which is one of the starting points of the renormalization-group theory of critical phenomena [6]. In the final section 5 we look in a more cursory manner at normal and superconducting many-body problems. The special case of small metallic particles, seen as zero-dimensional superconductors, leads us back to a local problem which finds a comparatively simple solution.

2. Coulomb repulsion between localized electrons Model.

The method of Gaussian averages can be explained best in applying it to a very simple example. This is the atomic one-orbital model. It deals with electrons in narrow orbitals of an atom or ion, for instance in the five 3d-orbitals of transition-metal atoms. The simplification consists of neglecting orbital degeneracy, that is, instead of five orbitals we consider only one "d-orbital" with energy ε_d. When two electrons of opposite spin are present in this narrow orbital they will feel an appreciable Coulomb repulsion $U > 0$ which cannot be neglected. The electronic energy is therefore described by the Hamiltonian

$$H_d = \varepsilon_d(n_\uparrow + n_\downarrow) + U n_\uparrow n_\downarrow \equiv \sum_s \varepsilon_d n_s + \frac{1}{2}\sum_s U n_s n_{-s} \quad (1)$$

where :

$$n_s = d_s^+ d_s ; \quad s = +1 = \uparrow \quad \text{or} \quad s = -1 = \downarrow \quad (2)$$

is the occupation number with values zero or one, and d_s^+ (d_s) creates (annihilates) an electron with spin s in the orbital.

H_d has four states : one with no electron present and energy $\varepsilon_o = 0$, two degenerate ones with one electron and energy $\varepsilon_1 = \varepsilon_d$, and one with two electrons present and energy $\varepsilon_2 = 2\,\varepsilon_d + U$. The partition function is therefore :

$$Z = \text{Trace}\ e^{-\beta H_d} = 1 + 2\,e^{-\beta\varepsilon_1} + e^{-\beta\varepsilon_2} . \quad (3)$$

When a magnetic field B is applied, the spin degeneracy is lifted by the Zeeman energy, and ε_d has to be replaced by $\varepsilon_{ds} = \varepsilon_d + \frac{1}{2} g\mu_B B_s$ in (1). The magnetic behavior is determined by the actual values of ε_d and U. Clearly, when ε_1 is larger than ε_o or ε_2 the atom is non magnetic, and its spin susceptibility would decrease exponentially with $T \to 0$. However, when $\varepsilon_1 = \varepsilon_d < \varepsilon_o = 0$, $\varepsilon_1 = \varepsilon_d < \varepsilon_2 = 2\,\varepsilon_d + U$, that is,

$$\varepsilon_d < 0, \quad \varepsilon_d + U > 0 , \quad (4)$$

only one electron is present in the ground state, and the atom is magnetic, with a Curie law for the susceptibility.

In spite of the apparent simplicity of the model, one has to realize that it describes an interacting system. Adding an electron needs an energy which depends on whether there is already another electron present, or not. In the magnetic case, for instance, (4) tells us that it is favorable to have only one electron in the orbital since the addition of a second one needs positive energy $\varepsilon_d + U$. The two-particle interaction is reflected also in the two-pole structure of the one-particle Green function

$$g_s(i\omega) = \frac{1 - \langle n_{-s} \rangle}{i\omega - \varepsilon_d} + \frac{\langle n_{-s} \rangle}{i\omega - (\varepsilon_d + U)} \tag{5}$$

where $\langle n_s \rangle$ is the thermal average of (2) with H_d.

Because the model is exactly solvable approximation schemes can be tested. Hartree-Fock, for intance, replace H_d by :

$$H_d^{HF} = (\varepsilon_d + U \langle n_\downarrow \rangle^{HF}) n_\uparrow + (\varepsilon_d + U \langle n_\uparrow \rangle^{HF}) n_\downarrow \tag{6}$$

whereby the analytical form of (5) is drastically changed into a one-pole structure reflecting the one-particle approximation :

$$g_s^{HF}(i\omega) = \frac{1}{i\omega - (\varepsilon_d + U \langle n_{-s} \rangle^{HF})} \quad . \tag{7}$$

Further, it is a quite non-trivial exercise to start with $U = 0$, $g^{(o)}(i\omega) = (i\omega - \varepsilon_d)^{-1}$, and to obtain (3), (5) by perturbation theory of infinite order with respect to U. The most compact relationship with an independent-particle system is however achieved by the introduction of an auxiliary field by means of a Gaussian average which we are going to discuss now.

Method. Consider first an operator A which appears squared in an exponential. Then the identity holds

$$e^{-A^2} = \int_{-\infty}^{\infty} dy \; e^{-\pi y^2 - 2\sqrt{\pi}\, i y A} \quad . \tag{8}$$

It tells us that by means of a Gaussian average a square in the exponential gets linearized. The application of a c-number formula to an operator function is justified by the spectral representation. (8) can easily be generalized to the case of a product of two operators A and B which commute with each other : AB = BA. We use then :

$$AB = \left(\frac{A + B}{2}\right)^2 - \left(\frac{A - B}{2}\right)^2 \tag{9}$$

and apply the identity of the type (8) twice to both squares in (9) by introducing a second Gaussian variable x :

$$e^{-AB} = \int_{-\infty}^{\infty} dx\, dy\, e^{-\pi(x^2+y^2)}\, e^{-\sqrt{\pi}\, x(A-B)\, -\sqrt{\pi}\, iy(A+B)}, \tag{10a}$$

with z = x + iy, dz = dx dy this takes the form :

$$e^{-AB} = \int_{-\infty}^{\infty} dz\, e^{-\pi\, |z|^2}\, e^{-\sqrt{\pi}\, Az\, +\, \sqrt{\pi}\, Bz^*}$$
$$= \langle e^{-\sqrt{\pi}\,(Az\, -\, Bz^*)} \rangle_{Gauss} \tag{10b}$$

The linearization of a product AB is achieved by a two-dimensional Gaussian average over z where A - B is coupled with Rez, and A + B with Imz, respectively.

The application of these simple manipulations on the atomic one-orbital model is immediate. In $\exp(-\beta H_d)$ we identify $\sqrt{\beta U}\, n_\uparrow$ with A and $\sqrt{\beta U}\, n_\downarrow$ with B, and obtain by the use of (10), and by performing the trace

$$Z = \langle \text{Trace } e^{-\beta\varepsilon_d(n_\uparrow + n_\downarrow) - \sqrt{\pi\beta U}\,(n_\uparrow z - n_\downarrow z^*)} \rangle_{Gauss}$$
$$= \langle Z(z) \rangle_{Gauss} \tag{11}$$

$$Z(z) = \text{Trace } e^{-\beta \sum_s E_s n_s} = (1 + e^{-\beta E_\uparrow})(1 + e^{-\beta E_\downarrow}), \tag{12}$$

$$E_\uparrow = \varepsilon_d + \sqrt{\frac{\pi U}{\beta}}\, z, \qquad E_\downarrow = \varepsilon_d - \sqrt{\frac{\pi U}{\beta}}\, z^* . \tag{13}$$

The partition function Z of interacting electrons is expressed by a Gaussian average over the partition function Z(z) which describes localized spin-up and spin-down electrons moving independently in an external field determined by the Gaussian variable z = x + iy. Writing (13) as :

$$E_s = \varepsilon_d + \sqrt{\frac{\pi U}{\beta}}\,(sx + iy) \tag{14}$$

we recognize that Rez = x (which couples with $n_\uparrow - n_\downarrow$) corresponds to a magnetic field whereas Imz = y (which couples with $n_\uparrow + n_\downarrow$) acts proportional to an electric potential.

The one-orbital model shows clearly the essence of the Gauss-average procedure : The direct two-body interaction (here Coulomb repulsion) is eliminated, and an auxiliary field z comes into play in which the particles move independently. The problem of dealing with an interaction is shifted to the problem of performing a Gaussian average over the auxiliary field. Due to the simplicity of the model these problems of course do not exist in the present case since the average in (11) is easily done and leads back to eq.(3).

Self-consistent field. It is worthwhile to consider the Hartree-Fock approximation from the viewpoint of auxiliary fields. We write (11) as :

$$Z = \int dz\, e^{-\beta F(z,\beta)}\,,$$
$$\beta F(z,\beta) = \pi |z|^2 - \log Z(z) \tag{15}$$

and obtain with (13) :

$$\frac{\partial \beta F}{\partial z^*} = \pi z - \sqrt{\pi\beta U}\,[\exp(\beta E_\downarrow) + 1]^{-1}\;;$$
$$\frac{\partial \beta F}{\partial z} = \pi z^* + \sqrt{\pi\beta U}\,[\exp(\beta E_\uparrow) + 1]^{-1}\;. \tag{16}$$

Putting the derivatives equal to zero determines the stationary point $\bar{z}$ of the integrand in (15). We insert this $\bar{z}$ into (13) and find the stationary-condition being equivalent to

$$E_s \to E_s^{HF} = \varepsilon_d + U < n_{-s} >^{HF}\;;$$
$$< n_{-s} >^{HF} = [\exp(\beta E_{-s}^{HF}) + 1]^{-1} \tag{17}$$

Thus the Hartree-Fock approximation selects one special value $\bar{z}(\beta)$ of the auxiliary field for which $F(z,\beta)$ is stationary (condition of self-consistency). The integral (15) over all values of z is approximated by the stationary value of its integrand

$Z^{HF} = \exp[-\beta F(\bar{z},\beta)]$, and $F(\bar{z}(\beta),\beta)$ is just the Hartree-Fock free energy $F^{HF}(\beta)$.

Vice versa, we may view the Gaussian average as a natural mathematical generalization of the approximative self-consistent mean field which gets replaced by a whole distribution of fields z. In other words : When we include all fluctuations of the mean field, both nearby and far-away from its self-consistent value, we will obtain the exact result.

Some additional properties should be collected. First, we note that the representation of Z by a Gaussian integral given in eq.(11), (15) is independent of the statistics of the particles. That we deal with fermions is used only afterwards in (12). Second, the fermion property $n_s^2 = n_s$ can, however, be applied at the beginning with the result:

$$H_d = (\varepsilon_d + \frac{U}{2})(n_\uparrow + n_\downarrow) - \frac{U}{2}(n_\uparrow - n_\downarrow)^2$$

$$= (\varepsilon_d - \frac{U}{2})(n_\uparrow + n_\downarrow) + \frac{U}{2}(n_\uparrow + n_\downarrow)^2. \qquad (18)$$

Here, only one square appears in H_d. Correspondingly, the above two-variable scheme x,y is replaced by a one-variable scheme with E_s of eq.(14) changed into

$$E_s = \varepsilon_d + \frac{U}{2} + \sqrt{\frac{2\pi U}{\beta}}\, sx \qquad \text{for } (n_\uparrow - n_\downarrow)^2, \text{ and}$$

$$E = \varepsilon_d - \frac{U}{2} + \quad i\sqrt{\frac{2\pi U}{\beta}}\, y \quad \text{for } (n_\uparrow + n_\downarrow)^2 \;,$$

respectively. Mathematically, these auxiliary fields give different representations of the same Z. But we emphasize that solely the original two-variable scheme is related to the Hartree-Fock approximation. Third, it is a general feature that Z(z) under the Gaussian integral in eqs.(11), (15) has a lower symmetry than Z. For the present model H_d possesses spin-rotation invariance which is not the case for $\Sigma\, E_s n_s$ entering Z(z). The symmetry is, of course, restored by the z-integration.

3. Functional integral for the Anderson model

Structure of models. Two-body forces can be eliminated by Gaussian averages in every interacting system. We will not list here all the models of solid state and statistical physics which have been treated this way. Rather we shall consider three groups of different structure and mention representative systems for each group.

Group I : In the statistical operator exp [- βH] the exponent is :

$$\beta H = \beta H_o + \beta AB \tag{1}$$

which has the same form as the model of the previous section (β being absorbed by A,B). The important difference is, however, that now A and B (which commute with each other) do not commute with H_o. We will see that this is responsible for the functional average. A representative system is the Anderson model of a magnetic impurity in a metal which is discussed mainly in this section. We note that also superconductivity in its most simplified version fits into the structure of (1).

Group II collects systems with

$$H = H_o + \sum_i A_i B_i \tag{2}$$

where all parts commute with each other. This is an obvious generalization of (2.10) since every term in the sum gets its Gaussian variable z_i, and the average (2.10) becomes a multi-dimensional integral. Both the Ising model and the classical gas belong to this group.

Group III is actually the combination of I and II. The systems have the form of eq.(2) but different parts do not commute with each other. As a consequence one faces a multi-dimensional functional average. All kinds of quantum-many-body systems are in this category. In addition, we mention the Heisenberg model of localized spins, and the multi-site Anderson and Hubbard model.

<u>Functional average</u>. Turning to group I, we start with the well-known formula :

$$e^{-\beta(H_o + AB)} = e^{-\beta H_o} T\, e^{-\int_0^\beta d\tau\, A(\tau)\, B(\tau)},$$

$$A(\tau) = e^{\tau H_o} A\, e^{-\tau H_o} \tag{3}$$

where the time-ordering symbol T arranges all operators in the series expansion of exp $[-\int d\tau\, AB]$ from left to right in order of decreasing τ in the interval $(\beta, 0)$. Therefore, before T does the work, we are allowed to apply the c-number formula (2.10) on every factor in :

$$E = e^{-\int_0^\beta d\tau\, A(\tau)\, B(\tau)} = \lim_{\substack{\Delta\tau\to 0\\ N\to\infty}} e^{-\sum_i^N \Delta\tau\, A(\tau_i) B(\tau_i)} \tag{4a}$$

with the result :

$$E = \lim_{\substack{\Delta\tau\to 0\\ N\to\infty}} \int \prod_i^N \left(\frac{\Delta\tau}{\beta} dz_i\right) e^{-\sum_i \Delta\tau\left[\frac{\pi}{\beta}|z_i|^2 + \sqrt{\frac{\pi}{\beta}}\, A(\tau_i) z_i - \sqrt{\frac{\pi}{\beta}}\, B(\tau_i) z_i^*\right]}$$

$$= \int Dz(\tau)\; e^{-\frac{\pi}{\beta}\int_0^\beta d\tau |z(\tau)|^2}\; e^{-\sqrt{\frac{\pi}{\beta}}\int_0^\beta d\tau\left[A(\tau) z(\tau) - B(\tau) z^*(\tau)\right]} \tag{4b}$$

The limit $\Delta\tau \to 0$, $N \to \infty$ defines here the functional integral with Gaussian measure. The measure function is :

$$Dz(\tau) = \lim_{\substack{\Delta\tau\to 0\\ N\to\infty}} \prod_i^N \left(\frac{\Delta\tau}{\beta}\, dz_i\right) . \tag{5}$$

The time-ordering operation T can be interchanged with the functional integration. Consequently, (3) becomes

$$e^{-\beta(H_o + AB)} = \int Dz\; e^{-\frac{\pi}{\beta}\int_0^\beta d\tau |z|^2}\; e^{-\beta H_o}\, T\, e^{-\sqrt{\frac{\pi}{\beta}}\int_0^\beta d\tau\left[A(\tau) z - B(\tau) z^*\right]}$$

$$= \left\langle e^{-\beta H_o}\, T\, e^{-\sqrt{\frac{\pi}{\beta}}\int_0^\beta d\tau\left[A(\tau)\, z(\tau) - B(\tau)\, z^*(\tau)\right]} \right\rangle_{\text{Gauss}} \tag{6}$$

This formula shows how the linearization of AB is achieved when the operators do not commute with H_o. There is an (dimensionless) auxiliary field $z(\tau) \overset{\circ}{=} x(\tau) + iy(\tau)$, dependent on time $\beta \geq \tau \geq 0$, which replaces the variable z of section 2. The normal Gauss average becomes an average over all functions

$z(\tau)$. From a formal view point it is worth noting that both exponentials in (6) can be recombined analogously to (3) :

$$e^{-\beta(H_o+AB)} = \langle T\, e^{-\int_0^\beta d\tau \{H_{o\tau} + \sqrt{\frac{\pi}{\beta}} [A_\tau z(\tau) - B_\tau z^*(\tau)]\}} \rangle_{Gauss} \qquad (7)$$

The subscript τ on the operators (Feynman ordering label) indicates that these are subject to ordering. Eq.(7) is sometimes called the "Stratonovich-Hubbard identity" since it was applied to quantum many-body problems by these authors in 1957-59 [4]. The former case of commuting operators is, of course, contained in (6), (7) because ordering is then superfluous and $\sqrt{\beta}$ A couples solely with :

$$z_o = \frac{1}{\beta} \int_0^\beta d\tau\ z(\tau) \ , \qquad (8)$$

whereas $\sqrt{\beta}$ B couples with z_o^*; the functional average shrinks to an average over the Gaussian variable z_o.

<u>Anderson model</u>. The above formulas will find an immediate application when we plant the one-orbital atom of section 2 as an impurity into a normal metal which is described by its extended conduction-band states

$$H_{band} = \sum_{ks} \varepsilon_k\ c^+_{ks}\ c_{ks} \qquad (9)$$

With H_d given by (2.1), $H_d + H_{band}$ would be an impurity isolated from the host metal. It is quite natural to assume that there is mixing by a matrix element V which allows for hopping of electrons from the impurity orbital to the extended states and vice versa

$$H_{Anderson} = H_d + W + H_{band} \ , \qquad (10a)$$

$$W = \sum_{ks} (V_k\ c^+_{ks}\ d_s + V^*_k\ d^+_s\ c_{ks}) \ . \qquad (10b)$$

We recognize that in the decomposition :

$$H_{Anderson} = H_{U=0} + Un_\uparrow n_\downarrow$$

$$H_{U=0} = \varepsilon_d\ (n_\uparrow + n_\downarrow) + W + H_{band} \qquad (11)$$

H_o does not commute with the Coulomb repulsion due to the mixing energy W. Therefore (7) rather than (2.11) has to be used for the partition function

$$Z = \text{Trace}\ e^{-\beta H_{\text{Anderson}}} = \langle Z[z] \rangle_{\text{Gauss}} ,$$

$$Z[z] = \text{Trace}\ T\ e^{-\int_o d\tau \{\sum_s E_s(\tau)\, n_{s\tau} + W_\tau + H_{\text{band}\tau}\}} \tag{12}$$

with :

$$E_s(\tau) = \varepsilon_d + \sqrt{\frac{\pi U}{\beta}}\,(sx(\tau) + iy(\tau)) . \tag{13}$$

<u>Particlesin a time-dependent field</u>. According to the above discussion, the physical meaning of Z[z] is quite clear. It describes free fermions moving in a time-dependent field contained in the d-electron energy $E_s(\tau)$. Since both spin directions are decoupled, we suppress for a moment the subscript s.

$$Z[z] = \text{Trace}\ T\ e^{-\int_o^\beta d\tau\, H_\tau} ,$$

$$H_\tau = E(\tau)\, n_\tau + W_\tau + H_{\text{band}\tau} . \tag{14}$$

Such a problem is solved formally by the use of one-particle Green functions :

$$G(\tau,\tau') = - \begin{cases} \langle d(\tau)\, d^+(\tau') \rangle & \tau > \tau' \\ - \langle d^+(\tau')\, d(\tau) & \tau < \tau' \end{cases} \tag{15}$$

where average and time dependence are made with the ordered exponential of eq.(14). The free (W = 0) Green function for the d-electron becomes :

$$G^{(o)}(\tau,\tau') = \begin{Bmatrix} -(1-f) \\ f \end{Bmatrix} e^{-\int_{\tau'}^{\tau} E(t)dt} ;$$

$$f = [\exp(\int_o^\beta E(t)\, dt) + 1]^{-1} . \tag{16}$$

Expressions similar to eq.(15) can be written down for

$< c_k(\tau)\ c_k^+(\tau') >$, $< c_k(\tau)\ d^+(\tau') >$. But due to the simple form of H_T in (14) the k-d Green function can be eliminated from the equations of motion. The mere effect of the mixing W upon the d-electron is that it introduces the d-electron self energy

$$\Sigma\ (\tau-\tau') = \sum_k |V_k|^2\ G_k(\tau-\tau') \tag{17}$$

where $G_k(\tau-\tau')$ is the propagator of band-electrons, obtained from (16) by $E(t) \to \varepsilon_k$. G of eq.(15) is then related to $G^{(o)}$ by the Dyson integral equation

$$G = G^{(o)} + G^{(o)}\ \Sigma G\ , \quad G = G^{(o)} + G\ \Sigma\ G^{(o)}\ . \tag{18}$$

In order to obtain the explicit form of the partition function (14), it is useful to do the Fourier transformation of the Green function. Here one has to remember that $E = E(\tau)$ breaks time-translation invariance and leads to two boundary conditions $G(0,\tau') = -G(\beta,\tau')$, $G(\tau,0) = -G(\tau,\beta)$. Therefore :

$$G(\tau,\tau') = \frac{1}{\beta} \sum_{nn'} G_{nn'}\ e^{-i\omega_n\tau + i\omega_{n'}\tau'} ,$$

$$\omega_n = \frac{\pi n}{\beta}\ ,\ n \text{ odd.} \tag{19}$$

G is a matrix in Fourier space reflecting the fact that energy $E(\tau)$ changes during propagation. The self energy (17) due to hopping back and forth to band states does not change and remains diagonal

$$\Sigma_{nn'} = \delta_{nn'}\ \Sigma(i\omega_n);\quad \Sigma\ (i\omega_n) = \sum_k \frac{|V_k|^2}{i\omega_n - \varepsilon_k}\ . \tag{20}$$

By means of the equations of motion one finds for the partition function (including spin $s = \pm 1$ again) :

$$Z\ [z] = Z_{band} \prod_s \left(1 + e^{-\int_0^\beta E_s(\tau)d\tau}\right) e^{\sum_n \{\log(1-G_s^{(o)}\ \Sigma\ \}_{nn}} \tag{21}$$

We see that Z[z] factorizes into the extensive band-electron system and the localized d-electrons (an intensive system) :

$$Z[z] = Z_{band} \cdot Z_d[z] \quad , \quad Z_{band} = \prod_{ks} (1 + e^{-\beta\varepsilon_k}) . \tag{22}$$

In $Z_d[z]$ the effect of band electrons is to produce Σ via the mixing. The special form of $Z_d[z]$ in (21) is also plausible from a diagrammatic viewpoint. The expansion :

$$\{\log (1 - G^{(o)} \Sigma)\}_{nn} = \sum_{p=1}^{\infty} \frac{-1}{p} \{(G^{(o)} \Sigma)^p\}_{nn} \tag{23}$$

represents the familiar loop expansion of the partition function which for the simple problem considered here reduces to a sum over ring diagrams.

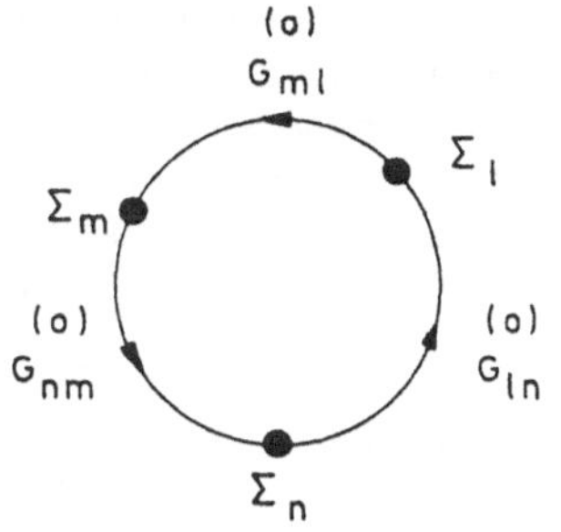

Fig.1 : Ring diagram of order p = 3 .

Whereas (21) is an expansion with respect to the mixing V contained in Σ, and the free Green function $G^{(o)}$ is made with $E(\tau)$ which depends on the auxiliary field $z(\tau)$ (13), there exists a somewhat different representation of $Z[z]$ which corresponds to an expansion in U :

$$Z[z] = Z_{U=0} \prod_s e^{\sum_n \{\log (1 - g^{(o)} v_s)\}_{nn}} \tag{24}$$

Here $Z_{U=0}$ is the partition function of the Hamiltonian (11), and $g^{(o)}$ is the corresponding d-electron Green function with the Fourier transform :

$$G^{(o)}_{nn'} = \delta_{nn'} \, g^{(o)}_n \, , \quad g^{(o)}_n = \frac{1}{i\omega_n - \varepsilon_d - \Sigma(i\omega_n)} \tag{25}$$

The energy v_s is essentially identical with the auxiliary field :

$$v_s(\tau) = E_s(\tau) - \varepsilon_d \rightarrow v_\uparrow(\tau) = \sqrt{\frac{\pi U}{\beta}} \, z(\tau) \, ,$$

$$v_\downarrow(\tau) = - \sqrt{\frac{\pi U}{\beta}} \, z^*(\tau) \, . \tag{26}$$

It follows in Fourier space :

$$z(\tau) = \sum_{\nu=-\infty}^{\infty} z_\nu \, e^{-i\frac{2\pi\nu}{\beta}\tau} \tag{27a}$$

$$(v_\uparrow)_{nn'} = \delta_{n,n'+2\nu}\sqrt{\frac{\pi U}{\beta}}\, z_\nu ,$$
$$(v_\downarrow)_{nn'} = -\,\delta_{n,n'+2\nu}\sqrt{\frac{\pi U}{\beta}}\, z_\nu^* . \tag{27b}$$

The graphical illustration of the terms in (24) gives us the same structure as above, but with a different meaning: the d-electrons propagate according to (25) with constant energy and get scattered by the time-dependent auxiliary field, thereby changing their energy (fig.2).

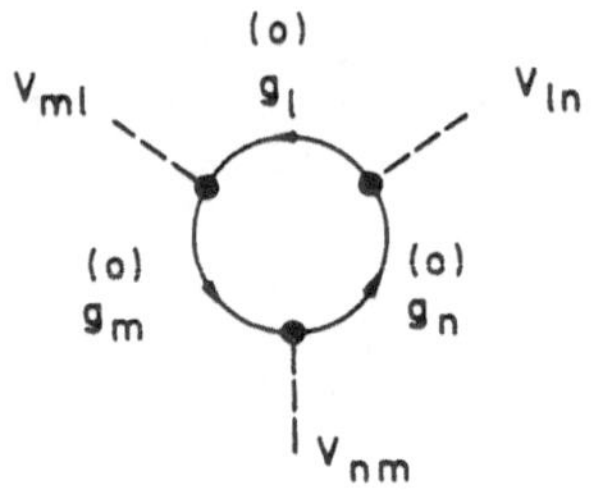

Fig.2 : Ring diagram of order p = 3.

Z_{band} factorizes out also in this representation since

$$Z_{U=0} = Z_{band}\, e^{-\sum_{ns} \log g_n^{(o)} \cdot e^{i\omega_n\tau}|_{\tau=0^+}} \tag{28}$$

Therefore, both forms (21) and (24) are at hand for the d-electron motion $Z_d[z]$ and may be used for approximations.

Static approximation. Anderson, in 1961, had studied his model in Hartree-Fock approximation which we already have written down in eq.(2.17). The only difference compared with (2.6) is that g^{HF} now includes the mixing W which changes g^{HF} of (2.7) into

$$g_s^{HF}(i\omega) = \frac{1}{i\omega - (\varepsilon_d + U < n_{-s} >^{HF}) - \Sigma(i\omega)} . \tag{29}$$

This then determines the self-consistency condition for $< n_s >^{HF}$. The solution (for $\beta^{-1} = k_B T = 0$) is in the textbooks [7,8] and shall not be discussed here. It gives a description of the local magnetic moment $< n_{\uparrow} >^{HF} - < n_{\downarrow} >^{HF}$ and its change by varying the model parameter ε_d, U, V. A fundamental problem of magnetism is certainly touched here since the explanation of the formation of local moments is a necessary step towards a true understanding of metallic magnetism.

In the functional integral for the partition function (12)

$$Z = \int Dz(\tau)\; e^{-\beta F[z]}$$

$$\beta F[z] = \frac{\pi}{\beta} \int_0^{\beta} d\tau\, |z(\tau)|^2 - \log Z[z] \qquad (30)$$

the H.F.-approximation, as was explained in section 2, corresponds to taking the time-independent stationary value $\bar{z}$ of F[z] and picking up the integrand solely at $\bar{z}$:

$$Z^{HF} = e^{-\beta F[\bar{z}(\beta)]} \quad . \qquad (31)$$

Clearly, from a mathematical point of view there is no justification for such a procedure, and we expect fluctuations of $z(\tau)$ around $\bar{z}$ to be of importance. The static approximation which was applied to the Anderson model in 1965, accounts at least for the time-independent fluctuations [5]. It replaces in (30) $z(\tau)$ by the variable z_0, thereby converting the functional integral into a normal (two-dimensional) integral :

$$Z^{stat} = \int dz_0\; e^{-\beta F(z_0)} \;,$$

$$\beta F(z_0) = \pi |z_0|^2 - \log Z^{stat}(z_0) \qquad (32)$$

Notice that $Z \rightarrow Z^{stat}$ is not a selection of a certain class of diagrams. The static approximation includes all diagrams. But we neglect in (27a) all Fourier components z_ν with $\nu \neq 0$; therefore $E_s(\tau)$ is kept constant, and no energy change takes place in both types of ring diagrams.

The further evaluation of Z^{stat} uses a technical simplification of the self energy Σ which implicitely was applied in Anderson's H.F.-approximation, too

$$\Sigma(i\omega_n) = -\,i\,\Gamma\ \mathrm{sign}\ \omega_n, \quad \Gamma = \pi\,\overline{|V|^2}\,N(o) \qquad (33)$$

This replacement is justified when the metal electrons have a broad band, and the matrix element V_k is so smooth that it can be taken at its value $\overline{V}$ at the Fermi energy where the band states have the density of states N(o). Γ is the width of d-electron energy ε_d due to mixing with the band states.

The two Gaussian integrations, $z_o = x_o + iy_o$, in (32) can be carried out by the use of Euler's beta function with the result [9]

$$Z^{stat} = Z_{band}\, Z_d^{stat}\ ,$$

$$Z_d^{stat} = (2\pi)^2\,[(\delta-1)!]^{-2} \int_o^1 dn_\uparrow dn_\downarrow e^{-\beta f(n_\uparrow, n_\downarrow)} \qquad (34a)$$

$$f(n_\uparrow, n_\downarrow) =$$

$$\varepsilon_d(n_\uparrow + n_\downarrow) + U n_\uparrow n_\downarrow - \frac{1}{\beta}(\delta-1)\log[4\sin(\pi n_\uparrow)\sin(\pi n_\downarrow)] \qquad (34b)$$

$$\delta = \frac{\beta\Gamma}{\pi}\ . \qquad (34c)$$

Note that the $n_\uparrow$, $n_\downarrow$ are continuous variables between 0 and 1, whereas before the same symbols were used for operators. Anderson's H.F.-solution follows from $\partial f/\partial n_\uparrow = \partial f/\partial n_\downarrow = 0$ and contains his magnetic state (two minima of f with $n_\uparrow \neq n_\downarrow$) for the proper choice of the model parameter. In contrast to Hartree-Fock, the static approximation preserves spin-up-down symmetry. Moreover, it becomes exact in the limiting cases $\eta \to 0, \infty$ of the relevant parameter

$$\eta = \frac{U}{\pi\Gamma} \qquad (35)$$

and gives the right first-order corrections to these limits. The impurity susceptibility $\chi_d = \beta\partial^2 \log Z_d/\partial B^2$ has a temperature dependence which shows a smooth transition from a Curie law to a temperature-independent Pauli susceptibility with decreasing η (Compare fig.3).

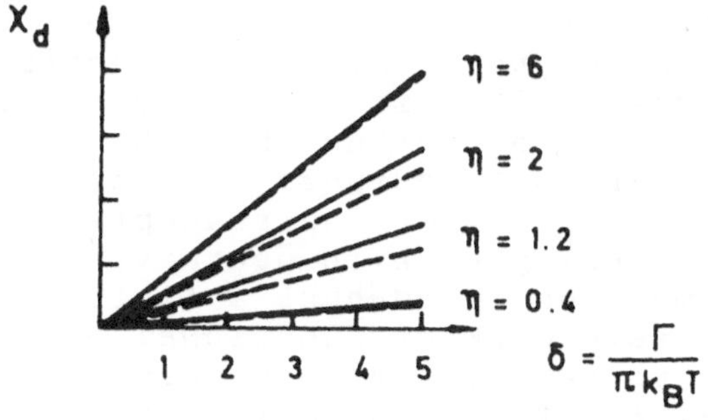

Fig.3 : Impurity susceptibility versus $\delta = \beta\Gamma/\pi$ for different values of $\eta = U/\pi\Gamma$. Solid curves represent the static approximation, dashed curves are for the harmonic approximation [9].

The functional integral (12) (30) suggests an extension of the static approximation which takes into account some dynamical fluctuations by including the Fourier components z_1 and z_{-1} in addition to the static z_o :

$$z(\tau) \to z_o + z_1 e^{-i\frac{2\pi}{\beta}\tau} + z_{-1} e^{+i\frac{2\pi}{\beta}\tau}. \qquad (36)$$

This "harmonic approximation" was studied in 1973 by Amit and Keiter [9]. They have calculated from the finite-dimensional integral the susceptibility. The influence of dynamical fluctuations is strongest for $\eta \approx 1$ and gives a diminished effective magnetic moment when compared with χ_d in static approximation (fig.3).

Further developments. Other work on the functional integral of the Anderson model has been described in an article by Hamann and Schrieffer [10]. We mention that the Kondo model (an impurity spin is coupled anti-ferro-magnetically to the band-electron spin density) is related to the magnetic Anderson model and should even take over its essential physical properties for large Coulomb repulsion U, that is for $\eta \gg 1$ (Schrieffer-Wolff transformation). Hamann has treated the functional integral for $\eta \gg 1$ by further approximations and finds results which also follow from the Kondo model. A diagrammatic analysis of $Z = \langle Z[z] \rangle_{Gauss}$ on the basis of eq.(24) with diagrams of the type of fig.2 has been made by Keiter [11].

A very interesting situation is created by the recent results about the impurity susceptibility of the Anderson model [12] which have been obtained using not functional-integral methods but the numerical renormalization-group techniques developed by Wilson [13] for the Kondo problem. The susceptibility, which develops Curie-like, goes with decreasing temperature through a maximum and approaches a finite value for $T \to 0$. In the frame work of functional averages, such a behavior can certainly not be understood by adding a few dynamical fluctuations to the static z_0. It remains a challenging question to find a somewhat more analytical explanation - with or without the functional integral.

4. Gaussian averages and the Ising model

Spins in a space-dependent field. Consider N spins $s_i = \pm 1$, $i = 1...N >> 1$ in an external magnetic field $b = (1/2) g \mu_B B$ which interact pairwise via the exchange energy $-I_{ij} < 0$. The energy of the system is then governed by the familiar Ising Hamiltonian of ferromagnetism

$$H = -b \sum_i s_i - \sum_{i<j} I_{ij} s_i s_j \tag{1}$$

It belongs to group II discussed in the previous section since all parts commute. Applying eq.(2.10) to every pair in (1) with the identification $A = \sqrt{\beta I_{ij}}\, s_i$, $B = -\sqrt{\beta I_{ij}}\, s_j$ gives :

$$e^{-\beta H} = e^{\beta b \sum_i s_i} \prod_{i<j} \int dz_{ij}\, e^{-\pi |z_{ij}|^2} e^{\sqrt{\pi \beta I_{ij}} (z_{ij} s_i + z^*_{ij} s_j)} \tag{2}$$

We define z_{ji} by $z_{ji} = z_{ij} = x_{ij} + i y_{ij}$ and obtain with $I_{ji} = I_{ij}$, $I_{ii} = 0$ the partition function as an $N(N-1)/2$-dimensional Gaussian average

$$Z = \mathrm{Trace}\, e^{-\beta H} = \int \prod_{i<j} dx_{ij}\, e^{-\pi \sum_{i<j} x_{ij}^2} Z(x) \tag{3a}$$

$$Z(x) = \text{Trace}\; e^{\sum_i [\beta b + \sum_j \sqrt{\pi\beta I_{ij}}\, x_{ij}]\, s_i}$$

$$= \prod_i 2 \cosh [\beta b + \sum_j \sqrt{\pi\beta I_{ij}}\, x_{ij}] \ . \qquad (3b)$$

As to be expected (3b) describes free spins in the magnetic field $b + b_i$ where

$$b_i = \sum_j \sqrt{\frac{\pi\ I_{ij}}{\beta}}\, x_{ij} \qquad (4)$$

is caused by the auxiliary variables x_{ij} attached to the bonds which connect the i-th spin with all the other ones.

So far we have assumed nothing about the arrangement of the spins and the form of I_{ij}. Regular or irregular distributed, the N spins could be arranged in any space dimension. It might be favorable to work with N(N-1/2 auxiliary variables x_{ij} when the explicite I_{ij}-dependence is of importance (for example in random systems where an average over a distribution $P(I_{ij})$ of exchange interactions is used which, in an approximative treatment, could be combined with the above Gaussian average). Usually, however, one goes over from N(N-1)/2 bond variables to N site variables by introducing the matrix $\underline{\underline{K}}$ by

$$\underline{\underline{K}}^2 = \underline{\underline{I}} \qquad (5)$$

We then take formula (2.8) for an N-dimensional vector $\underline{A} = i\sqrt{\beta/2}\ \underline{\underline{K}}\ \underline{s}$ and find the N-dimensional average

$$Z = \int d^N x\ e^{-\pi \sum_i x_i^2}\ Z(x) \ ,$$

$$Z(x) = \prod_i 2 \cosh [\beta b + \sum_j \sqrt{2\ \pi\beta}\ K_{ij}\ x_j] \ . \qquad (6)$$

The fluctuating field is now $b_i = \sqrt{2\ \pi/\beta}\ \sum_j K_{ij}\ x_j$.

Some conclusions can only be drawn in the following when the system is translational invariant. Therefore, when necessary, we shall assume that the spins occupy N lattice sites r_i with periodic boundary conditions. Fourier transformation gives :

$$I_{ij} = \frac{1}{N} \sum_p I(p)\, e^{ip(r_i - r_j)}, \quad s_i = \frac{1}{\sqrt{N}} \sum_p s(p)\, e^{ipr_i}, \qquad (7a)$$

$$H = -\sqrt{N}\, b\, s(o) - \frac{1}{2} I(o)\, s(o)^2 - \frac{1}{2} \sum_{p \neq 0} I(p)\, s(p)\, s(-p), \qquad (7b)$$

$$\sum_p s(p)\, s(-p) \equiv \sum_p |s(p)|^2 = \sum_i s_i^2 = N \quad . \qquad (7c)$$

p runs over the N vectors in the reciprocal lattice cell.

In the thermodynamic limit $N \to \infty$ Z is represented by an infinite-dimensional Gaussian integral which actually can be called a functional integral. Remember that the Anderson-functional integral of the last section can equally well be written as an infinite-dimensional integral by using the Fourier components z_ν of $z(\tau)$ (3.27) as integration variables. The difference is that Anderson deals with a small (intensive) system whereas here we have extensivity and a vanishing small parameter N^{-1}.

Infinite range. We apply the idea of the static approximation which, in the present case, obviously means : Neglect all Fourier components $p \neq 0$ in

$$x_i = \frac{1}{\sqrt{N}} \sum_{\bar{p}} x(p)\, e^{ipr_i} \quad , \qquad (8)$$

and take the same auxiliary field x(o) for all spins. With the substitution $x_o = [N\beta I(o)/2\pi]^{1/2} x$ it follows for (6) :

$$Z^{stat} = [N\beta I(o)/2\pi]^{1/2} \int dx\, e^{-Ng(x)} \quad ,$$
$$g(x) = \frac{1}{2} \beta I(o)\, x^2 - \log 2 \cosh \beta(b + I(o)x) . \qquad (9)$$

Using N^{-1} as expansion parameter the steepest-descents calculation gives the free energy

$$\beta F = -\log Z^{stat} = Ng(a) + \frac{1}{2} \log \frac{g''(a)}{\beta I(o)} + O(N^{-1}) \qquad (10)$$

where a is determined by the condition :

$$g'(a) = 0 \to a = \tanh \beta(b + I(o)\, a). \qquad (11)$$

Evaluating g"(a) we obtain for (10) :

$$\frac{F}{N} = \frac{F^{mf}}{N} + \frac{1}{2N\beta} \log [1 - \beta I(o)(1-a^2)] + O(N^{-2}) ,$$

$$\frac{F^{mf}}{N} = g(a) . \tag{12}$$

The magnetization becomes :

$$m = -\frac{1}{N}\frac{\partial F}{\partial b} = a + O(N^{-1}) . \tag{13}$$

We learn two things from these simple calculations. First, in contrast to the local system of Section 3 the static (spatially-constant) fluctuations are here suppressed due to $N >> 1$, and the stationary-value of the integrand in (9) is already the correct result for the integral (The N^{-1}-term in (12) blows up, however, for $b \to 0$, at the Curie temperature, $\beta_c I(o) = 1$, indicating the instability of the expansion). Second, the above molecular-field approximation is no approximation but becomes exact for infinite long range interactions. Clearly, in this case all spins interact with each other, and in (7a) only the zeroth component

$$I_{ij} = \frac{1}{N} I(o) > 0 \tag{14}$$

will survive. Consequently, Z^{stat} represents the full partition function.

It is obvious that translational invariance is not at all important for the latter result. We can obtain it equally well from (1) with the constant interaction (14). Therefore, molecular field is always exact for infinite range, independent of the arrangement of spins. We have here a system of the "Kac type". (Marc Kac obtained first the equivalent result for the classical gas [14]). Another system of the Kac type is the original BCS-model of superconductivity where the molecular field is replaced by the Bogoliubov pair field. Further, it will probably be discussed in the lectures by David Sherrington that the Ising-spin glass in replica form belongs to the same category.

Finite range, small fluctuations around the molecular field. The integral (6) has to be seen in full dimensionality when the range of exchange interactions I_{ij} is finite (reflecting physical reality). Indeed, when the spins are on a lattice, one usually considers only short range, e.g. nearest-neighbour coupling. The idea of searching for the stationary-value of the integrand in

$$Z = \int d^N x \; e^{-G(x)} ,$$

$$G(x) = \pi \sum_i x_i^2 - \sum_i \log 2 \cosh \left[\beta b + \sum_j \sqrt{2\pi\beta}\, K_{ij} x_j \right] \tag{15}$$

can be applied as before :

$$\frac{\partial G}{\partial x_i} = 2\pi\, x_i - \sum_\ell \sqrt{2\pi\beta}\, K_{\ell i} \tanh \left[\beta b + \sum_j \sqrt{2\pi\beta}\, K_{\ell j}\, x_j \right] \tag{16}$$

One realizes that, with translational invariance, the stationary point with all x_i being equal and given by $\bar{x}$:

$$\left. \frac{\partial G}{\partial x_i} \right|_{\bar{x}} = 2\pi\, \bar{x} - \sqrt{2\pi\beta I(o)} \tanh \left[\beta b + \sqrt{2\pi\beta I(o)}\; \bar{x} \right] = 0 \tag{17}$$

just reproduces the molecular-field condition (11) $a = \tanh \beta(b + I(o)a)$ where

$$\bar{x} = \sqrt{\beta I(o)/2\pi}\; a , \qquad \sum_j I_{ij} = I(o) . \tag{18}$$

As to be expected (already from our discussion in Section 2), the stationary-condition gives the molecular-field free energy $G(\bar{x}) = \beta F^{mf}$. One may, in addition, expand $G(x)$ around the stationary point $\bar{x}$:

$$Z = e^{-G(\bar{x})} \int d^N x \; e^{-\frac{1}{2} G''(\bar{x})(x-\bar{x})^2} = \frac{e^{-G(\bar{x})}}{\sqrt{G''(\bar{x})/2\pi}} . \tag{19}$$

Note that here a short-form notation is used, $G''(\bar{x})/2\pi$ actually stands for $\mathrm{Det}\,\{(1/2\pi)\, \partial^2 G/\partial x_i \partial x_j\}$. We assume in (19) that only small fluctuations of the auxiliary field around the fixed self-consistent value are important. Mathematically, there is nothing better since we are unable to integrate exponentials like $\exp -x^n$, $n > 2$. It follows from (16), (17) and (19)

$$\left. \frac{1}{2\pi} \frac{\partial^2 G}{\partial x_i\, \partial x_j} \right|_{\bar{x}} = \delta_{ij} - \beta I_{ij}\, (1-a^2) , \tag{20}$$

$$- \log Z = \beta F^{mf} + \sum_p \frac{1}{2} \log \left[1 - \beta I(p)(1-a^2) \right] . \tag{21}$$

For $I(p) = 0$, $p \neq 0$ one is back to (12), but in the

general case the sum gives an extensive contribution (It corresponds to the finite-temperature random-phase approximation, RPA, for general many-body systems). (21) is a good approximation for high enough temperatures but breaks down for $b \to 0$ at the mean-field Curie temperature, $\beta_c I(o) = 1$, $a \to 0$. The magnetization

$$m = \frac{1}{N\beta}\frac{\partial \log Z}{\partial b} = a - \frac{1}{N}\sum_p \frac{I(p)}{1-\beta I(p)(1-a^2)} \cdot a \frac{\partial a}{\partial b} \qquad (22)$$

blows up in zero field for $\beta \to \beta_c$ because of $\partial a/\partial b$. Notice also that the magnetization differs from the stationary-value a.

<u>Variation principle : Stationary-condition for both magnetization and correlation</u>. We describe here an approximation which is characterized by making the free energy itself stationary rather than expanding it at stationary points of G as was the case before [15]. This variation principle is similar to the one used by Feynman for the polaron problem [1]. Consider a trial function G_t with $G = G_t + (G - G_t)$ in eq.(15) and expand the exponential up to first order in $(G - G_t)$

$$\beta F = -\log Z \to \beta F_t = $$

$$= -\log \int d^N x\, e^{-G_t} + \frac{\int d^N x\, e^{-G_t}(G-G_t)}{\int d^N x\, e^{-G_t}} \qquad (23)$$

A set of variation parameters contained in G_t is determined by requiring $F_t = 0$. When both G and G_t are real, F_t is always an upper limit of the free energy. Obviously, the only trial function one can handle in a multi-dimensional integral is a general quadratic form

$$G_t = \sum_{ij} C_{ij}\,(x_i - \alpha_i)(x_j - \alpha_j) \qquad (24)$$

where the coefficients C_{ij} and the displacements α_i play the role of variation parameters. When these are determined by $\delta F_t = 0$, the "best" quadratic form has been found to approximate the exact G in a global manner.

For the Ising model with translational invariance the variational calculation can be worked out in detail. Before presenting the result a slight modification is useful. Remember the trivial sum rule

(7c) which expresses spin-number conservation. It makes the Fourier components S(p) dependent on one another. Since an approximation of the type (24) decouples different momenta p it is advisable to take care of spin-number conservation from the beginning by adding

$$\mu.(N - \sum_i s_i^2) \equiv 0 \tag{25}$$

to the Hamiltonian (1). The effect of this is the additional term μN to the free energy, and

$$I_{ij} \rightarrow v_{ij} = I_{ij} + 2\mu\,\delta_{ij}, \text{ Eigenvalues of } \underline{\underline{K}} : \quad \sqrt{v(p)} = \sqrt{I(p) + 2\mu} \quad . \tag{26}$$

It is then convenient to write the trial function (24) in the form :

$$G_t = \pi \sum_{ij} (\underline{\underline{S}}^{-2})_{ij} \,(x_i - \sqrt{\beta v(0)/2\pi}\; a)(x_j - \sqrt{\beta v(0)/2\pi}\; a) \tag{27}$$

where we assume the matrix S to be real, symmetric and translational invariant. Its eigenvalues S(p), the displacement a, and the Lagrange factor μ form the N + 2 variation parameter of the problem. Due to the comparatively simple structure of G one can perform the integration in (23) with the result for the trial free energy

$$\beta F_t = N\beta\mu + \sum_p [\tfrac{1}{2}\, S(p)^2 - \log S(p)] - \frac{N}{2} + \frac{N}{2}\,\beta v(0)\, a^2 - N\,\overline{\log 2 \cosh \beta\,[\,b + v(0)a]} \quad . \tag{28}$$

Here the average bar indicates a one-dimensional Gaussian integration defined by :

$$\overline{f(u)} = \int_{-\infty}^{\infty} dx\; e^{-\pi x^2}\; f(u + \sqrt{\frac{2\pi}{N} \sum_p \beta v(p)\; S(p)^2}\, x) . \tag{29}$$

The variational conditions can be summarized as follows :

$$\frac{\partial F_t}{\partial a} = 0 \;:\; a = \overline{\tanh \beta[\,b + v(0)a]} \quad (= -\frac{1}{N}\frac{\partial F_t}{\partial b} \equiv m) \tag{30a}$$

$$\frac{\partial F_t}{\partial S(p)} = 0 : \frac{1}{S(p)^2} = 1 - \beta v(p)\{1 - \overline{\tanh^2\beta[b + v(o)m]}\}, \tag{30b}$$

$$\frac{\partial F_t}{\partial \mu} = 0 : 1 - m^2 = \{1 - \overline{\tanh^2\beta[b+v(o)m]}\}.\frac{1}{N}\sum_p S(p)^2 . \tag{30c}$$

These coupled equations show that (in contrast to the previous approximation) the free energy is stationary with respect to the magnetization m = a (30a), and the spin number is conserved (30c). Further, by the use of $< S_i S_j > = - 2\ \partial F/\partial I_{ij}$ one obtains the relation :

$$< S_i S_j > - < S_i >< S_j > = (1-m^2)\frac{(S^2)_{ij}}{\frac{1}{N}\ \mathrm{Spur}\ \underline{\underline{S}}^2} \tag{31}$$

which brings the matrix S in close contact with the spin-correlation function. Therefore, eq.(30b) means physically that the free energy is stationary with respect to variations of the correlation function. This is clearly an important step beyond the molecular-field approach which solely optimizes the parameter a, i.e. the magnetization.

By combination of the eqs.(30) we find the Curie temperature $K_B T_c = \beta_c^{-1}$ putting b = 0 and m → 0

$$\beta_c = \frac{1}{N}\sum_p \frac{1}{I(o) - I(p)} \ . \tag{32}$$

T_c is identical with the Curie temperature of the socalled spherical model although F_t is not identical with the free energy of that model. $T_c = 0.66\ T_c^{mf}$ for nearest-neighbor interaction in the simple-cubic lattice.

A rather interesting feature of the variational treatment of the Ising model is connected with the observation that the self-consistency equations (30) are compatible with the condition

$$\sum_p \beta v(p)S(p)^2 = 0 \tag{33}$$

This implies that the average (29) becomes superfluous, and (40) reduces to (keep in mind $v(p) = I(p) + 2\mu$)

$$m = \tanh \beta[b + v(o)m] ,$$

$$1 = \frac{1}{N} \sum_p S(p)^2 , \quad \frac{1}{S(p)^2} = 1 - \beta v(p)(1 - m^2) . \qquad (34)$$

The corresponding free energy

$$\beta F_{Brout} = N\beta\mu + N [\frac{1}{2} \beta v(o)m^2 - \log 2 \cosh[\beta b + v(o)m]] + \sum_p \frac{1}{2} \log [1 - \beta v(p)(1 - m^2)] \qquad (35)$$

was first obtained by Brout [16], and is similar in form to (21), but with the important difference that (35) is stationary with respect to m and conserves the spin number.

Discussion, Landau functional integral. Both approximations for the partition function made above are of the Gaussian type and therefore include Gaussian fluctuations. In eq.(19) the quadratic form of second derivatives in the exponent is pinned on the molecular-field value $\bar{x}$ which (for magnetic field $b > 0$) is the absolute minimum of $G(x)$ (eq.(15)) in the N-dimensional space of the auxiliary variables x_i. The variational approximation (23), (24) contains more flexibility. It does not emphasize the curvature of G at its minimum $\bar{x}$ but offers G a general quadratic form G_t. The minimum position (magnetization) of G_t and its curvature (spin-correlation function) are determined by claiming that the free energy of the system should be stationary, leading to the coupled self-consistency equations (30).

The results discussed here within the frame of functional averages have been obtained in the early sixties also by means of rather elaborate diagrammatic methods by Brout and by Horwitz and Callen [16]. Compared with these techniques the Gaussian approximations are of great simplicity and allow for an easy physical interpretation. Concerning the variational principle for the functional integral, one must realize, however, that its succesful application to the Ising model rests upon the comparatively simple structure of the function G(x) which makes possible the explicite calculation of the second integral in eq.(23).

By means of the variational method, the Curie temperature is shifted from the "bare" molecular-field value T_c^{mf} to T_c of eq.(32) (or to some other value if one does not introduce the parameter μ which takes care of spin-number conservation). Approaching this new Curie temperature the lowest eigenvalue S(o) of the matrix S tends to infinity reflecting the long-range order of correlations at the phase transition. By the use of (34), one obtains for the spin-correlation function near T_c

$$< S_i S_j > - m^2 = \frac{\text{Const}}{r} e^{-r/\xi} ,$$

$$r = |r_i - r_j|, \ \xi \sim |T - T_c|^{-1/2} . \qquad (36)$$

The typical Ornstein-Zernike behavior is to be expected due to the Gaussian approximation.

To summarize the merits and limitations of the variation principle applied to the Ising model of ferromagnetism : It is an extension of the molecular-field approach in that it includes spin fluctuations which, together with the magnetization, are self-consistently determined. This gives a good approximation at high and at low temperatures. A value for the Curie temperature T_c is obtained (appreciably lower than the molecular-field value, but, of course, not comparable in precision with T_c-calculations by high-temperature expansions). In the critical region around T_c, the phenomenological Landau theory of second-order phase transitions is derived from the microscopic model.

Clearly, the great success of the theory of critical phenomena in recent years was to go beyond Landau by means of the renormalization-group approach. A starting point of this approach is the Landau functional integral of ϕ^4-theory. Let us show how it can be obtained from formula (6). Without loss of generality, we assume the interaction matrix I to be positive definite (which is achieved by adding a sufficiently large positive constant to all diagonal elements). New auxiliary variables ϕ_i are introduced by the linear transformation :

$$\underline{\phi} = \beta b + \sqrt{2\pi\beta}\ \underline{\underline{K}}\ \underline{x} \qquad (37)$$

which brings the partition function of the Ising model into the form :

$$Z = (2\pi\beta)^{-N/2} (\text{Det}\, I)^{-1/2} \int \phi d^N \, e^{-[\frac{1}{2\beta}\underline{\phi}\,\underline{\underline{I}}^{-1}\underline{\phi} - \underline{\phi}\,\underline{\underline{I}}^{-1}\underline{b} - \sum_i \log 2\cosh\phi_i]} \qquad (38)$$

The b^2-term has been dropped since the magnetic field is weak. We make three simplifications : 1) Expansion of the log-term : $\log\cosh\phi_i = 1/2\ \phi_i^2 - 1/12\ \phi_i^4$, 2) Transition from the discrete lattice with summations over i to a continuum with integration r over the lattice volume, lengths being dimensionless in units of lattice spacing, and 3) Short-range exchange interaction $I(p)^{-1} = I(o)^{-1}[1 + R^2p^2]$ of range R. Since p acts as $-i\nabla$ in space, it then follows with $\beta_o = I(o)^{-1}$ for the exponent in (38) :

$$[\ldots] = \int dr\ (\frac{1}{2}\ \frac{\beta_o}{\beta}\ \phi^2(r) + \frac{1}{2}\ \frac{\beta_o R^2}{\beta}\ (\nabla\phi(r)^2) -$$

$$- \beta_o b\ \phi(r) - \frac{1}{2}\ \phi^2(r) + \frac{1}{12}\ \phi^4(r)) , \qquad (39)$$

and (38) becomes the Landau functional integral

$$Z = A \int D\phi(r)\ e^{-\int H(r)dr} ,$$

$$H(r) = \frac{a}{2}\ \phi^2(r) + \frac{c}{2}\ (\nabla\phi(r))^2 + \frac{b}{4}\ \phi^4(r) - h\phi(r) \qquad (40)$$

with $h = \beta_o b$ being a dimensionless magnetic field.

We mention that (40) can be extracted also from the functional-integral representation of other many-body systems with phase transition, e.g. the van der Waals gas [2]. An interesting early study of (40) in connection with metastable states is due to Langer [2]. The connection with the Ising model was discovered repeatedly. It seems that the first derivation was given in 1967 by Zittartz [17] who applied (40) for space-dependent field h(r), studying the interfacial structure of an Ising-ferromagnet below T_o by means of the kink solution of the inhomogeneous "Ginzburg-Landau" equation $-c\nabla^2\phi + a\phi + b\phi^3 = h$.

The presentation of the general renormalization-group methods applied to the Landau functional integral is beyond our scope, and we refer to the Wilson-Kogut review [18], and especially to volume 6 (1976) of Domb and Green [6] as the most recent source of very detailed information.

5. Further applications of functional averages

In this final section we wish to sketch approximations, similar to the ones discussed above, to some other many-body systems. We consider here mainly fermions. Interacting bosons are treated by Wiegel in his review [2] and in his lectures.

Normal extended many-particle systems. The Hamiltonian in momentum space is :

$$H = H_o + H_1 = \sum_{ps} \varepsilon(p)\, c^+_{ps}\, c_{ps} + \frac{1}{2\,\Omega} \sum_q v(q)\, \rho^+_q\, \rho_q \qquad (1)$$

The chemical potential μ is included in $\varepsilon(p) = p^2/2m - \mu$, Ω is the volume, and $v(q)$, ρ_q are the Fourier transforms of the two-body interaction and of the particle-density operator, respectively. The system obviously belongs to group III of section 3 with the identification in (3.2)

$$i \to q,\; A_q = \sqrt{\frac{v(q)}{2\Omega}}\, \rho^+_q \;,\; B_q = \sqrt{\frac{v(q)}{2\Omega}}\, \rho_q \qquad (2)$$

and a multi-dimensional representation of the grand-canonical partition function according to (3.6), (3.7):

$$Z = \langle Z[z] \rangle_{Gauss} = \int \prod_q Dz_q(\tau)\, e^{-G[z]}, \qquad (3a)$$

$$G[z] = \frac{\pi}{\beta} \sum_q \int_0^\beta |z_q(\tau)|^2\, d\tau - \log Z[z] \;. \qquad (3b)$$

Zittartz [19] has made a thorough investigation of (3) for normal systems. An expansion at the stationary point $\bar{z}$ of G in complete analogy to eq.(4.19) gives with $G[\bar{z}]$ the Hartree approximation, and with the quadratic term the temperature-dependent RPA-contribution to the thermodynamic potential. A "best-quadratic form" approach for G by means of the variation principle (4.23) leads to a thermodynamic potential stationary with respect to density and two-particle correlation function. In terms of diagrams, this corresponds to a generalization of the well-known RPA-bubble diagram to a situation where v(q)-lines are inserted between particle propagators in a general manner. Physically, RPA becomes renormalized, and the variational condition corresponds to a self-consistent determination of the generalized dielectric function.

All these ideas apply also to the classical gas (without hard core). For the Coulomb case $v(q) = 4\pi e^2/q^2$, an elegant derivation of Debye-Hückel theory is a by-product.

Superconductivity. We do not enter into the famous history of this problem here but rather write down the Hamiltonian without any further comment

$$H = H_0 + H_1 = \sum_{ps} \varepsilon(p)\, c^+_{ps}\, c_{ps} - \frac{1}{\Omega} \sum_q g(q)\, b^+_q\, b_q \qquad (4)$$

where :

$$b^+_q = \sum_p c^+_{p+q\uparrow}\, c^+_{p\downarrow} \qquad (5)$$

is the creation operator of pairs with momentum q. The p-summation in (5) is confined to a shell of thickness $2\,\omega_D$ around the Fermi surface, and $g(q) > 0$ simulates the pair attraction caused by phonons. Two models of superconductivity are contained in (4) :

$$g(q) = V_{BCS}\, \delta_q \qquad \text{BCS model}\ , \qquad (6a)$$

$$g(q) = V_G \qquad \text{Gorkov model}\ . \qquad (6b)$$

The original BCS model considers only the $q = 0$ term in (4) leading to a long-range sub-extensive interaction between zero-momentum pairs. The corresponding functional integral has been studied a long time ago [20]. With the identification

$$A = \sqrt{\frac{V_{BCS}}{\Omega}}\, b^+_0\ , \qquad B = -\sqrt{\frac{V_{BCS}}{\Omega}}\, b_0 \qquad (7)$$

the BCS model belongs to group I of section 3 (A and B commute practically with each other for large volume Ω). As to be expected, the model is of the "Kac type" : The asymptotic expansion of the functional integral in the thermodynamic limit produces the BCS free energy with the energy gap Ψ as order parameter, proving that the self-consistent pair-field approach is exact.

The extension of the functional-integral treatment to the Gorkov model has been undertaken by several authors [21, 19, 22, 23]. We are here in group III of section 3 with operators A_q, B_q of the form (7) for every pair momentum q. Everts [19] has translated the approximation methods from the previous normal case to the superconducting case. One has now Nambu propagators, and the auxiliary variables $z_q(\tau)$ describe a space-time dependent pair field. The variation principle determines the energy gap and the pair-correlation function self-consistently. An explicite solution of the self-consistency equations as for the Ising model is not possible due to the more complicated structure of $G[z]$.

Pair-field fluctuations near the transition temperature have been studied in dependence of space dimension by Hassing and Wilkins [23]. These authors consider the functional integral within the frame of a Landau truncation, i.e. expansion of $G[z]$ up to fourth order in $z_q(\tau)$. They find that a static approximation $z_q(\tau) \to z_q$ is justified in the transition region.

Small metallic particles. The last observation stimulates us to write down more explicitely the functional integral of the BCS model which in static approximation becomes a two-dimensional integral :

$$Z^{stat} = \int dz\ e^{-\pi|z|^2}\ \mathrm{Trace}\ e^{-\sum_p \beta\varepsilon(p)(n_{p\uparrow}+n_{p\downarrow}) - \sqrt{\frac{\pi\beta V_{BCS}}{\Omega}}(zb_o^+ + z^* b_o)} \tag{8}$$

The trace can be performed, and so can the φ-integration in polar coordinates $z = |z| \exp(i\varphi)$ (restoring particle-number conservation which symmetry is violated by the integrand - keep in mind the remark at the end of section 2). The substitution

$$\Psi = \sqrt{\frac{\pi\ V_{BCS}}{\beta\Omega}}\ |z| \tag{9}$$

introduces the energy gap as order parameter, and (8) becomes :

$$Z^{stat} = Z_o \frac{\beta}{g\delta} \int_0^\infty d\Psi^2\ e^{-\beta f(\Psi^2,\beta)}\ , \tag{10a}$$

$$f(\Psi^2,\beta) = \frac{\Psi^2}{g\delta} - \frac{2}{\beta} \sum_{\substack{\alpha \\ |\varepsilon_\alpha| < \omega_D}} \log \frac{\cosh \beta E_\alpha/2}{\cosh \beta\varepsilon_\alpha/2}\ ,$$

$$E_\alpha = \sqrt{\varepsilon_\alpha^2 + \Psi^2} \ . \tag{10b}$$

We have defined a dimensionless coupling constant g by

$$\frac{V_{BCS}}{\Omega} = g\delta \ , \qquad \delta = \frac{1}{N(o)\ \Omega} \tag{11}$$

which measures the BCS coupling strength in (4) in units of δ, the reciprocal density of states at the Fermi energy. This is the important parameter for finite volume Ω since it is the average level spacing of one-electron states at the Fermi energy. Z_o in (10a) is the partition function of free electrons, and the subscript α in (10b) labels the one-electron states(previously α : p↑).

f is the BCS free-energy, and from

$$\frac{df}{d\Psi} = \frac{2\Psi}{\delta} \left[\frac{1}{g} - \sum_\alpha \delta \frac{1}{E_\alpha} \tanh \beta E_\alpha/2\right] \tag{12}$$

follows the well-known gap equation for the stationary point $\overline{\Psi}(\beta)$ of f. Clearly, in the thermodynamic limit $\Omega \to \infty$ ($\delta \to 0$) with δ^{-1} being the large expansion parameter in the integral (10a), only the minimal f will contribute and the mean-field BCS theory becomes exact, as was pointed out before.

When, however, the volume Ω is small ($\Omega^{1/3}$ << BCS-coherence length ξ), one must consider all static fluctuations of Ψ in (10). In other words, the BCS functional integral, in static approximation given by (10), is a simple starting point for studying size effects in minute superconducting particles [24]. By means of the transition temperature

$$K_B T_c = 1.14\ \omega_D\ e^{-1/g} \tag{13}$$

of the bulk system with coupling g, we introduce the dimensionless parameter

$$\overline{\delta} = \frac{\delta}{K_B T_c} \ . \tag{14}$$

Let C be an observable quantity of a superconductor (specific heat, susceptibility etc.). Then, by the use of (10) $C(T/T_c, \overline{\delta})$ can be obtained, and can be compared with the corresponding bulk behavior $C_{bulk} = C(T/T_c, \overline{\delta} = 0)$. Quite a number of experiments have been performed recently (on probes with many small particles of equal size, varying roughly in the range

30-150 Å), and have supported the simple model which is governed by the size parameter $\overline{\delta}$.

Multi-site models. We have discussed in sections 2 and 3 a local electronic system with Coulomb correlation. Imagine this system multiplied and arranged on N lattice sites r_i with periodic boundary conditions. Two models will emerge this way. The first is

$$H = \sum_i H_{di} + \sum_i W_i + H_{band} \quad . \tag{15}$$

H_{band} is given by (3.9), H_{di} by (2.1) with the subscript i on the operators, and

$$W_i = \sum_s [V\, c^+_{is}\, d_{is} + V^*\, d^+_{is}\, c_{is}] \; ;$$

$$c_{is} = \frac{1}{\sqrt{N}} \sum_k c_{ks}\, e^{ikr_i} \quad . \tag{16}$$

In W we have switched from extended band states k,s to local Wannier states i,s, and c^+_{is} is the corresponding creation operator. The functional integral for the partition function is obtained, in complete analogy to (3.12), (3.13),

$$Z = \text{Trace}\; e^{-\beta H} =$$

$$= \left\langle \text{Trace}\; T\; e^{-\int_0^\beta d\tau \{ \sum_{is} [E_{is}(\tau) n_{is\tau} + W_{i\tau}] + H_{band\tau} \}} \right\rangle_{Gauss} \tag{17}$$

with the Gaussian average being performed over N auxiliary fields $z_i(\tau)$.

We call (15) the periodic Anderson model. It is believed that this model is of some relevance for mixed-valence phenomena in rare-earth intermetallic compounds. Several approximations have been worked out for (15) but it seems that the functional integral (17) has not been studied so far; a Hartree-Fock calculation was made only recently for the half-filled system (2N electrons) [25].

The second model is of simpler structure. We discard the band states and multiply $H_d = \varepsilon_d(n_\uparrow + n_\downarrow) + U n_\uparrow n_\downarrow$ thereby assuming that ε_d on its own splits up to a band of width 2w with Fourier transform t_{ij} and U remains local :

$$H = H_{band} + H_U = \sum_{ij,s} t_{ij}\, d^{+}_{is}\, d_{js} + U \sum_i n_{i\uparrow}\, n_{i\downarrow} \;. \qquad (18)$$

This is the Hubbard model of a single band with local Coulomb correlations. It has attracted many researchers. Usually, the half-filled system is considered (N electrons). For U = 0 the half-filled band describes a metal. When U is increased at $U \gg k_B T$ w one faces a situation similar to N atomic-orbital models of section 2, almost isolated from each other. This is then an insulator. In between, at some critical value $U_c(T)/w$, a metal-insulator (Mott) transition should occur. The phase diagram in the T-U/w plane should, in addition, show transitions from anti-ferromagnetism to paramagnetism for increasing T.

The functional integral of the Hubbard model

$$Z_{Hubbard} = \left\langle \text{Trace}\; T\, e^{-\int_0^\beta d\tau \{\sum_{is} E_{is}(\tau) n_{is\tau} + H_{band_\tau}\}} \right\rangle_{Gauss} \qquad (19)$$

was studied by Kimball and Schrieffer and by others [26]. Without going into the details we mention only some steps of the calculations. Clearly, the one-body problem under the average in (19) can be solved formally quite analogously to the procedure for the one-site model section 3, eq.(3.24). In addition to the static approximation, the imaginary part y of the auxiliary fields which, according to (2.14) couples with $n_\uparrow + n_\downarrow$, is kept fixed. Therefore (19) becomes an N dimensional average over real static auxiliary fields with

$$E_{is}(\tau) \rightarrow E_{is} = \sqrt{\frac{\pi U}{\beta}}\, s x_i \;. \qquad (20)$$

The one-body Hamiltonian resembles a random-alloy where the electron sees a fluctuating potential (20) at the lattice sites. This then is the reason why, as a further step the coherent-potential approximation is used. We should remark that the static approximation appears doubtful when applied to the metal-insulator transition at T → 0 since all dynamical fluctuations are neglected [27]. The Hubbard model which treats Coulomb correlations in bands in the simplest possible form will certainly remain an interesting theoretical research object for quite a while.

Thanks are due to P.Entel and S.Krebs for their kind help in the preparation of the manuscript.

REFERENCES

1. R.P.Feynman and A.R.Hibbs, Quantum mechanics and path integrals, Mc Graw-Hill, New York (1965).
2. See for a review :
   F.W.Wiegel, Path integral methods in statistical mechanics. Phys.Reports 16C, 57 (1975).
3. N.N.Bogoliubov and D.U.Shirkov, Introduction to the theory of quantized fields. Interscience Publ. New York (1959).
4. R.L.Stratonovich, Dokl.Akad.nauk.SSSR 115 (1957). Translated Sov.Phys.Dokl. 2, 416 (1958); J.Hubbard, Phys.Rev.Lett. 3, 77 (1959). A general and more recent contribution which also includes Green functions and linear response is due to D.Sherrington, J.Phys. C4, 401 (1971).
5. For a short review see :
   B.Mühlschlegel, in Functional integration and its applications, ed.A.M.Arthurs, Clarendon Press, Oxford (1975).
6. Phase transitions and critical phenomena (eds. C.Domb and M.S.Green), volume 6, Academic Press, New York (1976).
7. C.Kittel, Quantum theory of solids, Wiley, New York (1963).
8. W.A.Harrison, Solid state theory, Mc Graw-Hill, New York (1970).
9. D.J.Amit and H.Keiter, J.Low Temp.Phys. 11, 603 (1973).
10. D.R.Hamann and J.R.Schrieffer, in Magnetism (ed. H.Suhl), Vol.V, p.237, Academic Press, New York (1973).
11. H.Keiter, Phys.Rev. B2, 3777 (1970).
12. H.R.Krishna-Murthy, K.B.Wilson and J.W.Wilkins, Phys.Rev.Lett. 35, 1101 (1975), and Int.Conf. on Valence Instabilities, Rochester, N.Y. 1976.

13. K.G.Wilson, Rev.Mod.Phys. 47, 773 (1975).

14. M.Kac, Phys.Fluids 2, 8 (1959) - For generalizations see L.W.J.den Ouden, H.W.Capel, J.H.H.Perk and P.A.J.Tindemans, Physica 85A, 51 and 425 (1976).

15. B.Mühlschlegel and J.Zittartz, Z.Physik 175, 533 (1963).

16. See, also for further references, R.Brout, Phase Transitions, Benjamin, New York (1965).

17. J.Zittartz, Phys.Rev. 154, 529 (1967).

18. K.G.Wilson and J.Kogut, The renormalization group and the ε-expansion, Phys.Reports 12, 75 (1974).

19. J.Zittartz, Z.Physik 180, 219 (1964); U.Everts, Z.Physik 199, 211 (1967).

20. B.Mühlschlegel, J.Math.Phys. 3, 522 (1962).

21. J.S.Langer, Phys.Rev. 134A, 553 (1964).

22. T.M.Rice, J.Math.Phys. 8, 1581 (1967).

23. R.F.Hassing and J.W.Wilkins, Phys.Rev. B7, 1890 (1973).

24. B.Mühlschlegel, D.J.Scalapino and R.Denton, Phys. Rev. B6, 1767 (1972).

25. H.J.Leder, Thesis, Universität Köln (1977).

26. See, also for other references : W.Weller, physica stat.sol. (b) 54, 611 (1972); 57, 593 (1973); 60, 783 (1973).

27. J.A.Hertz, Phys.Rev. B14, 1165 (1975).

THE INTERACTING BOSE FLUID: PATH INTEGRAL REPRESENTATIONS AND RENORMALIZATION GROUP APPROACH

F.W. Wiegel

Department of Applied Physics
Twente University of Technology
Enschede, The Netherlands

ABSTRACT

We review the various rigorous representations of the boson partition function as an integral over function space. Two approximation schemes for the evaluation of such path integrals are discussed: the saddle point method and the method of the renormalization group. These methods are applied to the critical behavior of the interacting Bose fluid. The saddle point method leads naturally to a description of the λ transition in terms of quantized vortex lines which create microturbulence in the superfluid. The renormalization group approach leads to the conclusion that the critical behavior of the Bose fluid is identical to that of a classical system of spins with two components.

1. INTRODUCTION

In these notes we review the equilibrium statistical mechanics of the interacting Bose fluid from the point of view of path integration. We shall also use this context to: (1) discuss the two most important calculational techniques for path integrals for their own sake; (2) collect some general remarks which belong to the folklore of "soaring" information on this subject. In those cases in which detailed derivations of results can be found in the literature we only outline the general idea and refer to the literature. If no good reference exists, however, we shall not hesitate to reconstruct the derivation in detail. Path integral methods in statistical mechanics form the subject of a recent review paper[1], and at several occasions the reader is referred to this paper and to the references quoted there.

The system under consideration consists of N spinless bosons, each of mass m, in a volume Ω. The particles interact with a pair interaction $V(\vec{r}-\vec{r}')$ which depends on their distance $|\vec{r}-\vec{r}'|$ only. This system has often been used as a model for ^{4}He. Its Hamiltonian is:

$$H_N = \sum_{i=1}^{N} \frac{\vec{p}_i^{\,2}}{2m} + \sum_{1<i<j<N} V(\vec{r}_i-\vec{r}_j), \tag{1.1}$$

in an obvious notation. The canonical partition function is usually represented by the trace of the density operator:

$$Z(N,\beta,\Omega) = \frac{1}{N!} \mathrm{Tr} \sum_P \exp(-\beta H_N)P. \tag{1.2}$$

Here $\beta = (k_B T)^{-1}$ with k_B denoting Boltzmann's constant and T the absolute temperature. P indicates a permutation operator in the Hilbert space of the N particles and the trace is extended over a complete orthonormal basis of this space. Note that Tr exp $(-\beta H_N)$ would give the partition function of quantum particles with Boltzmann statistics; the additional operator $(N!)^{-1} \sum_P P$ takes care of the Bose statistics.

In the first part of these notes the different C-number formalisms for this quantum system are reviewed. The advantage of an entirely operator free form of quantum statistics is that it enables one to apply approximation methods which were originally devised for classical systems. It turns out that several C-number formalisms exist, one of which seems the most convenient starting point for approximate calculations of the thermodynamic functions. These approximations form the subject of the second and third parts of the notes.

At this point it is appropriate to remark that the coherent state representation will play no role in what follows. For interacting bosons the coherent state representation is no doubt useful, but certain approximations have to be made to reduce the formalism to manageable proportions. For a recent detailed study of the coherent state formalism for interacting bosons the reader is referred to Ohanessian[2].

2. THE FEYNMAN PATH INTEGRAL

This integral expresses the partition function (1.2) as a sum of real contributions from sets of N trajectories $\vec{r}_1(\tau)$, $\vec{r}_2(\tau)$,..$\vec{r}_N(\tau)$, each of which connects the arbitrary initial positions $\vec{r}_1, \vec{r}_2, .. \vec{r}_N$, at reciprocal temperature 0, with final positions $\vec{r}_{P1}, \vec{r}_{P2}, ... \vec{r}_{PN}$, at reciprocal temperature β:

$$Z = \frac{1}{N!} \sum_P \int d^{3N}\vec{r} \prod_{n=1}^{N} \int_{(\vec{r}_n,0)}^{(\vec{r}_{Pn},\beta)} d[\vec{r}_n(\tau)]$$

$$\exp\{-\frac{m}{2\hbar^2} \int_0^\beta \sum_i \left(\frac{d\vec{r}_i}{d\tau}\right)^2 d\tau - \int_0^\beta \sum_{i<j} V(\vec{r}_i(\tau) - \vec{r}_j(\tau))\, d\tau\}. \tag{2.1}$$

The final positions $\vec{r}_{Pn}$ are permutions P of the initial positions $\vec{r}_n$. One has to sum over all trajectories, over all initial configurations and over all permutations P. A derivation of (2.1) from (1.2) can be found in reference 1.

The trajectories of different particles are coupled through the pair interaction term $V(\vec{r}_i(\tau) - \vec{r}_j(\tau))$. Also note that (2.1) is well defined if V has a hard core of diameter a: in that case those parts of the trajectories for which $|\vec{r}_i(\tau) - \vec{r}_j(\tau)| < a$ simply do not contribute to the integral.

This integral is particularly appropriate to study the classical limit of quantum statistics and the semi-classical correction terms: in the limit $(m/2\hbar^2) \to \infty$ the first term in the exponential suppresses the contribution from any set of trajectories, unless P=1 and $\dot{\vec{r}}_i=0$. This immediately gives the standard expression of the classical partition function.

3. COMPLEX GAUSSIAN RANDOM FUNCTIONS

In the future we shall often use random functions $\phi(x)$, where x denotes the pair $(\vec{r},\tau)$. Random functions are called Gaussian (with zero mean) if they have the decomposition property:

$$\langle\phi(x_1)\phi(x_2)\ldots\phi(x_{2\ell+1})\rangle = 0, \qquad (\ell=0,1,2,\ldots) \tag{3.1}$$

$$\langle\phi(x_1)\phi(x_2)\ldots\phi(x_{2\ell})\rangle = \Sigma' \prod_{\alpha=1}^{\ell} \langle\phi(x_{i_\alpha})\phi(x_{j_\alpha})\rangle, \quad (\ell=1,2,3,\ldots) \tag{3.2}$$

where the prime indicates a sum over all the different ways in which the 2ℓ indices $1,2,\ldots 2\ell$ can be subdivided into ℓ unordered pairs $(i_1,j_1), (i_2,j_2),\ldots (i_\ell,j_\ell)$.

One easily shows that the decomposition property implies a weight functional of the form:

$$P[\phi(x)] = N^{-1} \exp\{-\tfrac{1}{2} \int dx \int dx'\, \phi(x) \langle\phi(x)\phi(x')\rangle^{-1} \phi(x')\}. \tag{3.3}$$

Here the inverse of the covariance $\langle\phi(x)\phi(x')\rangle$ appears in the exponential, and the normalization integral is:

$$N = \int \exp\{-\tfrac{1}{2}\int dx \int dx' \phi(x)\langle\phi(x)\phi(x')\rangle^{-1}\phi(x')\}\, d[\phi(x)]. \qquad (3.4)$$

The integral over the space of functions $\phi(x)$ is defined as the limit of a multiple integral which is obtained either by dividing x-space in a large number (M) of cells and making $\phi(x)$ constant inside each cell, or by expanding $\phi(x)$ in the eigenfunctions of the kernel $\langle\phi(x)\phi(x')\rangle$ and using the first M expansion coefficients as stochastic variables.

4. THE ϕ-REPRESENTATION

For interacting bosons the grand canonical partition function

$$Z(z,\beta,\Omega) = \sum_{N=0}^{\infty} Z(N,\beta,\Omega) z^N \qquad (4.1)$$

has the representation:

$$Z(z,\beta,\Omega) = Z_o(\zeta,\beta,\Omega)\, \langle\exp\{-\mathrm{Tr}\,\ell n(1-\phi g)\}\rangle. \qquad (4.2)$$

Here Z_o denotes the grand partition function of the ideal Bose gas and the average is over Gaussian random functions with covariance:

$$\langle\phi(x)\phi(x')\rangle = -V(\vec{r}-\vec{r}\,')\delta(\tau-\tau'). \qquad (4.3)$$

The kernel $g(x)|x_o)$ which takes care of the Bose statistics, is defined (for a D-dimensional cubical volume) by the expressions:

$$g(x|x_o) = (2\pi)^{-D} \int (1-\zeta e^{-\beta\varepsilon_{\vec{k}}})^{-1} \exp\{-(\tau-\tau_o)\varepsilon_{\vec{k}}+i\vec{k}.(\vec{r}-\vec{r}_o)\}d^D\vec{k}$$

$$(0<\tau_o<\tau<\beta), \qquad (4.4a)$$

$$g(x|x_o) = (2\pi)^{-D} \int (1-\zeta e^{-\beta\varepsilon_{\vec{k}}})^{-1}$$

$$\zeta \exp\{-\beta\varepsilon_{\vec{k}} -(\tau-\tau_o)\varepsilon_{\vec{k}}+i\vec{k}.(\vec{r}-\vec{r}_o)\}d^D\vec{k} \quad (0<\tau\leq\tau_o<\beta). \qquad (4.4b)$$

Here $\varepsilon_{\vec{k}} = \hbar^2\vec{k}^2/2m$ and $\zeta = z\exp\{+\tfrac{1}{2}\beta V(\vec{0})\}$ is a renormalized activity. The derivations of this result and of the results in the next section have been reviewed in reference 1. For (4.2) to hold $V(\vec{r})$ must have a Fourier transform and $V(\vec{0})$ must be finite. A hard core in the interaction has therefore to be replaced by a core of finite "hardness"; from a physical point of view this should make no difference. Using the formula $\exp(\mathrm{Tr}A) = \det(e^A)$ one can also write (4.2) in the form:

$$\Xi(z,\beta,\Omega) = \Xi_o(\zeta,\beta,\Omega)\langle\det(1-\phi g)^{-1}\rangle, \tag{4.5}$$

but neither in this form nor in the form (4.2) has this representation found much practical application in the literature.

5. THE ψ-REPRESENTATION

Much more useful is the third form of a C-number formalism for interacting bosons with which our "exhibit" of path integrals ends. It has the form:

$$\Xi(z,\beta,\Omega) = \int \exp\{-L[\psi(x)]\}d[\psi(x)]. \tag{5.1}$$

The exponential is the sum of a second order term and a fourth order term:

$$L[\psi(x)] = \int \psi^*(x)\left[-\frac{\hbar^2}{2m}\Delta - \mu + \frac{\partial}{\partial\tau}\right]\psi(x)dx + \tfrac{1}{2}\int dx\int dx'|\psi(x)|^2 V(\vec{r}-\vec{r}')\delta(\tau-\tau')|\psi(x')|^2. \tag{5.2}$$

Here $\mu = k_BT\ell n\zeta$. As (5.1) will be used as the starting point for several approximation schemes the path integral symbol has to be defined precisely. First $\psi(x)$ is expanded in plane waves:

$$\psi(x) = (\beta\Omega)^{-\frac{1}{2}} \sum_{\vec{k},\ell} a_{\vec{k},\ell} \exp(i\vec{k}\vec{r} - 2\pi\ell\frac{\tau}{\beta}i). \tag{5.3}$$

In the summation ℓ runs through all integers and $\vec{k}$ is quantized because of the periodic boundary conditions on the surface of a cube with volume $\Omega = L^D$. For every $\vec{k},\ell$ one determines $\xi_{\vec{k}}$ as the unique positive real solution of:

$$2\sinh\tfrac{1}{2}\beta\xi_{\vec{k}} = \exp\{\tfrac{1}{2}\beta(\varepsilon_{\vec{k}}-\mu)\}. \tag{5.4}$$

Next one defines

$$\nu_{\vec{k},\ell} = (\xi_{\vec{k}} - \frac{2\pi\ell}{\beta}i)^{-1}. \tag{5.5}$$

The path integral symbol in (5.1) now stands for the repeated integral:

$$\int d[\psi(x)] \leftrightarrow \prod_{\vec{k},\ell} (\pi\nu_{\vec{k},\ell})^{-1} \int_{-\infty}^{+\infty} d(\mathrm{Re}\,a_{\vec{k},\ell}) \int_{-\infty}^{+\infty} d(\mathrm{Im}\,a_{\vec{k},\ell}). \tag{5.6}$$

Full details of the derivations can be found in references 3 and 4.

Let us very briefly outline a diagrammatic proof which has been worked out by Hirota[5] in his thesis but which to my knowledge was never published by this author. The diagrammatic proof proceeds by expanding the exponential of the fourth order term in (5.2) in a power series and by performing the path integral term by term. If the process of integration with a weight equal to the exponential of the second order term is indicated by square brackets the right hand side of (5.1) can be written as a series:

$$Z = [1] - \tfrac{1}{2}\int dx \int dx' [\psi^*(x)\psi(x)\psi^*(x')\psi(x')] V(\vec{r}-\vec{r}\,')\delta(\tau-\tau') + .. \quad (5.7)$$

Using an appropriate form of the decomposition property the averages can be written as sums of all possible matchings. Indicating a factor $V(\vec{r}-\vec{r}\,')\delta(\tau-\tau')$ by a wavy line and $[\psi(x)\psi^*(x')]$ by a solid line directed from vertex x' to vertex x (5.7) leads to a series of diagrams:

$Z = Z_o \{1 +$  $+$ 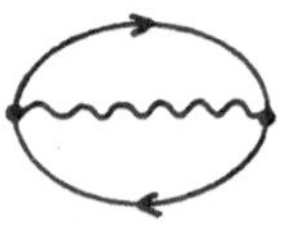 $+ \ldots\ldots\}$ (5.8)

A careful study shows that the diagram series is identical to the standard perturbation series for the partition function of the Bose fluid in powers of the interaction, thereby establishing (5.1) as an identity.

6. LANDAU-GINZBURG THEORY

In the next 6 sections we shall approximately evaluate the path integral (5.1) using increasingly more sophisticated forms of the saddle point method. The simplest approximation to evaluate the path integral consists of replacing the whole integral by the maximum of the integrand. This is the analog of the method of Laplace for ordinary integrals. The stationary points $\Phi(x)$ of $\exp(-L)$ are the solutions of $\delta L/\delta\Phi(x) = \delta L/\delta\Phi^*(x) = 0$. Substituting (5.2) one finds that the stationary functions do not depend on τ and that they are solutions of the non-linear equation:

$$\left[-\frac{\hbar^2}{2m}\Delta - \mu + \int_\Omega V(\vec{r}-\vec{r}\,')|\Phi(\vec{r}\,')|^2 d^D\vec{r}\,'\right]\Phi(\vec{r}) = 0 \quad (6.1)$$

under the boundary condition that $\Phi(\vec{r})$ vanishes if $\vec{r}$ approaches a hard wall.

We shall be interested in solutions of (6.1) which do not vary appreciably over a distance of the order a. For such solutions (6.1) can be replaced by the much simpler equation:

$$[-\frac{\hbar^2}{2m}\Delta - \mu + \tilde{V}(\vec{0})|\Phi(\vec{r})|^2]\Phi(\vec{r}) = 0, \tag{6.2}$$

where $\tilde{V}(\vec{0})$ is the $\vec{k}$=0 component of the Fourier transform of the interaction:

$$\tilde{V}(\vec{k}) = \int_\Omega V(\vec{r})\exp(i\vec{k}\vec{r})d^D\vec{k}. \tag{6.3}$$

It will be assumed that $\tilde{V}(\vec{0})$ is positive. It is now straightforward to prove the following theorem, which shows the special role played by the point μ=0 as far as the uniqueness or multiplicity of solutions of (6.2) is concerned:

Theorem. If μ<0 then (6.2) has only one solution:

$$\Phi_o(\vec{r}) = 0 \qquad (\mu<0). \tag{6.4}$$

If μ>0 then (6.2) has a great number of solutions, all satisfying the inequality:

$$|\Phi(\vec{r})| \leqslant (\mu/\tilde{V}(\vec{0}))^{\frac{1}{2}}. \qquad (\mu>0) \tag{6.5}$$

The contribution of a stationary solution is found by combining (6.2) with (5.2):

$$\exp\{-L[\Phi(\vec{r})]\} = \exp\{+\tfrac{1}{2}\beta\tilde{V}(\vec{0}) \int_\Omega |\Phi(\vec{r})|^4 \, d^D\vec{r}\}. \tag{6.6}$$

The optimal contribution to the path integral therefore comes in this case from the function:

$$\Phi_o(\vec{r}) = (\mu/\tilde{V}(0))^{\frac{1}{2}} \exp(i\alpha), \qquad (\mu>0) \tag{6.7}$$

where the phase α is arbitrary but constant.

The Landau-Ginzburg theory now simply amounts to the approximation:

$$Z \stackrel{\sim}{=} \exp\{-L[\Phi_o(\vec{r})]\}. \tag{6.8}$$

Using (6.4) and (6.7) this gives:

$$\beta p(\mu,\beta) = 0, \qquad (\mu<0) \tag{6.9a}$$

$$\beta p(\mu,\beta) = \frac{\beta\mu^2}{2\tilde{V}(\vec{0})}. \qquad (\mu>0) \tag{6.9b}$$

In this approximation the density is:

$$\rho(\mu,\beta) = \frac{\mu}{\tilde{V}(\vec{0})} \qquad (\mu>0) \qquad (6.10)$$

Hence the isotherm is given by $p(\rho,T) = \frac{1}{2}\tilde{V}(\vec{0})\rho^2$ and no phase transition develops. This approximation is obviously much too crude to give anything of interest, and we shall improve the calculation in the following sections.

Yet the reader should keep figure 1 in mind, which represents the contents of the theorem: the integrand of (5.1) is represented as a surface over function space. For $\mu<0$ this surface resembles a hill with a peak at the origin of function space, for $\mu>0$ the region of the surface near the absolute maximum resembles the ridge of a volcano.

7. THE BOGOLIUBOV THEORY

The failure of the method of Laplace (in the naive form used in the preceeding section) is caused by the fact that only the contribution of the maximum of the integrand of (5.1) was taken into account. The contribution of the vicinity of the maximum was neglected altogether. The simplest way to calculate the contribution of this vicinity to the partition function is by substituting:

$$\psi(x) = \Phi_0(\vec{r}) + \psi_1(x), \qquad (7.1)$$

where Φ_0 is the maximizing function (6.7), and by expanding L as given by (5.2) through second order in the "fluctuation" $\psi_1(x)$. This leads to a bilinear form which can be diagonalized. Calling the eigenvalues $\lambda_{\vec{k},\ell,\pm}$ and the expansion coefficients $c_{\vec{k},\ell,\pm}$ one finds an expression of the form:

$$Z = \exp\left(\frac{\beta\mu^2\Omega}{2\tilde{V}(\vec{0})}\right) \prod_{\vec{k},\ell} \int d^2c_{\vec{k},\ell,+} \int d^2c_{\vec{k},\ell,-}$$

$$\exp(-\lambda_{\vec{k},\ell,+}|c_{\vec{k},\ell,+}|^2 - \lambda_{\vec{k},\ell,-}|c_{\vec{k},\ell,-}|^2)$$

$$= \exp\left(\frac{\beta\mu^2\Omega}{2\tilde{V}(\vec{0})}\right) \prod_{\vec{k},\ell} \pi^2(\lambda_{\vec{k},\ell,+}\lambda_{\vec{k},\ell,-})^{-1}. \qquad (7.2)$$

The results of this calculation (for $\mu>0$) are as follows (details are in reference 6):

a) The eigenvalues are related to the Bogoliubov phonon spectrum by the relation:

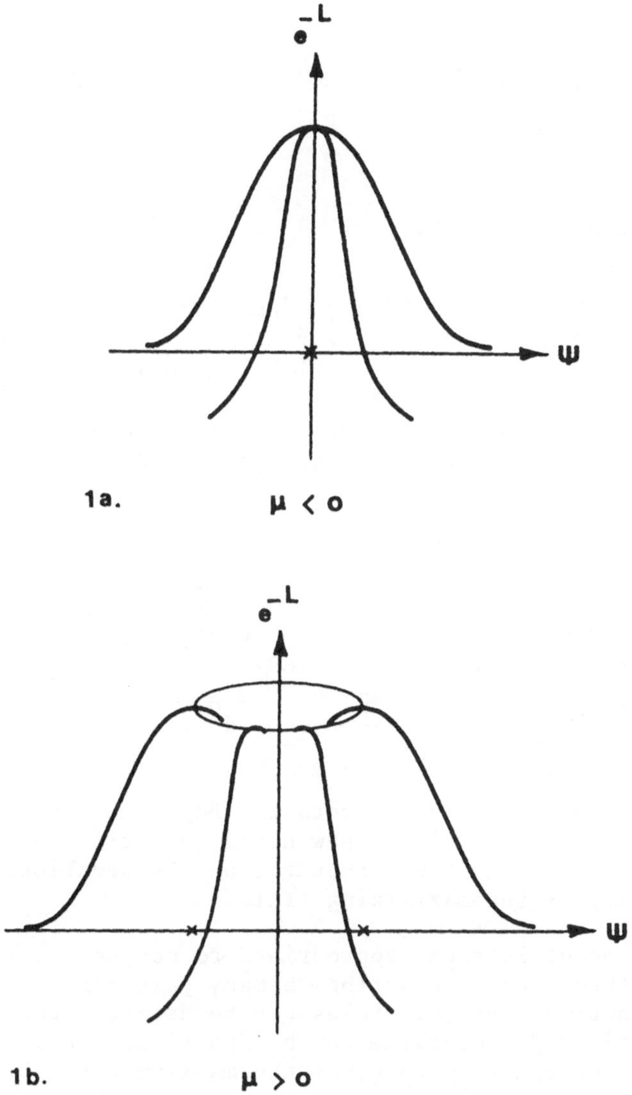

fig. 1: Qualitive behavior of the functional surface for $\mu<0$ and for $\mu>0$. The partition function (5.1) equals the total complex "volume" under the functional surface.

$$\lambda_{\vec{k},\ell,+}\,\lambda_{\vec{k},\ell,-} = E_{\vec{k}}^2 + \frac{4\pi^2\ell^2}{\beta^2}, \tag{7.3}$$

where

$$E_{\vec{k}} = \sqrt{\varepsilon_{\vec{k}}\,(\varepsilon_{\vec{k}}+2\mu\tilde{V}(\vec{k})/\tilde{V}(\vec{0}))} \tag{7.4}$$

is the phonon spectrum.

b) The pressure is given by:

$$\beta p(\mu,\beta) = \frac{\beta\mu^2}{2\tilde{V}(\vec{0})} + \tfrac{1}{2}(2\pi)^{-D}\int(\varepsilon_{\vec{k}}-\mu-E_{\vec{k}})d^D\vec{k} + (2\pi)^{-D}\int \ln(1-e^{-\beta E_{\vec{k}}})^{-1}\,d^D\vec{k}. \tag{7.5}$$

Note that the last term represents the pressure of a system of non-interacting phonons with energy spectrum $E_{\vec{k}}$.

c) The density is found to be:

$$\rho(\mu,\beta) = \frac{\mu}{\tilde{V}(\vec{0})} + \tfrac{1}{2}(2\pi)^{-D}\int\left(\varepsilon_{\vec{k}} + \mu\frac{\tilde{V}(\vec{k})}{\tilde{V}(\vec{0})} - E_{\vec{k}}\right)E_{\vec{k}}^{-1}\,d^D\vec{k} + (2\pi)^{-D}\int\left(\varepsilon_{\vec{k}} + \mu\frac{\tilde{V}(\vec{k})}{\tilde{V}(\vec{0})}\right)\frac{E_{\vec{k}}^{-1}\,e^{-\beta E_{\vec{k}}}}{1-e^{-\beta E_{\vec{k}}}}\,d^D\vec{k}. \tag{7.6}$$

Consequently one basically recovers the Bogoliubov theory. The spectrum of the quasiparticles now has a geometric interpretation: (7.3) relates the $E_{\vec{k}}$ to the curvatures of the functional surface in the vicinity of the maximizing field.

At this point it seems appropriate to conjecture that any approximate theory which describes a many particle system as a gas of non-interacting quasi particles can be derived from a path integral by fitting a Gaussian to the functional surface in that part of function space which gives the maximum contribution .

8. CRITIQUE OF THE METHOD OF LAPLACE

The procedure followed in sections 6, 7 is illustrated in figure 2 for the evaluation of an ordinary integral $\int_{-\infty}^{+\infty}\exp\{-f(\psi)\}d\psi$. In section 6 we replaced the whole integral by $\exp\{-f(\Phi_0)\}$; in section 7 a Gaussian was fitted to the integrand near to Φ_0. This procedure can be criticized for three reasons:

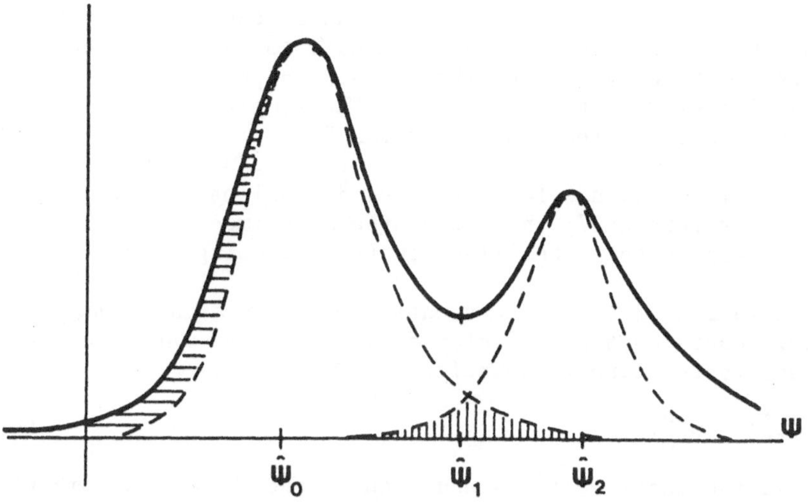

fig. 2: Evaluation of an ordinary integral with the method of Laplace. Discussion in section 8.

a) There is no large parameter in the path integral (5.1) which plays a role comparable to the large parameter in the standard saddle point method.

b) For $\mu>0$, according to the theorem of section 6, many secundary maxima exist (like $\hat{\Psi}_2$ in figure 2). Because of the absence of a large parameter the secundary maxima contribute to the thermodynamic functions, even in the thermodynamic limit $\Omega\to\infty$.

c) Even when all secundary maxima could be taken into account in a Gaussian approximation there are still contributions which are neglected (like the horizontally dashed area in figure 2) or which are counted twice (like the vertically dashed area in figure 2).

In the next three sections these points will be amplified by some further calculations.

9. SECONDARY MAXIMA, QUANTIZED VORTEX RINGS

For small positive values of μ (6.5) shows that all the maxima of exp(-L) become of comparable height. Therefore the method of Laplace now has to take the form:

$$Z \stackrel{\sim}{=} \sum_{\hat{\Psi}} \exp\{-L[\hat{\Psi}(\vec{r})]\}, \tag{9.1}$$

where the summation is over all solutions of the stationarity equation (6.2). This approximation amounts to replacing the curved functional surface by a step like surface which is flat in the vicinity of a stationary point. It is the simplest way to take care of the second objection of section 8.

A detailed analysis for the three dimensional case, which can be found in reference 7, shows that the most important secondary maxima correspond to vortex lines in the following sense:

a) Each solution of (6.2) represents a configuration in which the Bose fluid contains a set of closed vortex rings. The circulation around any vortex line is quantized in units h/m:

$$\oint \vec{v}.d\vec{\ell} = n \frac{h}{m}, \tag{9.2}$$

where $\vec{v}$ is the local velocity and n an integer. The most important vortices are those with $n = \pm 1$.

b) When cylindrical coordinates (r,θ,z) are choosen with the vortex line along the z axis the velocity equals:

$$v_\theta(r) = \frac{n\hbar}{mr}, \tag{9.3}$$

and the vortex flow pattern is restricted to a cylindrical region of a diameter of the order of the healing length:

$$\ell_o = (\hbar^2/2m\mu)^{\frac{1}{2}}. \tag{9.4}$$

c) A vortex filament may be curved provided the radius of curvature and the minimum distance between two vortex filaments are very large compared to ℓ_o.

d) A vortex filament with $|n|=1$ and length $N\ell_o$ contributes

$$\exp\{-L[\Phi(\vec{r})]\} = \exp(\frac{\beta\mu^2}{2\tilde{V}(\vec{0})}\Omega - c\mu^{\frac{1}{2}}N) \tag{9.5}$$

to the sum (9.1). Here

$$c = \frac{\beta}{\tilde{V}(\vec{0})} \left(\frac{\hbar^2}{2m}\right)^{3/2} c^*, \tag{9.6}$$

where c^* is a numerical constant. Note that the term (9.5) is non-analytic in the interaction and could not be found through perturbation theory.

10. THE λ-TRANSITION

Substituting (9.5) into (9.1) one finds for the grand canonical partition function an expression of the form:

$$\mathcal{Z}(\mu,\beta,\Omega) \cong \exp\left(\frac{\beta\mu^2}{2\tilde{V}(\vec{0})}\Omega\right) \sum_{N=0}^{\infty} W(N)\exp(-c\mu^{\frac{1}{2}}N), \tag{10.1}$$

where W(N) denotes the number of ways to draw directed, closed lines without intersections in the fluid such that their total circumference equals $N\ell_0$. It is often convenient to restrict the vortex rings to random walks along the bonds of a cubic lattice with a lattice constant of the order ℓ_0. Figure 3 shows a typical configuration in the analogous two-dimensional problem.

The combinatorial factor W(N) cannot be calculated rigorously. An approximate calculation is possible if one drops the constraints that different cycles should not intersect each other. That is: represent each vortex ring by a self-avoiding random walk on the lattice, but neglect the interactions between the rings. Figure 4 shows a typical configuration which is now permitted. If U_n denotes the number of self-avoiding random walks which return to a fixed starting point after n steps then it is straightforward to show that:

$$\mathcal{Z}(\mu,\beta,\Omega) = \exp\left(\frac{\beta\mu^2}{2\tilde{V}(\vec{0})}\Omega\right)\left\{1 + \frac{\Omega}{\ell_0^3}\sum_n \frac{U_n}{n} e^{-c\mu^{\frac{1}{2}}n} + \frac{1}{2!}\left(\frac{\Omega}{\ell_0^3}\sum_n \frac{U_n}{n} e^{-c\mu^{\frac{1}{2}}n}\right)^2 + \ldots\ldots\right\}. \tag{10.2}$$

The pressure is given by:

$$\beta p(\mu,\beta) = \frac{\beta\mu^2}{2\tilde{V}(\vec{0})} + \ell_0^{-3}\sum_n \frac{U_n}{n} e^{-c\mu^{\frac{1}{2}}n}. \tag{10.3}$$

Computer enumerations by Domb and coworkers have shown that U_n (in three dimensions) behaves asymptotically as:

$$U_n \sim A q_0^n n^{-a}; \quad a \cong 7/4 \qquad (n\to\infty). \tag{10.4}$$

It is now straightforward to combine the last two results and to show that the pressure has a singular part:

$$\beta p(\mu,\beta) = B(\mu-\mu_c)^a + \text{analytic function.} \quad (\mu>\mu_c) \tag{10.5}$$

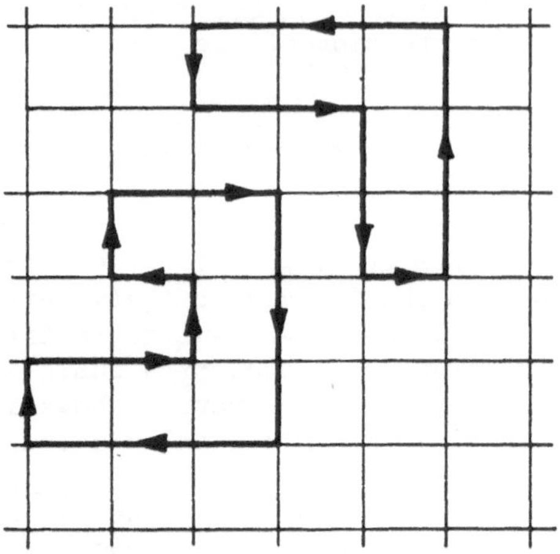

fig. 3: A configuration which is permitted in the two-dimensional analog of (10.1) and which is counted in W(26).

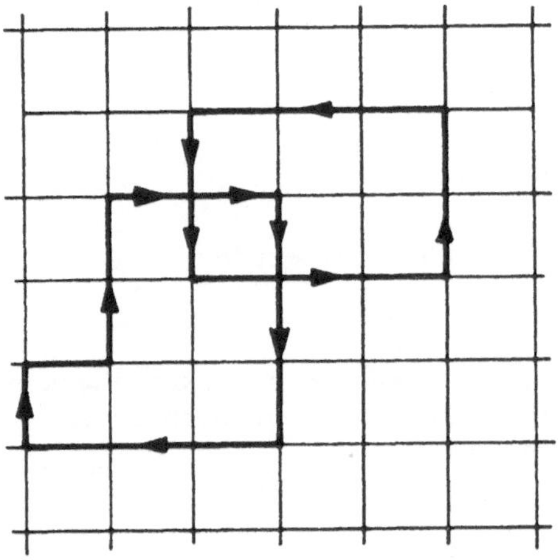

fig. 4: A configuration in an ideal gas of closed self-avoiding walks.

The critical value μ_c is given by:

$$\mu_c(T) = \left(\frac{\ell n q_o}{c}\right)^2 . \qquad (10.6)$$

The density is found by differentiating the pressure once:

$$\beta\rho(\mu,\beta) = Ba(\mu-\mu_c)^{a-1} + \text{analytic function}, \qquad (\mu>\mu_c) \qquad (10.7)$$

and the singular part of the specific heat by differentiating the pressure twice:

$$c_v(\mu,\beta) \sim (\mu-\mu_c)^{a-2}. \qquad (\mu\downarrow\mu_c) \qquad (10.8)$$

The last two equations give:

$$c_v(\rho,T) \sim (T_\lambda-T)^{(a-2)/(a-1)}. \qquad (T\uparrow T_\lambda) \qquad (10.9)$$

Hence in this approximation the specific heat critical exponent α is found to be:

$$\alpha = \frac{2-a}{a-1} \cong 0.33. \qquad (10.10)$$

This value is larger than the observed value which is closer to zero; it is possible that a more accurate determination of the combinatorial factor W(N) might improve the agreement with experiment.

11. ANALOGY WITH THE DROPLET MODEL OF CONDENSATION

The weights (10.2) also enable one to calculate the average number of vortex rings per unit of volume and the average length of a vortex ring. Both quantities are small if $\mu>>\mu_c$, they grow if μ decreases and they diverge in the limit $\mu\downarrow\mu_c$. The theory of section 10 thus leads to a model with two phases:

(a) A high density phase for $\mu>\mu_c$ in which the fluid consists of a uniform condensate (the N=0 term in (10.1)) with a few short vortex rings. This is the superfluid phase of the interacting Bose fluid.

(b) A low density phase for $\mu<\mu_c$ in which the system is instable for the formation of vortex filaments. The condensate is in a state of microturbulence which effectively destroys its long range coherence. Hence this phase is identified with the normal phase. The approximation (10.2) breaks down here and one needs a more accurate expression for the combinatorial factor W(N) to obtain the thermodynamic functions.

Note that this state of affairs is analogous to the results of the droplet model for the condensation of a classical gas: When a gas is compressed small droplets of the liquid phase appear. At a certain density the gas becomes instable for the formation of droplets: this is the density at which condensation sets in. Vortex filaments play the same role in the Bose fluid as droplets and bubbles in the classical gas. A more detailed discussion of this analogy can be found in reference 7.

12. THE RENORMALIZATION GROUP APPROACH

Using increasingly more sophisticated forms of the saddle point method we have developed in sections 6-11 an approximate theory of the Bose fluid. In the following sections another approximation scheme for the interacting Bose fluid will be discussed: the renormalization group approach. To date the renormalization group is the most powerful method to solve many body problems. Although several monographs exist which are devoted to the general ideas basic to this method [8, 9, 10] the application to the interacting Bose fluid can only be found in sometimes very sketchy letters. In the following sections we shall, therefore, present all derivations in much detail. It is hoped that these sections will serve as a pedestrian introduction to the renormalization group.

The great interest in the renormalization group derives from the fact that this technique provides a major step towards the establishment of a systematic calculus for path integrals. It is, therefore, of interest to formulate this calculational technique in as many different ways as possible, and to find that point of view from which the subject appears as simple as possible. In the following sections we shall focus on the explixit (although approximate) calculation of the infinitesimal generator of the group.

The first applications of the renormalization group to the interacting Bose fluid are due to Singh[11,12] and to Baldo, Catara and Lombardo[13]. These authors specifically showed that the critical exponents of the λ transition are the same as those of a classical model of spins with 2 components, and hence concluded that the universality hypothesis applies to phase transitions regardless of their classical or quantal nature

Singh's derivation starts from the expression of the boson partition function in second quantization. This approach has the drawback that the non-commuting nature of the creation- and annihilation operators has to be kept in mind. This not only leads to the use of a fairly complicated form of perturbation theory, but - as this author remarks - somewhat spoils the simplicity of the renormalization scheme because, unlike the case of classical spins, the creation and annihilation operators cannot be rescaled because of the commutation relations.

Baldo, Catara and Lombardo avoid this difficulty by using the coherent state representation. This representation had already been used by Ohanessian and Quattropani[14] to study the renormalization group approach to the ideal Bose gas. We shall not discuss this formalism but refer the interested reader to Ohanessian[2]. Two other papers on the renormalization group approach to the interacting Bose gas are those by Ma[15] and by Family and Stanley[16].

13. SHORT- AND LONG WAVELENGTHS

In order to set the scene for the renormalization group one introduces into (5.3) a cut-off Λ in $\vec{k}$ space by restricting the $\vec{k}$ values to the sphere $|\vec{k}| < \Lambda = 2\pi/a$, where a denotes the scattering length of the potential $V(\vec{r})$. The potential is assumed to be predominantly repulsive, so a>0. For hard spheres a equals the diameter of the spheres. With this cut-off the potential can be replaced (for a three dimensional problem) by the pseudopotential:

$$V(\vec{r}) = \frac{4\pi a\hbar^2}{m}\,\delta(\vec{r}). \qquad (D=3) \tag{13.1}$$

With the substitutions

$$\tau = \beta\theta, \quad \psi(\vec{r},\tau) = (m/\beta\hbar^2)^{\frac{1}{2}}\phi(\vec{r},\theta) \tag{13.2}$$

(5.1,2) can be written in the standard form:

$$Z(z,\beta,\Omega) = \int \exp\{-L[\phi(\vec{r},\theta)]\}d[\phi(\vec{r},\theta)], \tag{13.3}$$

$$L[\phi(\vec{r},\theta)] = \int_\Omega d^D\vec{r} \int_0^1 d\theta\{\tfrac{1}{2}|\vec{\nabla}\phi|^2 + b\phi^* \frac{\partial\phi}{\partial\theta} + U(\phi)\}, \tag{13.4}$$

where $U(\phi)$ has the special form:

$$U_o(\phi) = \tfrac{1}{2}r_o|\phi|^2 + s_o|\phi|^4. \tag{13.5}$$

The three constants b, r_o, s_o are specifically given by:

$$b = m/\beta\hbar^2, \tag{13.6a}$$

$$r_o = -2\mu m/\hbar^2, \tag{13.6b}$$

$$s_o = 2\pi am/\beta\hbar^2. \tag{13.6c}$$

The complex random field $\phi(\vec{r},\theta)$ has the expansion:

$$\phi(\vec{r},\theta) = \Omega^{-\frac{1}{2}} \sum_{\vec{k},\ell} c_{\vec{k},\ell} \exp\{i(\vec{k}.\vec{r}-2\pi\ell\theta)\} \tag{13.7}$$

and the integration symbol in (13.3) means:

$$\int d[\phi(r,\theta)] \leftrightarrow \prod_{\vec{k},\ell} (m/\pi\hbar^2\nu_{\vec{k},\ell}) \int_{-\infty}^{+\infty} d(\mathrm{Re}\,c_{\vec{k},\ell}) \int_{-\infty}^{+\infty} d(\mathrm{Im}\,c_{\vec{k},\ell}). \tag{13.8}$$

The canonical dimensions of these variables are $[\theta] = 1$; $[\phi] = [\text{length}]^{1-\frac{1}{2}D}$; $[c_{\vec{k},\ell}] = [\text{length}]$.

It will turn out to be convenient to write (13.3) as an average:

$$Z(z,\beta,\Omega) = N \;\langle \exp\{-\int U(\phi)d^D\vec{r}d\theta\}\rangle \tag{13.9}$$

over the complex valued weight functional:

$$W[\phi] \equiv \prod_{\vec{k},\ell} w(c_{\vec{k},\ell})$$

$$\equiv \prod_{\vec{k},\ell} \frac{1}{\pi} (\tfrac{1}{2}|\vec{k}|^2 - 2\pi\ell bi) \exp\{-(\tfrac{1}{2}|\vec{k}|^2 - 2\pi\ell bi)|c_{\vec{k},\ell}|^2\}. \tag{13.10}$$

The normalization constant equals:

$$N = \prod_{\vec{k},\ell} \frac{m}{\hbar^2\nu_{\vec{k},\ell}} (\tfrac{1}{2}|\vec{k}|^2 - 2\pi\ell bi)^{-1}. \tag{13.11}$$

Using (5.4) and (5.5) it is straightforward to calculate this constant, with the result:

$$N = \prod_{\vec{k}} e^{-\frac{1}{2}\beta\mu} (1 - e^{-\beta\varepsilon_{\vec{k}}})^{-1}. \tag{13.12}$$

This will add a term to the grand canonical pressure equal to:

$$\lim_{\Omega\to\infty} \frac{1}{\Omega} \ln N = -\tfrac{1}{2}\beta\mu(2\pi)^{-D} V_D(\Lambda) - (2\pi)^{-D} \int \ln(1 - e^{-\beta\varepsilon_{\vec{k}}})d^D\vec{k}, \tag{13.13}$$

where $V_D(\Lambda)$ denotes the volume of a D-dimensional sphere of radius Λ.

The sum over $\vec{k}$ in (13.7), which has a cut-off at $|\vec{k}| = \Lambda$, can be written as the sum of two terms:

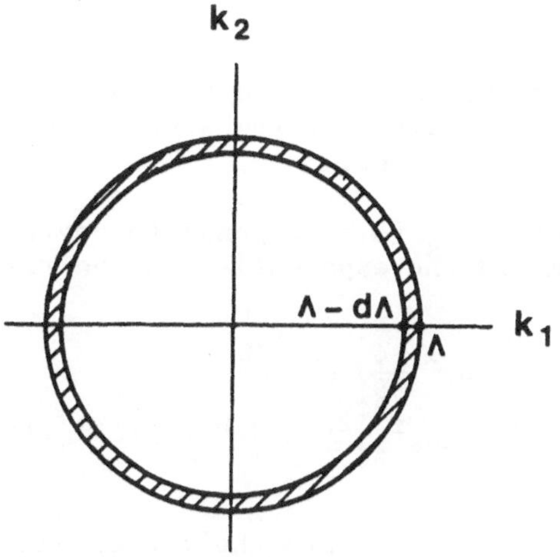

fig. 5: The dashed shell contains wave vectors which occur in ϕ_s; the inner sphere with radius $\Lambda-d\Lambda$ consists of wave vectors occuring in ϕ_ℓ.

$$\phi(\vec{r},\theta) = \phi_\ell(\vec{r},\theta) + \phi_s(\vec{r},\theta), \tag{13.14}$$

$$\phi_\ell(\vec{r},\theta) \equiv \Omega^{-\frac{1}{2}} \sum_{|\vec{k}|\leqslant\Lambda-d\Lambda} \sum_{\ell} c_{\vec{k},\ell} \exp\{i(\vec{k}.\vec{r} - 2\pi\ell\theta)\}, \tag{13.15}$$

$$\phi_s(\vec{r},\theta) = \Omega^{-\frac{1}{2}} \sum_{\Lambda-d\Lambda<|\vec{k}|\leqslant\Lambda} \sum_{\ell} c_{\vec{k},\ell} \exp\{i(\vec{k}.\vec{r} - 2\pi\ell\theta)\}. \tag{13.16}$$

Here $d\Lambda$ is a small positive quantity which will vanish in a later stage of the calculation. Hence ϕ_s is a sum of Fourier components with $\vec{k}$ values in the dashed region indicated in figure 5, and short wavelengths of order a. The function ϕ_ℓ contains all $|\vec{k}|$ values between 0 and $\Lambda-d\Lambda$; therefore the typical wavelengths in ϕ_ℓ are at least twice those in ϕ_s. We shall call ϕ_s the short wavelength part of the random field ϕ, and ϕ_ℓ the long wavelength part. In the following sections we shall often use the approximation in which the wavelengths in ϕ_ℓ are assumed to be much larger than those in ϕ_s.

14. THE CUMULANT EXPANSION

The average (13.9) over the random functions ϕ factorizes in a product of two averages:

$$< \; > = < < \; >_s >_\ell , \tag{14.1}$$

in which the two factors are defined by (13.10) with appropriate cut-offs in $\vec{k}$ space. In the next few sections we shall perform the integration over ϕ_s first.

Note that any function ϕ_s is orthogonal to any function ϕ_ℓ. Hence the first two terms in the exponential on the right side in (13.4) are:

$$\int d^D\vec{r} \int d\theta \left(\tfrac{1}{2}|\vec{\nabla}\phi|^2 + b\phi^* \frac{\partial\phi}{\partial\theta}\right) =$$
$$\int d^D\vec{r} \int d\theta \left(\tfrac{1}{2}|\vec{\nabla}\phi_\ell|^2 + b\phi_\ell^* \frac{\partial\phi_\ell}{\partial\theta}\right) + \int d^D\vec{r} \int d\theta \left(\tfrac{1}{2}|\vec{\nabla}\phi_s|^2 + b\phi_s^* \frac{\partial\phi_s}{\partial\theta}\right). \tag{14.2}$$

After splitting ϕ, ϕ_s and ϕ_ℓ into their real and imaginary parts:

$$\left.\begin{aligned} \phi &= a + ib \\ \phi_s &= a_s + ib_s \\ \phi_\ell &= a_\ell + ib_\ell \end{aligned}\right\} \tag{14.3}$$

$U(\phi)$ can be expanded in a Taylor series around the value ϕ_ℓ. One finds in an obvious notation:

$$U(\phi) = U(\phi_\ell) + F(\phi_s), \tag{14.4}$$

$$\begin{aligned} F(\phi_s) &= a_s(\partial U/\partial a)_\ell + b_s(\partial U/\partial b)_\ell + \\ &\quad \frac{1}{2!}\{a_s^2(\partial^2 U/\partial a^2)_\ell + 2a_s b_s(\partial^2 U/\partial a\partial b)_\ell + b_s^2(\partial^2 U/\partial b^2)_\ell\} + \ldots \\ &\equiv \sum_{n=1}^{\infty} F_n(\phi_s). \end{aligned} \tag{14.5}$$

In the last equation F_n denotes the sum of all terms of nth order in a_s and b_s.

The average over the short wavelength part ϕ_s of the random function thus leads to an expression of the form:

$$<\exp\{-\int U(\phi) d^D\vec{r} d\theta\}>_s = \exp\{-\int U(\phi_\ell) d^D\vec{r} d\theta\} <\exp\{-\int F(\phi_s) d^D\vec{r} d\theta\}>_s . \tag{14.6}$$

The average on the right hand side can be expanded in cumulants as follows:

$$<\exp\{-\int F(\phi_s)d^D\vec{r}d\theta\}>_s = \exp\Big[-\int<F(\phi_s)>_s d^D\vec{r}d\theta + \tfrac{1}{2}\int d^D\vec{r}d\theta d^D\vec{r}'d\theta'\{<F(\phi_s(\vec{r},\theta))F(\theta_s(\vec{r}',\theta')>_s - <F(\phi_s(\vec{r},\theta)>_s<F(\phi_s(\vec{r}',\theta')>_s\} + \ldots\Big]. \tag{14.7}$$

In these notes we shall use the approximation in which this expansion is terminated after the second cumulant. It can be shown that the corrections to the critical exponents due to the higher order cumulants are of order ε^2, where $\varepsilon = 4-D$.

15. PROPERTIES OF THE RAPID FLUCTUATIONS

Let us denote the pair $(\vec{r},\theta)$ by x and consider the fluctuations $\phi_s(x)$ with short wavelengths for their own sake. The complex weight functional of these fluctuations is given by (13.10) with $\vec{k}$ in the spherical shell $\Lambda-d\Lambda<|\vec{k}|\leqslant\Lambda$:

$$W_s[\phi_s] \equiv \prod_{\Lambda-d\Lambda<|\vec{k}|\leqslant\Lambda}\prod_{\ell} w(c_{\vec{k},\ell}). \tag{15.1}$$

The ϕ_s are Gaussian random functions and their properties are a generalization of those discussed in section 3. For example, one easily shows that:

$$<\phi_s(x)>_s = <\phi_s^*(x')>_s = 0, \tag{15.2a}$$

$$<\phi_s(x)\phi_s(x')>_s = <\phi_s^*(x)\phi_s^*(x')>_s = 0. \tag{15.2b}$$

A more interesting quantity is the covariance:

$$G(x|x') = <\phi_s(x)\phi_s^*(x')>_s. \tag{15.3}$$

Substituting (13.7) and (15.1) one finds:

$$G(x|x') = \Omega^{-1}\sum_{\vec{k}}{}''\sum_{\ell}(\tfrac{1}{2}|\vec{k}|^2 - 2\pi\ell bi)^{-1}\exp\{i\vec{k}.(\vec{r}-\vec{r}') - 2\pi\ell(\theta-\theta')i\}, \tag{15.4}$$

where the double prime indicates the constraints $\Lambda-d\Lambda<|\vec{k}|\leqslant\Lambda$.

It is instructive to perform the summation over ℓ explicitly. For $\theta>0$ this can be done by considering the complex integral:

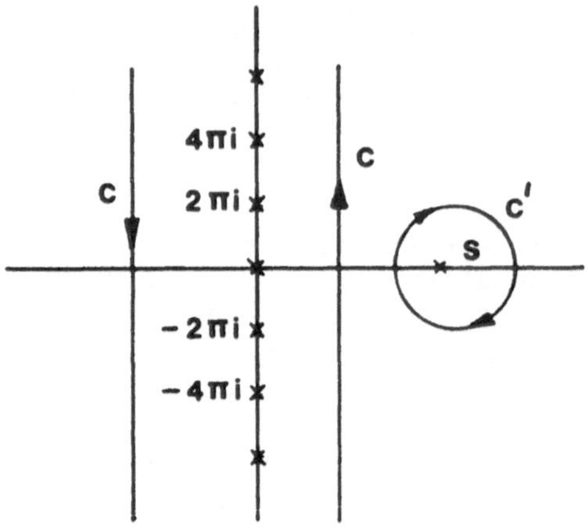

fig. 6: The contour of integration in (15.5).

$$F(S) \equiv \frac{1}{2\pi i} \oint_C (1-e^{-t})^{-1} (S-t)^{-1} e^{-\theta t} dt. \quad (S>0,\ 0<\theta<1) \qquad (15.5)$$

The contour of integration is shown in figure 6. The integrand has poles at $t_\ell = 2\pi\ell i$ on the imaginary t-axis. Applying Cauchy's theorem we find:

$$F(S) = \sum_\ell (S-2\pi\ell i)^{-1} e^{-2\pi\ell\theta i}. \qquad (15.6)$$

The integrand also has a pole at t=S. As it vanishes exponentially for $t\to\pm\infty$ C can be deformed into C' in figure 6. This gives:

$$F(S) = \frac{e^{-\theta S}}{1-e^{-S}} . \qquad (15.7)$$

For $\theta<0$ one considers the complex integral:

$$F(S) = \frac{1}{2\pi i} \oint_C e^{-t} (1-e^{-t})^{-1} (S-t)^{-1} e^{-\theta t} dt. \ (S>0,\ -1<\theta<0)$$

(15.8)

As the residues in t_ℓ are the same as before this integral is also equal to (15.6). But for $-1<\theta<0$ the integrand vanishes exponentially for $t \to \pm\infty$ and the pole at $t=S$ gives:

$$F(S) = \frac{e^{-S}}{1-e^{-S}} e^{-\theta S}. \tag{15.9}$$

Combining the last four equations one finds:

$$\sum_{\ell} (S-2\pi\ell i)^{-1} e^{-2\pi\ell\theta i} = \begin{cases} (1-e^{-S})^{-1} e^{-\theta S} & (0<\theta<1), \\ (1-e^{-S})^{-1} e^{-S} e^{-\theta S} & (-1<\theta<0). \end{cases} \tag{15.10}$$

The covariance (15.4) is thus given by:

$$G(x|x') = \begin{cases} \frac{1}{b} (2\pi)^{-D} d\Lambda \dfrac{e^{-(\theta-\theta')|\vec{k}|^2/2b}}{1 - e^{-|\vec{k}|^2/2b}} \oint e^{i\vec{k}.(\vec{r}-\vec{r}')} d^{D-1}S, & (\theta>\theta'), \\ \frac{1}{b} (2\pi)^{-D} d\Lambda \dfrac{e^{-|\vec{k}|^2/2b-(\theta-\theta')|\vec{k}|^2/2b}}{1 - e^{-|\vec{k}|^2/2b}} \oint e^{i\vec{k}.(\vec{r}-\vec{r}')} d^{D-1}S, & (\theta<\theta'), \end{cases} \tag{15.11}$$

where the integral extends over the surface of a D dimensional sphere of radius Λ. Note that G is just a cut-off part of g with $\mu=0$ (compare (4.4)). Obviously G is a function of $x-x_1'$ only and vanishes with a long oscillating tail when $|\vec{r}-\vec{r}'| >> \Lambda^{-1} = a/2\pi$. Also note that G does not depend on the volume Ω.

The random functions ϕ_s also have a decomposition property which now reads:

$$< \prod_{i=1}^{n} \{\phi_s(x_i)\} \prod_{j=1}^{m} \{\phi_s^*(x_j')\}>_s = 0 \text{ if } m \neq n,$$

$$= \sum_P \prod_{i=1}^{m} <\phi_s(x_i)\phi_s^*(x_{Pi}')>_s \quad \text{if } m=n. \tag{15.12}$$

Here the sum is over all permutations P of the m numbers 1, 2, ..m. The proof, which is somewhat tedious, can be found in reference 1.

16. INTEGRALS OF THE COVARIANCE

Integrating (15.4) over x' one finds immediately that the integral of the covariance vanishes:

$$\int G(x|x')dx' = 0. \tag{16.1}$$

The integral of G^2 follows in the same way:

$$\int G^2(x|x')dx' = \Omega^{-1} \sum_{\vec{k}}{}'' \sum_{\ell} (\tfrac{1}{2}|\vec{k}|^2 - 2\pi\ell bi)^{-1} (\tfrac{1}{2}|\vec{k}|^2 + 2\pi\ell bi)^{-1}. \tag{16.2}$$

Using the sum formula (which follows from (15.10)):

$$2S \sum_{\ell} (S^2 + 4\pi^2\ell^2)^{-1} = \text{ctgh}\ \tfrac{1}{2}S \tag{16.3}$$

one finds:

$$\int G^2(x|x')dx' = C_D d\Lambda, \tag{16.4}$$

$$C_D = \frac{1}{b\Lambda^2} \text{ctgh}\ (\frac{\Lambda^2}{4b}) (2\pi)^{-D}\ S_D(\Lambda). \tag{16.5}$$

Here the surface area of a D dimensional sphere of radius Λ is denoted by $S_D(\Lambda)$. The integral of a higher power of G can be estimated as follows. From (15.4) one has:

$$\int G^n(x|x')dx' = \Omega^{-n+1} \sum{}' \prod_{j=1}^{n} (\tfrac{1}{2}|\vec{k}_j|^2 - 2\pi\ell_j bi)^{-1}, \tag{16.6}$$

where the prime indicates a summation over $\vec{k}_1, \ell_1, \vec{k}_2, \ell_2, \ldots \vec{k}_n, \ell_n$ such that:

$$\sum_{j=1}^{n} \vec{k}_j = \vec{0}; \quad \sum_{j=1}^{n} \ell_j = 0 \tag{16.7}$$

in addition to the usual cut-offs $\Lambda - d\Lambda < |\vec{k}_j| \leq \Lambda$. These relations determine $\vec{k}_n$ and ℓ_n in terms of the $(n-1)^j$ other $\vec{k}_j, \ell_j$. As each $\vec{k}$ integration leads to one factor $d\Lambda$ we find the order of magnitude estimation:

$$\int G^n(x|x')dx' = \mathcal{O}((d\Lambda)^{n-1}). \tag{16.8}$$

The results of this section will now be used to calculate the first two cumulants in the expansion (14.7).

17. THE FIRST CUMULANT

The first cumulant follows from (14.5):

$$<F(\phi_s)>_s = \tfrac{1}{2}\{<a_s^2(x)>_s(\partial^2U/\partial a^2)_\ell + 2<a_s(x)b_s(x)>_s(\partial^2U/\partial a\partial b)_\ell + <b_s^2(x)>_s(\partial^2U/\partial b^2)_\ell\} + \ldots \quad (17.1)$$

The averages of the products of a's and b's are expressed in terms of averages of products of ϕ's and ϕ^*'s by making use of:

$$a_s(x) = \tfrac{1}{2}\{\phi_s(x) + \phi_s^*(x)\}; \quad b_s(x) = \frac{1}{2i}\{\phi_s(x) - \phi_s^*(x)\}. \quad (17.2)$$

Using (15.11) this gives:

$$<a_s(x)b_s(x)>_s = 0, \quad (17.3)$$

$$<a_s^2(x)>_s = <b_s^2(x)>_s = \tfrac{1}{4}C_D\Lambda^2 d\Lambda. \quad (17.4)$$

Because of the decomposition property and (16.8) the average of a higher order term in (17.1) is of at least second order in $d\Lambda$. Hence the first cumulant equals:

$$-\int<F(\phi_s(x))>_s dx = -\frac{1}{8}C_D\Lambda^2 d\Lambda\int\left(\frac{\partial^2U}{\partial a^2} + \frac{\partial^2U}{\partial b^2}\right)_\ell dx + \mathcal{O}((d\Lambda)^2). \quad (17.5)$$

As U depends on $|\phi_\ell| = \sqrt{a_\ell^2 + b_\ell^2}$ only this can also be written in the form:

$$-\int<F(\phi_s(x))>_s dx = -\frac{1}{8}C_D\Lambda^2 d\Lambda\int\left(\frac{d^2U}{d|\phi_\ell|^2} + \frac{1}{|\phi_\ell|}\frac{dU}{d|\phi_\ell|}\right)dx + \mathcal{O}((d\Lambda)^2). \quad (17.6)$$

Note that this has been derived without the use of the assumption that the scale of ϕ_ℓ is much larger than the scale of ϕ_s.

18. THE SECOND CUMULANT

The second cumulant is defined by the double integral:

$$+\tfrac{1}{2}\int dx\int dx'\{<F(\phi_s(x))F(\phi_s(x'))>_s - <F(\phi_s(x))>_s<F(\phi_s(x'))>_s\} \quad (18.1)$$

which appears on the right hand side of (14.7). This cumulant will be evaluated in the approximation in which $\phi_\ell(x)$ is assumed to vary infinitely slow in $\vec{r}$ space as compared to $\phi_s(x)$, i.e. it will be assumed that:

$$(\text{scale of } \phi_s) \ll (\text{scale of } \phi_\ell). \tag{18.2a}$$

In other words: we shall always replace integrals of the form:

$$\int G^n(x|x')F(\phi_\ell(x'))dx' \quad \text{by:} \quad F(\phi_\ell(x))\int G^n(x|x')dx', \tag{18.2b}$$

for those functions $F(\phi)$ which will be encountered. This approximation is somewhat crude, but has the advantage that it leads to a formalism which is still simple. It can be improved upon in a systematic way; see especially reference 9.

As $\langle F(\phi_s(x))\rangle_s$ is itself already of order $d\Lambda$ the second term in the integrand of (18.1) is $\mathcal{O}((d\Lambda)^2)$ and can be neglected. Substituting (14.5) the second culumant is given by the double series:

$$+\tfrac{1}{2}\int dx\int dx'\langle F(\phi_s(x))F(\phi_s(x'))\rangle_s =$$

$$= \tfrac{1}{2}\sum_{n,m=1}^{\infty} \int dx\int dx'\langle F_n(\phi_s(x))F_m(\phi_s(x'))\rangle_s + \mathcal{O}((d\Lambda)^2). \tag{18.3}$$

The first few terms are:

a) n=m=1. Substituting from (14.5) this term equals:

$$\begin{aligned}\langle F_1(\phi_s(x))F_1(\phi_s(x'))\rangle_s = {} & \langle a_s(x)a_s(x')\rangle_s \left(\frac{\partial U}{\partial a}\right)_x \left(\frac{\partial U}{\partial a}\right)_{x'} + \\ & \langle b_s(x)b_s(x')\rangle_s \left(\frac{\partial U}{\partial b}\right)_x \left(\frac{\partial U}{\partial b}\right)_{x'} + \\ & \langle a_s(x)b_s(x')\rangle_s \left(\frac{\partial U}{\partial a}\right)_x \left(\frac{\partial U}{\partial b}\right)_{x'} + \\ & \langle b_s(x)a_s(x')\rangle_s \left(\frac{\partial U}{\partial b}\right)_x \left(\frac{\partial U}{\partial a}\right)_{x'}.\end{aligned} \tag{18.4}$$

Integrating over x' and using (18.2) and (16.1) this is found to vanish:

$$\tfrac{1}{2}\int dx\int dx'\langle F_1(\phi_s(x))F_1(\phi_s(x'))\rangle_s = 0. \tag{18.5}$$

b) n=1, m=2 and n=2, m=1. These terms lead with (17.2) to averages of products of three factors ϕ_s, hence they vanish rigorously because of the decomposition property. For the same reason all terms vanish for which (n+m) is an odd number.

c) n=m=2. Substituting from (14.5) and using (17.2), (15.12) and the approximation (18.2) one finds after a straightforward but somewhat tedious calculation:

$$\int <F_2(\phi_s(x))F_2(\phi_s(x'))>_s dx' = \left[\frac{1}{16}\left(\frac{\partial^2 U}{\partial a^2}+\frac{\partial^2 U}{\partial b^2}\right)^2 + \frac{1}{16}\left\{\left(\frac{\partial^2 U}{\partial a^2}-\frac{\partial^2 U}{\partial b^2}\right)^2 + 4\left(\frac{\partial^2 U}{\partial a\partial b}\right)^2\right\}\right] . \int G^2(x|x')dx' + \mathcal{O}((d\Lambda)^2). \quad (18.6)$$

Upon integration over x and substitution from (16.4) this term contributes to the second cumulant:

$$\tfrac{1}{2}\int dx\int dx' <F_2(\phi_s(x))F_2(\phi_s(x'))>_s = \frac{1}{16}\, C_D d\Lambda . \int\left[\left(\frac{\partial^2 U}{\partial a^2}\right)^2 + \left(\frac{\partial^2 U}{\partial b^2}\right)^2 + 2\left(\frac{\partial^2 U}{\partial a\partial b}\right)^2\right]dx + \mathcal{O}((d\Lambda)^2). \quad (18.7)$$

d) n=1, m=3 or n=3, m=1. These two terms lead to contributions proportional to $\int G(x|x')dx'$ and vanish because of (16.1).

e) n+m≥6. These terms lead to contributions which are integrals of the form:

$$\int G^{\alpha}(x|x')dx' \left[G(x|x)\right]^{\beta}\left[G(x'|x')\right]^{\gamma}; \quad 2(\alpha+\beta+\gamma)\geq 6. \quad (18.8)$$

With (15.11) and (16.8) they are of order $(d\Lambda)^{\alpha-1+\beta+\gamma}$; hence they are at least of second order in $(d\Lambda)$.

Collecting all these results the second cumulant is found to equal:

$$\tfrac{1}{2}\int dx\int dx' <F(\phi_s(x))F(\phi_s(x'))>_s = \frac{1}{16}\, C_D d\Lambda\int\left[\left(\frac{\partial^2 U}{\partial a^2}\right)^2 + \left(\frac{\partial^2 U}{\partial b^2}\right)^2 + 2\left(\frac{\partial^2 U}{\partial a\partial b}\right)^2\right]_\ell dx + \mathcal{O}((d\Lambda)^2). \quad (18.9)$$

As $U = U(|\phi_\ell|)$ this can also be written in the form:

$$\tfrac{1}{2}\int dx \int dx' \langle F(\phi_s(x))F(\phi_s(x'))\rangle_s = \frac{1}{16} C_D d\Lambda \int \left[\left(\frac{d^2U}{d|\phi_\ell|^2}\right)^2 + \frac{1}{|\phi_\ell|^2}\left(\frac{dU}{d|\phi_\ell|}\right)^2 \right] dx + \mathcal{O}((d\Lambda)^2). \tag{18.10}$$

Note that this simple expression for the second cumulant only holds in the approximation (18.2).

19. SCALING

The result of averaging over the short-wavelength part of the random function is found by combination of (14.7), (17.6) and (18.10):

$$\langle \exp\{-\int_\Omega F(\phi_s)dx\}\rangle_s = \exp\{-\frac{1}{8} C_D d\Lambda \int_\Omega [\Lambda^2 (\frac{U'}{|\phi_\ell|} + U'') - \tfrac{1}{2}(\frac{U'^2}{|\phi_\ell|^2} + U''^2)] dx + \mathcal{O}((d\Lambda)^2)\}. \tag{19.1}$$

The primes denote differentation with respect to $|\phi_\ell|$. At this stage of the calculation the partition function is given by:

$$Z(z,\beta,\Omega) = N\langle \exp\{-\int_\Omega U_\ell(\phi_\ell)dx\}\rangle_\ell, \tag{19.2}$$

where the average is taken with respect to the weight functional

$$W_\ell[\phi_\ell] = \frac{\exp\{-\int_\Omega [\tfrac{1}{2}|\nabla\vec{\phi}_\ell|^2 + b\phi_\ell^* \frac{\partial \phi_\ell}{\partial \theta}] dx\}}{\int \exp\{-\int_\Omega [\tfrac{1}{2}|\nabla\vec{\phi}_\ell|^2 + b\phi_\ell^* \frac{\partial \phi_\ell}{\partial \theta}] dx\} d[\phi_\ell(x)]}, \tag{19.3}$$

and where:

$$U_\ell(\phi_\ell) = U(\phi_\ell) + \frac{1}{8} C_D d\Lambda [\Lambda^2 (\frac{U'}{|\phi_\ell|} + U'') - \tfrac{1}{2}(\frac{U'^2}{|\phi_\ell|^2} + U''^2)] + \mathcal{O}((d\Lambda)^2). \tag{19.4}$$

The average of ϕ_s thus only leads to an infinitesimal change in the function U and to a change in the cut-off in $\vec{k}$ space.

The similarity between these results and the original expressions

(13.9) and (13.10) can be improved by introducing scaled variables $\tilde{\vec{k}}$, $\tilde{\vec{r}}$ and $\tilde{\phi}$ which are defined by:

$$\vec{k} = (1-d\Lambda/\Lambda)\tilde{\vec{k}}, \tag{19.5}$$

$$\vec{r} = (1-d\Lambda/\Lambda)^{-1}\tilde{\vec{r}}, \tag{19.6}$$

$$\phi_{\ell} = (1-d\Lambda/\Lambda)^{\frac{1}{2}D-1}\tilde{\phi}. \tag{19.7}$$

In terms of these scaled variables one has the results:

$$Z(z,\beta,\Omega) = \tilde{N}\langle\exp\{-\int_{\tilde{\Omega}}\tilde{U}(\tilde{\phi})d\tilde{x}\}\rangle, \tag{19.8}$$

where the average is taken with respect to the weight functional:

$$\tilde{W}[\tilde{\phi}] = \frac{\exp\{-\int_{\tilde{\Omega}}[\frac{1}{2}|\tilde{\nabla}\tilde{\phi}|^2 + \tilde{b}\tilde{\phi}^* \frac{\partial\tilde{\phi}}{\partial\theta}]d\tilde{x}\}}{\int\exp\{-\int_{\tilde{\Omega}}[\frac{1}{2}|\tilde{\nabla}\tilde{\phi}|^2 + \tilde{b}\tilde{\phi}^* \frac{\partial\tilde{\phi}}{\partial\theta}]d\tilde{x}\}d[\tilde{\phi}(\tilde{x})]}, \tag{19.9}$$

and where:

$$\tilde{U} = U + \frac{d\Lambda}{\Lambda}[DU - (\tfrac{1}{2}D-1)|\tilde{\phi}|U' + \frac{1}{8}C_D\Lambda^3(\frac{U'}{|\tilde{\phi}|} + U'') -$$

$$\frac{1}{16}C_D\Lambda\,(\frac{U'^2}{|\tilde{\phi}|^2} + U''^2)] + \mathcal{O}((d\Lambda)^2). \tag{19.10}$$

Here the gradients are with respect to $\tilde{\vec{r}}$ and the primes denote differentation with respect to $|\tilde{\phi}|$. The arguments are $\tilde{x}$ and $\tilde{\phi}$ everywhere.

The effect of averaging over the short waves and rescaling the variables is thus threefold:

a) A decrease in the volume of the system from Ω to:

$$\tilde{\Omega} = (1-d\Lambda/\Lambda)^D\,\Omega. \tag{19.11}$$

b) An increase in the value of the constant b to:

$$\tilde{b} = (1-d\Lambda/\Lambda)^{-2}\,b. \tag{19.12}$$

c) A "renormalization" of the interaction from $U(\phi)$ to $\tilde{U}(\tilde{\phi})$. As both $\tilde{\phi}$ and $\tilde{x} \equiv (\tilde{\vec{r}},\theta)$ are integration variables we shall from here on omit the tilde on ϕ and x in (19.8), (19.9) and (19.10).

20. EQUIVALENCE WITH THE CLASSICAL SPIN SYSTEM

Equations (19.9) and (19.12) now lead to the following result. After a large number of renormalizations the effective value of b becomes large compared to unity. The weight (19.9) can be written in the spectral form:

$$\tilde{W}(\phi) = \prod_{|\vec{k}|\leqslant\Lambda} \prod_{\ell} w(c_{\vec{k},\ell}), \tag{20.1}$$

$$w(c_{\vec{k},\ell}) = \frac{1}{\pi}\,(\tfrac{1}{2}|\vec{k}|^2 - 2\pi\ell\tilde{b}i)\,\exp\{-(\tfrac{1}{2}|\vec{k}|^2 - 2\pi\ell\tilde{b}i)|c_{\vec{k},\ell}|^2\}. \tag{20.2}$$

But, as:

$$\lim_{\tilde{b}\to\infty} w(c_{\vec{k},\ell}) = \begin{cases} \delta(\mathrm{Re}\,c_{\vec{k},\ell})\,\delta(\mathrm{Im}\,c_{\vec{k},\ell}) & \text{if } \ell\neq 0 \quad (20.3)\\ \frac{1}{\pi}\,(\tfrac{1}{2}|\vec{k}|^2)\,\exp\{-\tfrac{1}{2}|\vec{k}|^2|c_{\vec{k},\ell}|^2\} & \text{if } \ell=0 \quad (20.4)\end{cases}$$

the only random fields which contribute in this case to the average (19.8) are those fields $\phi(\vec{r})$ which do not depend on θ. Hence, after a large number of renormalizations (19.8) has the form:

$$\Xi(z,\beta,\Omega) = (\text{constant})\,\frac{\int\exp\{-\int[\tfrac{1}{2}|\vec{\nabla}\phi|^2 + U(|\phi|)]d^D\vec{r}\}d[\phi(\vec{r})]}{\int\exp\{-\int[\tfrac{1}{2}|\vec{\nabla}\phi|^2]\,d^D\vec{r}\}d\,[\phi(\vec{r})]}. \tag{20.5}$$

As the critical exponents are determined by the way in which the ratio on the right hand side transforms under the renormalization group the critical behavior is the same as that of a D-dimensional system of classical spins with 2 components.

Note that this conclusion is quite general as it is based on (19.12) only; the explicit form (19.10) which relies on certain approximations, plays no role in this proof. For the sake of completeness we shall calculate the critical exponents ν and α using the approximation (19.10).

21. FIXED POINTS

The infinitesimal transformation

$$U(|\phi|) \to RU(|\phi|) \equiv \tilde{U}(|\phi|) \tag{21.1}$$

defines a flow in the space of interactions $U(|\phi|)$. We shall always define $U(|\phi|)$ in such a way that

$$U(0) = 0. \tag{21.2}$$

This implies that the constant part of U has to be absorbed in $\tilde{N}$:

$$\tilde{N} = N \exp\{-K\Omega\}, \tag{21.3}$$

$$K = K[U] \equiv \frac{d\Lambda}{\Lambda} \left[\frac{1}{8} C_D \Lambda^3 \left(\frac{U'}{|\phi|} + U''\right) - \frac{1}{16} C_D \Lambda \left(\frac{U'^2}{|\phi|^2} + U''^2\right)\right]_{|\phi|\to 0}. \tag{21.4}$$

The fixed points $U^*(|\phi|)$ of R are the solutions of the ordinary differential equation:

$$DU^* - (\tfrac{1}{2}D-1)|\phi|U^{*'} + \frac{1}{8} C_D \Lambda^3 \left(\frac{U^{*'}}{|\phi|} + U^{*''}\right) - \frac{1}{16} C_D \Lambda \left(\frac{U^{*'2}}{|\phi|^2} + U^{*''2}\right) = K^* \tag{21.5}$$

where

$$K^* = K[U^*], \tag{21.6}$$

is a constant which is determined by the differential equation. The boundary conditions are:

$$U^*(|\phi|) \to +\infty \quad \text{if} \quad |\phi| \to +\infty, \tag{21.7}$$

to guarantee the convergence of the average over ϕ-space, and

$$U^{*'}(0) = 0, \tag{21.8}$$

to guarantee that the solution stays finite if $|\phi| \to 0$.

The first boundary condition implies that U^* has the asymptotic behavior:

$$U^*(|\phi|) \cong A|\phi|^g \qquad (|\phi| \to +\infty) \tag{21.9}$$

where A and g have to be positive. Substituting into (21.5) and putting the sum of all terms of highest order equal to zero one finds:

$$g = 4, \tag{21.10}$$

$$A = \frac{4-D}{10 C_D \Lambda} \quad \text{or } A = 0. \tag{21.11}$$

Hence one finds the theorem: For D>4 only the trivial (Gaussian) fixed point $U^*=0$ exists; for D<4 a non-trivial fixed point exists.

22. THE NON-TRIVIAL FIXED POINT

For D near to 4 one puts:

$$D = 4-\varepsilon, \tag{22.1}$$

$$U^{*}(|\phi|) = \tfrac{1}{2}r^{*}|\phi|^{2} + s^{*}|\phi|^{4} + \ldots \tag{22.2}$$

The non-trivial fixed point can now be found by substituting the last equation into (21.5) and by assuming that $r = \mathcal{O}(\varepsilon)$, $s = \mathcal{O}(\varepsilon)$ and that all the higher order terms in (22.2) have coefficients which are at least $O(\varepsilon^2)$. This gives:

$$s^{*} = + \frac{\varepsilon}{10C_D\Lambda} + \mathcal{O}(\varepsilon^2), \tag{22.3}$$

$$r^{*} = - \frac{1}{5}\Lambda^{2}\varepsilon + \mathcal{O}(\varepsilon^2), \tag{22.4}$$

$$K^{*} = - \frac{1}{20} C_D\Lambda^{5}\varepsilon + \mathcal{O}(\varepsilon^2). \tag{22.5}$$

This is usually called the Ising fixed point. Note that (16.5) gives:

$$\lim_{b\to\infty} C_D(b) = 4\Lambda^{-4}(2\pi)^{-D} S_D(\Lambda). \tag{22.6}$$

Hence after many renormalizations C_D can be treated as a constant.

23. BEHAVIOR NEAR TO THE FIXED POINT

The behavior of the flow R in a vicinity of the Ising fixed point can be found by transforming:

$$U(|\phi|) = \tfrac{1}{2}r|\phi|^{2} + s|\phi|^{4} + \ldots \tag{23.1}$$

with:

$$r = r^{*} + \Delta r, \tag{23.2}$$

$$s = s^{*} + \Delta s, \tag{23.3}$$

where Δr and Δs are small. Substituting into (21.1) and (19.10) one finds that:

$$\tilde{\Delta}r = \Delta r + 2(\Delta r + 2C_D\Lambda^{3}\Delta s - 2C_D\Lambda r^{*}\Delta s - 2C_D\Lambda s^{*}\Delta r)\frac{d\Lambda}{\Lambda}, \tag{23.4}$$

$$\tilde{\Delta}s = \Delta s + (\varepsilon\Delta s - 20C_D\Lambda s^{*}\Delta s)\frac{d\Lambda}{\Lambda} . \tag{23.5}$$

Here all terms quadratic in Δr and Δs have been neglected. In a more algebraic notation this can be written as:

$$\begin{pmatrix}\tilde{\Delta}r\\ \tilde{\Delta}s\end{pmatrix} = R \begin{pmatrix}\Delta r\\ \Delta s\end{pmatrix} \tag{23.6}$$

where the explicit form of the matrix R is found from (22.3) and (22.4):

$$R = \begin{bmatrix} 1 + 2\,(1 - \frac{\varepsilon}{5})\,\frac{d\Lambda}{\Lambda} & 4C_D\Lambda^3\,(1 + \frac{\varepsilon}{5})\,\frac{d\Lambda}{\Lambda} \\ 0 & 1 - \varepsilon\,\frac{d\Lambda}{\Lambda} \end{bmatrix} \tag{23.7}$$

The eigenvalues of this matrix are:

$$\lambda_1 = 1 + 2\,(1 - \frac{\varepsilon}{5})\,\frac{d\Lambda}{\Lambda} , \tag{23.8}$$

$$\lambda_2 = 1 - \varepsilon\,\frac{d\Lambda}{\Lambda} . \tag{23.9}$$

After ℓ renormalizations:

$$U_o \xrightarrow{R} U_1 \xrightarrow{R} U_2 \xrightarrow{R} \ldots . \xrightarrow{R} U_\ell \tag{23.10}$$

the interaction $U_\ell(|\phi|)$ is characterized by:

$$r_\ell = r^{*} + \Delta r_o\,\lambda_1^\ell, \tag{23.11}$$

$$s_\ell = s^{*} + \Delta s_o\,\lambda_2^\ell. \tag{23.12}$$

As $\lambda_1>1$ and $\lambda_2<1$ the streamlines of the flow in (r,s) space are as drawn qualitatively in figure 7. For $\ell>>1$ one thus has:

$$r_\ell = r^{*} + \Delta r_o\,\lambda_1^\ell, \tag{23.13}$$

$$s_\ell \cong s^{*}. \tag{23.14}$$

In the language of the renormalization group r is a relevant variable and s an irrelevant variable. Note that b is also a relevant variable.

24. THE CRITICAL EXPONENT ν

The constant Δr_o is still a function $f(\mu,T)$ of temperature and chemical potential. Let $T_c(\mu)$ denote the temperature where this

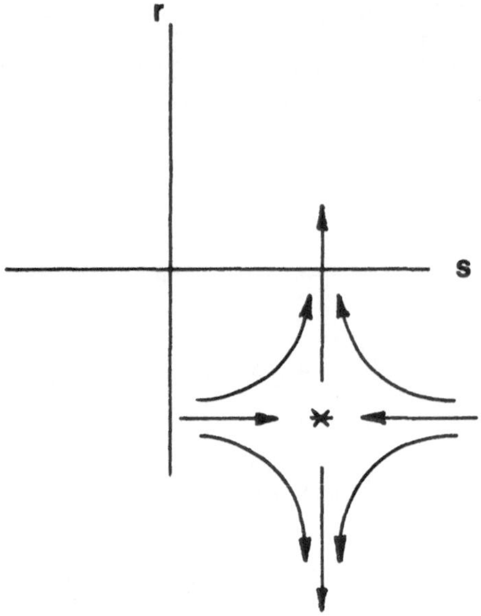

fig. 7: Qualitative behavior of R in the vicinity of the Ising fixed point.

function vanishes:

$$f(\mu,T_c(\mu)) = 0. \tag{24.1}$$

For T near to T_c one has:

$$f(\mu,T) = (T-T_c)\left(\frac{\partial f}{\partial T}\right)_{T=T_c(\mu)}. \qquad (|T-T_c|<<T_c) \tag{24.2}$$

In combination with equation (23.13) this shows that:

$$U_\ell(\mu,T=T_c+\tau) = U_{\ell+1}(\mu,T=T_c+\frac{\tau}{\lambda_1}). \quad (\ell>>1) \tag{24.3}$$

But after ℓ renormalizations the length scale has increased with a factor $(1-d\Lambda/\Lambda)^{-\ell}$. Consequently the correlation length $\xi(\mu,T)$ has the property:

$$(1-\frac{d\Lambda}{\Lambda})^{\ell}\xi(\mu,T=T_c+\tau) = (1-\frac{d\Lambda}{\Lambda})^{\ell+1}\xi(\mu,T=T_c+\frac{\tau}{\lambda_1}). \tag{24.4}$$

This shows that ξ must diverge for $\tau\downarrow 0$, thereby identifying T_c with the critical temperature. Substituting the asymptotic formula:

$$\xi(T) \cong (T-T_c)^{-\nu} \qquad (\mu \text{ fixed, } T\downarrow T_c) \qquad (24.5)$$

and (23.8) one finds:

$$\nu = \tfrac{1}{2} + \frac{\varepsilon}{10} + \mathcal{O}(\varepsilon^2) \qquad (24.6)$$

for the critical exponent of the correlation length.

25. THE CRITICAL EXPONENT α

The relation (24.5) implies that also

$$\xi(\mu,T) \cong |\mu-\mu_c(T)|^{-\nu} \qquad (T \text{ fixed, } \mu\uparrow\mu_c(T)). \qquad (25.1)$$

The scaling hypothesis states that the singular part of any thermodynamic function depends on μ and T through $\xi(\mu,T)$ only. Hence the singular part of $\beta p(\mu,T)$, which has the same dimension as ξ^{-D}, must have the form:

$$\beta p(\mu,T) \cong |\mu-\mu_c(T)|^{+\nu D} + \text{analytic function.} \qquad (25.2)$$

Just like in section 10 this implies a specific heat critical exponent:

$$\alpha = \frac{2-\nu D}{\nu D-1} \,. \qquad (25.3)$$

Substituting from (24.6) one finds:

$$\alpha = \frac{\varepsilon}{10} + O(\varepsilon^2). \qquad (25.4)$$

For D=3 this gives $\alpha \cong 0.10$ by extrapolation.

26. CONCLUDING REMARKS

The two approximation schemes for the Boson partition function are somewhat complementary: the saddle point method (section 6-11) seems more appropriate for the system on that side of the λ-line where a condensate is present; the renormalization group approach (sections 12-25) is more appropriate at the other side of the λ-line. However, in the second paper by Singh[12] it is demonstrated how the renormalization scheme can be applied to the condensed phase too; in the same way a more accurate treatment of the combinatorics of the vortex lines should also lead to a description of the high-temperature side of the λ-line.

ACKNOWLEDGMENTS

The material of sections 2-11 is based in part on seminars presented at the National Bureau of Standards and the Universities of Maryland, Toronto and Montréal.

The author is indebted to Mrs. C.J. Cohn for typing the manuscript and Mr. P.A. Kops for making the line drawings.

REFERENCES

1) F.W. Wiegel, Phys. Rep. 16 (1975) 57.
2) S.P. Ohanessian, "The interacting Bose gas: formalism in coherent states and applications", Thesis, Ecole Polytechnique Fédérale de Lausanne (1976).
3) F.W. Wiegel and J. Hijmans, Proc. Kon. Ned. Acad. B77 (1974) 177.
4) F.W. Wiegel and J. Hijmans, Proc. Kon. Ned. Acad. B77 (1974) 189.
5) R. Hirota, Thesis, Northwestern University (1961).
6) F.W. Wiegel and J.B. Jalickee, Physica 57 (1972) 317.
7) F.W. Wiegel, Physica 65 (1973) 321.
8) G. Toulouse and P. Pfeuty: "Introduction au groupe de renormalisation et a ses applications", Presses Univ. de Grenoble (1975).
9) S.K. Ma: "Moderne theory of critical phenomena", Benjamin (1976).
10) C. Domb and M.S. Green: "Phase transitions and critical phenomena", Vol. 6, Academic Press (1976).
11) K.K. Singh, Phys. Lett. 51A (1975) 27.
12) K.K. Singh, Phys. Lett. 57A (1976) 309.
13) M. Baldo, E. Catara and U. Lombardo, Lett. Nuov. Cim. 15 (1976) 214.
14) S.P. Ohanessian and A. Quattropani, Helv. Phys. Acta 46 (1973) 473.
15) S.K. Ma, Phys. Rev. Lett. 29 (1972) 1311.
16) F. Family and H.E. Stanley, Phys. Lett. 53A (1975) 111.

FUNCTIONAL EQUATIONS AND THEIR TREATMENT IN NON-EQUILIBRIUM STATISTICAL MECHANICS

Iwao Hosokawa

Dept. Mech. Engrg., Iwate University

Morioka 020, Japan

ABSTRACT

The classical BBGKY hierarchy is reformulated into a single functional equation. A suitable modification of this (neglecting the anti-irreversible or entropy-destructive process) gives an irreversible dynamics of the many-particle system evolution, which is called the functional random-walk model.

Then, the basic (functional-differential) equation governs formally the time development of the probability density functional of the particle-number density in the one-body phase space. Some topics, mathematical as well as physical, around this equation are presented. For instance, the integral representation of the equation expresses an ensemble of generalized Brownian paths in the function space of the particle-number density; any state of the system approaches irreversibly to the ultimate equilibrium state (if the system is closed), for which the probabilistic mode in the function space is nothing other than of canonical form, etc.

I. INTRODUCTION

There have been many discussions upon the irreversible approach to the ultimate equilibrium of a closed many-body system since Boltzmann. The Boltzmann kinetic equation met with the first success in explaining this matter, so as to derive the Maxwell-Boltzmann distribution of molecular velocity at equilibrium. However, the exact Liouville dynamics of a many-body system is time reversible. Thus, many disputes have been centered on how to achieve the Boltzmann-type kinetic equation with irreversibility as an appropriate first approximation to the Liouville dynamics [1]. "Coarse-graining" has been a key word of disputes on this problem.

On the other hand, it was a serious matter for Boltzmann and others to understand an equilibrium state on the exact mechanical basis. They required the ergodic hypothesis to replace the long-time average by the phase average of summable functions in phase space, and justified the establishment of microcanonical distribution on an energy-constant surface in phase space. Then, it seemed to be immediate to derive canonical distribution (including the Maxwell-Boltzmann distribution) on the basis of microcanonical distribution. But it should be commented that the process of derivation is not purely mechanical but necessitates to use the other discipline in statistics, called the central limit theorem [2]. This is acknowledged as a kind of statistical operation, which randomly mixes many independent dynamical paths in a large ensemble.

Anyway, it is apparent that to introduce some kind of operation beyond pure many-body mechanics is essential in order to formulate an irreversible dynamics of a many-body system, as well as to derive canonical distribution that allowed Gibbs so much to develop the thermodynamics of equilibrium grounded on microscopic mechanics.

In this lecture, the author would like to present a functional formalism of non-equilibrium statistical mechanics, which holds the irreversible perspective of the Liouville dynamics and predicts the canonical distribution for one body in a self-consistent way at the equilibrium state of a closed system. The so-called, some kind of operation to be required for getting a final equilibrium in this formalism is not anything conventional, at least, on surface; but a formally rather simple mathematical one. After the formulation is made in Chapters II and III, the interrelation of the present operation to the concept of "coarse-graining" will be explained in Chapter IV.

We call this irreversible dynamics the functional random-walk model, because the basic functional equation is of the Fokker-Planck type, defined in some Riemannian space. A time-dependent solution can generally be expressed in terms of a kind of path integral. A steady-state solution may be found by the method of entropy, which is developed in Chapter V. From this solution, it is deduced in Chapter VI that the one-body distribtion should be canonical in a self-consistent way for a closed system. Some interesting properties of our entropy introduced above are shown in Chapter VII and VIII, including the derivation of Onsager's reciprocity relation.

In this lecture, all the discussion is confined to a one-component, classical many-body system, for simplicity. The readers who are interested in application to a multi-component, special-relativistic particle and radiation system are referred to the other work of the author [3], of which the mathematical presentation should be somewhat corrected in the present way but the essence of argument shall be unchanged. The Addendum concerns the Gibbs theorem.

II. FUNCTIONAL FORMALISM OF THE BBGKY HIERARCHY

An idea of using a functional to describe a statistical state of a many-body system was initiated by Bogoliubov [4], who introduced it in 1946 to reform the Bogoliubov-Born-Green-Kirkwood-Yvon (BBGKY) hierarchy into a single closed equation for the generating functional. The generating functional is defined, for a one-component system, as

$$L(y,t) = 1 + \sum_{s=1}^{\infty} \frac{1}{s!} \int_X \cdots \int_X F_s(x_1,\ldots,x_s,t) y(x_1)\cdots y(x_s) dx_1 \cdots dx_s . \tag{2.1}$$

Here t is the time variable, X the one-body phase space, $x \in X$, $y(x)$ a proper complex function and F_s the s-body generic distribution function. We treat the limiting state in which both the number of particles N and the spatial volume V containing them tend to infinity, holding the relation $N/V = \text{const}$, say n. Then, F_s should satisfy

$$F_s(x_1,\ldots,x_s,t) = \lim_{V\to\infty} \int_X F_{s+1}(x_1,\ldots x_s, x_{s+1},t) dx_{s+1}/V \tag{2.2}$$

and $F_0 \equiv 1$. The BBGKY hierarchy of equations governing $\{F_s\}$ is, thus, transformed into the Bogoliubov closed-form, linear functional-differential equation:

$$\partial L(y,t)/\partial t = L_B\, L(y,t). \tag{2.3}$$

Once this equation could be solved, we might have all the information of $\{F_s\}$.

However, Bogoliubov questioned to himself about convergence of the function Taylor series (2.1) and considered the condition on y for convergence. As a result, it becomes hard to solve (2.3) in the function space of y.

Avoiding this difficulty, the author proposed to limit all y to imaginary functions [5]. In other words, the functional is redefined as

$$\psi(y,t) = 1 + \sum_{s=1}^{\infty} \frac{i^s}{s!} \int_X \cdots \int_X F_s(x_1,\ldots,x_s,t) y(x_1) \cdots y(x_s) dx_1 \cdots dx_s , \tag{2.4}$$

where y should be an arbitrary real-valued function. And we expect that all ψ having a finite (maximum) norm:

$$\|\psi(y,t)\| = \max_{y\varepsilon A}|\psi(y,t)| < \infty \tag{2.5}$$

corresponds to a physically meaningful state of the system, where A is the function space of y(x); A may be understood as the M-dimensional Euclidean space, say A^M, in the limit $M\to\infty$, when y(x) is mode-analyzed by the first M functions of a complete orthonormal function set in X.

For example, consider the limit situation in which all particles are independent; then

$$F_s(x_1,..x_s) = F_1(x_1)..F_1(x_s) \tag{2.6}$$

so that

$$\psi(y) = \exp[i \int_X F_1(x)y(x)dx] \tag{2.7}$$

which gives

$$\|\Psi(y)\| = 1. \tag{2.8}$$

Moreover, consider the integrated form of (2.3):

$$\Psi(y,t) = \int_A K(y,t/y',t')\psi(y',t')\delta y', \quad t > t' \tag{2.9}$$ *

where K is the Green functional which is usually absolutely integrable. Then,

$$\|\psi(y;t)\| \leq \max_{y\varepsilon A} \int_A |K|\delta y'\|\psi(y',t')\|. \tag{2.10}$$

This means that if the initial functional is given with a finite norm, at any time later the functional continues having a finite norm. Thus, the problem of convergence has gone out of sight, so long as we deal initially with functionals with a finite norm. We call such a functional a state functional. It may be added that unity is the lower limit of the norm, since always

$$\|\psi(y,t)\| \geq 1, \tag{2.11}$$

* The functional integration may be understood as the limit at $M\to\infty$ of the integration with the M-dimensional Euclidean measure divided by $(2\pi)^{M/2}$, when A is approached by A^M; there, the amplitudes in mode-analysis of y(x) constitute R^M. (See Ref. 5.)

as is known from (2.4)

Now, it is easy to rewrite (2.3) for the state functional ψ. That is, in the explicit form,

$$\frac{\partial\psi}{\partial t} = i\int_X y(x)[H_1(x); \frac{\delta\psi}{i\delta y(x)}]dx - \frac{1}{2}\iint_{XX}\{y(x)y(x')-iny(x)-iny(x')\}$$

$$\times [\phi(|q-q'|); \frac{\delta^2\psi}{i^2\delta y(x)\delta y(x')}]dxdx', \qquad (2.12)^{\star}$$

where $H_1(x)$ is the one-body Hamiltonian, $\phi(|q-q'|)$ the potential of interaction between particles, $x = (q,p)$, q and p being the position and momentum vector, repectively, and [;] denotes the Poisson bracket. To identify the equivalence of the ψ equation (2.12) and the BBGKY hierarchy, one may directly substitute (2.4) into (2.12).

The condition (2.2) is rewritten in terms of ψ as

$$\psi = \lim_{V\to\infty} \int_X \frac{\delta\psi}{i\delta y(x)} dx/V \qquad (2.13)$$

Since (2.12) as well as (2.13) are linear equations, we have to make the solution for the state functional unique by the condition of normalization [which comes from the definition (2.4)]:

$$\psi(0,t) = 1 \qquad (2.14)$$

Once a state functional satisfies (2.13) and (2.14) at an initial time, it is guaranteed that (2.12) governs all the development of the state functional at a later time [5].

III. BASIC EQUATION OF THE FUNCTIONAL RANDOM-WALK MODEL

A. Fourier-transformed State Functional

★ $\delta/\delta y(x)$ denotes the functional differentiation, which may be defined as

$$\frac{\delta G}{\delta y(x)} = \lim_{\varepsilon\to 0} \frac{G[y(x')+\varepsilon\delta(x-x')] - G[y(x')]}{\varepsilon} .$$

If y(x) is mode-analyzed in M dimensions as $y(x) = \sum_{i=1}^{M} a_i s_i(x)$, where $\{s_i(x)\}$ is a complete orthonormal function set in X, the functional derivative may be approached by $\frac{\delta G}{\delta y(x)} = \sum_{i=1}^{M} s_i(x)\frac{\partial G}{\partial a_i}$.

First, we try to Fourier-transform the state functional ψ. In order to make safe the mathematical treatment, we adopt the M-dimensional Euclidean approach for the function space, $_M A^M$, as was already introduced in Chapter II; there we may write y^M for y and ψ^M for $\psi(y^M)$. In this convention, the Fourier transform is represented by

$$\psi^M(y,t) = \int_{A^M} \rho^M(z,t)\exp[\, i \int_X y^M(x) z^M(x)dx]\, \delta z^M. \tag{3.1}$$

Attachement of the superscript M as this, however, is rather tedious in every equation. Then, it will be omitted hereafter unless necessary in particular, with the understanding of this approach.*

As a result, the basic equation (2.12) transforms to the following equation:

$$\frac{\partial \rho}{\partial t} = - \int_X \frac{\delta}{\delta z(x)} \{Qz(x)\rho\} + \frac{1}{2} \iint_{XX} \frac{\delta^2}{\delta z(x)\delta z(x')}$$

$$\times \{[\phi(|q-q'|);\ z(x)z(x')]\,\rho\}dxdx', \tag{3.2}$$

where Q denotes the nonlinear operator such that

$$Qz(x) = [H_1(x);\ z(x)] + n\int_X dx'[\phi(|q-q'|);\ z(x)z(x')]. \tag{3.3}$$

It is noted that $\partial z/\partial t = Qz$ is nothing but the self-consistent Vlasov equation, therefore, Q may be called the Vlasov operator.

Corresponding to (2.14) and (2.13), we have two conditions on ρ;

$$\int_A \rho\delta z = 1 \text{ and} \tag{3.4}$$

$$\int_A \rho[\lim_{V\to\infty} \int z(x)\frac{dx}{V} - 1] \exp[\, i\int_X y(x)z(x)dx]\, \delta z = 0. \tag{3.5}$$

From (3.5), ρ should vanish for all z unless

$$\lim_{V\to\infty} \int_X z(x)dx/V = 1, \tag{3.6}$$

---

* Although M should always be finite in discussion of ρ in order for (3.1) to be mathematically meaningful, it is possible to assume the convergence of $\{\psi^M\}$. In fact, $\{\psi^M\}$ can make a Cauchy sequence, if y(x) belongs to the Banach space L_1 ($\int|y|dx < \infty$), and if the derivative of ψ^M is uniformly bounded[17].

which defines a hyperplane in A.

Since our basic equations so far obtained are equivalent to the BBGKY hierarchy or the Liouville equation in the limit $N = nV \to \infty$ (n finite), any dynamical property of the particle system is preserved in our formalism. It is easy to verify the time reversibility of the basic equation (3.2): if $\rho[z(q,p),t]$ is a solution, then $\rho[z(q,-p),-t]$ becomes another solution. Also, it is possible to derive the conservation laws of total mass, momentum, angular momentum and energy on the direct basis of (3.2) if the system is closed[6]. In this case, on comparing (2.4) with the series expression in (3.1) in powers of y, we have

$$F_1(x) = \int_A \rho(z,t)z(x)\delta z, \tag{3.7}$$

$$F_2(x,x') = \int_A \rho(z,t)z(x)z(x')\delta z, \tag{3.8}$$

which can be used to express the above-described physical quantities.

B. From the Dynamical Process to the Markovian Process

The basic equation (3.2) is very close in form to the generalized Fokker-Planck or Kolmogorov equation which describes a Markov stochastic process. The first term of the right-hand side is the analog of the so-called friction or drift term and the second resembles the diffusion term. However, there is a difference in principle between (3.2) and the Fokker-Planck or Kolmogorov equation; the former is time reversible, as was mentioned already, while the latter should be irreversible. This reflects the fact that the coefficient function $[\phi;z(x)z(x')]$ inside the second-order derivative is not positive definite. This similarity and difference may contain a key to solving the noted historical question: how to bridge the microscopic reversibility and the macroscopic irreversibility in the system evolution in a general way.

Now, if the coefficient function is considered as the sum of the positive definite and negative definite parts at each local point in A [exactly saying, the hyperplane defined by (3.6)], the former plays a role of creating an irreversible process, while the latter that of destroying it or creating an anti-irreversible process so that both roles may offset each other to cause the reversible process exactly. So, in order to extract a purely irreversible process from the dynamical process, evidently it is necessary and sufficient to retain only the former part of the coefficient function, so that (3.2) reduces to a true Fokker-Planck or Kolmogorov equation. This is the simplest universal way of introducing the positive time-arrow in the system evolution. There is no other assumption in this procedure, such as weak interaction, diluteness, etc. Therefore, its application will be free from any such restriction on physical conditions.

Let the Fokker-Planck or Kolmogorov equation thus obtained be written as

$$\frac{\partial\hat{\rho}}{\partial t} = L\hat{\rho} \equiv - \int_X \frac{\delta}{\delta z(x)} [Qz(x)\hat{\rho}]\,dx + \frac{1}{2}\int\int_{XX} \frac{\delta^2}{\delta z(x)\delta z(x')}$$

$$\times \{P[\phi; z(x)z(x')]\ \hat{\rho}\}dx\,dx', \qquad (3.9)$$

where the symbol ρ has been replaced by $\hat{\rho}$ to distinguish its approximate nature because of the above procedure, and P is the operator on the subsequent symmetric function to make it positive definite in the way just described. Together with the condition (3.4), now it is possible to interpret $\hat{\rho}$ as the probability density in the hyperplane (3.6)] in A.

If the diffusion term is neglected out of (3.9), it can be found that the equation governs the time evolution of the probability on an ensemble of the trajectories in A, which develop from various initial values according to the Vlasov equation, $\partial z/\partial t = Qz$. (Consider the characteristic curves in A.) This fact is obvious from the perfect analogy with the Hopf equation in turbulence mechanics[7], if we go back to the equation for the characteristic functional $\hat{\Psi}$ which corresponds to $\hat{\rho}$. In this case, we really deal with nothing but a so-called turbulent field which is basically governed by the Vlasov equation instead of the Navier-Stokes fluid-dynamic equation. If the diffusion term is included in (3.9), the motion of z is always affected by some *internal* random force implied by $P[\phi; zz]$, as is well expected from the Langevin equation for Brownian motion (whose stochastic process may also be governed by the Fokker-Planck equation). Thus, *what we may call a turbulence* of the field z goes away from the Hopf turbulence, getting more irregular and more random. For this case, there is the analogy with the Novikiv equation in the turbulence mechanics with random force action[8] (though there is a slight difference in situation in that the random force in the Novikov equation is not internal but externally given independently of the field z). From all these facts, it is quite reasonable to interpret z(x) as the *stochastic* particle-number density in X, normalized in the sense of (3.6). Obviously, z(x) is not a macroscopic observable [as (3.7) and (3.8) show] nor is it a microscopic density such as considered by Klimontovich[9], since it can be a regular function; but it may be understood as having an intermediate property. [As for the essential nonnegativeness of z(x), see the appendix of Ref. 6.]

Let us see if (3.9) is consistent with the conservation laws in spite of the modification by P. It is known from a simple functional calculus that the term with a second-order functional derivative in the basic equation has no effect on the conservation laws of the quantities expressed only by $\hat{F}_1$ (which is linear in z). Here, let $\hat{F}_1, \hat{F}_2, \ldots, \hat{F}_s$ be the approximations to $F_1, F_2, \ldots, F_s$, when ρ is replaced by $\hat{\rho}$. [See (3.14).] Then, it is evident that the conservation laws of total mass, momentum, and angular momentum hold also for (3.9). For the energy case, however, we have

$$\frac{\partial}{\partial t}\Big[\int_X nH_1(x)\hat{F}_1(x)dx + \iint_{XX}\frac{1}{2}n^2\phi(q-q')\hat{F}_2(x,x')dxdx'\Big]$$

$$= \frac{1}{2}n^2\int_A\iint_{XX}\phi P[\phi;z(x)z(x')]dxdx'\,\hat{\rho}\delta z, \tag{3.10}$$

which cannot vanish in general. This shows that (3.9) does not yield the conservation of total energy, without any special condition on its usage. Such a condition can be made by confining the space of z into a manifold in which the conservation of total energy is strictly insured. Indeed, since the total energy found in the left-hand side of (3.10) is rewritten in terms of (3.7) and (3.8) as

$$\int_A \hat{\rho}\Big[\int_X nH_1 z(x)dx + \iint_{XX}\frac{1}{2}n^2\phi z(x)z(x')dxdx'\Big]\delta z,$$

we can impose the condition on z,

$$\lim_{V\to\infty}\Big[\int_X nH_1 z(x)dx + \iint_{XX}\frac{1}{2}n^2\phi z(x)z(x')dxdx'\Big]/V = \text{const}, \tag{3.11}$$

when the system is closed.

It is thus essential that our basic equation (3.9) should be redefined in such a Riemannian space as prescribed by (3.11) for a closed system. There is no special difficulty in dealing with a Fokker-Planck or Kolmogorov equation in a Riemannian space[10]. Hence, our equation governs a generalized Brownian motion of the field z in the special Riemannian space, say B, prescribed by the condition: $z \geq 0$, (3.6) and (3.11). This is the reason why our irreversible dynamics was named as the functional random-walk model. If the system is not closed, the condition (3.11) may be replaced by other conditions required according to the physical situation.

A form of the initial value of $\hat{\rho}$ is simply given, when initially the correlations among particles are negelcted. From (2.7), we may have

$$\hat{\Psi}(y,0) = \exp\Big[\, i\int_X y(x)\hat{F}_1(x,0)dx\Big], \tag{3.12}$$

the inverse Fourier transform of which is symbolically written as

$$\mathfrak{p}(z,0) = \delta[z(x)-\hat{F}_1(x,0)]. \tag{3.13}$$

(See Ref. 5 for the delta functional, $\delta[\ ]$, and Ref. 17 for its symbolic sense.) This means that our random walk begins from the single point $F_1(x)$ in the Riemannian space B. Then, the total of the Brownian trajectories beginning from this point provides all the physical information of the system evolution. Indeed, all F_s are given by the correlations of z(x), as are obtained in a similar way to (3.7) and (3.8);

$$\hat{F}_s(x_1,\ldots,x_s,t) = \int_A z(x_1)\ldots z(x_s)\mathfrak{p}(z,t)\delta z, \tag{3.14}$$

where $\mathfrak{p}\delta z$ vanishes outside B; in this sense, the change of notation, $\mathfrak{p} \to \mathfrak{p}\delta[B]$, is pertinent, as will be used in the next section. As time goes on, we expect that all trajectories finally cover the whole B with a unique density, as a result of the property of our Markovian process. The $\mathfrak{p}\,\delta[B]$ which expresses this state, apparently, corresponds to microcanonical distribution in phase space in ergodic theory.

C. Basic Equation and Integral Representation

We note, first, that z(x) is a Euclidean coordinate in A. In order to reformulate (3.9) in B, it is convenient to retain the integral equation for weak solution of (3.9). That is

$$\int_0^\infty\int_A \left(\frac{\partial W}{\partial t} - L^\star W \right)\mathfrak{p}\delta[B]\,\delta z dt - \int_A W(z,0)\mathfrak{p}(z,0)\delta\,[B]\,\delta z = 0, \tag{3.15}$$

where W(z,t) is an arbitrary, second-order differentiable test functional which vanishes outside a finite region of the z,t-space; $\delta[B]\delta z$ denotes the differential measure in A such that it vanishes outside B but is the Riemannian measure in B, and $L^\star$ is the functional differential operator adjoint to L in (3.9).

We may introduce the curvilinear coordinate $\{\mu_i\}$ in B. Then, we can set the measure equality in the coordinate transformation,

$$\mathfrak{p}(z)\ \delta[B]\,\delta z = \mathfrak{p}_B(\mu)g^{1/2}\prod_i d\mu_i. \tag{3.16}$$

where g is the determinant of the metric tensor g_{ij}, which is defined as

$$g_{ij} = \int_X \frac{\partial z(x)}{\partial \mu_i}\frac{\partial z(x)}{\partial \mu_j}\,dx. \tag{3.17}$$

From (3.15) and (3.16) and assuming that any W vanishes at the boundary of B, we have the equation in the Riemannian coordinate,

$$\frac{\partial \hat{\rho}_B g^{1/2}}{\partial t} = -\frac{\partial}{\partial \mu_i}\left[\int_X \frac{\delta \mu_i}{\delta z(x)} Qz(x)dx\, \hat{\rho}_B g^{1/2}\right] + \frac{1}{2}\frac{\partial^2}{\partial \mu_i \partial \mu_j}\left[\int_X\int_X \frac{\delta \mu_i}{\delta z(x)} \frac{\delta \mu_j}{\delta z(x')} D(z;x,x')dxdx'\hat{\rho}_B g^{1/2}\right], \tag{3.18}$$

where

$$D(z;x,x') \equiv \mathcal{P}[\phi;z(x)z(x')] \tag{3.19}$$

and the subscript i, j obeys the summation convention.

In the same right, it is possible to make the equation for $\hat{\rho}\delta[B]$ from (3.15). Namely,

$$\frac{\partial \hat{\rho}\delta[B]}{\partial t} = -\int_X \frac{\delta}{\delta z(x)}\{Qz(x)\hat{\rho}\delta[B]\}dx + \frac{1}{2}\int_X\int_X \frac{\delta^2}{\delta z(x)\delta z(x')}\{D(z;x,x')\hat{\rho}\delta[B]\}dxdx'. \tag{3.20}$$

This offers the equation in a seemingly Euclidean coordinate, but z(x) is guaranteed to be always in B; also $\delta/\delta z(x)$ is non-trivially active only in B. This equation is sometimes easier to manipulate than (3.18). In this case, (3.4) is changed to be

$$\int_A \hat{\rho}\delta[B]\delta z = 1. \tag{3.21}$$

A general solution for $\hat{\rho}(z,t)\delta[B]$ may be given in terms of the functional Green kernel G. That is,

$$\hat{\rho}(z,t)\delta[B] = \int_A G(z,t/z',0)\hat{\rho}(z',0)\delta[B]\,\delta z', \tag{3.22}$$

$$G(z,t/z',t') = \lim_{\Delta t\to 0}\int_A..\int_A P_{\Delta t}(z^L/z^{L-1})..P_{\Delta t}(z^1/z^0)\prod_{k=1}^{L-1}\delta z^k, \tag{3.23}$$

where $z = z^L$, $z' = z^0$, $\Delta t = (t-t')/L$, and the superscripts indicate the order of the time-subintervals. The infinitesimal Green kernel $P_{\Delta t}$ is explicitly obtained from (3.20) as

$$P_{\Delta t}(z^{k+1}/z^k) = \int_A \exp\{i \int_X y^k(x)\, [z^k(x)-z^{k+1}(x) + \Delta t Q z^k(x)]\, dx$$

$$- \frac{1}{2}\Delta t \int_X \int_X y^k(x) y^k(x') D(z^k;x,x')dxdx'\}\delta y^k \delta[B]^{k+1}, \quad (3.24)$$

where $\delta[B]^k$ denotes $\delta[B]$ for z^k. Obviously, $P_{\Delta t}(z_k^{k+1}/z^k)$ is Gaussian for z^{k+1} in B, whose parameters depend on z^k. Hence, it may be said that (3.23)-(3.24) provide a generalization of the Wiener measure. Accordingly, (3.22) may be understood as a kind of generalized path integral due to a generalized Brownian motion in the space B. For direct computation of an integral with such a measure, a high-speed computer is expected to be useful in future, with the aid of Monte-Carlo quadrature [6], etc.

IV. QUALITY OF "COARSE-GRAINING"

Our way from the dynamical process to the Markovian process, developed in Chapter III, B, is purely mathematically motivated. It is worthwhile to pick up the physical implication latent in this simple mathematical operation, which is represented by $\mathcal{P}$.

Let us examine the change by $\mathcal{P}$ of the quadratic form, $\int\int y(x) y(x')[\phi;z(x)z(x')]dxdx'$, in some detail. If we *coarse-grain* X by such a finite small cell Δx, that it can always contain, at least, two interacting particles, the quadratic form may be approximated as

$$\int_X \int_X y(x)y(x')[\phi;z(x)z(x')]\, dxdx'$$

$$= \sum_i y(x_i)y(x_i')[\phi(|q-q'|);z(x_i)z(x_i')]\,\Delta x_i^2. \quad (4.1)$$

where both x_i and x_i' are randomly chosen in Δx_i. We assume that ϕ has eventually the nature of short-range interaction so that particles in different cells cannot interact each other, since its long-range effect is taken full account of by the Q term in the basic equation. On putting

$$A_{12} \equiv [\phi;z(x_i)z(x_i')] = a_i/2 = A_{21} \text{ for } x_i \neq x_i'$$

$$A_{11} \equiv [\phi;z(x_i)z(x_i)] = 0 = A_{22} \quad (4.2)$$

(remember $\partial\phi/\partial q|_{q=q'} = 0.$), we can easily find the principal axes for the matrix A.

After all, we have

$$\sum_i y(x_i)\, y(x_i') \; [\phi; z(x_i) z(x_i')]\, \Delta x_i^2 = \sum_i (\eta_i^2 - \zeta_i^2)\, |a_i|\, \Delta x_i^2/2, \qquad (4.3)$$

where

$$\eta_i = [y(x_i) + y(x_i')]/2^{1/2}$$

$$\zeta_i = [y(x_i) - y(x_i')]/2^{1/2}. \qquad (4.4)$$

Therefore, our operator P causes

$$\sum_i y(x_i) y(x_i') P[\phi; z(x_i) z(x_i')]\, \Delta x_i^2 = \sum_i \eta_i^2 |a_i| \Delta x_i^2/2. \qquad (4.5)$$

What is meant by (4.5) is that ζ_i in (4.4) is completely neglected in our functional random-walk model. This is quite natural, if y is a slowly varying function in Δx; and this assumption is a fundamental premise of the coarse-graining of X we have just started from, or of the finite difference approach to functions y and z. Thus, the coarse-graining nature in our modelling is clear.

Somewhere else [6,12], other physical arguments on this nature were presented (though they were rather inaccurate discussions). But the present one looks easiest to grasp.

V. PROOF OF ENTROPY PRODUCTION

According to Shannon [13], it might seem natural to define the H function (per degree of freedom) as

$$H_s = \lim_{M\to\infty} \int_A \hat{\rho}^M \log \hat{\rho}^M \delta[B]\, \delta z^M / M. \qquad (5.1)$$

However, we should take into account that there are many possible permutations of particles which yield the same density field z(x). Then, the total number of permutations makes the weight for the field z(x), which is calculated as

$$\frac{N!}{\prod_i [n z(x_i) \Delta x]!} = \exp\{- \sum_i n z(x_i) \log z(x_i) \Delta x + \mathrm{const}\}, \qquad (5.2)$$

using the Stirling formula for $N = nV \to \infty$.

If the weight now calculated is considered, the suitable form of our H function should be

$$H = \int_A \hat{\rho} \log \left[\hat{\rho} \exp\left(n\int_X z\log z dx\right)\right] \delta[B]\, \delta z. \tag{5.3}$$

Here, all M to be attached as in (5.1) have been omitted according to our convention described in Chapter III, A. The limit of H (as $M \to \infty$)is assumed to exist.

With the aid of (3.20), the time rate of H is calculated as follows.

$$\frac{dH}{dt} = \int_A \frac{\partial\hat{\rho}}{\partial t} \left(\log \hat{\rho} + n \int_X z\log z dx\right) \delta[B]\, \delta z$$

$$= \int_A \left(- \int_X \frac{\delta}{\delta z(x)} \{Qz(x)\hat{\rho}\delta[B]\} dx + \frac{1}{2} \int\int_{XX} \frac{\delta^2}{\delta z(x)\delta z(x')} \{D\hat{\rho}\delta[B]\} dxdx'\right)$$

$$\times \left(\log\hat{\rho} + n \int_X z\log z dx\right) \delta z.$$

By partial integration, we have

$$\frac{dH}{dt} = \int_A \left[- \int_X \frac{\delta}{\delta z(x)} (Qz)dx + n \int_X (\log z + 1)Qzdx\right] \hat{\rho}\delta[B]\, \delta z$$

$$+ \frac{1}{2} \int_A \int_X \int_X D\left[\frac{\delta^2\hat{\rho}}{\delta z(x)\delta z(x')} - \frac{1}{\hat{\rho}} \frac{\delta\hat{\rho}}{\delta z(x)} \frac{\delta\hat{\rho}}{\delta z(x')} + \frac{n\delta(x-x')}{z(x)} \hat{\rho}\right]$$

$$dxdx'\delta[B]\, \delta z, \tag{5.4}$$

noting that $\delta/\delta z(x)\delta[B] = 0$ in B. The first functional integral vanishes on account of the functional form of Q as well as the boundary condition on z(x): $z(x) \to 0$ as $|x| \to \infty$ or the condition of no entropy gain:

$$- \int_X \frac{\partial H_1}{\partial p} \frac{\partial}{\partial q} z\log z\, dx = 0. \tag{5.5}$$

For a closed system, both conditions are fulfilled.

If we make the substitution:

$$\hat{\rho} = C \exp\left[-n \int_X z\log z\, dx - \Psi(z)\right] \tag{5.6}$$

with C as the normalization constant, (5.4) reduces to

$$\frac{dH}{dt} = -\frac{1}{2}\int_A \int_X \int_X D \frac{\delta^2 \Psi}{\delta z(x)\delta z(x')} dx dx' \hat{\rho}\, \delta[B]\delta z. \tag{5.7}$$

Ψ can be considered as an analytic functional like

$$\Psi(z) = \frac{1}{2}\int_X \int_X A_2(x_1,x_2)\ [z(x_1)-z_0(x_1)][z(x_2)-z_0(x_2)]\,dx_1 dx_2$$

$$+ \text{ higher-order terms in } z-z_0 \text{ with } A_n(x_1,\ldots,x_n). \tag{5.8}$$

Here z_0, $A_2,\ldots,A_n$ are certain difinte functions in $X, X^2,\ldots, X^n$, respectively. A_n is symmetric with respect to interchange of arguments without loss of generality, and A_2 and $\int\int A_n(z-z_0)\ldots(z-z_0)dx_3 \ldots dx_n$ are naturally assumed to be nonnegative definite functions of x_1 and x_2 in order for $\hat{\rho}$ to be integrable [to give (3.21)]. Accordingly, $\delta^2\Psi/\delta z(x)\delta z(x')$ must be nonnegative definite.

Thus, the double integral over X^2 in (5.7) makes the trace of the product of two nonnegative definite functions, which can be proved to be nonnegative definite[12]. Hence, it is clear that

$$dH/dt \leq 0. \tag{5.9}$$

We have the conclusion that entropy as the minus H function is always produced except at the only one state, where all $A_n = 0$ so that $\Psi = 0$. $\hat{\rho}$ for that state is expressed as

$$\hat{\rho}_\infty = C_\infty \exp(-n\int_X z\log z dx), \tag{5.10}$$

for which H must be steady and minimum. In fact, it is easy to see, by functional variation, that (5.10) gives the minimum of H [defined by (5.3)] under the condition of (3.21). What we have just obtained is a general H theorem for a closed system.

VI. STEADY-STATE SOLUTION AND CANONICAL DISTRIBUTION

It is no doubt that (5.10) represents the maximum-entropy state of a closed system, and then it is natural to expect that it is the steady solution of (3.20) representing the ultimate equilibrium of the system. However, a sufficient condition for that is that (5.10) directly satisfies the steady case of (3.20). This is very difficult to prove, because of the complication of the D term. Here, we present just an indirect argument for it as follows.

If (5.10) is not a steady solution of (3.20), this state should readily be succeeded by another state at the next instant, so that entropy should increase endlessly so long as the assumption made on A_p holds. We could not expect any steady state to exist in this case, since entropy must stay constant for a steady state. However, this fact contradicts a general property of the Fokker-Planck or Kolomogorov equation: there should be a unique steady state which all states go towards [10,14]. Hence, it may be reasoned that $\hat{\rho}_\infty \delta[B]$ should be the steady solution of (3.20) finally approached by all $\hat{\rho}\delta[B]$. And then, $\hat{\rho}_\infty \delta[B]$ corresponds to microcanonical distribution in ergodic theory, although all points in the space B would never be equally realizable.

It is desirable to know the most probable value, i.e. the mode of z(x) at the maximum-entropy state. It is given as the z(x) in B such that the exponent of (5.10) is maximum. Then, it is solved by the standard variation method applied to the quantity $-n\int z \log z dx$ under the subsidiary conditions (3.6) and (3.11). It is easy to see that the implicit solution is a canonical (or Maxwell-Boltzmann) distribution in the one-body phase space with the particle interaction included in a self-consistent way:

$$z_B(x) = \exp\{-\lambda - \beta[H_1(x) + n\int_X \phi(|q-q'|) z_B(x')dx']\} \qquad (6.1)$$

where $e^{-\lambda}$ and β are the constants to be determined by the conditions (3.6) and (3.11). It must be noted that $z_B(x)$ is also a particular (Maxwellian-type) solution of the steady Vlasov equation;

$$Qz_B = 0, \qquad (6.2)$$

as is easily verified.

The exponent of (5.10) may be rewritten in terms of the functional Taylor expansion in the neighborhood of z_B;

$$\hat{\rho}_\infty = C_\infty \exp\left[-n\int_X z_B \log z_B dx - \frac{n}{2}\int_X \frac{1}{z_B}(z-z_B)^2 dx - ..\right]. \qquad (6.3)$$

If n is large enough, this may be considered as Gaussian in B. Therefore, the mode z_B may actually be the average of all z. From this result, the significance of the particular steady-state solution of the Vlasov equation is larger than usually known. One can say that $z_B(x)$ is irreversibly approached by the one-body distribution $\hat{F}_1(x,t)$, despite the fact that the Vlasov equation itself is time reversible! This being the case, z_B is widely applicable to an equilibrium state of any classical system: gas, liquid, and even crystalline state [15]. When we calculate the correlation of z, that is $\hat{F}_2$, etc from (5.10), care should be more taken of the effect of curving of the space B; in such a case, it would be safer to choose a curvilinear coordinate $\{\mu_i\}$ in B.

It is known from (6.3) that $\hat{\rho}_\infty$ is sharp nearly like $\delta[z - z_B]$, if n is sufficiently large. If so, what happens when $\hat{\rho}_\infty$ is inserted into (3.20)? By virtue of (6.2), the first term of the right-hand side of (3.20) nearly vanishes. Next, let us see the second term. Approximately, we have

$$D\hat{\rho}_\infty \cong D(z_B; x, x')\hat{\rho}_\infty \tag{6.4}$$

and

$$D(z_B; x, x') = \frac{\beta}{m} P \left[\frac{\partial\phi}{\partial q} (p'-p) z_B(x) z_B(x')\right] \tag{6.5}$$

(m: the mass of a particle).
If the system is cold so that $z_B(x) \sim \delta(p)$, D is nearly zero. On the contrary, if it is hot, β tends to vanish so that D approaches to zero. Then, D is presumed to vanish for $\hat{\rho}_\infty$ to the first approximation, at least. This result suggests that dH/dT vanishes for $\hat{\rho}_\infty$ for the stronger reason than argued in the preceding chapter, as is seen in (5.7). In consequence, it is plausible that $\hat{\rho}_\infty$ [B] would directly satisfy the steady case of (3.20).

VII. COMPARISON WITH THE BOLTZMANN ENTROPY

It is at this equilibrium state that our entropy can be compared with those of Boltzmann and Gibbs. Our steady-state solution (6.1) for $\hat{F}_1$ is obviously an extension of the Maxwell-Boltzmann distribution to the case with the particle interaction present. It is canonical-type, so that $\hat{F}_1$ is conjectured to be also close to the one-body distribution to be derived from the Gibbs canonical distribution in the many-body phase space. All the theories are coincident in the limit when the particle interaction tends to vanish.

It is interesting to see that our theory derives a canonical distribution of $\hat{F}_1$ so naturally, without appealing neither to ergodic theory nor to the central limit theorem. Instead, calculation of the mode z_B is obviously on the same technical basis as the information-theoretical approach of Jaynes [16].

As is known from (5.3), H is $\log C_\infty$ at $\hat{\rho} = \hat{\rho}_\infty$. In order to compare our entropy with Boltzmann's original form, a new definition, changed in level:

$$S = -kH + k \int_A \hat{\rho}_\infty \log\hat{\rho}_\infty \delta[B]\,\delta z \tag{7.1}$$

is most suitable, where k is the Boltzmann constant. In fact, at $\hat{\rho} = \hat{\rho}_\infty$ this entropy is related to the original Boltzmann entropy, as follows [Cf. (6.3)],

$$S_\infty = -kn \int_A \int_X z\log z dx \hat{\rho}_\infty \delta[B]\delta z \leq -kn \int_X z_B \log z_B dx. \tag{7.2}$$

The Boltzmann entropy gives an upper bound of S_∞.

By virtue of (6.1), we have

$$S_\infty \leq k[N\lambda + n\beta \int_X E(x) z_B(x)dx] \tag{7.3}$$

with

$$E(x) = H_1(x) + n\int_X \phi(|q-q'|) z_B(x')dx'. \tag{7.4}$$

If $N\lambda$ is read as thermodynamic potential, β as $1/kT$ (T: absolute temperature), and $n\int E(x)z_B(x)dx$ as total energy, then the right-hand side of (7.3) gives the usual thermodynamic entropy [2]. It is noted, however, that the particle interaction energy is counted twice in the total energy so read; so that the correspondence of the right-hand side of (7.3) with thermodynamic entropy is not perfect by that amount. This fact is not so serious for the case where the particle interaction energy is relatively small. However, our entropy S_∞ may have a better value compensating this discrepancy, if the integral in B is estimated correctly.

VIII. RELAXATION OF A NEARLY EQUILIBRIUM SYSTEM

An irreversible process in a nearly equilibrium system can be discussed within the frame of the functional random-walk model.

Let us assume that some small general forces $\{f_i\}$ act at $t = 0$ on a closed system which was at an equilibrium state, $\hat{\rho}_0\delta[B]$. The forces are either constant (e.g. electric field) or impulsive (e.g. initially given gradient of temperature). Then, the system will relax itself to the final equilibrium state, $\hat{\rho}_\infty\delta[B+\Delta B]$, as $t \to \infty$, where B denotes the deviation of B due to some of $\{f_i\}$. This irreversible dynamics is governed by (3.20) with the proper Hamiltonian $H_1+\Delta H_1$ which has the extra effect of some constant forces, say $\{f_i\}_1$, and with the proper initial condition which includes the effect of the other forces, say $\{f_i\}_2$. An intermediate state in the relaxation process may be written as

$$(\hat{\rho}_0 + \Delta\hat{\rho})\delta[B+\Delta B] = (\hat{\rho}_\infty + \delta\hat{\rho})\delta[B+\Delta B]. \tag{8.1}$$

First, application of (3.20) to the left-hand side of (8.1) results in the equation for $\Delta\hat{\rho}\delta[B+\Delta B]$ with the inhomogeneous terms which are in proportion to $\{f_i\}_1$. Of course, $\{f_i\}_2$ acts linearly on the initial condition on $\Delta\hat{\rho}$. Then, it is obvious that $\Delta\hat{\rho}\delta[B+\Delta B]$ can be formulated in terms of the same Green kernel as (3.23), except for the change in H_1 and in B, in such a way that $\Delta\hat{\rho}$ is in proportion to all f_i. Next, we have $\hat{\rho}_0 \cong \hat{\rho}_\infty$ because of (5.10). Then, the similar formula as for $\Delta\hat{\rho}$ is given to $\delta\hat{\rho}$. (Note $\hat{\rho}_0-\hat{\rho}_\infty \sim \Delta B \sim\{f_i\}_1$.)

With this fact in mind, let us calculate the deviation of the H function from the final value:

$$\delta H = \int_A \delta\hat{\rho}[(\log\hat{\rho}_\infty + 1) + n\int_X z\log z dx]\ \delta[B+\Delta B]\delta z + O(\delta\hat{\rho}^2). \tag{8.2}$$

It is natural that

$$\int_A \delta\hat{\rho}\delta[B+\Delta B]\delta z = 0. \tag{8.3}$$

Hence and by virtue of (5.10), we have

$$\delta H = O(\delta\hat{\rho}^2) = \frac{1}{2}\int_A \frac{\delta\hat{\rho}^2}{\hat{\rho}}\ \delta[B+\Delta B]\delta z > 0. \tag{8.4}$$

Since $\delta\hat{\rho}$ is in proportion to all f_i, δH must be a *positive definite* quadratic form of $\{f_i\}$. Then, we may write

$$\delta H = \sum_{i,j} \gamma_{ij} f_i f_j > 0, \tag{8.5}$$

where γ_{ij} is considered to be time-dependent and symmetric without loss of generality.

Hence,

$$\frac{d}{dt}\delta S = -k\frac{d}{dt}\delta H = -k\sum_{i,j}\frac{d\gamma_{ij}}{dt} f_i f_j. \tag{8.6}$$

γ_{ij} will asymptotically decrease to zero as time goes on, so as to achieve $\delta H = 0$. Therefore, $d\gamma_{ij}/dt \to 0$ as $t \to \infty$. Since the final state should have the maximum entropy according to the general result of Chapter V, $d\gamma_{ij}/dt$ must be *negative definite*.

If we define

$$J_i \equiv -kT_\infty \sum_j d\gamma_{ij}/dt f_j \tag{8.7}$$

(where T_∞ is the absolute temperature of a reference state, which may be either the initial or final equilibrium state), (8.6) is rewritten as

$$d\delta S/dt = \sum_i J_i f_i / T_\infty. \tag{8.8}$$

Then, J_i can be interpreted as the flow conjugate to f_i, such that $J_i f_i$ makes an energy dissipation in the system.

In general, flows are caused through the relation (8.7) by all forces. The transport coefficients in this process are defined as

$$\Gamma_{ij} \equiv -kT_\infty d\gamma_{ij}/dt. \tag{8.9}$$

Their reciprocity is clear from the symmetry of γ_{ij}. This is consistent with Onsager's well known principle for the irreversible process. In order to obtain the explicit form of Γ_{ij}, it is necessary to formulate γ_{ij} on the basis of (8.4). $\delta\hat{\rho}$ can be linearly related to $\{f_i\}$ by the formula just discussed. But eventually, a computer-work will be necessary for executing the pretty complicated form of functional integrals for this purpose.

In this chapter, we have dealt with a closed system, which finally goes to equilibrium where all flows disappear. Therefore, it is impossible here to obtain constant transport coefficients, such as expected for a steady non-equilibrium state. But, even if so, it may be practical to know the properly time-averaged value $\overline{\Gamma}_{ij}$ of the transient Γ_{ij} and to predict the linear relations of $\{f_i\}$ and $\{\overline{J}_i\}$, the properly time-averaged values of transient flows. Otherwise, we have to investigate (3.20) for an open system.

IX. CONCLUSION

We began with the functional formalism of statistical mechanics, which was initiated by Bogoliubov. The mathematical shape of the formalism naturally led us to make the functional random-walk model as an irreversible dynamics of the system evolution. In contrast to a usual way of deriving an irreversible dynamics, the concept of coarse-graining became the thing to be explained later in our theory; a mathematical make-up was first set forth in place of a physical assumption, and then the physical content of the former was pursued to be open in various ways. This is a unique point of our theory.

It is interesting to note that, for the steady state of a closed system, canonical distribution is naturally deduced by the same mathematical routine as in Jaynes' information-theoretical method. In his theory, this routine was the first principle we have to believe. In our theory, it is not such. Then, it may be said that Jaynes' method for an equilibrium state was theoretically grounded by our theory to some degree. For an open system, however, this routine might not be valid. This problem will be an interesting future target.

The concept of entropy in our theory turned out to be useful for deriving canonical distribution, as well as to be really important in that it has the thermodynamic meaning, comparable to the Boltzmann entropy or even beyond. In fact, the reciprocity relation of transport coefficients by Onsager is established on the basis of our entropy principle for a nearly equilibrium system.

These results were more than first expected when the author set out the research for the functional formalism, but now they may be considered as the evidences which show the reality of the functional random-walk model. Some mathematical problems in our theory to be investigated in future are the functional integration in the Riemannian space for estimating $\hat{F}_2$, etc at equilibrium, and the path integral expressed by the Green kernel (3.23)-(3.24) for a time-dependent solution for $\hat{\rho}$ and also for $\delta\hat{\rho}$ which is necessary for estimating transport coefficients. A high-speed computer will, of course, play an essential role for solving these problems.

REFERENCES

1. For example, see I. Prigogine, *Non-Equilibrium Statistical Mechanics* (Interscience Publishers, Inc., New York, 1962); T. Y. Wu, *Kinetic Equations of Gases and Plasmas* (Addison-Wesley Publ. Co., Inc., Reading, Mass., 1966).
2. A. I. Khinchin, *Mathematical Foundations of Statistical Mechanics* (Dover Publ., Inc., New York, 1949).
3. I. Hosokawa, Prog. Theor. Phys. (Kyoto) 52, 1513 (1974).
4. N. N. Bogoliubov, in *Studies in Statistical Mechanics*, J. de Boer and G. E. Uhlenbeck, Eds. (North-Holland Publ. Co., Amsterdam, 1962), Vol. 1.
5. I. Hosokawa, J. Math. Phys. 8, 221 (1967).
6. I. Hosokawa, J. Math. Phys. 11, 657 (1970).
7. E. Hopf, J. Ratl. Mech. Anal. 1, 87 (1952).
8. E. A. Novikov, Zh. Exsp. Teor. Fiz. 47, 1919 (1964) [Soviet-Phys. - JETP 20, 1290 (1965)];I. Hosokawa, J. Phys. Soc. Japan 25, 271 (1968).
9. Iu. L. Klimontovich, Zh. Exsp. Teor. Fiz. 33, 982 (1957)[Soviet-Phys. - JETP 6, 753 (1958)]; *The Statistical Theory of Non-Equilibrium Processes in a Plasma* (Pergamon Press, Inc., New York, 1967).
10. A. Kolmogorov, Math. Ann. 108, 149 (1933).

11. R. D. Richtmyer and K. W. Morton, *Difference Methods for Initial-Value Problems* (Interscience Publishers, Inc., New York, 1967).
12. I. Hosokawa, J. Math. Phys. 14, 1374 (1973).
13. C. E. Shannon and W. Weaver, *The Mathematical Theory of Communication* (University of Illinois Press, Urbana, 1959).
14. A. H. GRAY, Jr., J. Math. Phys. 6, 644 (1965).
15. A. A. Vlasov, *Many-Particle Theory* (Moscow-Leningrad, 1950, trans. AEC-tr-3406, Washington, D. C., 1959), Part 2, Chap.1; also the book by Wu in Ref. 1, Chap. 5.
16. E. T. Jaynes, Phys. Rev. 106, 620 (1957).
17. I. Hosokawa, J. Stat. Phys. 15, 87 (1976).

ADDENDUM

The Generalized Particle Formalism - Proof of Gibbs' Theorem

Our functional treatment has been confined to the particle system, each particle of which is located within X, that is, the six-dimensional phase space. We can extend the theory by considering the X in more than six dimensions. In this case, our particle becomes a generalized particle; where

$$x = (q_1, \ldots, q_m, p_1, \ldots, p_m), \qquad m > 3 \qquad (A.1)$$

Extra variables such as q_4, p_4, etc. may indicate any internal degrees of freedom of a particle. Or they may be usual degrees of freedom of other particles, when the generalized particle represents a cluster of particles.

In this extension to the theory, some care should be taken in interpreting the starting formulas (2.12)-(2.13). H_1 is no longer the one-body Hamiltonian but the Hamiltonian relevant to all the degrees of freedom of a generalized particle, and ϕ should be the sum of all the interaction potentials between two generalized particles; V is not really three-dimensional but should be the m-dimensional total volume. Then, n is defined as the average density of generalized particles in this volume. With this understanding, all the formal process in the text is acceptable.

Particularly, (6.1) is interesting. Consider the generalized particle as a gigantic cluster of particles and assume that the inter-cluster interaction is negligibly weak compared to H_1. Then, we have

$$z_B(x) = \exp\{-\lambda - \beta H_1(x)\} \qquad (A.2)$$

This is the Gibbs canonical distribution for m/3 particles. Here, the assumption preset above is just equivalent to the traditional assumption that a cluster of particles has a weak interaction with the heat bath (understood as all of the other clusters) so that the Hamiltonians of both are completely separable. The thermodynamic limit in our cluster is achieved when $V \to \infty$, $m \to \infty$. Keeping $(m/3)/V^{-m/3}$ (particle-number density in a cluster) to be constant.

It is worth noting that the Gibbs theorem given above - the important foundation of equilibrium thermodynamics - was derived not by way of the ergodic theorem and microcanonical distribution, but through the other discipline, the functional random-walk model with a generalized particle, which offers itself as the foundation of the irreversible non-equilibrium thermodynamics.

Part III

Seminars

SEMI-CLASSICAL APPROXIMATION TO PATH INTEGRALS - PHASES AND CATASTROPHES

Shimon Levit

Dept. of Nuclear Physics, Weizmann Institute of Science

Rehovot, Israel

INTRODUCTION

The semi-classical limit of quantum mechanics has recently gained much interest in various fields of physics, (e.g. Ref.[1-7]). The main feature of the semi-classical approach is that, while the dynamics of the system is determined by the rules of classical mechanics, the quantal effects are introduced through the superposition principle.

The general equations in this approach are easily written down. However, when one attempts to apply them to realistic cases aiming at the comparison with experimental data or with another type of calculation, many semi-intuitive prescriptions often fail to reproduce correct answers. A notable example is the phase of the semi-classical amplitudes, a problem known already from the WKB connection formulae. The correct definition of this phase is in turn related to the divergences (catastrophes) of the simple-minded ("primitive") semi-classical expressions.

We encountered the problems of phases and catastrophes trying to apply the classical S-matrix theory [1,2] to the scattering phenomena in nuclear physics [4-6]. The path integral formulation provided a suitable basis for the treatment of these and related problems. Within conventional mathematical language it was possible [8-12] to give practical prescriptions and discuss their limitations. Since the semi-classical (stationary phase) approximation is commonly used in any application of the path integral method, our results are not restricted to the scattering problems and may be of general interest.

The reported work has been done in collaboration with Prof. U. Smilansky, Dr. K. Möhring and Dr. T. Dreyfus.

I. PRIMITIVE SEMI-CLASSICAL APPROXIMATION

For simplicity we discuss a system with one degree of freedom described by a coordinate x , the corresponding momentum p and a Hamiltonian H(p,x) . The propagator $K(x'',t''; x',t') = \langle x'',t''|\exp(-iHt)|x',t'\rangle$ ($\hbar$ is one in our units) is given [13,14] by the path integral

$$K(x'',t''; x',t') = \int \exp\left\{iT[x(t),p(t)]\right\} DxDp. \tag{1.1}$$

The derivation of this expression was discussed during this school by several lecturers. $T[x(t), p(t)]$ is the action functional and the integration in (1.1) is extended over all paths satisfying

$$x(t') = x' \; ; \quad x(t'') = x'' \; . \tag{1.2}$$

In order to take a semi-classical (SC) limit to (1.1) it is first necessary to find all the stationary, i.e. classical trajectories satisfying the boundary conditions (1.2). In general there exist several such trajectories and the SC propagator is given by the coherent sum over all of them

$$K \cong \sum_j \tilde{K}^{(j)} \exp[i\, T_{cl}^{(j)}] \; . \tag{1.3}$$

Here $\tilde{K}^{(j)}$ are the gaussian path integrals of the type

$$\tilde{K} = \int \exp\,[(i/2)\delta^2 T]\; DxDp \tag{1.4}$$

Various methods to evaluate the absolute value of (1.4) exist [1,7,8,11,14],the simplest is probably using the unitarity property of the propagator. The result is

$$|\tilde{K}| = (2\pi)^{-1/2}|\partial x(t'')/\partial p(t')|^{-1/2} \tag{1.5}$$

On the other hand the determination of the phase of $\tilde{K}$ is one of the difficult points of the entire SC approach. These phases ensure the correct quantal interference between the various contributing trajectories. To describe their calculation we discuss first the concept of a focal point.

The quantity $\partial x(t)/\partial p(t')$ appearing in (1.5) defines the response of the given trajectory to an infinitesimal change of the initial momentum. The point along the trajectory where this quantity vanishes is called a focal point [15]. Here neighbouring trajectories focus. It is obvious [8] that the phase of $\tilde{K}$ may change only at a focal point. The usually accepted prescription [14] states that the phase of $\tilde{K}$ is equal to $(-\pi/2)$ times the number of focal points along the trajectory. This prescription is actually based on the result of the Morse focal point theorem [15] which relates the number of focal points along the trajectory to the number of negative eigenvalues of the operator which governs the second variation $\delta^2 T$ in (1.4).

Recent studies showed that the mentioned prescription is in general incorrect, e.g. in the important case of the momentum representation [11,12]. The reason is that it is based on the assumption of the positive definiteness of the mass in the Hamiltonian. In the momentum representation the role of the mass is played by the second derivative of the potential. In most cases this will not be definite along the trajectory. The generalized prescription to calculate the phase of $\tilde{K}$ for any number of degrees of freedom is derived in Ref. [12]. It turns out to be only slightly more complicated than the old prescription.

The notion of the focal point is crucial not only for the calculation of the phases. Suppose that the end point of the propagator is a focal point. The result (1.5) diverges in this case and the approximation (1.3) is not valid. Physically this corresponds to the situation where two or more trajectories in (1.3) coalesce.

As an illustration we discuss briefly a simple example of the motion in a repulsive $1/x^2$ potential. We consider the propagator $K(x,t;\, x'=1,\, t'=0)$ as a function of the final point (x,t). The family of all classical trajectories which start at $x' = 1$, $t' = 0$ is shown on Fig. 1. This family has an envelope on which $\partial x(t,p')/\partial p' = 0$. The focal points lie on the envelope which is called caustic [16]. (This line would "burn" if the trajectories were light rays.) The caustic divides the (x,t) plane into classically allowed and forbidden regions. If the final point (x,t) is in the allowed region, it can be reached by two trajectories of which one hits the caustic. The expression (1.3) contains therefore two terms. As the final point moves towards the caustic the two trajectories coincide and (1.3) diverges. The region beyond the caustic cannot be reached by real-valued classical paths. Allowing for complex values of the initial momenta provides the correct treatment of the forbidden region.

This situation is typical for the applications of SC approximation. One is usually interested in the dependence of the propagator on various parameters such as the position of the end points, the propagation time or any parameter in the Hamiltonian.

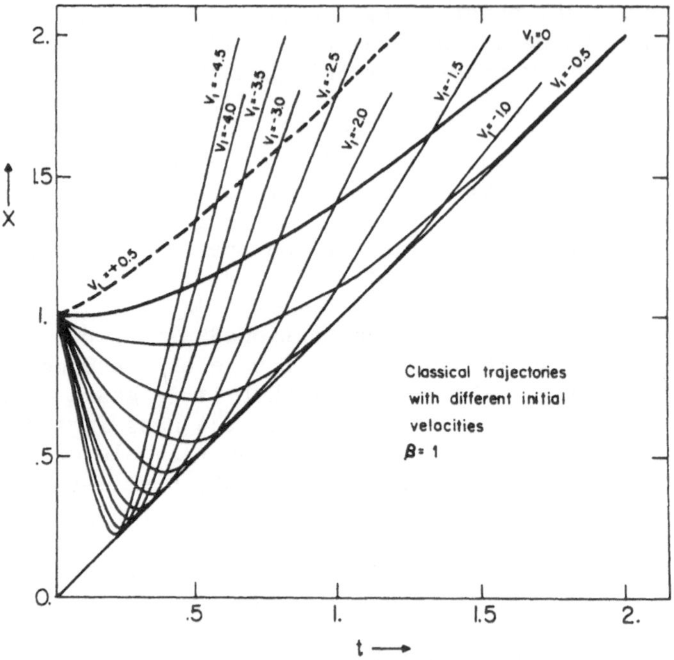

Fig. 1. Classical trajectories satisfying the initial conditions $x(t=0) = 1$ and $dx/dt(t=0) = v_1$ for various values of v_1. The caustic $x_c = {}^1t_c$ is clearly constructed as the envelope of the classical trajectories with $v_1 < 0$ (from Ref. [10]).

The positions of the classical paths depend on these parameters and it often happens that for a certain value of the parameters some of the paths are nearly coincident. Under these conditions the primitive SC expression (1.3) diverges. This situation is referred to as a catastrophe. Further variation of the parameters may cause the transition to become classically inexcessible ("forbidden") and the complex valued trajectories should be used [1,2]. What one needs is a uniform approximation which is valid for all values of the relevant parameters including the region of catastrophes.

Catastrophic phenomena are encountered already in the asymptotic evaluation of finite dimensional integrals [17, 18]. In this case the concept of an "unfolding of a singularity" [20] was used in order to develope the uniform technique. We review now this concept for the ordinary integrals and then show how it can be extended to path integrals.

II. UNIFORM TREATMENT

2.1. Ordinary Integrals [17-19]

Consider the simplest example of a one dimensional integral

$$I(\alpha) = \int \exp[i\ f(x,\alpha)]\ dx \tag{2.1}$$

which depends on the parameter α . Let the stationary phase condition $\partial f/\partial x = 0$ have two solutions $x_1(\alpha)$ and $x_2(\alpha)$ such that as $\alpha \to \alpha_0$ they coincide. The SPA result

$$I(\alpha) \sim \sum_j 1/\sqrt{f''(x_j)}\ \exp[i\ f(x_j,\alpha)] \tag{2.2}$$

diverges as $\alpha \to \alpha_0$. This is the simplest catastrophe. To obtain the uniform approximation one should "unfold the singularity".

Suppose that we have a function $\phi(y,\beta)$ of the variable y and parameter β such that: 1) $\phi(y,\beta)$ has the same number and topology of stationary points $y_1(\beta)$ and $y_2(\beta)$ as $f(x,\alpha)$ (by topology we mean the order and the sign of the first nonvanishing derivative at the stationary points); 2) $\phi(y,\beta)$ is simple so that the integrals

$$I_0(\beta) = \int \exp[i\ \phi(y,\beta)]\ dy \tag{2.3a}$$

$$I_1(\beta) = -i\partial I/\partial\beta = \int (\partial f/\partial\beta)\ \exp[i\phi(y,\beta)]\ dy \tag{2.3b}$$

are known numerically as functions of β .

Change now the integration variables by the following mapping

$$f(x,\alpha) = \phi(y,\beta) + A \tag{2.4}$$

with additive parameter A . This mapping is one-to-one if $dx/dy = (\partial\phi/\partial y)/(\partial f/\partial x) \neq 0$ and is finite. This requires that the stationary points of f and ϕ coincide

$$f(x_i,\alpha) = \phi(y_i,\beta) + A\ , \quad i = 1,\ 2 \quad . \tag{2.5}$$

The two equations (2.5) fix the parameters β and A . The original integral (2.1) is now

$$I = \exp(iA) \int \exp[i\phi(y,\beta)]\ (dx/dy)dy \quad . \tag{2.6}$$

This expression is still exact but now the whole complexity of the integral is in the Jacobian dx/dy . In the SPA spirit we approximate

$$G(y) = dx/dy \cong g_0 + g_1 (\partial\phi/\partial\beta) \quad , \tag{2.7}$$

where g_0 and g_1 are constants which are fixed by the requirement that (2.7) is exact at the stationary points.

$$G(y) = \left| \frac{(\partial^2\phi/\partial y^2)^{(i)}}{(\partial^2 g/\partial x^2)^{(i)}} \right|^{1/2} = g_0 + g_1(\partial\phi/\partial\beta)^{(i)} \qquad (i = 1,2) \tag{2.8}$$

This defines g_0 and g_1 and the integral $I(\alpha)$ reduces to the known integrals (2.3)

$$I(\alpha) \cong \exp(iA) \; [g_0 I_0 + g_1 I_1] \tag{2.9}$$

It is important to notice that the information used to get (2.9) is the same as in the primitive SPA result (2.2), namely the values of $f(x_i)$ and $f''(x_i)$. The mapping function in this simple case is $\phi(y,\beta) \doteq (1/3)\; y^3 - \beta y$ and gives a well known Airy approximation [19]. If more than two stationary points are present the mapping function is more complicated and depends on more parameters β . S-1 parameters are necessary to bring the number S of the stationary points in coalescence.

2.2 Path Integrals [10,11]

The uniform technique can be extended to the domain of the path integration in Eq. (1.1). For this purpose it is convenient to work in the so-called path expansion scheme [10,11], which we now briefly describe.

Let $\{X(t), P(t)\}$ be an arbitrary path satisfying the boundary conditions (1.2). The path variations from this path form a linear space of 2-component vector functions with upper components vanishing at $t=t'$ and $t=t''$. Using any orthonormal basis $(u^{(\alpha)}, v^{(\alpha)})$ in this space one can expand every path satisfying (1.2)

$$(x(t), p(t)) = (X(t), P(t) + \sum_{\alpha=1}^{\infty} a_\alpha (u^{(\alpha)}(t), v^{(\alpha)}(t)) \tag{2.10}$$

The integration over all paths in this scheme means the integration over the coefficients $\{a_\alpha\}$. The action $T[x(t), p(t)]$ is now $T[X(t), P(t), a_1, \ldots, a_N, \ldots]$. Truncating the series (2.10) to the first N terms we may define the path integral (1.1) as

$$K = \lim_{N\to\infty} J_N \int \exp(i\ T_N[X(t),\ P(t),\ a_1,\ \ldots\ a_N])\ d^N a \qquad (2.11)$$

The normalization factor J_N is necessary to provide a proper convergence of the limit (2.11) and can be easily found [10,11].

We show now how the uniform semi-classical approximation is derived for the integral in (2.11). We are interested in the situation where S stationary points of the function T_N exist and may coalesce when the parameters of which propagator depends are varied.

It is clear that the number and the topology of the S stationary points do not change when N increases, once N is sufficiently large. It is then possible to find a function F of ℓ variables $b = (b_1, \ldots, b_\ell)$ which is simpler in form than T_N, but identical as far as the number of stationary points and their topology is concerned. The number ℓ is independent of N . Since there are S points involved in the coalesence, the function F depends on S-1 parameters $A = (A_1, \ldots, A_{S-1})$. As in the one-dimensional example we wish now to change the variables $b = b(a)$ using the function F . We define in addition a quadratic form

$$L_N = \sum_{j=\ell+1}^{N} b_j^2$$

and construct the mapping

$$T_N = F(A,\ b_1,\ \ldots,\ b_\ell) + \frac{1}{2} L_N + F_0 \quad . \qquad (2.10)$$

Obviously L_N does not add new stationary points. The parameters $A_1, \ldots, A_{S-1}$ and F_0 are found in analogy to (2.5) from (2.10) taken at the stationary points (where $L_N = 0$)

$$T_N^{(i)} = F^{(i)}(A,\ b_1^{(i)},\ \ldots,\ b_\ell^{(i)}) + F_0 \quad , \quad (i = 1,\ldots,S) \quad (2.11)$$

In the new integration variables (2.11) is written exactly

$$K = \lim_{N\to\infty} \exp(i\ F_0) \int \det(\partial a_\alpha/\partial b_\beta) \exp\left\{i(F+\frac{1}{2}L_N)\right\} d^N b \qquad (2.12)$$

Now comes the approximation

$$\det(\partial a_\alpha/\partial b_\beta) \cong g_0 + \sum_{k=1}^{S-1} g_k \cdot \partial F/\partial A_k \tag{2.13}$$

Requiring this to be exact at the SP we find g_k using

$$\det(\partial a_\alpha/\partial b_\beta)^{(i)} = \left|\frac{\det(\partial^2 F/\partial b_\alpha \partial b_\beta)^{(i)}}{\det(\partial^2 T_N/\partial a_\gamma \partial a_\mu)^{(i)}}\right|^{1/2} = \tag{2.14}$$

$$= g_0 + \sum_{k=1}^{S-1} g_k(\partial F/\partial A_k)^{(i)}, \quad (i = 1, \ldots S)$$

Using (2.13) in (2.12) one can perform the integration over the variables of the quadratic form L_N. The result is expressed through the integrals

$$I^{(0)}(A) = (2\pi i)^{-\ell/2} \int \exp[i\, F(A,b)]\, d^\ell b \quad ; \tag{2.15a}$$

$$I^{(i)}(A) = -i\partial I^{(0)}/\partial A_i \quad , \quad i = 1, \ldots, S \tag{2.15b}$$

which are assumed to be known.

Taking the limit of $N\to\infty$ we get after some simple transformations [11] the final uniform expression

$$K = \sum_{k=1}^{S} \tilde{K}^{(i)}\, \tilde{I}^{(i)}(A, F_0) \tag{2.16}$$

where $\tilde{K}^{(i)}$ were already discussed in connection to Eq. (1.3) and the functions $\tilde{I}^{(i)}(A, F_0)$ are certain linear combinations [11] of the integrals (2.15). When $N\to\infty$ the functions $T_N^{(i)}$ in Eq. (2.11) are replaced by the actions $T_{cl}^{(i)}$ along the contributing classical paths.

We wish to emphasise again that the information needed for the application of (2.16) is the same as in the simple SPA (1.3). When the paths are well separated the result (2.16) approaches (1.3). It stays finite near a caustic and gives the transitional approximation discussed by Schulman [20] and DeWitt [21]. The crucial point for the application of (2.16) is to choose the mapping function $F(A, b_1, \ldots, b_\ell)$. The action functional T depends on an infinite number of variables. However, only the dependence on a finite and usually small number of variables is essential to define the required mapping. It is the main advantage of the path expansion scheme that with the proper choice of the reference path $\{X(t), P(t)\}$ and the basis $(u^{(\alpha)}, v^{(\alpha)})$ the relevant variables are the lowest expansion coefficients $\{a_\alpha\}$ in (2.10).

Thom's classification of the elementary catastrophes [17, 18] often helps to choose the mapping. In the example of $1/x^2$ potential which was discussed in Section I the simple cubic mapping $F(A,b) = (1/3)b^3+Ab$ is sufficient. The detailed calculations are found in Ref. [10]. For the treatment of situations with more complicated catastrophes we refer the reader to Ref. [17, 18].

2.3 The Initial Value Representation

Under certain conditions the physical system itself provides us with the natural mapping function. In these circumstances the problem of finding the uniform approximation may be considerably simplified.

We remind that we are looking for the propagation from x' to x'' in time $T = t''-t'$. Consider all the classical trajectories starting at x' at $t = t'$ with all possible initial momenta p_i. Calculating at the final time $t = t''$ the functions $x^f(p')$, $p^f(p')$ and the classical action $S(p') = S(x^f(p'), x')$, construct the function

$$F(p') = S(p') - p^f(p')\,(x^f(p') - x'') \tag{2.17}$$

It is easily checked [11] that under the condition

$$\partial p^f(p')/\partial p' \neq 0 \tag{2.18}$$

the function $F(p')$ provides the correct mapping. Following the procedure similar to that of section 2.2 one obtains [11] the following integral representation for the propagator

$$K \cong \int (\partial p^f(p')/\partial p') \exp[i\, F(p')]\, dp' \tag{2.19}$$

This Initial Value Representation (IVR) is known in optics [17] under the name diffraction integral and was used in recent studies [1-6] of the semi-classical S-matrix.

In the IVR the propagator is approximated by an ordinary integral. This suggests a relatively simple practical way of calculations, especially in the case of several degrees of freedom. Unfortunately the condition (2.18) severely restricts the possible applications [11]. It is highly desired to remove this restriction.

CONCLUSION

The presented discussion of the phases and catastrophes in the semi-classical approximations was restricted to the systems with one degree of freedom for the sake of simplicity. It is the main advantage of the path integral formulation that this discussion is easily extended to the (really interesting) cases of several degrees of freedom. This is in contrast to WKB solutions of the Schrödinger equation.

In practice one is usually not interested in the propagator in coordinate-time representation. For the calculation of the S-matrix for example the relevant quantities are the matrix elements of the energy dependent propagator. Even though different exact quantal representations are related to each other by simple transformations, the corresponding connection between the SC expressions is not as simple. This is due to the fact that the structure of the caustics is entirely different in the different representations. The derivation of the uniform approximations in the energy representation should therefore use the exact path integral expression as the starting point, rather that performing Fourier transforms on the expressions derived in the present lecture.

REFERENCES

[1] W.H. Miller, Adv. Chem. Phys. 25 (1974) 69, 30 (1975) 77.
[2] R.A. Marcus, J. Chem. Phys. 54 (1971) 3965.
[3] J. Knoll, R. Schaeffer, Ann. Phys. 97 (1976) 307.
[4] S. Levit, U. Smilansky and D. Pelte, Phys. Lett. B53 (1974) 39.
[5] M.W. Guidry, H. Massman, R. Donangelo and J.O. Rasmussen, to be published in Nucl. Phys. A.
[6] P. Fröbrich, Q.K.K. Liu and K. Möhring, to be published in Nucl. Phys. A.
[7] R. Rajaraman, Phys. Rev. 21C (1975) 227 and the references therein.
[8] S. Levit and U. Smilansky, Ann. Phys. 103 (1977) 198.
[9] S. Levit and U. Smilansky, Proc. Amer. Math. Soc. (1977), in press.
[10] S. Levit and U. Smilansky, preprint, Weizmann Institute of Science, WIS-76/22-Ph, Rehovot, Israel, 1976.
[11] S. Levit and U. Smilansky, Ann. Phys. (1977) in press.
[12] S. Levit, K. Möhring, U. Smilansky and T. Dreyfus, preprint, Weizmann Institute of Science, WIS-77/23-Ph, Rehovot, Israel, 1977, submitted to Ann. of Phys.
[13] C. Garrod, Rev. Mod. Phys. 38 (1966) 483.
[14] M.C. Gutzwiller, present school and the references therein.
[15] M. Morse, "Variational Analysis", Wiley, New York, 1973.
[16] M.V. Berry, K.E. Mount, Rep. Progr. Phys. 35 (1972) 315.
[17] M.V. Berry, Advan. Physics 25 (1976) 1 and the references therein.
[18] J.N.L. Connor, Mol. Phys. 31 (1976) 33.
[19] C. Chester, B. Friedman and F. Ursell, Proc. Camb. Phil. Soc. 53 (1957) 599.

[20] L.S. Schulman, in Functional integration and its application, Proc. Int. Conf., Ed. A.M. Arthurs, London, April 1974.
[21] C. De Witt-Morette, Ann. Phys. 97 (1976) 367.

CURRENT SATURATION IN CdS

Stephen J. Nettel

Department of Physics, Rensselaer Polytechnic Institute
Troy, New York 12181, U.S.A.

A calculation of the high electric fields found in domains beyond the onset of current saturation in piezoelectric semi-conductors is outlined. The calculation is a first principles one, deriving from Feynman's exact polaron mobility formula. Special attention is given to explaining the motives behind an iteration scheme, that uses a classical like, constant drift velocity path as a zero-order approximation. The electrons are found to bunch in drifting localized potentials.

1. INTRODUCTION

When the drift velocity of electrons in semi-conductors approaches the velocity of phonons some interesting effects can be expected. Quite a long time ago von Hippel suggested the existence of a "sound barrier", which if broken by the application of sufficient electric field, would lead to electric breakdown [1]. The treatment of this sound barrier has long been a challenge to theoreticians. A Boltzmann equation approach by Devreese and Evrard has been given earlier in the lectures of Professor Devreese [2].

The notion that a "resonance" takes place when an electron approaches an average velocity equal to that of a phase velocity of a crystal mode of vibration is easily understood in classical physics.

This is the reason we turned to the method of the Feynman path integral, where we know that classical trajectories are easily identified, and play a special role. It is going to take some deliberation to get at this classical aspect of the problem in a first principle way when dealing with as light and wave-like particle as the electron. It is not our purpose in this short write up to go into quantitative detail, especially as a careful description is already given in the literature [3,4]. Rather we would like to give an explanatory account, emphasizing those points which were brought out in the discussions at the seminar here. We see this as an opportunity to make the forest appear for all the trees.

2. THE PHENOMENA IN CdS

The high electric field "resonance" phenomena appear particularly clearly in the piezoelectric semi-conductor cadmium sulphide. The questions that one would like answered are :

a) The basic fact, observed, for example, by Hutson et. al.[5], that the electric current saturates when the electron drift velocity approaches the velocity of the crystal vibrations; (there is a "knee" in the current-voltage characteristic).

b) The saturation does not occur until the phase velocity of the longest shear waves is reached, rather than at a lower velocity associated with shorter waves.

c) Upon reaching saturation the emission of a flux of phonons of frequencies up to about 4×10^9 cycles/sec is observed [6].

d) The saturation is found to disappear at conduction-electron densities less than 5×10^{13} cms^{-3} [5].

e) Haydl and Quate among others have found that beyond the onset of saturation there appear domains with anomalously high electric fields (of the order of 5×10^4 V/cm) [6].

f) Piezoelectric materials appear to favor the saturation effects.

3. ANALYTIC TREATMENT

Starting with Feynman's expression for the mobility of an electron we first derived a Schrödinger like integro-differential-equation for the density matrix ρ [3]. We did this with a view to applying iteration methods. The equation is generalized to include all conduction electrons (see (d) of the previous section). Coulomb interactions between these electrons were not included. The effects of neglecting them are discussed below.

As our "zero-order" iteration we assumed for each electron a classical (Dirac delta) path

$$\rho = \delta(\bar{r}(t) - \bar{v}t) \tag{1}$$

where $\bar{r}$ is the position of the electron, $\bar{v}$ corresponds to its constant drift velocity, and t is the time. The substitution of eqs.(1) into the integro-differential equation leads to a Feynman influence function with two terms :

$$NV_{p.v.}(\vec{r}) = \frac{2N}{\hbar} \sum_k |C_k|^2 \frac{k_z v + \omega_k}{(k_z v + \omega_k)^2 + \Gamma^2} \cos \vec{k}.\vec{r} \tag{2a}$$

and

$$NV_\delta(r) = \frac{-2N}{\hbar} \sum_k |C_k|^2 \pi\delta(k_z v - \omega_k) \sin \vec{k}.\vec{r} \tag{2b}$$

Here N is the number of conduction electrons, the sum is over all lattice modes k, C_k in the electron coupling to the mode k, the drift velocity v is in the z direction, ω_k are the mode frequencies, and Γ is the reciprocal life-time of an oscillator mode (see below). The quasi-potentials in eqs.(2a) and (2b) have opposite spatial symmetry, and are different in their significance. Equation (2a) is a local potential well that travels with the drifting electrons, (i.e. its center is at $\vec{v}t$). Upon detailed calculation it is found that such a well created by a single electron will not hold the electron. But when a number of electrons combine or "bunch", they will cooperatively produce a potential which is binding. We were able to estimate this number at about 100 [3] after drawing on the experimental fact that the effect disappears below conduction electron concentrations of 5×10^{13} cm^{-3}, see (d) of Section 2. That the electrons travel in bunches under conditions of saturation corresponds to what is observed, and has been taken note of theoretically before [7]. By N in eqs.2 is meant just the number of conduction electrons bunched together, and the extent of the well is directly proportional to this number.

It is an interesting fact that the potential well, (2a), cannot be evaluated without a knowledge of the reciprocal life-time Γ, even though the dependence turns out to be weak. Actually, in the original formulae of Feynman et. al. no Γ appears [8]. One must go to complex oscillator frequencies $(\omega_k + i\Gamma_k)$. The legality of introducing dissipation into the Feynman mobility formula worried us so much that we explicitly re-did the Feynman polaron calculation, eliminating lattice-oscillator coordinates for a system in which in addition the phonons could collide with one another, that is, by supplying a damping mechanism. The interested reader is referred to the literature [4].

The second term, eq.(2b), is the back action of the lattice, connected with the emission of real phonons. To get a self-consistent solution of the differential equation governing ρ it was found necessary to balance this term with the influence of the applied electric field, E.

Physically, how do we arrive at the argument in the delta function of eq.(2b)? We suppose that an electron of mass m and initial momentum $(m\vec{v} + \hbar\vec{k}/2)$ emits a phonon of energy $\hbar\omega_k$, momentum $\hbar\vec{k}$, giving a final electron momentum $(m\vec{v} - \hbar\vec{k}/2)$. The average momentum is $m\vec{v}$. We have :

initial kinetic - final kinetic energy - phonon energy

$$= \vec{k}.\vec{v} - \hbar\omega_k .$$

Why was it useful to characterize a recoiling electron by an average velocity v even in a quantum-mechanical calculation? The electron motions in our analysis are quantum-mechanical average motions, an averaging over all possible final crystal oscillator (phonon)states. This averaging is a consequence of Feynman's formula, in which lattice coordinates have been eliminated by taking expectation values. (We find in practice that at every step one always first takes a summation over all phonon modes k). This means that just as in a macroscopic calculation of diffusion one does not look at the details of the motion of the solvent molecules, here individual electron-phonon collisions are not of first concern. When at current saturation electron-phonon resonance sets in, phonon emission takes place more rapidly and systematically than in the ohmic region; witness the emission of a steady phonon flux. Consequently, one is led as a zero-order substitution, to a simplified motion in which recoils from the emission of individual phonons have been obliterated. By choosing an average path characterized by a steady-state velocity $\vec{v}$ a steady force exerted by the lattice on the electrons as a result of the average phonon emission can be isolated in the quantum calculation, and balanced against the forces of the applied electric field, just as happens in classical physics. By this procedure the largest perturbations in the problem are eliminated.

The actual motion of the electrons is not, of course, as given by eq.(1). A self-consistent solution of our governing equation is obtained only when we superimpose on the steady drift motion of the electrons an oscillatory motion in the drifting well of eq.(2a). When such a compound motion is substituted into our integro-differential equation as a second iteration, the terms in equations (2a) and (2b) are modified. Factors appear containing matrix elements of $\exp.i\vec{k}.\vec{r}$ between localized electronic states of the well, a, i.e. elements $< a|\exp.i\vec{k}.\vec{r}|a >$.

As already mentioned the electric field under saturation conditions is obtained by using the electric potential eE_z as a balance to the now modified influence function of equation (2b). Only with this procedure do we get a self-consistent solution of our governing differential equation, i.e. an overall situation drifting at the velocity v. An independent calculation of the total power absorbed by all the phonon modes gave the same result for E when energy conservation was postulated [9]. The final expression for E was found to be :

$$eE = (\sum_b |<b + 1|z|b>|^2)^{-1} \sum_k |C_k|^2 \sum_{a,b} e^{ik\rho}_{\beta\beta}\, e^{-i\vec{k}.\vec{r}}_{aa} \tag{3}$$

$$\times <b + 1|z|b>\, i e^{ik_z z}_{b+1,b} \left\{ \frac{\hbar\Gamma/2}{(\omega_k + k_z v)^2 + (\Gamma/2)^2} - \frac{\hbar\Gamma/2}{(\omega_k - k_z v)^2 + (\Gamma/2)^2} \right\}$$

where (x,y,z) are the components of $\vec{r}$, b is the z-direction quantum number of the three-dimensional oscillator state b, and β the other two quantum numbers for b. The sum on b is over the states on the Fermi-surface only, that is the state b is filled, but the state b + 1 is empty. The sum on a is over the N lowest oscillator states, i.e. the filled ones. $[\exp - i\vec{k}.\vec{r}]_{aa}$ are matrix elements, and k_ρ means $k_{xx} + k_{yy},.|$ and ω_o is the frequency associated with the electron oscillator states, a,b. A numerical evaluation of equation (3) was found possible [3,10].

4. RESULTS

We discuss the results of the calculation with respect to the six observations for CdS listed in Section 2.

a) We find that the "resonance" leads to very high electric fields. As the voltage increases to beyond where the electron drift velocity reaches the sound velocity an increasingly broader section of the crystal will be at resonance, and display "high field domains", see (e). The drift velocity does not increase beyond the resonance value, current saturation occurs.

b) The matrix element $< a|\exp i\vec{k}.\vec{r}|a >$ between oscillator states of electrons localized in the travelling well limits the interaction to crystal oscillations of wave-length λ longer than $1/r_o$, where r_o is a characteristic measure of the extension of the well. At wave-lengths longer than this the oscillating lattice sees the electron as a point charge. At shorter wave-length of crystal vibration the oscillator motion of the electron affects the electron-phonon interaction. The matrix elements $< a|\exp i\vec{k}.\vec{r}|a >$ become small, and cause a cut-off in lattice models that have the resonant interaction with the drifting electrons. Thus, saturation is delayed until the phase velocity of the long waves is reached.

c) A small increase in the order of magnitude of r_o, the well extension (from about 10^{-6} cms to 2×10^{-5} cms), was found to be warranted to take due account of the neglected coulomb repulsion among electrons in one well (about 100 electrons). The cut-off mechanism discussed then limits the frequency of generated phonons to just about 4×10^9 cycles/sec., as observed by Haydl and Quate [5].

d) For low conduction-electron densities a self-consistent binding well does not form, presumably because there is not time for the cooperation to occur. As mentioned, we have actually used the observed limit in the density to get at the numerical characteristics of the well [3].

e) The electric field E calculated from equation (3) decreased from 10^5 V/cm to around the observed value of 3×10^4 V/cm when r_o was increased to 2×10^{-5} cms.

Besides the terms listed in equation (3) for E we found as a result of the repeated iteration a number of other terms. These generally turned out to be small (vanish for the delta functions) because the energy separation between adjacent oscillator levels in the localized well are a factor of 10 to 100 greater than the energy of those long-wave phonons which are not cut-off in their interaction. What this means is that the bunched electrons in the collective well have a certain stability against scattering by phonons. Only the resonant interaction is significant. Because the new terms are small, we can expect convergence in our iterative solution of the governing Schrödinger-Feynman equation for .

f) The long range of piezoelectric interactions, ($C_k \propto 1/\sqrt{k}$), is essential for a binding potential well.

In conclusion, our knowledge of piezoelectric electron-phonon coupling is too poor [3], and we have had to make too many approximations to make the quantitative agreement we have found very significant by itself. We feel that it is above all the simple, qualitative explanation of all the observed aspects of the saturation phenomenon which make our calculation warrant additional consideration. In the meantime experimental interest continues [11].

ACKNOWLEDGEMENTS

It is a pleasure to acknowledge the contribution of my colleagues and former students at Rensselaer, Ping-Kong Lai and Paul O. Massicot. They are the co-authors of our previous papers. I wish also to thank Dr. T.D.Schultz at IBM who gave so much of his time to discussion and review.

The probing questions of Dr. R.von Baltz, Dr. F.Beleznay, Professor J.T.Devreese, Professor H.Reik and Dr. K.K.Thornber during the seminar have helped to shape this particular account of our long work. Their interest and encouragement is indeed appreciated.

REFERENCES

1. A.von Hippel, Dielectrics and Waves, p.248, Wiley, New York (1954).

2. Besides the lectures of Prof. Devreese in this monograph, see J.T. Devreese and R.Evrard, Phys. Stat.Sol. (B) 78, 85 (1976).

3. P.K. Lai, P.O. Massicot and S.J. Nettel, J.Phys. Chem.Solids 35, 1703 (1974).

4. P.K. Lai, P.O. Massicot and S.J. Nettel, Z.Physik (B) 23, 97 (1976).

5. A.R. Hutson, J.H. McFee and D.L. White, Phys. Rev. Letters 7, 237 (1961).

6. W.H. Haydl and C.F. Quate, Appl.Phys.Letters 7, 45 (1965). See also W.H. Haydl, K. Harker and C.F. Quate, J.Appl.Phys. 38, 4295 (1967).

7. P.K. Tien, Phys.Rev. 171, 970 (1968).

8. R.P.Feynman, R.W. Hellwarth, C.K. Iddings and P.M. Platzman, Phys.Rev. 127, 1004 (1962).

9. C. Kourkoumelis, B.S. Thesis, Dept. of Physics, Rensselaer Polytechnic Institute (1972).

10. P.K. Lai, Ph.D. Thesis, Dept. of Physics, Rensselaer Polytechnic Institute (1972).

11. K. Junker and H. Heinz, Phys.Stat.Sol. (A), 21, 451(1974).

LIST OF CONTRIBUTORS

J.T. Devreese
Rijksuniversitair Centrum Antwerpen, Groenenborgerlaan 171, B-2020 Antwerpen, Belgium and Universitaire Instelling Antwerpen, Universiteitsplein 1, B-2610 Wilrijk, Belgium.

S.F. Edwards
Science Research Council, State House, High Holborn, London WC1 4TA, Great Britain.

M.C. Gutzwiller
Thomas J. Watson Research Center, I.B.M., P.O. Box 218, Yorktown Heights, N.Y. 10598, U.S.A.

I. Hosokawa
Department of Mechanical Engineering, Iwate University, Morioka 020, Japan.

J.R. Klauder
Bell Laboratories, 600 Mountain Avenue, Murray Hill, N.J. 07974, U.S.A.

S. Levit
Department of Physics, The Weizmann Institute of Science, Rehovot, Israel.

J.M. Luttinger
Department of Physics, Columbia University, 538 West 120th Street, New York, N.Y. 10027, U.S.A.

B. Mühlschlegel
Institut für Theoretische Physik, Universität zu Köln, Zülpicher Strasse, 5000 Köln 41, B.R.D.

S.J. Nettel
Department of Physics, Rensselaer Polytechnic Institute, Troy, N.Y. 12181, U.S.A.

G.J. Papadopoulos
Division of Mechanics, Department of Physics, University of Athens, Panepistimiopolis, Athens 621, Greece.

G. Rosen
Department of Physics, Drexel University, Philadelphia, Pa. 19104, U.S.A.

D. Sherrington
Imperial College of Science and Technology, Prince Consort Road, London SW7 2BZ, Great Britain.

K.K. Thornber
Bell Laboratories, 600 Mountain Avenue, Murray Hill, N.J. 07974, U.S.A.

F.W. Wiegel
Department of Applied Physics, Twente University of Technology, P.O. Box 217, Enschede, the Netherlands.

AUTHOR INDEX

SUBJECT INDEX